SOLUTIONS
TO BLACK EXERCISES

ROXY WILSON
University of Illinois, Urbana-Champaign

CHEMISTRY THE CENTRAL SCIENCE 11E

BROWN LeMAY BURSTEN MURPHY

PEARSON

Prentice
Hall

Upper Saddle River, NJ 07458

Project Editor: Jennifer Hart
Senior Editor: Andrew Gilfillan
Editor-in-Chief, Science: Nicole Folchetti
Assistant Managing Editor, Science: Gina M. Cheselka
Supplement Cover Manager: Paul Gourhan
Supplement Cover Designer: Victoria Colotta
Operations Specialist: Amanda A. Smith
Senior Operations Supervisor: Alan Fischer

© 2009 Pearson Education, Inc.
Pearson Prentice Hall
Pearson Education, Inc.
Upper Saddle River, NJ 07458

Printed in the United States of America

10 9 8 7 6 5 4 3 2 1

ISBN-13: 978-0-13-600324-3
ISBN-10:   0-13-600324-9

Pearson Education Ltd., *London*
Pearson Education Australia Pty. Ltd., *Sydney*
Pearson Education Singapore, Pte. Ltd.
Pearson Education North Asia Ltd., *Hong Kong*
Pearson Education Canada, Inc., *Toronto*
Pearson Educación de Mexico, S.A. de C.V.
Pearson Education—Japan, *Tokyo*
Pearson Education Malaysia, Pte. Ltd.

Contents

Introduction

Chemistry: The Central Science, 11th edition, contains nearly 2600 end-of-chapter exercises. Considerable attention has been given to these exercises because one of the best ways for students to master chemistry is by solving problems. Grouping the exercises according to subject matter is intended to aid the student in selecting and recognizing particular types of problems. Within each subject matter group, similar problems are arranged in pairs. This provides the student with an opportunity to reinforce a particular kind of problem. There are also a substantial number of general exercises in each chapter to supplement those grouped by topic. Visualizing Concepts, general exercises which require students to analyze visual data in order to formulate conclusions about chemical concepts, and Integrative Exercises, which require students to integrate concepts from several chapters, are continuing features of the 11th edition. Answers to the odd numbered topical exercises plus selected general exercises, about 1200 in all, are provided in the text. These appendix answers help to make the text a useful self-contained vehicle for learning.

This manual, **Solutions to Black Exercises** in **Chemistry: The Central Science, 11th edition**, was written to enhance the end-of-chapter exercises by providing documented solutions for those problems <u>not</u> answered in the appendix of the text. The manual assists the instructor by saving time spent generating solutions for assigned problem sets and aids the student by offering a convenient independent source to check their understanding of the material. Most solutions have been worked in the same detail as the in-chapter sample exercises to help guide students in their studies.

When using this manual, keep in mind that the numerical result of any calculation is influenced by the precision of the numbers used in the calculation. In this manual, for example, atomic masses and physical constants are typically expressed to four significant figures, or at least as precisely as the data given in the problem. If students use slightly different values to solve problems, their answers will differ slightly from those listed in the appendix of the text or this manual. This is a normal and a common occurrence when comparing results from different calculations or experiments.

Rounding methods are another source of differences between calculated values. In this manual, when a solution is given in steps, intermediate results will be rounded to the correct number of significant figures; however, unrounded numbers will be used in subsequent calculations. By following this scheme, calculators need not be cleared to re-enter rounded intermediate results in the middle of a calculation sequence. The final answer will appear with the correct number of significant figures. This may result in a small discrepancy in the last significant digit between student-calculated answers and those given in this manual. Variations due to rounding can occur in any analysis of numerical data.

The first step in checking your solution and resolving differences between your answer and the listed value is to look for similarities and differences in problem-solving methods. Ultimately, resolving the small numerical differences described above is less important than understanding the general method for solving a problem. The goal of this manual is to provide a reference for sound and consistent problem-solving methods in addition to accurate answers to text exercises.

Extraordinary efforts have been made to keep this manual as error-free as possible. All exercises were worked and proof-read by at least three chemists to ensure clarity in methods and accuracy in mathematics. The work and advice of Dr. Angela Manders Cannon and Ms. Renee Rice have been invaluable to this project. However, in a written work as technically challenging as this manual, typos and errors inevitably creep in. Please help us find and eliminate them. We hope that both instructors and students will find this manual accurate, helpful and instructive.

Roxy B. Wilson
University of Illinois
School of Chemical Sciences
505 S. Mathews Ave., Box 49-1
Urbana, IL 61801
rbwilson@uiuc.edu

1 Introduction: Matter and Measurement

Visualizing Concepts

1.2   After a *physical change*, the identities of the substances involved are the same as their identity before the change. That is, molecules retain their original composition. During a *chemical change*, at least one new substance is produced; rearrangement of atoms into new molecules occurs.

     The diagram represents a **chemical change**, because the molecules after the change are different than the molecules before the change.

1.3   (a)   time   (b) density   (c) length   (d) area   (e) temperature

          (f)   volume   (g)   temperature

1.5   Measurements (darts) that are close to each other are *precise*. Measurements that are close to the "true value" (the bull's eye) are *accurate*.

     (a)   Figure ii represents data that are both accurate and precise. The darts are close to the bull's eye and each other.

     (b)   Figure i represents data that are precise but inaccurate. The darts are near each other but their center point (average value) is far from the bull's eye.

     (c)   Figure iii represents data that are imprecise but their average value is accurate. The darts are far from each other, but their average value, or geometric center point, is close to the bull's eye.

1.7   The determined age of the artifact, 1,900 years, has two significant figures. There is uncertainty in the hundreds place, indicating that the minimum uncertainty in age is 100 years. The 20-year period since the age was determined is not significant relative to the determined age.

1.9   In order to cancel units, the conversion factor must have the unit being canceled opposite the starting position. For example, if the unit cm starts in the numerator, then the conversion factor must have cm in its denominator. However, if the unit cm starts in the denominator, the conversion factor must have cm in the numerator. Ideally, this will lead to the desired units in the appropriate location, numerator or denominator. However, the inverse of the answer can be taken when necessary.

1.10   Given: mi/hr Find: km/s

Classification and Properties of Matter

1.12   (a)   homogeneous mixture

(b)   heterogeneous mixture (particles in liquid)

(c)   pure substance

(d)   heterogeneous mixture

1.14   (a)   C   (b) N   (c) Br   (d) Zn   (e) Fe   (f) phosphorus

(g)   calcium   (h) helium   (i) lead   (j) silver

1.16   Before modern instrumentation, the classification of a pure substance as an element was determined by whether it could be broken down into component elements. Scientists subjected the substance to all known chemical means of decomposition, and if the results were negative, the substance was an element. Classification by negative results was somewhat ambiguous, since an effective decomposition technique might exist but not yet have been discovered.

1.18   *Physical properties*: silver-grey (color); melting point = 420°C; hardness = 2.5 Mohs; density = 7.13 g/cm^3 at 25°C. *Chemical properties*: metal; reacts with sulfuric acid to produce hydrogen gas; reacts slowly with oxygen at elevated temperatures to produce ZnO.

1.20   (a)   chemical

(b)   physical

(c)   physical (The production of H_2O is a chemical change, but its **condensation** is a physical change.)

(d)   physical (The production of soot is a chemical change, but its **deposition** is a physical change.)

1.22   Take advantage of differences in physical properties to separate the components of a mixture. First heat the liquid to 100°C to evaporate the water. This is conveniently done in a distillation apparatus (Figure 1.13) so that the water can be collected. After the water is completely evaporated and if there is a residue, measure the physical properties of the residue such as color, density, and melting point. Compare the observed properties of the residue to those of table salt, NaCl. If the properties match, the colorless liquid contained table salt. If the properties don't match, the liquid contained a different dissolved solid. If there is no residue, no dissolved solid is present.

Units and Measurement

1.24   (a) $6.35 \times 10^{-2}\,L \; \times \; \dfrac{1\,mL}{1 \times 10^{-3}\,L} = 63.5\,mL$

(b) $6.5 \times 10^{-6}\,s \; \times \; \dfrac{1\,\mu s}{1 \times 10^{-6}\,s} = 6.5\,\mu s$

(c) $9.5 \times 10^{-4}\,m \; \times \; \dfrac{1\,mm}{1 \times 10^{-3}\,m} = 0.95\,mm$

(d) $4.23 \times 10^{-9} \, m^3 \times \dfrac{1^3 \, mm^3}{(1 \times 10^{-3})^3 \, m^3} = 4.23 \, mm^3$

$4.23 \, mm^3 \times \dfrac{(10^{-1})^3 \, cm^3}{1^3 \, mm^3} \times \dfrac{1 \, mL}{1 \, cm^3} \times \dfrac{1 \times 10^{-3} \, L}{1 \, mL} \times \dfrac{1 \, \mu L}{1 \times 10^{-6} \, L} = 4.23 \, \mu L$

(e) $12.5 \times 10^{-8} \, kg \times \dfrac{1 \times 10^3 \, g}{1 \, kg} \times \dfrac{1 \, mg}{1 \times 10^{-3} \, g} = 0.125 \, mg \, (125 \, \mu g)$

(f) $3.5 \times 10^{-10} \, g \times \dfrac{1 \, ng}{1 \times 10^{-9} \, g} = 0.35 \, ng$

(g) $6.54 \times 10^9 \, fs \times \dfrac{1 \times 10^{-15} \, s}{1 \, fs} \times \dfrac{1 \, \mu s}{1 \times 10^{-6} \, s} = 6.54 \, \mu s$

1.26 (a) $°C = 5/9 \, (87°F - 32°) = 31°C$

(b) $K = 25°C + 273 = 298 \, K$; $°F = 9/5 \, (25°C) + 32 = 77°F$

(c) $°C = 5/9 \, (175°F - 32°) = 79.444 = 79.4°C$

$K = °C + 273.15 = 79.444°C + 273.15 = 352.6 \, K$

(d) $°F = 9/5 \, (755°C) + 32 = 1391°F$; $K = 755°C + 273.15 = 1028 \, K$

(It could be argued that the result of 9/5 (755) has 3 sig figs, so the final Fahrenheit temperature should have 3 sig figs, 1390°F.)

(e) melting point = $-248.6°C + 273.15 = 24.6 \, K$
boiling point = $-246.1°C + 273.15 = 27.1 \, K$

1.28 (a) volume = length^3 (cm^3); density = mass/volume (g/cm^3)

volume = $(1.500)^3 \, cm^3 = 3.375 \, cm^3$

density = $\dfrac{76.31 \, g}{3.375 \, cm^3} = 22.61 \, g/cm^3$ osmium

(b) $125.0 \, mL \times \dfrac{1 \, cm^3}{1 \, mL} \times \dfrac{4.51 \, g}{1 \, cm^3} = 563.75 = 564 \, g$ titanium

(c) $0.1500 \, L \times \dfrac{1 \, mL}{1 \times 10^{-3} \, L} \times \dfrac{0.8787 \, g}{1 \, mL} = 131.8 \, g$ benzene

1.30 (a) $\dfrac{21.95 \, g}{25.0 \, mL} = 0.878 \, g/mL$

The tabulated value has four significant figures, while the experimental value has three. The tabulated value rounded to three figures is 0.879. The values agree within one in the last significant figure of the experimental value; the two results agree. The liquid could be benzene.

(b) $15.0 \, g \times \dfrac{1 \, mL}{0.7781 \, g} = 19.3 \, mL$ cyclohexane

(c) $r = d/2 = 5.0 \, cm/2 = 2.5 \, cm$

$V = 4/3 \, \pi \, r^3 = 4/3 \times \pi \times (2.5)^3 \, cm^3 = 65 \, cm^3$

$65.4498 \, cm^3 \times \dfrac{11.34 \, g}{cm^3} = 7.4 \times 10^2 \, g$

(The answer has two significant figures because the diameter had only two figures.)

Note: This is the first exercise where "intermediate rounding" occurs. In this manual, when a solution is given in steps, the intermediate result will be rounded to the correct number of significant figures. However, the **unrounded** number will be used in subsequent calculations. The final answer will appear with the correct number of significant figures. That is, calculators need not be cleared and new numbers entered in the middle of a calculation sequence. This may result in a small discrepancy in the last significant digit between student-calculated answers and those given in the manual. These variations occur in any analysis of numerical data.

For example, in this exercise the volume of the sphere, $65.4498 \, cm^3$, is rounded to $65 \, cm^3$, but 65.4498 is retained in the subsequent calculation of mass, $7.4 \times 10^2 \, g$. In this case, $65 \, cm^3 \times 11.34 \, g/cm^3$ also yields 7.4×10^2 g. In other exercises, the correctly rounded results of the two methods may not be identical.

1.32   Calculate the volume of the rod:

$2.17 \, kg \times \dfrac{1000 \, g}{1 \, kg} \times \dfrac{1 \, cm^3}{2.33 \, g} = 931.3 = 931 \, cm^3$

$V = \pi \, r^2 h; \; d = 2r, r = d/2; \quad V = \pi \left(\dfrac{d}{2}\right)^2 h; \quad d^2 = \dfrac{4 \, V}{\pi \, h}; d = \left(\dfrac{4 \, V}{\pi \, h}\right)^{1/2}$

$d = \left(\dfrac{4 \, (931.3) \, cm^3}{\pi \, (16.8) \, cm}\right)^{1/2} = 8.401 = 8.40 \, cm$

Uncertainty in Measurement

1.34   Exact: (b), (e) (The number of students is exact on any given day.)

1.36   (a)   4   (b) 3   (c) 4   (d) 5   (e) 6

1.38   (a) 7.93×10^3 mi   (b) 4.001×10^4 km

1.40   (a) $[320.5 - 6104.5/2.3] = -2.3 \times 10^3$ (The intermediate result has two significant figures, so only the thousand and hundred places in the answer are significant.)

         (b) $[285.3 \times 10^5 - 0.01200 \times 10^5] \times 2.8954 = 8.260 \times 10^7$ (Since subtraction depends on decimal places, both numbers must have the same exponent to determine decimal places/sig figs. The intermediate result has 1 decimal place and 4 sig figs, so the answer has 4 sig figs.)

(c) $\underset{\text{2 sig figs / 0 dec pl}}{(0.0045 \times 20{,}000.0)}$ + $\underset{\text{2 sig figs / first 2 digits}}{(2813 \times 12)}$ $= 3.4 \times 10^4$

(d) $\underset{\text{3 sig figs}}{863} \quad \times \quad \underset{\text{[0 dec pl/3 sig figs]}}{[1255 \quad - \quad \underset{\text{(3 sig figs / 0 dec pl)}}{(3.45 \times 108)]}}$ $= 7.62 \times 10^5$

                                $= 3$ sig figs

Dimensional Analysis

1.42   In each conversion factor, the old unit appears in the denominator, so it cancels, and the new unit appears in the numerator.

(a) $\mu m \rightarrow mm:$ $\dfrac{1 \times 10^{-6}\,m}{1\,\mu m} \times \dfrac{1\,mm}{1 \times 10^{-3}\,m} = 1 \times 10^{-3}\,mm/\mu m$

(b) $ms \rightarrow ns:$ $\dfrac{1 \times 10^{-3}\,s}{1\,ms} \times \dfrac{1\,ns}{1 \times 10^{-9}\,s} = 1 \times 10^6\,ns/ms$

(c) $mi \rightarrow km:$ $1.6093\,km/mi$

(d) $ft^3 \rightarrow L:$ $\dfrac{(12)^3\,in^3}{1\,ft^3} \times \dfrac{(2.54)^3\,cm^3}{1\,in^3} \times \dfrac{1\,L}{1000\,cm^3} = 28.3\,L/ft^3$

1.44   (a) $\dfrac{2.998 \times 10^8\,m}{s} \times \dfrac{1\,km}{1000\,m} \times \dfrac{60\,s}{1\,min} \times \dfrac{60\,min}{1\,hr} = 1.079 \times 10^9\,km/hr$

(b) $1454\,ft \times \dfrac{1\,yd}{3\,ft} \times \dfrac{1\,m}{1.0936\,yd} = 443.18 = 443.2\,m$

(c) $3{,}666{,}500\,m^3 \times \dfrac{1^3\,dm^3}{(1 \times 10^{-1})^3\,m^3} \times \dfrac{1\,L}{1\,dm^3} = 3.6665 \times 10^9\,L$

(d) $\dfrac{232\,mg\ cholesterol}{100\,mL\ blood} \times \dfrac{1\,mL}{1 \times 10^{-3}\,L} \times 5.2\,L \times \dfrac{1 \times 10^{-3}\,g}{1\,mg} = 12\,g\ cholesterol$

1.46   (a) $0.105\,in \times \dfrac{2.54\,cm}{in} \times \dfrac{1 \times 10^{-2}\,m}{cm} \times \dfrac{1\,mm}{1 \times 10^{-3}\,m} = 2.667 = 2.67\,mm$

(b) $0.650\,qt \times \dfrac{1\,L}{1.057\,qt} \times \dfrac{1\,mL}{1 \times 10^{-3}\,L} = 614.94 = 615\,mL$

(c) $\dfrac{8.75\,\mu m}{s} \times \dfrac{1 \times 10^{-6}\,m}{1\,\mu m} \times \dfrac{1\,km}{1 \times 10^3\,m} \times \dfrac{60\,s}{1\,min} \times \dfrac{60\,min}{1\,hr} = 3.15 \times 10^{-5}\,km/hr$

(d) $1.955\,m^3 \times \dfrac{(1.0936)^3\,yd^3}{1\,m^3} = 2.55695 = 2.557\,yd^3$

(e) $\dfrac{\$3.99}{lb} \times \dfrac{2.205\,lb}{1\,kg} = 8.798 = \$8.80/kg$

(f) $\dfrac{8.75\,\text{lb}}{\text{ft}^3} \times \dfrac{453.59\,\text{g}}{1\,\text{lb}} \times \dfrac{1\,\text{ft}^3}{12^3\,\text{in}^3} \times \dfrac{1\,\text{in}^3}{2.54^3\,\text{cm}^3} \times \dfrac{1\,\text{cm}^3}{1\,\text{mL}} = 0.140\,\text{g/mL}$

1.48   (a) $1486\,\text{mi} \times \dfrac{1\,\text{km}}{0.62137\,\text{mi}} \times \dfrac{\text{charge}}{225\,\text{km}} = 10.6\,\text{charges}$

Since charges are integral events, 11 charges are required.

(b) $\dfrac{14\,\text{m}}{\text{s}} \times \dfrac{1\,\text{km}}{1\times10^3\,\text{m}} \times \dfrac{1\,\text{mi}}{1.6093\,\text{km}} \times \dfrac{60\,\text{s}}{1\,\text{min}} \times \dfrac{60\,\text{min}}{1\,\text{hr}} = 31\,\text{mi/hr}$

(c) $450\,\text{in}^3 \times \dfrac{(2.54)^3\,\text{cm}^3}{1\,\text{in}^3} \times \dfrac{1\,\text{mL}}{1\,\text{cm}^3} \times \dfrac{1\times10^{-3}\,\text{L}}{1\,\text{mL}} = 7.37\,\text{L}$

(d) $2.4\times10^5\,\text{barrels} \times \dfrac{42\,\text{gal}}{1\,\text{barrel}} \times \dfrac{4\,\text{qt}}{1\,\text{gal}} \times \dfrac{1\,\text{L}}{1.057\,\text{qt}} = 3.8\times10^7\,\text{L}$

1.50 $9.0\,\text{ft} \times 14.5\,\text{ft} \times 18.8\,\text{ft} = 2453.4 = 2.5\times10^3\,\text{ft}^3$

$2453.4\,\text{ft}^3 \times \dfrac{(1\,\text{yd})^3}{(3\,\text{ft})^3} \times \dfrac{(1\,\text{m})^3}{(1.094\,\text{yd})^3} \times \dfrac{48\,\mu\text{g CO}}{1\,\text{m}^3} \times \dfrac{1\times10^{-6}\,\text{g}}{1\,\mu\text{g}} = 3.3\times10^{-3}\,\text{g CO}$

1.52   Select a common unit for comparison, in this case the kg.

1 kg > 2 lb, 1 L ≈ 1 qt

5 lb potatoes < 2.5 kg

5 kg sugar = 5 kg

1 gal = 4 qt ≈ 4 L. 1 mL H_2O = 1 g H_2O. 1 L = 1000 g, 4 L = 4000 g = 4 kg

The order of mass from lightest to heaviest is 5 lb potatoes < 1 gal water < 5 kg sugar.

1.54   A wire is a very long, thin cylinder of volume, $V = \pi r^2 h$, where h is the length of the wire and πr^2 is the cross-sectional area of the wire.

Strategy: 1) Calculate total volume of copper in cm^3 from mass and density

         2) $h\,(\text{length in cm}) = \dfrac{V}{\pi r^2}$

         3)   Change cm → ft

$150\,\text{lb Cu} \times \dfrac{453.6\,\text{g}}{1\,\text{lb Cu}} \times \dfrac{1\,\text{cm}^3}{8.94\,\text{g}} = 7610.7 = 7.61\times10^3\,\text{cm}^3$

$r = d/2 = 8.25\,\text{mm} \times \dfrac{1\,\text{cm}}{10\,\text{mm}} \times \dfrac{1}{2} = 0.4125 = 0.413\,\text{cm}$

$h = \dfrac{V}{\pi r^2} = \dfrac{7610.7\,\text{cm}^3}{\pi(0.4125)^2\,\text{cm}^2} = 1.4237\times10^4 = 1.42\times10^4\,\text{cm}$

$1.4237\times10^4\,\text{cm} \times \dfrac{1\,\text{in}}{2.54\,\text{cm}} \times \dfrac{1\,\text{ft}}{12\,\text{in}} = 467\,\text{ft}$

(too difficult to estimate)

Additional Exercises

1.56   (a)   A gold coin is probably a *solid solution*. Pure gold (element 79) is too soft and too valuable to be used for coinage, so other metals are added. However, the simple term "gold coin" does not give a specific indication of the other metals in the mixture.

A cup of coffee is a *solution* if there are no suspended solids (coffee grounds). It is a heterogeneous mixture if there are grounds. If cream or sugar is added, the homogeneity of the mixture depends on how thoroughly the components are mixed.

A wood plank is a *heterogeneous mixture* of various cellulose components. The different domains in the mixture are visible as wood grain or knots.

(b)   The ambiguity in each of these examples is that the name of the substance does not provide a complete description of the material. We must rely on mental images, and these vary from person to person.

1.57   (a)   A *hypothesis* is a possible explanation for certain phenomena based on preliminary experimental data. A *theory* may be more general, and has a significant body of experimental evidence to support it; a theory has withstood the test of experimentation.

(b)   A scientific *law* is a summary or statement of natural behavior; it tells how matter behaves. A *theory* is an explanation of natural behavior; it attempts to explain why matter behaves the way it does.

1.59   (a)   I. $(22.52 + 22.48 + 22.54)/3 = 22.51$

II. $(22.64 + 22.58 + 22.62)/3 = 22.61$

Based on the average, set I is more accurate. That is, it is closer to the true value of 22.52%.

(b)   Average deviation = $\sum$ | value – average | /3

I. | 22.52 – 22.51 | + | 22.48 – 22.51 | + | 22.54 – 22.51 | /3 = 0.02

II. | 22.64 – 22.61 | + | 22.58 – 22.61 | + | 22.62 – 22.61 | /3 = 0.02

The two sets display the same precision, even though set I is more accurate.

1.60   (a)   Inappropriate. The circulation of a widely-read publication like *National Geographic* would vary over a year's time, and could simply not be counted to the nearest single subscriber. Probably about four significant figures would be appropriate.

(b)   Inappropriate. In a county with 5 million people, the population fluctuates with moves, births, and deaths. The population cannot be known precisely to the nearest person, even over the course of a day. There would be uncertainty in at least the tens, probably the hundreds place in the population.

(c)   Appropriate. The percentage has three significant figures. In a population as large as the United States, the number of people named Brown can surely be counted by census data or otherwise to a precision of three significant figures.

1.62   (a) $\dfrac{m}{s^2}$ (b) $\dfrac{kg - m}{s^2}$ (c) $\dfrac{kg - m}{s^2} \times m = \dfrac{kg - m^2}{s^2}$

       (d) $\dfrac{kg - m}{s^2} \times \dfrac{1}{m^2} = \dfrac{kg}{m - s^2}$ (e) $\dfrac{kg - m^2}{s^2} \times \dfrac{1}{s} = \dfrac{kg - m^2}{s^3}$

1.63   (a) $2.4 \times 10^5 \text{ mi} \times \dfrac{1.609 \text{ km}}{1 \text{ mi}} \times \dfrac{1000 \text{ m}}{1 \text{ km}} = 3.9 \times 10^8 \text{ m}$

       (b) $2.4 \times 10^5 \text{ mi} \times \dfrac{1.609 \text{ km}}{1 \text{ mi}} \times \dfrac{1 \text{ hr}}{350 \text{ km}} \times \dfrac{60 \text{ min}}{1 \text{ hr}} \times \dfrac{60 \text{ s}}{1 \text{ min}} = 4.0 \times 10^6 \text{ s}$

1.65   (a) $\dfrac{\$1950}{acre - ft} \times \dfrac{1 \text{ acre}}{4840 \text{ yd}^2} \times \dfrac{3 \text{ ft}}{1 \text{ yd}} \times \dfrac{(1.094 \text{ yd})^3}{(1 \text{ m})^3} \times \dfrac{(1 \text{ m})^3}{(10 \text{ dm})^3} \times \dfrac{(1 \text{ dm})^3}{1 \text{ L}} =$

          $\$1.583 \times 10^{-3}/L$ or $0.1583 \text{ ¢}/L$ $(0.158 \text{ ¢}/L$ to 3 sig figs)

       (b) $\dfrac{\$1950}{acre - ft} \times \dfrac{1 \text{ acre} - ft}{2 \text{ households} - year} \times \dfrac{1 \text{ year}}{365 \text{ days}} \times 1 \text{ household} = \dfrac{\$2.671}{day} = \dfrac{\$2.67}{day}$

1.66   There are 209.1 degrees between the freezing and boiling points on the Celsius (C) scale and 100 degrees on the glycol (G) scale. Also, –11.5°C = 0°G. By analogy with °F and °C,

      $°G = \dfrac{100}{209.1}(°C + 11.5)$ or $°C = \dfrac{209.1}{100}(°G) - 11.5$

These equations correctly relate the freezing point (and boiling point) of ethylene glycol on the two scales.

      f.p. of H_2O: $°G = \dfrac{100}{209.1} (0°C + 11.5) = 5.50°G$

1.68   Density is the ratio of mass and volume. For substances with different densities, the greater the density the smaller the volume of substance that will contain a certain mass. Since volume is directly related to diameter ($V = 4/3 \, \pi \, r^3 = 1/6 \, \pi \, d^3$), the more dense the substance, the smaller the diameter of a ball that contains a certain mass. The order of the sphere sizes (diameters) is the reverse order of densities: Pb < Ag < Al.

1.69   The mass of water in the bottle does not change with temperature, but the density (ratio of mass to volume) does. That is, the amount of volume occupied by a certain mass of water changes with temperature. Calculate the mass of water in the bottle at 25°C, and then the volume occupied by this mass at –10°C.

       (a)   25°C: $1.50 \text{ L } H_2O \times \dfrac{1000 \text{ cm}^3}{1 \text{ L}} \times \dfrac{0.997 \text{ g } H_2O}{1 \text{ cm}^3} = 1.4955 \times 10^3 = 1.50 \times 10^3 \text{ g } H_2O$

          –10°C: $1.4955 \times 10^3 \text{ g } H_2O \times \dfrac{1 \text{ cm}^3}{0.917 \text{ g } H_2O} \times \dfrac{1 \text{ L}}{1000 \text{ cm}^3} = 1.6309 = 1.63 \text{ L}$

       (b)   If the soft-drink bottle is completely filled with 1.50 L of water, the 1.63 L of ice **cannot** be contained in the bottle. The extra volume of ice will push through any opening in the bottle, or crack the bottle to create an opening.

1.71   (a)   density $= (35.66 \text{ g} - 14.23 \text{ g})/4.59 \text{ cm}^3 = 4.67 \text{ g/cm}^3$

(b) $34.5 \text{ kg} \times \dfrac{1000 \text{ g}}{1 \text{ kg}} \times \dfrac{1 \text{ mL}}{13.6 \text{ g}} \times \dfrac{1 \text{ L}}{1000 \text{ mL}} = 2.54 \text{ L}$

(c) $V = 4/3 \, \pi \, r^3 = 4/3 \, \pi \, (28.9 \text{ cm})^3 = 1.0111 \times 10^5 = 1.01 \times 10^5 = 1.01 \times 10^5 \text{ cm}^3$

$1.011 \times 10^5 \text{ cm}^3 \times \dfrac{19.3 \text{ g}}{\text{cm}^3} = 1.95 \times 10^6 \text{ g}$

The sphere weighs 1950 kg or 4300 pounds. The student is unlikely to be able to carry the sphere.

1.72 $0.500 \text{ L battery acid} \times \dfrac{1000 \text{ mL}}{\text{L}} \times \dfrac{1.28 \text{ g}}{\text{mL}} = 640 \text{ g battery acid}$

$640 \text{ g battery acid} \times \dfrac{38.1 \text{ g sulfuric acid}}{100 \text{ g battery acid}} = 243.84 = 244 \text{ g sulfuric acid}$

1.74   (a)   Calculate the volume of the coin. It is a cylinder,

$V = \pi \, r^2 h, \;\; r = \dfrac{d}{2}, \;\; V = \pi \left(\dfrac{d}{2}\right)^2 h = \dfrac{\pi d^2 h}{4}$

Then use density of pure gold to calculate mass.

$V = \pi \times \dfrac{(2.2)^2 \text{ cm}^2}{4} \times 3.0 \text{ mm} \times \dfrac{1 \text{ cm}}{10 \text{ mm}} = 1.140 = 1.1 \text{ cm}^3$

$1.140 \text{ cm}^3 \times \dfrac{19.3 \text{ g gold}}{1 \text{ cm}^3} = 22.01 = 22 \text{ g pure gold}$

(b) $22.01 \text{ g gold} \times \dfrac{1 \text{ tr oz}}{31.1 \text{ g}} \times \dfrac{\$640}{\text{tr oz}} = \$452.93 = 4.5 \times 10^2 \ (\$450)$

1.75 $8.0 \text{ oz} \times \dfrac{1 \text{ lb}}{16 \text{ oz}} \times \dfrac{453.6 \text{ g}}{\text{lb}} \times \dfrac{1 \text{ cm}^3}{2.70 \text{ g}} = 84.00 = 84 \text{ cm}^3$

$\dfrac{84 \text{ cm}^3}{50 \text{ ft}^2} \times \dfrac{1^2 \text{ ft}^2}{12^2 \text{ in}^2} \times \dfrac{1^2 \text{ in}^2}{2.54^2 \text{ cm}^2} \times \dfrac{10 \text{ mm}}{1 \text{ cm}} = 0.018 \text{ mm}$

1.77   (a)   Let x = mass of Au in jewelry

9.85 - x = mass of Ag in jewelry

The total volume of jewelry = volume of Au + volume of Ag

$0.675 \text{ cm}^3 = x \text{ g} \times \dfrac{1 \text{ cm}^3}{19.3 \text{ g}} + (9.85 - x)\text{g} \times \dfrac{1 \text{ cm}^3}{10.5 \text{ g}}$

$0.675 = \dfrac{x}{19.3} + \dfrac{9.85 - x}{10.5}$ (To solve, multiply both sides by (19.3)(10.5))

$$0.675\,(19.3)(10.5) = 10.5\,x + (9.85 - x)(19.3)$$

$$136.79 = 10.5\,x + 190.105 - 19.3\,x$$

$$-53.315 = -8.8\,x$$

$$x = 6.06\ \text{g Au};\ 9.85\ \text{g total} - 6.06\ \text{g Au} = 3.79\ \text{g Ag}$$

$$\text{mass \% Au} = \frac{6.06\ \text{g Au}}{9.85\ \text{g jewelry}} \times 100 = 61.5\%\ \text{Au}$$

(b)   24 carats $\times$ 0.615 = 15 carat gold

1.78   A solution can be separated into components by physical means, so separation would be attempted. If the liquid is a solution, the solute could be a solid or a liquid; these two kinds of solutions would be separated differently. Therefore, divide the liquid into several samples and do different tests on each. Try evaporating the solvent from one sample. If a solid remains, the liquid is a solution and the solute is a solid. If the result is negative, try distilling a sample to see if two or more liquids with different boiling points are present. If this result is negative, the liquid is probably a pure substance, but negative results are never entirely conclusive. We might not have tried the appropriate separation technique.

1.80   The densities are:

carbon tetrachloride (methane, tetrachloro) – 1.5940 g/cm^3

hexane – 0.6603 g/cm^3

benzene – 0.87654 g/cm^3

methylene iodide (methane, diiodo) – 3.3254 g/cm^3

Only methylene iodide will separate the two granular solids. The undesirable solid (2.04 g/cm^3) is less dense than methylene iodide and will float; the desired material is more dense than methylene iodide and will sink. The other three liquids are less dense than both solids and will not produce separation.

1.82   Study (a) is likely to be both precise and accurate, because the errors are carefully controlled. The secondary weight standard will be resistant to chemical and physical changes, the balance is carefully calibrated, and weighings are likely to be made by the same person. The relatively large number of measurements is likely to minimize the effect of random errors on the average value. The accuracy and precision of study (b) depend on the veracity of the participants' responses, which cannot be carefully controlled. It also depends on the definition of "comparable lifestyle." The percentages are not precise, because the broad definition of lifestyle leads to a range of results (scatter). The relatively large number of participants improves the precision and accuracy. In general, controlling errors and maximizing the number of data points in a study improves precision and accuracy.

2 Atoms, Molecules, and Ions

Visualizing Concepts

2.2   (a)   % abundance $= \dfrac{\text{\# of mass number} \times \text{particles}}{\text{total number of particles}} \times 100$

12 red ^{293}Nv particles

8 blue ^{295}Nv particles

20 total particles

% abundance ^{293}Nv $= \dfrac{12}{20} \times 100 = 60\%$

% abundance ^{295}Nv $= \dfrac{8}{20} \times 100 = 40\%$

(b)   Atomic weight (AW) is the same as average atomic mass.

Atomic weight (average atomic mass) $= \sum$ fractional abundance $\times$ mass of isotope

AW of Nv $= 0.60(293.15) + 0.40(295.15) = 293.95$

(Since % abundance was calculated by counting exact numbers of particles, assume % abundance is an exact number. Then, the number of significant figures in the AW is determined by the number of sig figs in the masses of the isotopes.)

2.3   In general, metals occupy the left side of the chart, and nonmetals the right side.

metals: red and green   *nonmetals*: blue and yellow

alkaline earth metal: red   *noble gas*: yellow

2.5   In a solid, particles are close together and their relative positions are fixed. In a liquid, particles are close but moving relative to each other. In a gas, particles are far apart and moving. All ionic compounds are solids because of the strong forces among charged particles. Molecular compounds can exist in any state: solid, liquid, or gas.

Since the molecules in *ii* are far apart, *ii* must be a molecular compound. The particles in *i* are near each other and exist in a regular, ordered arrangement, so *i* is likely to be an ionic compound.

2.7   See Figure 2.22. yellow box: 1+ (group 1A); blue box: 2+ (group 2A)

          black box: 3+ (a metal in Group 3A); orange box: 2– (a nonmetal in group 6A);

          green box: 1– (a nonmetal in group 7A)

2.8   Cations (red spheres) have positive charges; anions (blue spheres) have negative charges. There are twice as many anions as cations, so the formula has the general form CA_2. Only $Ca(NO_3)_2$, calcium nitrate, is consistent with the diagram.

Atomic Theory and the Discovery of Atomic Structure

2.10   (a)   6.500 g compound – 0.384 g hydrogen = 6.116 g sulfur

          (b)   *Conservation of mass*

          (c)   According to postulate 3 of the atomic theory, atoms are neither created nor destroyed during a chemical reaction. If 0.384 g of H are recovered from a compound that contains only H and S, the remaining mass must be sulfur.

2.12   (a)   1: $\dfrac{3.56 \text{ g fluorine}}{4.75 \text{ g iodine}} = 0.749 \text{ g fluorine/1 g iodine}$

                    2:     $\dfrac{3.43 \text{ g fluorine}}{7.64 \text{ g iodine}} = 0.449 \text{ g fluorine/1 g iodine}$

                    3:     $\dfrac{9.86 \text{ g fluorine}}{9.41 \text{ g iodine}} = 1.05 \text{ g fluorine/1 g iodine}$

          (b)   To look for integer relationships among these values, divide each one by the smallest.

               If the quotients aren't all integers, multiply by a common factor to obtain all integers.

               1: 0.749/0.449 = 1.67; 1.67 × 3 = 5

               2: 0.449/0.449 = 1.00; 1.00 × 3 = 3

               3: 1.05/0.449 = 2.34; 2.34 × 3 = 7

               The ratio of g fluorine to g iodine in the three compounds is 5:3:7. These are in the ratio of small whole numbers and, therefore, obey the *law of multiple proportions*. This integer ratio indicates that the combining fluorine "units" (atoms) are indivisible entities.

2.14   Since the unknown particle is deflected in the opposite direction from that of a negatively charged beta (β) particle, it is attracted to the (–) plate and repelled by the (+) plate. The unknown particle is positively charged. The magnitude of the deflection is less than that of the β particle, or electron, so the unknown particle has greater mass than the electron. The unknown is a positively charged particle of greater mass than the electron.

2.16   (a)   The droplets carry different total charges because there may be 1, 2, 3, or more electrons on the droplet.

(b) The electronic charge is likely to be the lowest common factor in all the observed charges.

(c) Assuming this is so, we calculate the apparent electronic charge from each drop as follows:

A: $1.60 \times 10^{-19} / 1 = 1.60 \times 10^{-19}$ C

B: $3.15 \times 10^{-19} / 2 = 1.58 \times 10^{-19}$ C

C: $4.81 \times 10^{-19} / 3 = 1.60 \times 10^{-19}$ C

D: $6.31 \times 10^{-19} / 4 = 1.58 \times 10^{-19}$ C

The reported value is the average of these four values. Since each calculated charge has three significant figures, the average will also have three significant figures.

$(1.60 \times 10^{-19} \text{ C} + 1.58 \times 10^{-19} \text{ C} + 1.60 \times 10^{-19} \text{ C} + 1.58 \times 10^{-19} \text{ C}) / 4 = 1.59 \times 10^{-19}$ C

Modern View of Atomic Structure; Atomic Weights

2.18 (a) $r = d/2 ; r = \dfrac{2.8 \times 10^{-8} \text{ cm}}{2} \times \dfrac{1 \text{ Å}}{1 \times 10^{-8} \text{ cm}} = 1.4$ Å

$r = \dfrac{2.8 \times 10^{-8} \text{ cm}}{2} \times \dfrac{1 \text{ m}}{100 \text{ cm}} = 1.4 \times 10^{-10}$ m

(b) Aligned Sn atoms have **diameters** touching. $d = 2.8 \times 10^{-8}$ cm $= 2.8 \times 10^{-10}$ m

$6.0 \, \mu\text{m} \times \dfrac{1 \times 10^{-6} \text{ m}}{1 \, \mu\text{m}} \times \dfrac{1 \text{ Sn atom}}{2.8 \times 10^{-10} \text{ m}} = 2.1 \times 10^{4}$ Sn atoms

(c) $V = 4/3 \, \pi \, r^{3} ; r = 1.4 \times 10^{-10}$ m

$V = (4/3)[(\pi(1.4 \times 10^{-10})^{3}] \text{ m}^{3} = 1.149 \times 10^{-29} = 1.1 \times 10^{-29} \text{ m}^{3}$

2.20 (a) The nucleus has most of the mass **but occupies very little** of the volume of an atom.

(b) True

(c) The number of electrons in an atom is equal to the number of **protons** in the atom.

(d) True

2.22 (a) $^{31}_{16}$X and $^{32}_{16}$X are isotopes of the same element, because they have identical atomic numbers.

(b) These are isotopes of the element sulfur, S, atomic number = 16.

2.24 (a) ^{32}P has 15 p, 17 n   (b) ^{51}Cr has 24 p, 27 n

(c) ^{60}Co has 27 p, 33 n   (d) ^{99}Tc has 43 p, 56 n

(e) ^{131}I has 53 p, 78 n   (f) ^{201}Tl has 81 p, 120 n

2.26

Symbol	^{65}Zn	^{101}Ru	^{87}Sr	^{108}Ag	^{235}U
Protons	30	44	38	47	92
Neutrons	35	57	49	61	143
Electrons	30	44	38	47	92
Mass No.	65	101	87	108	235

2.28 Since the two nuclides are atoms of the same element, by definition they have the same number of protons, 54. They differ in mass number (and mass) because they have different numbers of neutrons. ^{129}Xe has 75 neutrons and ^{130}Xe has 76 neutrons.

2.30 (a) 12 amu

(b) The atomic weight of carbon reported on the front-inside cover of the text is the abundance-weighted average of the atomic masses of the two naturally occurring isotopes of carbon, ^{12}C, and ^{13}C. The mass of a ^{12}C atom is exactly 12 amu, but the atomic weight of 12.011 takes into account the presence of some ^{13}C atoms in every natural sample of the element.

2.32 Atomic weight (average atomic mass) = Σ fractional abundance $\times$ mass of isotope

Atomic weight = 0.7215(84.9118) + 0.2785(86.9092) = 85.4681 = 85.47

(The result has 2 decimal places and 4 sig figs because each term in the sum has 4 sig figs and 2 decimal places.)

2.34 (a) The purpose of the magnet in the mass spectrometer is to change the path of the moving ions. The magnitude of the deflection is inversely related to mass, which is the basis of the discrimination by mass.

(b) The atomic weight of Cl, 35.5, is an average atomic mass. It is the average of the masses of two naturally occurring isotopes, weighted by their abundances.

(c) The single peak at mass 31 in the mass spectrum of phosphorus indicates that the sample contains a single isotope of P, and the mass of this isotope is 31 amu.

2.36 (a) Three peaks: $^{1}H - {}^{1}H$, $^{1}H - {}^{2}H$, $^{2}H - {}^{2}H$

(b) $^{1}H - {}^{1}H = 2(1.00783) = 2.01566$ amu

$^{1}H - {}^{2}H = 1.00783 + 2.01410 = 3.02193$ amu

$^{2}H - {}^{2}H = 2(2.01410) = 4.02820$ amu

The mass ratios are 1 : 1.49923 : 1.99845 or 1 : 1.5 : 2.

(c) $^{1}H - {}^{1}H$ is largest, because there is the greatest chance that two atoms of the more abundant isotope will combine.

$^{2}H - {}^{2}H$ is the smallest, because there is the least chance that two atoms of the less abundant isotope will combine.

The Periodic Table; Molecules and Ions

2.38   (a)   calcium (metal)   (b) titanium (metal)   (c) gallium (metal)

(d)   thorium (metal)   (e) platinum (metal)   (f) selenium (nonmetal)

(g)   krypton (nonmetal)

2.40   C, carbon, nonmetal; Si, silicon, metalloid; Ge, germanium, metalloid; Sn, tin, metal;

Pb, lead, metal

2.42   Compounds with the same empirical but different molecular formulas differ by the integer number of empirical formula units in the respective molecules. Thus, they can have very different molecular structure, size, and mass, resulting in very different physical properties.

2.44   A molecular formula contains all atoms in a molecule. An empirical formula shows the simplest ratio of atoms in a molecule or elements in a compound.

(a)   molecular formula: C_6H_6; empirical formula: CH

(b)   molecular formula: $SiCl_4$; empirical formula: $SiCl_4$ (1:4 is the simplest ratio)

(c)   molecular: B_2H_6; empirical: BH_3

(d)   molecular: $C_6H_{12}O_6$; empirical: CH_2O

2.46   (a)   4   (b) 8   (c) 9

2.48   (a) C_2H_6O (b) C_2H_6O

(c) CH_4O (d) PF_3

2.50

Symbol	$^{31}P^{3-}$	$^{80}Br^-$	$^{115}In^{3+}$	$^{197}Au^{3+}$
Protons	15	35	49	79
Neutrons	16	45	66	118
Electrons	18	36	46	76
Net Charge	3–	1–	3+	3+

2.52   (a) Ga^{3+} (b) Sr^{2+} (c) As^{3-} (d) Br^- (e) Se^{2-}

2.54   (a)   AgI   (b) Ag_2S (c)   AgF

2.56   (a) $CuBr_2$ (b) Fe_2O_3 (c) Hg_2CO_3 (d) $Ca_3(AsO_4)_2$ (e) $(NH_4)_2CO_3$

2.58

Ion	Na^+	Ca^{2+}	Fe^{2+}	Al^{3+}
O^{2-}	Na_2O	CaO	FeO	Al_2O_3
NO_3^-	$NaNO_3$	$Ca(NO_3)_2$	$Fe(NO_3)_2$	$Al(NO_3)_3$
SO_4^{2-}	Na_2SO_4	$CaSO_4$	$FeSO_4$	$Al_2(SO_4)_3$
AsO_4^{2-}	Na_3AsO_4	$Ca_3(AsO_4)_2$	$Fe_3(AsO_4)_2$	$AlAsO_4$

2.60   Molecular (all elements are nonmetals):

     (a) PF_5 (c)   SCl   (h) N_2O_4

     Ionic (formed from ions, usually contains a metal cation):

     (b)   NaI   (d) $Ca(NO_3)_2$ (e) $FeCl_3$ (f)   LaP   (g) $CoCO_3$

Naming Inorganic Compounds; Organic Molecules

2.62   (a)   selenate   (b)   selenide   (c)   hydrogen selenide (biselenide)

     (d)   hydrogen selenite (biselenite)

2.64   (a)   copper, 2+; sulfide, 2–   (b)   silver, 1+; sulfate, 2–

     (c)   aluminum, 3+; chlorate, 1–   (d)   cobalt, 2+; hydroxide, 1–

     (e)   lead, 2+; carbonate, 2–

2.66   (a)   potassium oxide   (b)   sodium chlorite

     (c)   strontium cyanide   (d)   cobalt(II) hydroxide (cobaltous hydroxide)

     (e)   iron(III) carbonate (ferric carbonate)

     (f)   chromium(III) nitrate (chromic nitrate)

     (g)   ammonium sulfite   (h)   sodium dihydrogen phosphate

     (i)   potassium permanganate   (j)   silver dichromate

2.68   (a) Na_3PO_4 (b) $Zn(NO_3)_2$ (c) $Ba(BrO_3)_2$ (d) $Fe(ClO_4)_2$

     (e) $Co(HCO_3)_2$ (f) $Cr(CH_3COO^-)_3$ (g) $K_2Cr_2O_7$

2.70   (a)   HBr   (b) H_2S (c) HNO_2

     (d)   carbonic acid   (e)   chloric acid   (f)   acetic acid

2.72 (a) dinitrogen monoxide (b) nitrogen monoxide (c) nitrogen dioxide

(d) dinitrogen pentoxide (e) dinitrogen tetroxide

2.74 (a) $NaHCO_3$ (b) $Ca(ClO)_2$ (c) HCN

(d) $Mg(OH)_2$ (e) SnF (f) CdS, H_2SO_4, H_2S

2.76 (a) *-ane*

(b) **Hexane** has 6 carbons in its chain.

molecular: C_6H_{14}

empirical: C_3H_7

2.78 (a) They both have two carbon atoms in their molecular backbone, or chain.

(b) In 1-propanol one of the H atoms on an outer (terminal) C atom has been replaced by an $-OH$ group.

Additional Exercises

2.80 (a) Most of the volume of an atom is empty space in which electrons move. Most alpha particles passed through this space. The path of the massive alpha particle would not be significantly altered by interaction with a "puny" electron.

(b) Most of the mass of an atom is contained in a very small, dense area called the nucleus. The few alpha particles that hit the massive, positively charged gold nuclei were strongly repelled and essentially deflected back in the direction they came from.

(c) The Be nuclei have a much smaller volume and positive charge than the Au nuclei; the charge repulsion between the alpha particles and the Be nuclei will be less, and there will be fewer direct hits because the Be nuclei have an even smaller volume than the Au nuclei. Fewer alpha particles will be scattered in general and fewer will be strongly back scattered.

2.81 (a) Droplet D would fall most slowly. It carries the most negative charge, so it would be most strongly attracted to the upper (+) plate and most strongly repelled by the lower (–) plate. These electrostatic forces would provide the greatest opposition to gravity.

(b) Calculate the lowest common factor.

A: 3.84×10^{-8} / 2.88×10^{-8} = 1.33; $1.33 \times 3 = 4$

B: 4.80×10^{-8} / 2.88×10^{-8} = 1.67; $1.67 \times 3 = 5$

C: 2.88×10^{-8} / 2.88×10^{-8} = 1.00; 1.00 × 3 = 3

D: 8.64×10^{-8} / 2.88×10^{-8} = 3.00; 3.00 × 3 = 9

The total charge on the drops is in the ratio of 4:5:3:9. Divide the total charge on each drop by the appropriate integer and average the four values to get the charge of an electron in warmombs.

A: 3.84×10^{-8} / 4 = 9.60×10^{-9} wa

B: 4.80×10^{-8} / 5 = 9.60×10^{-9} wa

C: 2.88×10^{-8} / 3 = 9.60×10^{-9} wa

D: 8.64×10^{-8} / 9 = 9.60×10^{-9} wa

The charge on an electron is 9.60×10^{-9} wa

(c)   The number of electrons on each drop are the integers calculated in part (b). A has 4 e^-, B has 5 e^-, C has 3 e^- and D has 9 e^-.

(d) $$\frac{9.60 \times 10^{-9}\text{ wa}}{1\,e^-} \times \frac{1\,e^-}{1.60 \times 10^{-16}\text{ C}} = 6.00 \times 10^{7}\text{ wa/C}$$

2.83   (a)   2 protons and 2 neutrons

        (b)   the nuclear strong force

        (c)   The charge of an α particle is twice the magnitude of the charge of an electron, with the opposite sign. That is, 2 $(+1.6022 \times 10^{-19})$ C = $+3.2044 \times 10^{-19}$ C.

        (d) $$\frac{3.2044 \times 10^{-19}\text{ C}}{4.8224 \times 10^{4}\text{ g/C}} = 6.6448 \times 10^{-24}\text{ g}$$

            $$6.6448 \times 10^{-24}\text{ g} \times \frac{1\,\text{amu}}{1.66054 \times 10^{-24}\text{ g}} = 4.0016\,\text{amu}$$

        (e)   The sum of the particle masses in an α particle is 2(1.0073) amu and 2(1.0087) amu = 4.0320 amu. The actual particle mass, 4.0016 amu, is less than the sum of the masses of the components. The difference is the nuclear binding energy, the energy released when protons and neutrons combine to form a nucleus. Mass and energy are interchangeable according to the Einstein relationship $E = mc^2$.

2.84   (a)   Calculate the mass of a single gold atom, then divide the mass of the cube by the mass of the gold atom.

            $$\frac{197.0\,\text{amu}}{\text{gold atom}} \times \frac{1\,\text{g}}{6.022 \times 10^{23}\,\text{amu}} = 3.2713 \times 10^{-22} = 3.271 \times 10^{-22}\text{ g/gold atom}$$

            $$\frac{19.3\,\text{g}}{\text{cube}} \times \frac{1\,\text{gold atom}}{3.271 \times 10^{-22}\text{ g}} = 5.90 \times 10^{22}\text{ Au atoms in the cube}$$

        (b)   The shape of atoms is spherical; spheres cannot be arranged into a cube so that there is no empty space. The question is, how much empty space is there? We can calculate the two limiting cases, no empty space and maximum empty space. The true diameter will be somewhere in this range.

No empty space: volume cube/number of atoms = volume of one atom

$V = 4/3 \pi r^3; r = (3\pi V/4)^{1/3}; d = 2r$

$$\text{vol. of cube} = (1.0 \times 1.0 \times 1.0) = \frac{1.0 \, cm^3}{5.90 \times 10^{22} \, Au \, atoms} = 1.695 \times 10^{-23}$$

$$= 1.7 \times 10^{-23} \, cm^3$$

$r = [\pi (1.695 \times 10^{-23} \, cm^3)/4]^{1/3} = 3.4 \times 10^{-8} \, cm; d = 2r = 6.8 \times 10^{-8} \, cm$

Maximum empty space: assume atoms are arranged in rows in all three directions so they are touching across their diameters. That is, each atom occupies the volume of a cube, with the atomic diameter as the length of the side of the cube. The number of atoms along one edge of the gold cube is then

$(5.90 \times 10^{22})^{1/3} = 3.893 \times 10^7 = 3.89 \times 10^7$ atoms/1.0 cm.

The diameter of a single atom is 1.0 cm/3.89×10^7 atoms = 2.569×10^{-8}

$$= 2.6 \times 10^{-8} \, cm.$$

The diameter of a gold atom is between 2.6×10^{-8} cm and 6.8×10^{-8} cm (2.6 – 6.8 Å).

(c)   Some atomic arrangement must be assumed, since none is specified. The solid state is characterized by an orderly arrangement of particles, so it isn't surprising that atomic arrangement is required to calculate the density of a solid. A more detailed discussion of solid-state structure and density appears in Chapter 11.

2.86   (a)   diameter of nucleus = 1×10^{-4} Å; diameter of atom = 1 Å

$V = 4/3 \pi r^3; r = d/2; r_n = 0.5 \times 10^{-4}$ Å; $r_a = 0.5$ Å

volume of nucleus = $4/3 \pi (0.5 \times 10^{-4})^3$ Å^3

volume of atom = $4/3 \pi (0.5)^3$ Å^3

$$\text{volume fraction of nucleus} = \frac{\text{volume of nucleus}}{\text{volume of atom}} = \frac{4/3 \pi (0.5 \times 10^{-4})^3 \, Å^3}{4/3 \pi (0.5)^3 \, Å^3} = 1 \times 10^{-12}$$

diameter of atom = 5 Å, r_a = 2.5 Å

$$\text{volume fraction of nucleus} = \frac{4/3 \pi (0.5 \times 10^{-4})^3 \, Å^3}{4/3 \pi (2.5)^3 \, Å^3} = 8 \times 10^{-15}$$

Depending on the radius of the atom, the volume fraction of the nucleus is between 1×10^{-12} and 8×10^{-15}, that is, between 1 part in 10^{12} and 8 parts in 10^{15}.

(b)   mass of proton = 1.0073 amu

1.0073 amu × 1.66054×10^{-24} g/amu = 1.6727×10^{-24} g

diameter = 1.0×10^{-15} m, radius = 0.50×10^{-15} m × $\dfrac{100 \, cm}{1 \, m}$ = 5.0×10^{-14} cm

Assuming a proton is a sphere, $V = 4/3 \pi r^3$.

$$\text{density} = \frac{g}{cm^3} = \frac{1.6727 \times 10^{-24}\ g}{4/3\pi\,(5.0 \times 10^{-14})^3\ cm^3} = 3.2 \times 10^{15}\ g/cm^3$$

2.87   The integer on the lower left of a nuclide is the atomic number; it is the number of protons in any atom of the element and gives the element's identity. The number of neutrons is the mass number (upper left) minus atomic number.

(a)   As, 33 protons, 41 neutrons

(b)   I, 53 protons, 74 neutrons

(c)   Eu, 63 protons, 89 neutrons

(d)   Bi, 83 protons, 126 neutrons

2.89 $F = k\,Q_1 Q_2/d^2$; $k = 9.0 \times 10^9\ N\,m^2/C^2$; $d = 0.53 \times 10^{-10}\ m$;

Q (electron) $= -1.6 \times 10^{-19}\ C$; Q (proton) $= -Q$ (electron) $= 1.6 \times 10^{-19}\ C$

$$F = \frac{\dfrac{9.0 \times 10^9\ N\,m^2}{C^2} \times -1.6 \times 10^{-19}\ C \times 1.6 \times 10^{-19}\ C}{(0.53 \times 10^{-10})^2\ m^2} = 8.202 \times 10^{-8} = 8.2 \times 10^{-8}\ N$$

2.90   Atomic weight (average atomic mass) $= \Sigma$ fractional abundance $\times$ mass of isotope

Atomic weight $= 0.014(203.97302) + 0.241(205.97444) + 0.221(206.97587) +$
$$0.524(207.97663) = 207.22 = 207\ amu$$

(The result has 0 decimal places and 3 sig figs because the fourth term in the sum has 3 sig figs and 0 decimal places.)

2.92   (a)   There are 24 known isotopes of Ni, from ^{51}Ni to ^{74}Ni.

(b)   The five most abundant isotopes are

^{58}Ni, 57.935346 amu, 68.077%

^{60}Ni, 59.930788 amu, 26.223%

^{62}Ni, 61.928346 amu, 3.634%

^{61}Ni, 60.931058 amu, 1.140%

^{64}Ni, 63.927968 amu, 0.926%

Data from *Handbook of Chemistry and Physics*, 74th Ed. [Data may differ slightly in other editions.]

2.93   (a)   A Br_2 molecule could consist of two atoms of the same isotope or one atom of each of the two different isotopes. This second possibility is twice as likely as the first. Therefore, the second peak (twice as large as peaks 1 and 3) represents a Br_2 molecule containing different isotopes. The mass numbers of the two isotopes are determined from the masses of the two smaller peaks. Since $157.836 \approx 158$, the first peak represents a ^{79}Br$-^{79}$Br molecule. Peak 3, $161.832 \approx 162$, represents a ^{81}Br$-^{81}$Br molecule. Peak 2 then contains one atom of each isotope, ^{79}Br$-^{81}$Br, with an approximate mass of 160 amu.

(b) The mass of the lighter isotope is 157.836 amu/2 atoms, or 78.918 amu/atom. For the heavier one, 161.832 amu/2 atoms = 80.916 amu/atom.

(c) The relative size of the three peaks in the mass spectrum of Br_2 indicates their relative abundance. The average mass of a Br_2 molecule is

0.2569(157.836) + 0.4999(159.834) + 0.2431(161.832) = 159.79 amu.

(Each product has four significant figures and two decimal places, so the answer has two decimal places.)

(d) $\dfrac{159.79 \text{ amu}}{\text{avg. } Br_2 \text{ molecule}} \times \dfrac{1\,Br_2 \text{ molecule}}{2\,Br \text{ atoms}} = 79.895 \text{ amu}$

(e) Let x = the abundance of ^{79}Br, 1 – x = abundance of ^{81}Br. From (b), the masses of the two isotopes are 78.918 amu and 80.916 amu, respectively. From (d), the mass of an average Br atom is 79.895 amu.

x(78.918) + (1 – x)(80.916) = 79.895, x = 0.5110

^{79}Br = 51.10%, ^{81}Br = 48.90%

2.95 (a) an alkali metal: K   (b) an alkaline earth metal: Ca   (c) a noble gas: Ar

(d) a halogen: Br   (e) a metalloid: Ge   (f) a nonmetal in 1A: H

(g) a metal that forms a 3+ ion: Al   (h) a nonmetal that forms a 2– ion: O

(i) an element that resembles Al: Ga

2.96 (a) $^{266}_{106}Sg$ has 106 protons, 160 neutrons and 106 electrons

(b) Sg is in Group 6B (or 6) and immediately below tungsten, W. We expect the chemical properties of Sg to most closely resemble those of W.

2.97 (a) chlorine gas, Cl_2: ii   (b) propane, C_3H_8: v   (c) nitrate ion, NO_3^-: i

(d) sulfur trioxide, SO_3: iii   (e) methylchloride, CH_3Cl: iv

2.99 (a) IO_3^- (b) IO_4^- (c) IO (d) HIO (e) HIO_4 or (H_5IO_6)

2.100 (a) perbromate ion   (b) selenite ion

(c) AsO_4^{3-} (d) $HTeO_4^-$

2.101 Carbonic acid: H_2CO_3; the cation is H^+ because it is an acid; the anion is carbonate because the acid reacts with lithium hydroxide to form lithium carbonate.
Lithium hydroxide: LiOH; lithium carbonate: Li_2CO_3

2.103 (a) potassium nitrate   (b) sodium carbonate   (c) calcium oxide

(d) hydrochloric acid   (e) magnesium sulfate   (f) magnesium hydroxide

2.105   (a)   In an alkane, all C atoms have 4 single bonds, so each C in the partial structure needs 2 more bonds. All alkanes are hydrocarbons, so 2 H atoms will bind to each C atom in the ring.

    (b)   The molecular formula of cyclohexane is C_6H_{12}; the molecular formula of *n*-hexane is C_6H_{14} (see Solution 2.64(d)). Cyclohexane can be thought of as *n*-hexane in which the two outer (terminal) C atoms are joined to each other. In order to form this C–C bond, each outer C atom must lose 1 H atom. The number of C atoms is unchanged, and each C atom still has 4 single bonds. The resulting molecular formula is $C_6H_{14-2} = C_6H_{12}$.

    (c)   On the structure in part (a), replace 1 H atom with an OH group.

2.106   Elements are arranged in the periodic table by increasing atomic number and so that elements with similar chemical and physical properties form a vertical column or group. By its position in the periodic chart, we know whether an element is a metal, nonmetal, or metalloid, and the common charge of its ion. Members of a group have the same common ionic charge and combine in similar ways with other elements.

3 Stoichiometry: Calculations with Chemical Formulas and Equations

Visualizing Concepts

3.2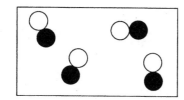

Write the balanced equation for the reaction.

$$2H_2 + CO \rightarrow CH_3OH$$

The combining ratio of H_2: CO is 2:1. If we have 8 H_2 molecules, 4 CO molecules are required for complete reaction. Alternatively, you could examine the atom ratios in the formula of CH_3OH, but the balanced equation is most direct.

3.4 The box contains 4 C atoms and 16 H atoms, so the empirical formula of the hydrocarbon is CH_4.

3.6 *Analyze.* Given: 4.0 mol CH_4. Find: mol CO and mol H_2

Plan. Examine the boxes to determine the CH_4:CO mol ratio and CH_4:H_2O mole ratio.

Solve. There are 2 CH_4 molecules in the reactant box and 2 CO molecules in the product box. The mole ratio is 2:2 or 1:1. Therefore, 4.0 mol CH_4 can produce 4.0 mol CO. There are 2 CH_4 molecules in the reactant box and 6 H_2 molecules in the product box. The mole ratio is 2:6 or 1:3. So, 4.0 mol CH_4 can produce 12:0 mol H_2.

Check. Use proportions. 2 mol CH_4/2 mol CO = 4 mol CH_4/4 mol CO;
2 mol CH_4/6 mol H_2 = 4 mol CH_4/12 mol H_2.

3.8 (a) $2NO + O_2 \rightarrow 2NO_2$, $O_2 =$ 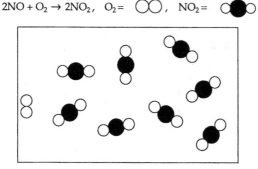, $NO_2 =$

Each NO molecule reacts with 1 O atom (1/2 of an O_2 molecule) to produce 1 NO_2 molecule. Eight NO molecules react with 8 O atoms (4 O_2 molecules) to produce 8 NO_2 molecules. One O_2 molecule doesn't react (is in excess). NO is the limiting reactant.

(b) $$\% \text{ yield} = \frac{\text{actual yield}}{\text{theoretical yield}} \times 100; \quad \text{actual yield} = \frac{\% \text{ yield}}{100} \times \text{theoretical yield}$$

The theoretical yield from part (a) is 8 NO_2 molecules. If the percent yield is 75%, then 0.75(8) = 6 NO_2 would appear in the products box.

Balancing Chemical Equations

3.10 (a) In a CO molecule, there is one O atom bound to C. 2CO indicates that there are **two CO molecules**, each of which contains one C and one O atom. Adding a subscript 2 to CO to form CO_2 means that there are **two O atoms** bound to one C in a CO_2 molecule. The composition of the different molecules, CO_2 and CO, is different and the physical and chemical properties of the two compounds they constitute are very different. The subscript 2 changes molecular composition and thus properties of the compound. The prefix 2 indicates how many molecules (or moles) of the original compound are under consideration.

(b) Yes. There are the same number and kinds of atoms on the reactants side and the products side of the equation.

3.12 (a) $6Li(s) + N_2(g) \rightarrow 2Li_3N(s)$

(b) $La_2O_3(s) + 3H_2O(l) \rightarrow 2La(OH)_3(aq)$

(c) $2NH_4NO_3(s) \rightarrow 2N_2(g) + O_2(g) + 4H_2O(g)$

(d) $Ca_3P_2(s) + 6H_2O(l) \rightarrow 3Ca(OH)_2(aq) + 2PH_3(g)$

(e) $3Ca(OH)_2(aq) + 2H_3PO_4(aq) \rightarrow Ca_3(PO_4)_2(s) + 6H_2O(l)$

(f) $2AgNO_3(aq) + Na_2SO_4(aq) \rightarrow Ag_2SO_4(s) + 2NaNO_3(aq)$

(g) $4CH_3NH_2(g) + 9O_2(g) \rightarrow 4CO_2(g) + 10H_2O(g) + 2N_2(g)$

3.14 (a) $SO_3(g) + H_2O(l) \rightarrow H_2SO_4(aq)$

(b) $B_2S_3(s) + 6H_2O(l) \rightarrow 2H_3BO_3(aq) + 3H_2S(g)$

(c) $Pb(NO_3)_2(aq) + 2\,NaI(aq) \rightarrow 2\,NaNO_3(aq) + PbI_2(s)$

(d) $2Hg(NO_3)_2(s) \overset{\Delta}{\rightarrow} 2HgO(s) + 4NO_2(g) + O_2(g)$

(e) $Cu(s) + 2H_2SO_4(aq) \rightarrow CuSO_4(aq) + SO_2(g) + 2H_2O(l)$

Patterns of Chemical Reactivity

3.16   (a)   Neutral Ca atom loses $2e^-$ to form Ca^{2+}. Neutral O_2 molecule gains $4e^-$ to form $2O^{2-}$. The formula of the product will be CaO, because the cationic and anionic charges are opposite and equal. $2Ca(s) + O_2(g) \rightarrow 2CaO$

          (b)   The products are $CO_2(g)$ and $H_2O(l)$. $C_3H_6O(l) + 4O_2(g) \rightarrow 3CO_2(g) + 3H_2O(l)$

3.18   (a) $2Al(s) + 3O_2(g) \rightarrow Al_2O_3(s)$

          (b) $Cu(OH)_2(s) \overset{\Delta}{\rightarrow} CuO(s) + H_2O(g)$

          (c) $C_7H_{16}(l) + 11O_2(g) \rightarrow 7CO_2(g) + 8H_2O(l)$

          (d) $2C_5H_{12}O(l) + 15O_2(g) \rightarrow 10CO_2(g) + 12H_2O(l)$

3.20   (a) $2C_3H_6(g) + 9O_2(g) \rightarrow 6CO_2(g) + 6H_2O(l)$ combustion

          (b) $NH_4NO_3(s) \rightarrow N_2O(g) + 2H_2O(l)$ decomposition

          (c) $C_5H_6O(l) + 6O_2(g) \rightarrow 5CO_2(g) + 3H_2O(l)$ combustion

          (d) $N_2(g) + 3H_2(g) \rightarrow 2NH_3(g)$ combination

          (e) $K_2O(s) + H_2O(l) \rightarrow 2KOH(aq)$ combination

Formula Weights

3.22   Formula weight in amu to 1 decimal place.

          (a) N_2O: FW = $2(14.0) + 1(16.0) = 44.0$ amu

          (b) $HC_7H_5O_2$: $7(12.0) + 6(1.0) + 2(16.0) = 122.0$ amu

          (c) $Mg(OH)_2$: $1(24.3) + 2(16.0) + 2(1.0) = 58.3$ amu

          (d) $(NH_2)_2CO$: $2(14.0) + 4(1.0) + 1(12.0) + 1(16.0) = 60.0$ amu

          (e) $CH_3CO_2C_5H_{11}$: $7(12.0) + 14(1.0) + 2(16.0) = 130.0$ amu

3.24   (a) C_2H_2: FW = $2(12.0) + 2(1.0) = 26.0$ amu

$$\%\,C = \frac{2(12.0)\,\text{amu}}{26.0\,\text{amu}} \times 100 = 92.3\%$$

          (b) $HC_6H_7O_6$: FW = $6(12.0) + 8(1.0) + 6(16.0) = 176.0$ amu

$$\%\,H = \frac{8(1.0)\,\text{amu}}{176.0\,\text{amu}} \times 100 = 4.5\%$$

(c) $(NH_4)_2SO_4$: FW = 2(14.0) + 8(1.0) + 1(32.1) + 4(16.0) = 132.1 amu

$$\% \, H = \frac{8(1.0)\,amu}{132.1\,amu} \times 100 = 6.1\%$$

(d) $PtCl_2(NH_3)_2$: FW = 1(195.1) + 2(35.5) + 2(14.0) + 6(1.0) = 300.1 amu

$$\% \, Pt = \frac{1(195.1)\,amu}{300.1\,amu} \times 100 = 65.01\%$$

(e) $C_{18}H_{24}O_2$: FW = 18(12.0) + 24(1.0) + 2(16.0) = 272.0 amu

$$\% \, O = \frac{2(16.0)\,amu}{272.0\,amu} \times 100 = 11.8\%$$

(f) $C_{18}H_{27}NO_3$: FW = 18(12.0) + 27(1.0) + 1(14.0) + 3(16.0) = 305.0 amu

$$\% \, C = \frac{18(12.0)\,amu}{305.0\,amu} \times 100 = 70.8\%$$

3.26   (a) CO_2: FW = 1(12.0) + 2(16.0) = 44.0 amu

$$\% \, C = \frac{12.0\,amu}{44.0\,amu} \times 100 = 27.3\%$$

(b) CH_3OH: FW = 1(12.0) + 4(1.0) + 1(16.0) = 32.0 amu

$$\% \, C = \frac{12.0\,amu}{32.0\,amu} \times 100 = 37.5\%$$

(c) C_2H_6: FW = 2(12.0) + 6(1.0) = 30.0 amu

$$\% \, C = \frac{2(12.0)\,amu}{30.0\,amu} \times 100 = 80.0\%$$

(d) $CS(NH_2)_2$: FW = 1(12.0) + 1(32.1) + 2(14.0) + 4(1.0) = 76.1 amu

$$\% \, C = \frac{12.0\,amu}{76.1\,amu} \times 100 = 15.8\%$$

Avogadro's Number and the Mole

3.28   (a)   <u>exactly</u> 12 g   (b) 6.0221421×10^{23}, Avogadro's number

3.30 3.0×10^{23} H_2O_2 molecules contains (4 atoms × 0.5 mol) = 2 mol atoms

       32 g O_2 contains (2 atoms × 1 mol) = 2 mol atoms

       2.0 mol CH_4 contains (5 atoms × 2 mol) = 10 mol atoms

3.32   300 million = 300×10^6 = 3.00×10^8 or 3×10^8 people

       (The number 300 million has an ambiguous number of sig figs.)

$$\frac{6.022 \times 10^{23}\,¢}{3.00 \times 10^8\,people} \times \frac{\$1}{100\,¢} = \frac{\$6.022 \times 10^{21}}{3.00 \times 10^8\,people} = 2.007 \times 10^{13} = \$2.01 \times 10^{13}/person$$

$$\$13.5 \, trillion = \$1.35 \times 10^{13} \qquad \frac{\$2.007 \times 10^{13}}{\$1.35 \times 10^{13}} = 1.487 = 1.49 \, or \, 1$$

Each person would receive an amount that is 1.49 (or 1) times the dollar amount of the national debt.

3.34   (a)   molar mass = 1(112.41) + 1(32.07) = 144.48 g

$$5.76 \times 10^{-3} \text{ mol CdS} \times \frac{144.48 \text{ g}}{1 \text{ mol}} = 0.832 \text{ g CdS}$$

(b)   molar mass = 1(14.01) + 4(1.008) + 1(35.45) = 53.49 g/mol

$$112.6 \text{ g NH}_4\text{Cl} \times \frac{1 \text{ mol}}{53.49 \text{ g}} = 2.1051 = 2.11 \text{ mol NH}_4\text{Cl}$$

(c) $1.305 \times 10^{-2} \text{ mol C}_6\text{H}_6 \times \dfrac{6.02214 \times 10^{23} \text{ molecules}}{1 \text{ mol}} = 7.859 \times 10^{21} \text{ C}_6\text{H}_6 \text{ molecules}$

(d) $4.88 \times 10^{-3} \text{ mol Al(NO}_3)_3 \times \dfrac{9 \text{ mol O}}{1 \text{ mol Al(NO}_3)_3} \times \dfrac{6.022 \times 10^{23} \text{ O atoms}}{1 \text{ mol}}$

$$= 2.64 \times 10^{22} \text{ O atoms}$$

3.36   (a) $\text{Fe}_2(\text{SO}_4)_3$ molar mass = 2(55.845) + 3(32.07) + 12(16.00) = 399.900 = 399.9 g/mol

$$0.0714 \text{ mol Fe}_2(\text{SO}_4)_3 \times \frac{399.9 \text{ g Fe}_2(\text{SO}_4)_3}{1 \text{ mol}} = 28.553 = 28.6 \text{ g Fe}_2(\text{SO}_4)_3$$

(b) $(\text{NH}_4)_2\text{CO}_3$ molar mass = 2(14.007) + 8(1.008) + 12.011 + 3(15.9994) = 96.0872

$$= 96.087 \text{ g/mol}$$

$$8.776 \text{ g (NH}_4)_2\text{CO}_3 \times \frac{1 \text{ mol}}{96.087 \text{ g (NH}_4)_2\text{CO}_3} \times \frac{2 \text{ mol NH}_4^+}{1 \text{ mol (NH}_4)_2\text{CO}_3} = 0.1827 \text{ mol NH}_4^+$$

(c) $\text{C}_9\text{H}_8\text{O}_4$ molar mass = 9(12.01) + 8(1.008) + 4(16.00) = 180.154 = 180.2 g/mol

$$6.52 \times 10^{21} \text{ molecules} \times \frac{1 \text{ mol}}{6.022 \times 10^{23} \text{ molecules}} \times \frac{180.2 \text{ g C}_9\text{H}_8\text{O}_4}{1 \text{ mol aspirin}} = 1.95 \text{ g C}_9\text{H}_8\text{O}_4$$

(d) $\dfrac{15.86 \text{ g Valium}}{0.05570 \text{ mol}} = 284.7 \text{ g Valium/mol}$

3.38   (a) $\text{C}_{14}\text{H}_{18}\text{N}_2\text{O}_5$ molar mass = 14(12.01) + 18(1.008) + 2(14.01) + 5(16.00)

$$= 294.30 \text{ g/mol}$$

(b) $1.00 \text{ mg aspartame} \times \dfrac{1 \times 10^{-3} \text{ g}}{1 \text{ mg}} \times \dfrac{1 \text{ mol}}{294.3 \text{ g}} = 3.398 \times 10^{-6} = 3.40 \times 10^{-6} \text{ mol aspartame}$

(c) $3.398 \times 10^{-6} \text{ mol aspartame} \times \dfrac{6.022 \times 10^{23} \text{ molecules}}{1 \text{ mol}} = 2.046 \times 10^{18}$

$$= 2.05 \times 10^{18} \text{ aspartame molecules}$$

(d) $2.046 \times 10^{18} \text{ aspartame molecules} \times \dfrac{18 \text{ H atoms}}{1 \text{ aspartame molecule}} = 3.68 \times 10^{19} \text{ H atoms}$

3.40   (a) 7.08×10^{20} H atoms $\times \dfrac{19 \, \text{C atoms}}{28 \, \text{H atoms}} = 4.80 \times 10^{20}$ C atoms

      (b) 7.08×10^{20} H atoms $\times \dfrac{1 \, C_{19}H_{28}O_2 \, \text{molecule}}{28 \, \text{H atoms}} = 2.529 \times 10^{19}$

                               $= 2.53 \times 10^{19} \, C_{19}H_{28}O_2$ molecules

      (c) $2.529 \times 10^{19} \, C_{19}H_{28}O_2$ molecules $\times \dfrac{1 \, \text{mol}}{6.022 \times 10^{23} \, \text{molecules}} = 4.199 \times 10^{-5}$

                               $= 4.20 \times 10^{-5} \, \text{mol} \, C_{19}H_{28}O_2$

      (d) $C_{19}H_{28}O_2$ molar mass $= 19(12.01) + 28(1.008) + 2(16.00) = 288.41 = 288.4$ g/mol

          $4.199 \times 10^{-5} \, \text{mol} \, C_{19}H_{28}O_2 \times \dfrac{288.4 \, \text{g} \, C_{19}H_{28}O_2}{1 \, \text{mol}} = 0.0121 \, \text{g} \, C_{19}H_{28}O_2$

3.42 $25 \times 10^{-6} \, \text{g} \, C_{21}H_{30}O_2 \times \dfrac{1 \, \text{mol} \, C_{21}H_{30}O_2}{314.5 \, \text{g} \, C_{21}H_{30}O_2} = 7.95 \times 10^{-8} = 8.0 \times 10^{-8} \, \text{mol} \, C_{21}H_{30}O_2$

          $7.95 \times 10^{-8} \, \text{mol} \, C_{21}H_{30}O_2 \times \dfrac{6.022 \times 10^{23} \, \text{molecules}}{1 \, \text{mol}} = 4.8 \times 10^{16} \, C_{21}H_{30}O_2$ molecules

Empirical Formulas

3.44   (a)   Calculate the simplest ratio of moles.

          0.104 mol K / 0.052 = 2

          0.052 mol C / 0.052 = 1

          0.156 mol O / 0.052 = 3

          The empirical formula is K_2CO_3.

      (b)   Calculate moles of each element present, then the simplest ratio of moles.

          $5.28 \, \text{g} \, Sn \times \dfrac{1 \, \text{mol} \, Sn}{118.7 \, \text{g} \, Sn} = 0.04448 \, \text{mol} \, Sn; \; 0.04448 / 0.04448 = 1$

          $3.37 \, \text{g} \, F \times \dfrac{1 \, \text{mol} \, F}{19.00 \, \text{g} \, FSn} = 0.1774 \, \text{mol} \, F; \, 0.1774 / 0.04448 \approx 4$

          The integer ratio is 1 Sn : 4 F; the empirical formula is SnF_4.

      (c)   Assume 100 g sample, calculate moles of each element, find the simplest ratio of moles.

          $87.5\% \, N = 87.5 \, \text{g} \, N \times \dfrac{1 \, \text{mol} \, N}{14.01 \, \text{g}} = 6.25 \, \text{mol} \, N; \; 6.25 / 6.25 = 1$

          $12.5\% \, H = 12.5 \, \text{g} \, H \times \dfrac{1 \, \text{mol}}{1.008 \, \text{g}} = 12.4 \, \text{mol} \, H; \; 12.4 / 6.25 \approx 2$

          The empirical formula is NH_2.

3.46   See Solution 3.45 for stepwise problem-solving approach.

(a) $55.3 \text{ g K} \times \dfrac{1 \text{ mol K}}{39.10 \text{ g K}} = 1.414 \text{ mol K}; \ 1.414/0.4714 \approx 3$

$14.6 \text{ g P} \times \dfrac{1 \text{ mol P}}{30.97 \text{ g P}} = 0.4714 \text{ mol P}; \ 0.4714/0.4714 = 1$

$30.1 \text{ g O} \times \dfrac{1 \text{ mol O}}{16.00 \text{ g O}} = 1.881 \text{ mol O}; \ 1.881/0.4714 \approx 4$

The empirical formula is K_3PO_4.

(b) $24.5 \text{ g Na} \times \dfrac{1 \text{ mol Na}}{22.99 \text{ g Na}} = 1.066 \text{ mol Na}; \ 1.066/0.5304 \approx 2$

$14.9 \text{ g Si} \times \dfrac{1 \text{ mol Si}}{28.09 \text{ Si}} = 0.5304 \text{ mol si}; \ 0.5304/0.5304 = 1$

$60.6 \text{ g F} \times \dfrac{1 \text{ mol F}}{19.00 \text{ g F}} = 3.189 \text{ mol F}; \ 3.189/0.5304 \approx 6$

The empirical formula is Na_2SiF_6.

(c) $62.1 \text{ g C} \times \dfrac{1 \text{ mol C}}{12.01 \text{ g C}} = 5.17 \text{ mol C}; \ 5.17 / 0.864 \approx 6$

$5.21 \text{ g H} \times \dfrac{1 \text{ mol H}}{1.008 \text{ g H}} = 5.17 \text{ mol O}; \ 5.17 / 0.864 \approx 6$

$12.1 \text{ g N} \times \dfrac{1 \text{ mol N}}{14.01 \text{ g N}} = 0.864 \text{ mol N}; \ 0.864 / 0.864 = 1$

$20.7 \text{ g O} \times \dfrac{1 \text{ mol O}}{16.00 \text{ g O}} = 1.29 \text{ mol O}; \ 1.29 / 0.864 \approx 1.5$

Multiplying by two, the empirical formula is $C_{12}H_{12}N_2O_3$.

3.48   (a)   FW $HCHO_2 = 12.01 + 1.008 + 2(16.00) = 45.0$ $\dfrac{MM}{FW} = \dfrac{90.0}{45.0} = 2$

The molecular formula is $H_2C_2O_4$.

(b)   FW $C_2H_4O = 2(12) + 4(1) + 16 = 44.$ $\dfrac{MM}{FW} = \dfrac{88}{44} = 2$

The molecular formula is $C_4H_8O_2$.

3.50   Assume 100 g in the following problems.

(a) $75.69 \text{ g C} \times \dfrac{1 \text{ mol C}}{12.01 \text{ g C}} = 6.30 \text{ mol C}; \ 6.30/0.969 = 6.5$

$8.80 \text{ g H} \times \dfrac{1 \text{ mol H}}{1.008 \text{ g H}} = 8.73 \text{ mol H}; \ 8.73/0.969 = 9.0$

$15.51 \text{ g O} \times \dfrac{1 \text{ mol O}}{16.00 \text{ g O}} = 0.969 \text{ mol O}; \ 0.969/0.969 = 1$

Multiply by 2 to obtain the integer ratio 13:18:2. The empirical formula is $C_{13}H_{18}O_2$, FW = 206 g. Since the empirical formula weight and the molar mass are equal (206 g), the empirical and molecular formulas are $C_{13}H_{18}O_2$.

(b) $58.55 \text{ g C} \times \dfrac{1 \text{ mol C}}{12.01 \text{ g C}} = 4.875 \text{ mol C}; \quad 4.875/1.956 \approx 2.5$

$13.81 \text{ g H} \times \dfrac{1 \text{ mol H}}{1.008 \text{ g H}} = 13.700 \text{ mol H}; \quad 13.700/1.956 \approx 7.0$

$27.40 \text{ g N} \times \dfrac{1 \text{ mol N}}{14.01 \text{ g N}} = 1.956 \text{ mol N}; \quad 1.956/1.956 = 1.0$

Multiply by 2 to obtain the integer ratio 5:14:2. The empirical formula is $C_5H_{14}N_2$; FW = 102. Since the empirical formula weight and the molar mass are equal

(102 g), the empirical and molecular formulas are $C_5H_{14}N_2$.

(c) $59.0 \text{ g C} \times \dfrac{1 \text{ mol C}}{12.01 \text{ g C}} = 4.91 \text{ mol C}; \quad 4.91/0.550 \approx 9$

$7.1 \text{ g H} \times \dfrac{1 \text{ mol H}}{1.008 \text{ g H}} = 7.04 \text{ mol H}; \quad 7.04/0.550 \approx 13$

$26.2 \text{ g O} \times \dfrac{1 \text{ mol O}}{16.00 \text{ g O}} = 1.64 \text{ mol O}; \quad 1.64/0.550 \approx 3$

$7.7 \text{ g N} \times \dfrac{1 \text{ mol N}}{14.01 \text{ g N}} = 0.550 \text{ mol N}; \quad 0.550/0.550 = 1$

The empirical formula is $C_9H_{13}O_3N$, FW = 183 amu (or g). Since the molecular weight is approximately 180 amu, the empirical formula and molecular formula are the same, $C_9H_{13}O_3N$.

3.52   (a)   *Plan.* Calculate mol C and mol H, then g C and g H; get g O by subtraction.

Solve.

$6.32 \times 10^{-3} \text{ g CO}_2 \times \dfrac{1 \text{ mol CO}_2}{44.01 \text{ g CO}_2} \times \dfrac{1 \text{ mol C}}{1 \text{ mol CO}_2} = 1.436 \times 10^{-4} = 1.44 \times 10^{-4} \text{ mol C}$

$2.58 \times 10^{-3} \text{ g H}_2\text{O} \times \dfrac{1 \text{ mol H}_2\text{O}}{18.02 \text{ g H}_2\text{O}} \times \dfrac{2 \text{ mol H}}{1 \text{ mol H}_2\text{O}} = 2.863 \times 10^{-4} = 2.86 \times 10^{-4} \text{ mol H}$

$1.436 \times 10^{-4} \text{ mol C} \times \dfrac{12.01 \text{ g C}}{1 \text{ mol C}} = 1.725 \times 10^{-3} \text{ g C} = 1.73 \text{ mg C}$

$2.863 \times 10^{-4} \text{ mol H} \times \dfrac{1.008 \text{ g H}}{1 \text{ mol H}} = 2.886 \times 10^{-4} \text{ g H} = 0.289 \text{ mg H}$

mass of O = 2.78 mg sample – (1.725 mg C + 0.289 mg H) = 0.77 mg O

$0.77 \times 10^{-3} \text{ g O} \times \dfrac{1 \text{ mol O}}{16.00 \text{ g O}} = 4.81 \times 10^{-5} \text{ mol O. Divide moles by } 4.81 \times 10^{-5}.$

$$C: \frac{1.44 \times 10^{-4}}{4.81 \times 10^{-5}} \approx 3; \quad H: \frac{2.86 \times 10^{-4}}{4.81 \times 10^{-5}} \approx 6; \quad O: \frac{4.81 \times 10^{-5}}{4.81 \times 10^{-5}} = 1$$

The empirical formula is C_3H_6O.

(b)   *Plan.* Calculate mol C and mol H, then g C and g H. In this case, get N by subtraction.   *Solve.*

$$14.242 \times 10^{-3} \text{ g } CO_2 \times \frac{1 \text{ mol } CO_2}{44.01 \text{ g } CO_2} \times \frac{1 \text{ mol } C}{1 \text{ mol } CO_2} = 3.2361 \times 10^{-4} \text{ mol C}$$

$$4.083 \times 10^{-3} \text{ g } H_2O \times \frac{1 \text{ mol } H_2O}{18.02 \text{ g } H_2O} \times \frac{2 \text{ mol } H}{1 \text{ mol } H_2O} = 4.5136 \times 10^{-4} = 4.532 \times 10^{-4} \text{ mol H}$$

$$3.2361 \times 10^{-4} \text{ g mol C} \times \frac{12.01 \text{ g } C}{1 \text{ mol } H} = 3.8866 \times 10^{-3} \text{ g C} = 3.8866 \text{ mg C}$$

$$4.532 \times 10^{-4} \text{ mol H} \times \frac{1.008 \text{ g } H}{1 \text{ mol } H} = 0.45683 \times 10^{-3} \text{ g H} = 0.4568 \text{ mg H}$$

mass of N = 5.250 mg sample – (3.8866 mg C + 0.4568 mg H) = 0.9066

$$= 0.907 \text{ mg N}$$

$$0.9066 \times 10^{-3} \text{ g N} \times \frac{1 \text{ mol } N}{14.01 \text{ g } N} = 6.47 \times 10^{-5} \text{ mol N. Divide moles by } 6.47 \times 10^{-5}.$$

$$C: \frac{3.24 \times 10^{-4}}{6.47 \times 10^{-5}} \approx 5; \quad H: \frac{4.53 \times 10^{-4}}{6.47 \times 10^{-5}} \approx 7; \quad N: \frac{6.47 \times 10^{-5}}{6.47 \times 10^{-5}} = 1$$

The empirical formula is C_5H_7N, FW = 81. A molar mass of 160 ± 5 indicates a factor of 2 and a molecular formula of $C_{10}H_{14}N_2$.

3.54   The reaction involved is $MgSO_4 \cdot xH_2O(s) \rightarrow MgSO_4(s) + xH_2O(g)$. First, calculate the number of moles of product $MgSO_4$; this is the same as the number of moles of starting hydrate.

$$2.472 \text{ g } MgSO_4 \times \frac{1 \text{ mol } MgSO_4}{120.4 \text{ g } MgSO_4} \times \frac{1 \text{ mol } MgSO_4 \cdot xH_2O}{1 \text{ mol } MgSO_4} = 0.02053 \text{ mol } MgSO_4 \cdot xH_2O$$

Thus, $\dfrac{5.061 \text{ g } MgSO_4 \cdot xH_2O}{0.02053} = 246.5 \text{ g/mol} = \text{FW of } MgSO_4 \cdot xH_2O$

FW of $MgSO_4 \cdot xH_2O$ = FW of $MgSO_4$ + x(FW of H_2O).

246.5 = 120.4 + x(18.02). x = 6.998. The hydrate formula is $MgSO_4 \cdot \underline{7}H_2O$.

Alternatively, we could calculate the number of moles of water represented by weight loss: (5.061 – 2.472) = 2.589 g H_2O lost.

$$2.589 \text{ g } H_2O \times \frac{1 \text{ mol } H_2O}{18.02 \text{ g } H_2O} = 0.1437 \text{ mol } H_2O; \quad \frac{\text{mol } H_2O}{\text{mol } MgSO_4} = \frac{0.1437}{0.02053} = 7.000$$

Again the correct formula is $MgSO_4 \cdot \underline{7}H_2O$.

Calculations Based on Chemical Equations

3.56   The **integer coefficients** immediately preceding each molecular formula in a chemical equation give information about relative numbers of moles of reactants and products involved in a reaction.

3.58 $C_6H_{12}O_6(aq) \rightarrow 2C_2H_5OH(aq) + 2CO_2(g)$

         (a) $0.400 \text{ mol } C_6H_{12}O_6 \times \dfrac{2 \text{ mol } CO_2}{1 \text{ mol } C_6H_{12}O_6} = 0.800 \text{ mol } CO_2$

         (b) $7.50 \text{ g } C_2H_5OH \times \dfrac{1 \text{ mol } C_2H_5OH}{46.07 \text{ g } C_2H_5OH} \times \dfrac{1 \text{ mol } C_6H_{12}O_6}{2 \text{ mol } C_2H_5OH} \times \dfrac{180.2 \text{ g } C_6H_{12}O_6}{1 \text{ mol } C_6H_{12}O_6}$

                                                                  $= 14.7 \text{ g } C_6H_{12}O_6$

         (c) $7.50 \text{ g } C_2H_5OH \times \dfrac{1 \text{ mol } C_2H_5OH}{46.07 \text{ g } C_2H_5OH} \times \dfrac{2 \text{ mol } CO_2}{2 \text{ mol } C_2H_5OH} \times \dfrac{44.01 \text{ g } CO_2}{1 \text{ mol } CO_2} = 7.16 \text{ g } CO_2$

3.60   (a) $Fe_2O_3(s) + 3CO(g) \rightarrow 2Fe(s) + 3CO_2(g)$

         (b) $0.150 \text{ kg } Fe_2O_3 \times \dfrac{1000 \text{ g}}{1 \text{ kg}} \times \dfrac{1 \text{ mol } Fe_2O_3}{159.688 \text{ g } Fe_2O_3} = 0.9393 = 0.939 \text{ mol } Fe_2O_3$

                $0.9393 \text{ mol } Fe_2O_3 \times \dfrac{3 \text{ mol } CO}{1 \text{ mol } Fe_2O_3} \times \dfrac{28.01 \text{ g } CO}{1 \text{ mol } CO} = 78.929 = 78.9 \text{ g } CO$

         (c) $0.9393 \text{ mol } Fe_2O_3 \times \dfrac{2 \text{ mol } Fe}{1 \text{ mol } Fe_2O_3} \times \dfrac{55.845 \text{ g } Fe}{1 \text{ mol } Fe} = 104.914 = 105 \text{ g } Fe$

                $0.9393 \text{ mol } Fe_2O_3 \times \dfrac{3 \text{ mol } CO_2}{1 \text{ mol } Fe_2O_3} \times \dfrac{44.01 \text{ g } CO_2}{1 \text{ mol } CO_2} = 124.015 = 124 \text{ g } CO_2$

         (d)   reactants: $150 \text{ g } Fe_2O_3 + 78.9 \text{ g } CO = 228.9 = 229 \text{ g}$

               products: $104.9 \text{ g } Fe + 124.0 \text{ g } CO_2 = 228.9 = 229 \text{ g}$

               Mass is conserved.

3.62   (a) $CaH_2(s) + 2H_2O(l) \rightarrow Ca(OH)_2(aq) + 2H_2(g)$

         (b) $8.500 \text{ g } H_2 \times \dfrac{1 \text{ mol } H_2}{2.016 \text{ g } H_2} \times \dfrac{1 \text{ mol } CaH_2}{2 \text{ mol } H_2} \times \dfrac{42.10 \text{ g } CaH_2}{1 \text{ mol } CaH_2} = 88.75 \text{ g } CaH_2$

3.64 $2C_8H_{18}(l) + 25O_2(g) \rightarrow 16CO_2(g) + 18H_2O(l)$

         (a) $1.25 \text{ mol } C_8H_{18} \times \dfrac{25 \text{ mol } O_2}{2 \text{ mol } C_8H_{18}} = 15.625 = 15.6 \text{ mol } O_2$

         (b) $10.0 \text{ g } C_8H_{18} \times \dfrac{1 \text{ mol } C_8H_{18}}{114.2 \text{ g } C_8H_{18}} \times \dfrac{25 \text{ mol } O_2}{2 \text{ mol } C_8H_{18}} \times \dfrac{32.00 \text{ g } O_2}{1 \text{ mol } O_2} = 35.0 \text{ g } O_2$

         (c) $1.00 \text{ gal } C_8H_{18} \times \dfrac{3.7854 \text{ L}}{1 \text{ gal}} \times \dfrac{1000 \text{ mL}}{1 \text{ L}} \times \dfrac{0.692 \text{ g}}{1 \text{ mL}} = 2619.5 = 2.62 \times 10^3 \text{ g } C_8H_{18}$

$$2.6195 \times 10^3 \text{ g C}_8\text{H}_{18} \times \frac{1 \text{ mol C}_8\text{H}_{18}}{114.2 \text{ g C}_8\text{H}_{18}} \times \frac{25 \text{ mol O}_2}{2 \text{ mol C}_8\text{H}_{18}} \times \frac{32.00 \text{ g O}_2}{1 \text{ mol O}_2} = 9{,}175.1 \text{ g}$$

$$= 9.18 \times 10^3 \text{ g O}_2$$

3.66 (a) *Plan.* Calculate a "mole ratio" between nitroglycerine and total moles of gas produced. $(12 + 6 + 1 + 10) = 29$ mol gas; 4 mol nitro: 29 total mol gas. *Solve.*

$$2.00 \text{ mL nitro} \times \frac{1.592 \text{ g}}{\text{mL}} \times \frac{1 \text{ mol nitro}}{227.1 \text{ g nitro}} \times \frac{29 \text{ mol gas}}{4 \text{ mol nitro}} = 0.10165 = 0.102 \text{ mol gas}$$

(b) $$0.10165 \text{ mol gas} \times \frac{55 \text{ L}}{\text{mol}} = 5.5906 = 5.6 \text{ L}$$

(c) $$2.00 \text{ mL nitro} \times \frac{1.592 \text{ g}}{\text{mL}} \times \frac{1 \text{ mol nitro}}{227.1 \text{ g nitro}} \times \frac{6 \text{ mol N}_2}{4 \text{ mol nitro}} \times \frac{28.01 \text{ g N}_2}{1 \text{ mol N}_2} = 0.589 \text{ g N}_2$$

Limiting Reactants; Theoretical Yields

3.68 (a) *Theoretical yield* is the maximum amount of product possible, as predicted by stoichiometry, assuming that the limiting reactant is converted entirely to product.

Actual yield is the amount of product actually obtained, less than or equal to the theoretical yield. *Percent yield* is the ratio of (actual yield to theoretical yield) × 100.

(b) No reaction is perfect. Not all reactant molecules come together effectively to form products; alternative reaction pathways may produce secondary products and reduce the amount of desired product actually obtained, or it might not be possible to completely isolate the desired product from the reaction mixture. In any case, these factors reduce the actual yield of a reaction.

(c) No, 110% actual yield is not possible. Theoretical yield is the maximum possible amount of pure product, assuming all available limiting reactant is converted to product, and that all product is isolated. If an actual yield of 110% if obtained, the product must contain impurities which increase the experimental mass.

3.70 (a) $$40{,}875 \text{ L beverage} \times \frac{1 \text{ bottle}}{0.355 \text{ L}} = 115{,}140.85 = 1.15 \times 10^5 \text{ portions of beverage}$$

(The uncertainty in 355 mL limits the precision of the number of portions we can reasonably expect to deliver to three significant figures.)

121,515 bottles; 122,500 caps; 1.15×10^5 bottles can be filled and capped.

(b) 122,500 caps – 115,141 portions = 7,359 = 7×10^3 caps remain

121,515 empty bottles – 115,141 portions = 6374 = 6×10^3 bottles remain

(Uncertainty in the number of portions delivered limits the results to 1 sig fig.)

(c) The volume of beverage limits production.

3.72 $$0.500 \text{ mol Al(OH)}_3 \times \frac{3 \text{ mol H}_2\text{SO}_4}{2 \text{ mol Al(OH)}_3} = 0.750 \text{ mol H}_2\text{SO}_4 \text{ needed for complete reaction}$$

Only 0.500 mol H_2SO_4 available, so H_2SO_4 limits.

$$0.500 \text{ mol } H_2SO_4 \times \frac{1 \text{ mol } Al_2(SO_4)_3}{3 \text{ mol } H_2SO_4} = 0.1667 = 0.167 \text{ mol } Al_2(SO_4)_3 \text{ can form}$$

$$0.500 \text{ mol } H_2SO_4 \times \frac{2 \text{ mol } Al(OH)_3}{3 \text{ mol } H_2SO_4} = 0.3333 = 0.333 \text{ mol } Al(OH)_3 \text{ react}$$

0.500 mol $Al(OH)_3$ initial – 0.333 mol react = 0.167 mol $Al(OH)_3$ remain

3.74 $4NH_3(g) + 5O_2(g) \rightarrow 4NO(g) + 6H_2O(g)$

(a)   Follow the approach in Sample Exercise 3.19.

$$1.50 \text{ g } NH_3 \times \frac{1 \text{ mol } NH_3}{17.03 \text{ g } NH_3} = 0.08808 = 0.0881 \text{ mol } NH_3$$

$$2.75 \text{ g } O_2 \times \frac{1 \text{ mol } O_2}{32.00 \text{ g } O_2} = 0.08594 = 0.0859 \text{ mol } O_2$$

$$0.08594 \text{ mol } O_2 \times \frac{4 \text{ mol } NH_3}{5 \text{ mol } O_2} = 0.06875 = 0.0688 \text{ mol } NH_3 \text{ required}$$

More than 0.0688 mol NH_3 is available, so O_2 is the limiting reactant.

(b) $$0.08594 \text{ mol } O_2 \times \frac{4 \text{ mol } NO}{5 \text{ mol } O_2} \times \frac{30.01 \text{ g } NO}{1 \text{ mol } NO} = 2.063 = 2.06 \text{ g } NO \text{ produced}$$

$$0.08594 \text{ mol } O_2 \times \frac{6 \text{ mol } H_2O}{5 \text{ mol } O_2} \times \frac{18.02 \text{ g } H_2O}{1 \text{ mol } H_2O} = 1.8583 = 1.86 \text{ g } H_2O \text{ produced}$$

(c)   0.08808 mol NH_3 – 0.06875 mol NH_3 reacted = 0.01933 = 0.0193 mol NH_3 remain

$$0.01933 \text{ mol } NH_3 \times \frac{17.03 \text{ g } NH_3}{1 \text{ mol } NH_3} = 0.32919 = 0.329 \text{ g } NH_3 \text{ remain}$$

(d)   mass products = 2.06 g NO + 1.86 g H_2O + 0.329 g NH_3 remaining = 4.25 g products

mass reactants = 1.50 g NH_3 + 2.75 g O_2 = 4.25 g reactants

(For comparison purposes, the mass of excess reactant can be either added to the products, as above, or subtracted from reactants.)

3.76   *Plan.* Write balanced equation; determine limiting reactant; calculate amounts of excess reactant remaining and products, based on limiting reactant.

Solve. $H_2SO_4(aq) + Pb(C_2H_3O_2)_2(aq) \rightarrow PbSO_4(s) + 2HC_2H_3O_2(aq)$

$$7.50 \text{ g } H_2SO_4 \times \frac{1 \text{ mol } H_2SO_4}{98.09 \text{ g } H_2SO_4} = 0.07646 = 0.0765 \text{ mol } H_2SO_4$$

$$7.50 \text{ g } Pb(C_2H_3O_2)_2 \times \frac{1 \text{ mol } Pb(C_2H_3O_2)_2}{325.3 \text{ g } Pb(C_2H_3O_2)_2} = 0.023056 = 0.0231 \text{ mol } Pb(C_2H_3O_2)_2$$

1 mol H_2SO_4:1 mol $Pb(C_2H_3O_2)_2$, so $Pb(C_2H_3O_2)_2$ is the limiting reactant.

0 mol $Pb(C_2H_3O_2)_2$, $(0.07646 - 0.023056) = 0.0534$ mol H_2SO_4, 0.0231 mol $PbSO_4$,

$(0.023056 \times 2) = 0.0461$ mol $HC_2H_3O_2$ are present after reaction

0.053405 mol $H_2SO_4 \times 98.09$ g/mol $= 5.2385 = 5.24$ g H_2SO_4

0.023056 mol $PbSO_4 \times 303.3$ g/mol $= 6.9928 = 6.99$ g $PbSO_4$

0.046111 mol $HC_2H_3O_2 \times 60.05$ g/mol $= 2.7690 = 2.77$ g $HC_2H_3O_2$

Check. The initial mass of reactants was 15.00 g; and the final mass of excess reactant and products is 15.00 g; mass is conserved.

3.78   (a) $C_2H_6 + Cl_2 \rightarrow C_2H_5Cl + HCl$

$$125 \text{ g } C_2H_6 \times \frac{1 \text{ mol } C_2H_6}{30.07 \text{ g } C_2H_6} = 4.157 = 4.16 \text{ mol } C_2H_6$$

$$255 \text{ g } Cl_2 \times \frac{1 \text{ mol } Cl_2}{70.91 \text{ g } Cl_2} = 3.596 = 3.60 \text{ mol } Cl_2$$

Since the reactants combine in a 1:1 mole ratio, Cl_2 is the limiting reactant. The theoretical yield is:

$$3.596 \text{ mol } Cl_2 \times \frac{1 \text{ mol } C_2H_5Cl}{1 \text{ mol } Cl_2} \times \frac{64.51 \text{ g } C_2H_5Cl}{1 \text{ mol } C_2H_5Cl} = 231.98 = 232 \text{ g } C_2H_5Cl$$

(b) $\% \text{ yield} = \dfrac{206 \text{ g } C_2H_5Cl \text{ actual}}{232 \text{ g } C_2H_5Cl \text{ theoretical}} \times 100 = 88.8\%$

3.80 $H_2S(g) + 2NaOH(aq) \rightarrow Na_2S(aq) + 2H_2O(l)$

$$1.50 \text{ g } H_2S \times \frac{1 \text{ mol } H_2S}{34.08 \text{ g } H_2S} = 0.04401 = 0.0440 \text{ mol } H_2S$$

$$2.00 \text{ g } NaOH \times \frac{1 \text{ mol } NaOH}{40.00 \text{ g } NaOH} = 0.0500 \text{ mol } NaOH$$

By inspection, twice as many mol NaOH as H_2S are needed for exact reaction, but mol NaOH given is less than twice mol H_2S, so NaOH limits.

$$0.0500 \text{ mol } NaOH \times \frac{1 \text{ mol } Na_2S}{2 \text{ mol } NaOH} \times \frac{78.05 \text{ g } Na_2S}{1 \text{ mol } Na_2S} = 1.95125 = 1.95 \text{ g } Na_2S \text{ theoretical}$$

$$\frac{92.0\%}{100} \times 1.95125 \text{ g } Na_2S \text{ theoretical} = 1.7951 = 1.80 \text{ g } Na_2S \text{ actual}$$

Additional Exercises

3.82   The formulas of the fertilizers are NH_3, NH_4NO_3, $(NH_4)_2SO_4$ and $(NH_2)_2CO$. Qualitatively, the more heavy, non-nitrogen atoms in a molecule, the smaller the mass % of N. By inspection, the mass of NH_3 is dominated by N, so it will have the greatest % N, $(NH_4)_2SO_4$ will have the least. In order of increasing % N:

$(NH_4)_2SO_4 < NH_4NO_3 < (NH_2)_2CO < NH_3$.

Check by calculation:

$(NH_4)_2SO_4$: FW = $2(14.0) + 8(1.0) + 1(32.1) + 4(16.0) = 132.1$ amu

% N = $[2(14.0)/132.1] \times 100 = 21.2\%$

NH_4NO_3: FW = $2(14.0) + 4(1.0) + 3(16.0) = 80.0$ amu

% N = $[2(14.0)/80.0] \times 100 = 35.0\%$

$(NH_2)_2CO$: FW = $2(14.0) + 4(1.0) = 1(12.0) + 1(16.0) = 60.0$ amu

% N = $[2(14.0)/60.0] \times 100 = 46.7\%$ N

NH_3: FW = $1(14.0) + 3(1.0) = 17.0$

% N = $[14.0/17.0] \times 100 = 82.4$ % N

3.83   (a) $1.25 \text{ carat} \times \dfrac{0.200 \text{ g}}{1 \text{ carat}} \times \dfrac{1 \text{ mol C}}{12.01 \text{ g C}} = 0.020816 = 0.0208 \text{ mol C}$

              $0.020816 \text{ mol C} \times \dfrac{6.022 \times 10^{23} \text{ C atoms}}{1 \text{ mol C}} = 1.25 \times 10^{22} \text{ C atoms}$

      (b) $0.500 \text{ g C}_9\text{H}_8\text{O}_4 \times \dfrac{1 \text{ mol C}_9\text{H}_8\text{O}_4}{180.2 \text{ g C}_9\text{H}_8\text{O}_4} = 2.7747 \times 10^{-3} = 2.77 \times 10^{-3} \text{ mol HC}_9\text{H}_7\text{O}_4$

              $0.0027747 \text{ mol C}_9\text{H}_8\text{O}_4 \times \dfrac{6.022 \times 10^{23} \text{ molecules}}{1 \text{ mol}} = 1.67 \times 10^{21} \text{ HC}_9\text{H}_7\text{O}_4 \text{ molecules}$

3.84   (a) $\dfrac{5.342 \times 10^{-21} \text{ g}}{1 \text{ molecule penicillin G}} \times \dfrac{6.0221 \times 10^{23} \text{ molecules}}{1 \text{ mol}} = 3217 \text{ g/mol penicillin G}$

      (b)   1.00 g hemoglobin (hem) contains 3.40×10^{-3} g Fe.

              $\dfrac{1.00 \text{ g hem}}{3.40 \times 10^{-3} \text{ g Fe}} \times \dfrac{55.85 \text{ g Fe}}{1 \text{ mol Fe}} \times \dfrac{4 \text{ mol Fe}}{1 \text{ mol hem}} = 6.57 \times 10^4 \text{ g/mol hemoglobin}$

3.86   *Plan.* Assume 100 g, calculate mole ratios, empirical formula, then molecular formula from molar mass.   *Solve.*

      $68.2 \text{ g C} \times \dfrac{1 \text{ mol C}}{12.01 \text{ g C}} = 5.68 \text{ mol C}; \ 5.68/0.568 \approx 10$

      $6.86 \text{ g H} \times \dfrac{1 \text{ mol H}}{1.008 \text{ g H}} = 6.81 \text{ mol H}; \ 6.81/0.568 \approx 12$

      $15.9 \text{ g N} \times \dfrac{1 \text{ mol N}}{14.01 \text{ g N}} = 1.13 \text{ mol N}; \ 1.13/0.568 \approx 2$

      $9.08 \text{ g O} \times \dfrac{1 \text{ mol O}}{16.00 \text{ g O}} = 0.568 \text{ mol O}; \ 0.568/0.568 = 1$

The empirical formula is $C_{10}H_{12}N_2O$, FW = 176 amu (or g). Since the molar mass is 176, the empirical and molecular formula are the same, $C_{10}H_{12}N_2O$.

3.87   *Plan.* Assume 1.000 g and get mass O by subtraction.   *Solve.*

(a) $0.7787 \, g \, C \times \dfrac{1 \, mol \, C}{12.01 \, g \, C} = 0.06484 \, mol \, C$

$0.1176 \, g \, H \times \dfrac{1 \, mol \, H}{1.008 \, g \, H} = 0.1167 \, mol \, H$

$0.1037 \, g \, O \times \dfrac{1 \, mol \, C}{16.00 \, g \, O} = 0.006481 \, mol \, O$

Dividing through by the smallest of these values we obtain $C_{10}H_{18}O$.

(b)   The formula weight of $C_{10}H_{18}O$ is 154. Thus, the empirical formula is also the molecular formula.

3.88   Since all the C in the vanillin must be present in the CO_2 produced, get g C from g CO_2.

$2.43 \, g \, CO_2 \times \dfrac{1 \, mol \, CO_2}{44.01 \, g \, CO_2} \times \dfrac{12.01 \, g \, C}{1 \, mol \, C} = 0.6631 = 0.663 \, g \, C$

Since all the H in vanillin must be present in the H_2O produced, get g H from g H_2O.

$0.50 \, g \, H_2O \times \dfrac{1 \, mol \, H_2O}{18.02 \, g \, H_2O} \times \dfrac{2 \, mol \, H}{1 \, mol \, H_2O} \times \dfrac{1.008 \, g \, H}{1 \, mol \, H} = 0.0559 = 0.056 \, g \, H$

Get g O by subtraction. (Since the analysis was performed by combustion, an unspecified amount of O_2 was a reactant, and thus not all the O in the CO_2 and H_2O produced came from vanillin.) 1.05 g vanillin – 0.663 g C – 0.056 g H = 0.331 g O

$0.6631 \, g \, C \times \dfrac{1 \, mol \, C}{12.01 \, g \, C} = 0.0552 \, mol \, C; \ 0.0552 \, / \, 0.0207 = 2.67$

$0.0559 \, g \, H \times \dfrac{1 \, mol \, H}{1.008 \, g \, H} = 0.0555 \, mol \, C; \ 0.0555 \, / \, 0.0207 = 2.68$

$0.331 \, g \, O \times \dfrac{1 \, mol \, O}{16.00 \, g \, O} = 0.0207 \, mol \, O; \ 0.0207 \, / \, 0.0207 = 1.00$

Multiplying the numbers above by **3** to obtain an integer ratio of moles, the empirical formula of vanillin is $C_8H_8O_3$.

3.90   The mass percentage is determined by the relative number of atoms of the element times the atomic weight, divided by the total formula mass. Thus, the mass percent of bromine in $KBrO_x$ is given by $0.5292 = \dfrac{79.91}{39.10 + 79.91 + x(16.00)}$. Solving for x, we obtain x = 2.00. Thus, the formula is $KBrO_2$.

3.91   (a)   Let AW = the atomic weight of X.

According to the chemical reaction, moles XI_3 reacted = moles XCl_3 produced

$$0.5000 \text{ g XI}_3 \times 1 \text{ mol XI}_3 / (AW + 380.71) \text{ g XI}_3$$

$$= 0.2360 \text{ g XCl}_3 \times \frac{1 \text{ mol XCl}_3}{(AW + 106.36) \text{ g XCl}_3}$$

$0.5000 (AW + 106.36) = 0.2360 (AW + 380.71)$

$0.5000 \text{ AW} + 53.180 = 0.2360 \text{ AW} + 89.848$

$0.2640 \text{ AW} = 36.67; \text{ AW} = 138.9 \text{ g}$

(b)   X is lanthanum, La, atomic number 57.

3.92 $C_2H_5OH(l) + 3O_2(g) \rightarrow 2CO_2(g) + 3H_2O(g)$

$C_3H_8(g) + 5O_2(g) \rightarrow 3CO_2(g) + 4H_2O(g)$

$CH_3CH_2COCH_3(l) + 11/2 \, O_2(g) \rightarrow 4CO_2(g) + 4H_2O(l)$

In a combustion reaction, all H in the fuel is transformed to H_2O in the products. The reactant with most mol H/mol fuel will produce the most H_2O. C_3H_8 and $CH_3CH_2COCH_3$ (C_4H_8O) both have 8 mol H/mol fuel, so 1.5 mol of either fuel will produce the same amount of H_2O. 1.5 mol C_2H_5OH will produce less H_2O.

3.94 $2NaCl(aq) + 2H_2O(l) \rightarrow 2NaOH(aq) + H_2(g) + Cl_2(g)$

Calculate mol Cl_2 and relate to mol H_2, mol NaOH.

$$1.5 \times 10^6 \text{ kg} \times \frac{1000 \text{ g}}{1 \text{ kg}} \times \frac{1 \text{ mol Cl}_2}{70.91 \text{ g Cl}_2} = 2.115 \times 10^7 = 2.1 \times 10^7 \text{ mol Cl}_2$$

$$2.115 \times 10^7 \text{ mol Cl}_2 \times \frac{1 \text{ mol H}_2}{1 \text{ mol Cl}_2} \times \frac{2.016 \text{ g H}_2}{1 \text{ mol H}_2} = 4.26 \times 10^7 \text{ g H}_2 = 4.3 \times 10^4 \text{ kg H}_2$$

$$4.3 \times 10^7 \text{ g} \times \frac{1 \text{ metric ton}}{1 \times 10^6 \text{ g (1 Mg)}} = 43 \text{ metric tons H}_2$$

$$2.115 \times 10^7 \text{ mol Cl}_2 \times \frac{2 \text{ mol NaOH}}{1 \text{ mol Cl}_2} \times \frac{40.0 \text{ g NaOH}}{1 \text{ mol NaOH}} = 1.69 \times 10^9 = 1.7 \times 10^9 \text{ g NaOH}$$

1.7×10^9 g NaOH = 1.7×10^6 kg NaOH = 1.7×10^3 metric tons NaOH

3.95 $2C_{57}H_{110}O_6 + 163O_2 \rightarrow 114CO_2 + 110H_2O$

molar mass of fat = 57(12.01) + 110(1.008) + 6(16.00) = 891.5

$$1.0 \text{ kg fat} \times \frac{1000 \text{ g}}{1 \text{ kg}} \times \frac{1 \text{ mol fat}}{891.5 \text{ g fat}} \times \frac{110 \text{ mol H}_2O}{2 \text{ mol fat}} \times \frac{18.02 \text{ g H}_2O}{1 \text{ mol H}_2O} \times \frac{1 \text{ kg}}{1000 \text{ g}} = 1.1 \text{ kg H}_2O$$

3.96   (a)   *Plan.* Calculate the total mass of C from g CO and g CO_2. Calculate the mass of H from g H_2O. Calculate mole ratios and the empirical formula.   *Solve.*

$$0.467 \text{ g CO} \times \frac{1 \text{ mol CO}}{28.01 \text{ g CO}} \times \frac{1 \text{ mol C}}{1 \text{ mol CO}} \times 12.01 \text{ g C} = 0.200 \text{ g C}$$

$$0.733 \text{ g CO}_2 \times \frac{1 \text{ mol CO}_2}{44.01 \text{ g CO}_2} \times \frac{1 \text{ mol C}}{1 \text{ mol CO}_2} \times 12.01 \text{ g C} = 0.200 \text{ g C}$$

Total mass C is 0.200 g + 0.200 g = 0.400 g C.

$$0.450 \text{ g H}_2\text{O} \times \frac{1 \text{ mol H}_2\text{O}}{18.02 \text{ g H}_2\text{O}} \times \frac{2 \text{ mol H}}{1 \text{ mol H}_2\text{O}} \times \frac{1.008 \text{ g H}}{1 \text{ mol H}} = 0.0503 \text{ g H}$$

(Since hydrocarbons contain only the elements C and H, g H can also be obtained by subtraction: 0.450 g sample – 0.400 g C = 0.050 g H.)

$$0.400 \text{ g C} \times \frac{1 \text{ mol C}}{12.01 \text{ g C}} = 0.0333 \text{ mol C}; \ 0.0333/0.0333 = 1.0$$

$$0.0503 \text{ g H} \times \frac{1 \text{ mol H}}{1.008 \text{ g H}} = 0.0499 \text{ mol H}; \ 0.0499/0.0333 = 1.5$$

Multiplying by a factor of 2, the empirical formula is C_2H_3.

(b) Mass is conserved. Total mass products – mass sample = mass O_2 consumed.

0.467 g CO + 0.733 g CO_2 + 0.450 g H_2O – 0.450 g sample = 1.200 g O_2 consumed

(c) For complete combustion, 0.467 g CO must be converted to CO_2.

$$2CO(g) + O_2(g) \rightarrow 2CO_2(g)$$

$$0.467 \text{ g CO} \times \frac{1 \text{ mol CO}}{28.01 \text{ g C}} \times \frac{1 \text{ mol O}_2}{2 \text{ mol CO}} \times \frac{32.00 \text{ g O}_2}{1 \text{ mol O}_2} = 0.267 \text{ g O}_2$$

The total mass of O_2 required for complete combustion is

1.200 g + 0.267 g = 1.467 g O_2.

3.98 All of the O_2 is produced from $KClO_3$; get g $KClO_3$ from g O_2. All of the H_2O is produced from $KHCO_3$; get g $KHCO_3$ from g H_2O. The g H_2O produced also reveals the g CO_2 from the decomposition of $NaHCO_3$. The remaining CO_2 (13.2 g CO_2– g CO_2 from $NaHCO_3$) is due to K_2CO_3 and g K_2CO_3 can be derived from it.

$$4.00 \text{ g O}_2 \times \frac{1 \text{ mol O}_2}{32.00 \text{ g O}_2} \times \frac{2 \text{ mol KClO}_3}{3 \text{ mol O}_2} \times \frac{122.6 \text{ g KClO}_3}{1 \text{ mol KClO}_3} = 10.22 = 10.2 \text{ g KClO}_3$$

$$1.80 \text{ g H}_2\text{O} \times \frac{1 \text{ mol H}_2\text{O}}{18.02 \text{ g H}_2\text{O}} \times \frac{2 \text{ mol KHCO}_3}{1 \text{ mol H}_2\text{O}} \times \frac{100.1 \text{ g KHCO}_3}{1 \text{ mol KHCO}_3} = 20.00 = 20.0 \text{ g KHCO}_3$$

$$1.80 \text{ g H}_2\text{O} \times \frac{1 \text{ mol H}_2\text{O}}{18.02 \text{ g H}_2\text{O}} \times \frac{2 \text{ mol CO}_2}{1 \text{ mol H}_2\text{O}} \times \frac{44.01 \text{ g CO}_2}{1 \text{ mol CO}_2} = 8.792 = 8.79 \text{ g CO}_2 \text{ from KHCO}_3$$

13.20 g CO_2 total – 8.792 CO_2 from $KHCO_3$ = 4.408 = 4.41 g CO_2 from K_2CO_3

$$4.408 \text{ g CO}_2 \times \frac{1 \text{ mol CO}_2}{44.01 \text{ g CO}_2} \times \frac{1 \text{ mol K}_2\text{CO}_3}{1 \text{ mol CO}_2} \times \frac{138.2 \text{ g K}_2\text{CO}_3}{1 \text{ mol K}_2\text{CO}_3} = 13.84 = 13.8 \text{ g K}_2\text{CO}_3$$

100.0 g mixture – 10.22 g $KClO_3$ – 20.00 g $KHCO_3$ – 13.84 g K_2CO_3 = 56.0 g KCl

3.99 (a) $2C_2H_2(g) + 5O_2(g) \rightarrow 4CO_2(g) + 2H_2O(g)$

(b) Following the approach in Sample Exercise 3.18,

$$10.0 \text{ g } C_2H_2 \times \frac{1 \text{ mol } C_2H_2}{26.04 \text{ g } C_2H_2} \times \frac{5 \text{ mol } O_2}{2 \text{ mol } C_2H_2} \times \frac{32.00 \text{ g } O_2}{1 \text{ mol } O_2} = 30.7 \text{ g } O_2 \text{ required}$$

Only 10.0 g O_2 are available, so O_2 limits.

(c)   Since O_2 limits, 0.0 g O_2 remain.

Next, calculate the g C_2H_2 consumed and the amounts of CO_2 and H_2O produced by reaction of 10.0 g O_2.

$$10.0 \text{ g } O_2 \times \frac{1 \text{ mol } O_2}{32.00 \text{ g } O_2} \times \frac{2 \text{ mol } C_2H_2}{5 \text{ mol } O_2} \times \frac{26.04 \text{ g } C_2H_2}{1 \text{ mol } C_2H_2} = 3.26 \text{ g } C_2H_2 \text{ consumed}$$

10.0 g C_2H_2 initial – 3.26 g consumed = 6.74 = 6.7 g C_2H_2 remain

$$10.0 \text{ g } O_2 \times \frac{1 \text{ mol } O_2}{32.00 \text{ g } O_2} \times \frac{4 \text{ mol } CO_2}{5 \text{ mol } O_2} \times \frac{44.01 \text{ g } CO_2}{1 \text{ mol } CO_2} = 11.0 \text{ g } CO_2 \text{ produced}$$

$$10.0 \text{ g } O_2 \times \frac{1 \text{ mol } O_2}{32.00 \text{ g } O_2} \times \frac{2 \text{ mol } H_2O}{5 \text{ mol } O_2} \times \frac{18.02 \text{ g } H_2O}{1 \text{ mol } H_2O} = 2.25 \text{ g } H_2O \text{ produced}$$

3.100   (a) $1.5 \times 10^5 \text{ g } C_9H_8O_4 \times \dfrac{1 \text{ mol } C_9H_8O_4}{180.2 \text{ g } C_9H_8O_4} \times \dfrac{1 \text{ mol } C_7H_6O_3}{1 \text{ mol } C_9H_8O_4} \times \dfrac{138.1 \text{ g } C_7H_6O_3}{1 \text{ mol } C_7H_6O_3}$

$$= 1.1496 \times 10^5 \text{ g} = 1.1 \times 10^2 \text{ kg } C_7H_6O_3$$

(b)   If only 80 percent of the acid reacts, then we need 1/0.80 = 1.25 times as much to obtain the same mass of product: $1.25 \times 1.15 \times 10^2$ kg = 1.4×10^2 kg $C_7H_6O_3$

(c)   Calculate the number of moles of each reactant:

$$1.25 \times 10^5 \text{ g } C_4H_6O_3 \times \frac{1 \text{ mol } C_4H_6O_3}{102.1 \text{ g } C_4H_6O_3} = 1.224 \times 10^3 = 1.22 \times 10^3 \text{ mol } C_4H_6O_3$$

We see that $C_4H_6O_3$ limits, because equal numbers of moles of the two reactants are consumed in the reaction.

$$1.224 \times 10^3 \text{ mol } C_4H_6O_3 \times \frac{1 \text{ mol } C_9H_8O_4}{1 \text{ mol } C_7H_6O_3} \times \frac{180.2 \text{ g } C_9H_8O_4}{1 \text{ mol } C_9H_8O_4} = 2.206 \times 10^5$$

$$= 2.21 \times 10^5 \text{ g } C_9H_8O_4$$

(d)   percent yield $= \dfrac{1.82 \times 10^5 \text{ g}}{2.206 \times 10^5 \text{ g}} \times 100 = 82.5\%$

Integrative Exercises

3.102   (a)   *Plan.* volume of Ag cube $\xrightarrow{\text{density}}$ mass of Ag $\rightarrow$ mol Ag $\rightarrow$ Ag atoms

Solve. $(1.000)^3 \text{ cm}^3 \text{ Ag} \times \dfrac{10.5 \text{ g Ag}}{1 \text{ cm}^3 \text{ Ag}} \times \dfrac{1 \text{ mol Ag}}{107.87 \text{ g Ag}} \times \dfrac{6.022 \times 10^{23} \text{ atoms}}{1 \text{ mol}}$

$$= 5.8618 \times 10^{22} = 5.86 \times 10^{22} \text{ Ag atoms}$$

(b)   1.000 cm^3 cube volume, 74% is occupied by Ag atoms

0.74 cm^3 = volume of 5.86×10^{22} Ag atoms

$$\frac{0.7400 \text{ cm}^3}{5.8618 \times 10^{22} \text{ Ag atoms}} = 1.2624 \times 10^{-23} = 1.3 \times 10^{-23} \text{ cm}^3 / \text{Ag atom}$$

Since atomic dimensions are usually given in Å, we will show this conversion.

$$1.2624 \times 10^{-23} \text{ cm}^3 \times \frac{(1 \times 10^{-2})^3 \text{ m}^3}{1 \text{ cm}^3} \times \frac{1 \text{Å}^3}{(1 \times 10^{-10})^3 \text{ m}^3} = 12.62 = 13 \text{ Å}^3 / \text{Ag atom}$$

(c) $V = 4/3 \, \pi \, r^3$; $r^3 = 3V/4\pi$; $r = (3V/4\pi)^{1/3}$

       $r_A = (3 \times 12.62 \text{ Å}^3 / 4\pi)^{1/3} = 1.444 = 1.4 \text{ Å}$

3.103   (a)   *Analyze.* Given: gasoline = C_8H_{18}, density = 0.69 g/mL, 20.5 mi/gal, 225 mi.

Find: kg CO_2.

Plan. Write and balance the equation for the combustion of octane. Change mi $\rightarrow$ gal octane $\rightarrow$ mL $\rightarrow$ g octane. Use stoichiometry to calculate g and kg CO_2 from g octane.

Solve. $2C_8H_{18}(l) + 25O_2(g) \rightarrow 16CO_2(g) + 18H_2O(l)$

$$225 \text{ mi} \times \frac{1 \text{ gal}}{20.5 \text{ mi}} \times \frac{3.7854 \text{ L}}{1 \text{ gal}} \times \frac{1 \text{ mL}}{1 \times 10^{-3} \text{ L}} \times \frac{0.69 \text{ g octane}}{1 \text{ mL}} = 2.8667 \times 10^4 \text{ g}$$

$$= 29 \text{ kg octane}$$

$$2.8667 \times 10^4 \text{ g } C_8H_{18} \times \frac{1 \text{ mol } C_8H_{18}}{114.2 \text{ g } C_8H_{18}} \times \frac{16 \text{ mol } CO_2}{2 \text{ mol } C_8H_{18}} \times \frac{44.01 \text{ g } CO_2}{1 \text{ mol } CO_2} = 8.8382 \times 10^4 \text{ g}$$

$$= 88 \text{ kg } CO_2$$

Check. $\left(\dfrac{225 \times 4 \times 0.7}{20}\right) \times 10^3 = (45 \times 0.7) \times 10^3 = 30 \times 10^3 \text{ g} = 30 \text{ kg octane}$

$\dfrac{44}{114} \approx \dfrac{1}{3}$; $\dfrac{30 \text{ kg} \times 8}{3} \approx 80 \text{ kg } CO_2$

(b)   *Plan.* Use the same strategy as part (a). *Solve.*

$$225 \text{ mi} \times \frac{1 \text{ gal}}{5 \text{ mi}} \times \frac{3.7854 \text{ L}}{1 \text{ gal}} \times \frac{1 \text{ mL}}{1 \times 10^{-3} \text{ L}} \times \frac{0.69 \text{ g octane}}{1 \text{ mL}} = 1.1754 \times 10^5$$

$$= 1 \times 10^2 \text{ kg octane}$$

$$1.1754 \times 10^5 \text{ g } C_8H_{18} \times \frac{1 \text{ mol } C_8H_{18}}{114.2 \text{ g } C_8H_{18}} \times \frac{16 \text{ mol } CO_2}{2 \text{ mol } C_8H_{18}} \times \frac{44.01 \text{ g } CO_2}{1 \text{ mol } CO_2} = 3.624 \times 10^5 \text{ g}$$

$$= 4 \times 10^2 \text{ kg } CO_2$$

Check. Mileage of 5 mi/gal requires ~4 times as much gasoline as mileage of 20.5 mi/gal, so it should produce ~4 times as much CO₂. 90 kg CO₂ [from (a)] $\times$ 4 = 360 = 4×10^2 kg CO₂ [from (b)].

3.104   *Plan.* We can proceed by writing the ratio of masses of Ag to $AgNO_3$, where y is the
atomic mass of nitrogen.   *Solve.*

$$\frac{Ag}{AgNO_3} = 0.634985 = \frac{107.8682}{107.8682 + 3(15.9994) + y}$$

Solve for y to obtain y = 14.0088. This is to be compared with the currently accepted
value of 14.0067.

3.106   *Analyze.* Given: 2.0 in × 3.0 in boards, 5000 boards, 0.65 mm thick Cu; 8.96 g/cm^3 Cu;
85% Cu removed; 97% yield for reaction. Find: mass $Cu(NH_3)_4Cl_2$, mass NH_3.

Plan. vol Cu/board × density → mass Cu/board → 5000 boards × 85% = total Cu
removed = actual yield; actual yield/0.97 = theoretical yield Cu.

mass Cu → mol Cu → mol $Cu(NH_3)_4$ Cl or NH_3 → desired masses.

Solve. $2.0 \, in \times 3.0 \, in \times \dfrac{(2.54)^2 \, cm^2}{in^2} \times 0.65 \, mm \times \dfrac{1 \, cm}{10 \, mm} = 2.516 = 2.5 \, cm^3$ Cu/board

$\dfrac{2.516 \, cm^3 \, Cu}{board} \times \dfrac{8.96 \, g}{cm^3} \times 5000 \, boards \times 0.85 \, removed = 95,814 \, g = 96 \, kg$ Cu removed

$\dfrac{95,814 \, g \, Cu \, actual \, yield}{0.97} = 98,777 \, g = 99 \, kg$ Cu theoretical

$98,777 \, g \, Cu \times \dfrac{1 \, mol \, Cu}{63.546 \, g \, Cu} \times \dfrac{1 \, mol \, Cu(NH_3)_4Cl_2}{1 \, mol \, Cu} \times \dfrac{202.575 \, g}{1 \, mol \, Cu(NH_3)_4Cl_2} = 314,887 \, g$

$= 3.1 \times 10^2 \, kg \, Cu(NH_3)_4Cl_2$

$98,777 \, g \, Cu \times \dfrac{1 \, mol \, Cu}{63.546 \, g \, Cu} \times \dfrac{4 \, mol \, NH_3}{1 \, mol \, Cu} \times \dfrac{17.03 \, g \, NH_3}{mol \, NH_3} = 105,891 \, g = 1.1 \times 10^2 \, kg \, NH_3$

3.107   (a)   *Plan.* Calculate the kg of air in the room and then the mass of HCN required to
produce a dose of 300 mg HCN/kg air.   *Solve.*

12 ft × 15 ft × 8.0 ft = 1440 = $1.4 \times 10^3 \, ft^3$ of air in the room

$1440 \, ft^3 \, air \times \dfrac{(12 \, in)^3}{1 \, ft^3} \times \dfrac{(2.54 \, cm)^3}{1 \, in^3} \times \dfrac{0.00118 \, g \, air}{1 \, cm^3 \, air} \times \dfrac{1 \, kg}{1000 \, g} = 48.12 = 48 \, kg$ air

$48.12 \, kg \, air \times \dfrac{300 \, mg \, HCN}{1 \, kg \, air} \times \dfrac{1 \, g}{1000 \, mg} = 14.43 = 14 \, g$ HCN

(b) $2NaCN(s) + H_2SO_4(aq) \rightarrow Na_2SO_4(aq) + 2HCN(g)$

The question can be restated as: What mass of NaCN is required to produce 14 g
of HCN according to the above reaction?

$14.43 \, g \, HCN \times \dfrac{1 \, mol \, HCN}{27.03 \, g \, HCN} \times \dfrac{2 \, mol \, NaCN}{2 \, mol \, HCN} \times \dfrac{49.01 \, g \, NaCN}{1 \, mol \, NaCN} = 26.2 = 26 \, g$ NaCN

(c) $12 \, ft \times 15 \, ft \times \dfrac{1 \, yd^2}{9 \, ft^2} \times \dfrac{30 \, oz}{1 \, yd^2} \times \dfrac{1 \, lb}{16 \, oz} \times \dfrac{454 \, g}{1 \, lb} = 17,025$

$= 1.7 \times 10^4 \, g$ acrilan in the room

50% of the carpet burns, so the starting amount of CH_2CHCN is $0.50(17,025)$
$= 8,513 = 8.5 \times 10^3$ g

$$8,513 \, g \, CH_2CHCN \times \frac{50.9 \, g \, HCN}{100 \, g \, CH_2CHCH} = 4333 = 4.3 \times 10^3 \, g \, HCN \, possible$$

If the actual yield of combustion is 20%, actual g HCN = 4,333(0.20) = 866.6
$= 8.7 \times 10^2$ g HCN produced. From part (a), 14 g of HCN is a lethal dose. The fire produces much more than a lethal dose of HCN.

3.108 (a) $N_2(g) + O_2(g) \rightarrow 2NO(g); \; 2NO(g) + O_2(g) \rightarrow 2NO_2(g)$

(b)   1 million = 1×10^6

$$19 \times 10^6 \, tons \, NO_2 \times \frac{2000 \, lb}{1 \, ton} \times \frac{453.6 \, g}{1 \, lb} = 1.724 \times 10^{13} = 1.7 \times 10^{13} \, g \, NO$$

(c)   *Plan.* Calculate g O_2 needed to burn 500 g octane. This is 85% of total O_2 in the engine. 15% of total O_2 is used to produce NO_2, according to the second equation in part (a).

Solve. $2C_8H_{18}(l) + 25O_2(g) \rightarrow 16CO_2(g) + 18H_2O(l)$

$$500 \, g \, C_8H_{18} \times \frac{1 \, mol \, C_8H_{18}}{114.2 \, g \, C_8H_{18}} \times \frac{25 \, mol \, O_2}{2 \, mol \, C_8H_{18}} \times \frac{32.00 \, g \, O_2}{mol \, O_2} = 1751 = 1.75 \times 10^3 \, g \, O_2$$

$$\frac{1751 \, g \, O_2}{total \, g \, O_2} = 0.85; \; 2060 = 2.1 \times 10^3 \, g \, O_2 \, total \, in \, engine$$

2060 g O_2 total × 0.15 = 309.1 = 3.1×10^9 g O_2 used to produce NO_2. One mol O_2 produces 2 mol NO. Then 2 mol NO react with a second mol O_2 to produce 2 mol NO_2. Two mol O_2 are required to produce 2 mol NO_2; one mol O_2 per mol NO_2.

4 Aqueous Reactions and Solution Stoichiometry

Visualizing Concepts

4.2   Although CH_3OH and HCl are both molecular compounds, HCl is an acid and strong electrolyte. Strong electrolytes exist in solution almost completely as ions, so an aqueous HCl solution conducts electricity. CH_3OH is a nonelectrolyte that exists as neutral molecules in aqueous solution. Since there are no charge carriers, aqueous solutions of nonelectrolytes such as CH_3OH do not conduct electricity.

4.4   The brightness of the bulb in Figure 4.2 is related to the number of ions per unit volume of solution. If 0.1 M $HC_2H_3O_2$ has about the same brightness of 0.001 M HBr, the two solutions have about the same number of ions. Since 0.1 M $HC_2H_3O_2$ has 100 times more solute than 0.001 M HBr, HBr must be dissociated to a much greater extent than $HC_2H_3O_2$. HBr is one of the few molecular acids that is a strong electrolyte. $HC_2H_3O_2$ is a weak electrolyte; if it were a nonelectrolyte, the bulb in Figure 4.2 wouldn't glow.

4.6   Certain pairs of ions form precipitates because their attraction is so strong that they cannot be surrounded and separated by solvent molecules. That is, the attraction between solute particles is greater than the stabilization offered by interaction of individual ions with solvent molecules.

4.8   Use the difference in reactivities with SO_4^{2-} to identify Pb^{2+}(aq) and Mg^{2+}(aq). Test a portion of each solution with H_2SO_4(aq). Pb^{2+}(aq) is an exception to the soluble sulfates rule, so $Pb(NO_3)_2$(aq) will form a precipitate, while $Mg(NO_3)_2$(aq) will not.

4.10   Concentration is a ratio of amount of solute to amount of solution or solvent. Thus there are two ways to double the concentration of a solution: double the amount of solute, keeping volume constant or reduce the volume of solution by half, keeping the amount of solute the same.

Electrolytes

4.12   When CH_3OH dissolves, neutral CH_3OH molecules are dispersed throughout the solution. These electrically neutral particles do not carry charge and the solution is nonconducting. When CH_3COOH dissolves, mostly neutral molecules are dispersed throughout the solution. A few of the dissolved molecules ionize to form H^+(aq) and CH_3COO^-(aq). These few ions carry some charge and the solution is weakly conducting.

4.14   Ions are hydrated when they are surrounded by H_2O molecules in aqueous solution.

4.16   (a) $MgI_2(aq) \rightarrow Mg^{2+}(aq) + 2I^-(aq)$

            (b) $Al(NO_3)_3(aq) \rightarrow Al^{3+}(aq) + 3NO_3^-(aq)$

            (c) $HClO_4(aq) \rightarrow H^+(aq) + ClO_4^-(aq)$

            (d) $NaCH_3COO(aq) \rightarrow Na^+(aq) + CH_3COO^-(aq)$

4.18   (a)   acetone (nonelectrolyte): $CH_3COCH_3(aq)$ molecules only; hypochlorous acid (weak electrolyte): $HClO(aq)$ molecules, $H^+(aq)$, ClO^- (aq); ammonium chloride (strong electrolyte): $NH_4^+(aq)$, $Cl^-(aq)$

            (b) NH_4Cl, 0.2 mol solute particles; $HClO$, between 0.1 and 0.2 mol particles; CH_3OCH_3, 0.1 mol of solute particles

Precipitation Reactions and Net Ionic Equations

4.20   According to Table 4.1:

            (a) $Ni(OH)_2$: insoluble

            (b) $PbBr_2$: insoluble;

            (c) $Ba(NO_3)_2$: soluble

            (d) $AlPO_4$: insoluble

            (e) $AgCH_3COO$: soluble

4.22   In each reaction, the precipitate is in bold type.

            (a) $Ni(NO_3)_2(aq) + 2NaOH(aq) \rightarrow$ **$Ni(OH)_2(s)$** $+ 2NaNO_3(aq)$

            (b)   No precipitate, and, therefore, no reaction. There is no chemical change to any of the reactant ions.

            (c) $Na_2S(aq) + CuCH_3COO(aq) \rightarrow$ **$CuS(s)$** $+ 2NaCH_3COO(aq)$

4.24   Spectator ions are those that do not change during reaction.

            (a) $2Cr^{3+}(aq) + 3CO_3^{2-}(aq) \rightarrow Cr_2(CO_3)_3(s)$; spectators: NH_4^+, SO_4^{2-}

            (b) $Ba^{2+}(aq) + SO_4^{2-}(aq) \rightarrow BaSO_4(s)$; spectators: K^+, NO_3^-

            (c) $Fe^{2+}(aq) + 2OH^-(aq) \rightarrow Fe(OH)_2(s)$; spectators: K^+, NO_3^-

4.26 Br^- and NO_3^- can be ruled out because the Ba^{2+} salts are soluble. (Actually all NO_3^- salts are soluble.) CO_3^{2-} forms insoluble salts with the three cations given; it must be the anion in question.

4.28   (a) $Pb(CH_3COO)_2(aq) + Na_2S(aq) \rightarrow PbS(s) + 2NaCH_3COO(aq)$

                    net ionic: $Pb^{2+}(aq) + S^{2-}(aq) \rightarrow PbS$

                    $Pb(CH_3COO)_2(aq) + CaCl_2(aq) \rightarrow (PbCl_2)s + Ca(CH_3COO)_2(aq)$

                    net ionic: $Pb^{2+}(aq) + 2Cl^-(aq) \rightarrow PbCl_2(s)$

                    $Na_2S(aq) + CaCl_2(aq) \rightarrow CaS(aq) + 2NaCl(aq)$

                    net ionic: no reaction

            (b)   Spectator ions: Na^+, Ca^{2+}, CH_3COO^-

Acid-Base Reactions

4.30 $NH_3(aq)$ is a weak base, while KOH and $Ca(OH)_2$ are strong bases. $NH_3(aq)$ is only slightly ionized, so even 0.6 M NH_3 is less basic than 0.150 M KOH. $Ca(OH)_2$ has twice as many OH^- per moles as KOH, so 0.100 M $Ca(OH)_2$ is more basic than 0.150 M KOH. The most basic solution is 0.100 M $Ca(OH)_2$.

4.32   (a) NH_3 produces OH^- in aqueous solution by reacting with H_2O (hydrolysis): $NH_3(aq) + H_2O(l) \rightleftharpoons NH_4^+(aq) + OH^-(aq)$. The OH^- causes the solution to be basic.

         (b)   The term "weak" refers to the tendency of HF to dissociate into H^+ and F^- in aqueous solution, not its reactivity toward other compounds.

         (c) H_2SO_4 is a **diprotic** acid; it has two ionizable hydrogens. The first hydrogen completely ionizes to form H^+ and HSO_4^-, but HSO_4^- only **partially** ionizes into H^+ and SO_4^{2-} (HSO_4^- is a weak electrolyte). Thus, an aqueous solution of H_2SO_4 contains a mixture of H^+, HSO_4^- and SO_4^{2-}, with the concentration of HSO_4^- greater than the concentration of SO_4^{2-}.

4.34   All soluble ionic hydroxides from Table 4.1 are listed as strong bases in Table 4.2. Insoluble hydroxides like $Cd(OH)_2$ are not listed as strong bases. "Insoluble" means that less than 1% of the base molecules exist as separated ions and are dissolved. Thus, insoluble hydroxide salts produce too few $OH^-(aq)$ to be considered strong bases.

4.36   Since the solution does conduct some electricity, but less than an equimolar NaCl solution (a strong electrolyte), the unknown solute must be a weak electrolyte. The weak electrolytes in the list of choices are NH_3 and H_3PO_3; since the solution is acidic, the unknown must be **H_3PO_3**.

4.38   (a) $HClO_4$: strong   (b) HNO_3: strong   (c) NH_4Cl: strong

         (d) CH_3OCH_3: non   (e) $CoSO_4$: strong   (f) $C_{12}H_{22}O_{11}$: non

4.40   (a) $HC_2H_3O_2(aq) + KOH(aq) \rightarrow KC_2H_3O_2(aq) + H_2O(l)$

                $HC_2H_3O_2(aq) + OH^-(aq) \rightarrow C_2H_3O_2^-(aq)\ H_2O(l)$

         (b) $Cr(OH)_3(s) + 3HNO_3(aq) \rightarrow Cr(NO_3)_3(aq) + 3H_2O(l)$

                $Cr(OH)_3(s) + 3H^+(aq) \rightarrow 3H_2O(l) + Cr^{3+}(aq)$

         (c) $Ca(OH)_2(aq) + 2HClO(aq) \rightarrow Ca(ClO)_2(aq) + 2H_2O(l)$

                $HClO(aq) + OH^-(aq) \rightarrow ClO^-(aq) + H_2O(l)$

4.42   (a) $FeO(s) + 2H^+(aq) \rightarrow H_2O(l) + Fe^{2+}(aq)$

         (b) $NiO(s) + 2H^+(aq) \rightarrow H_2O(l) + Ni^{2+}(aq)$

4.44 $K_2O(aq) + H_2O(l) \rightarrow 2KOH(aq)$, molecular; $O^{2-}(aq) + H_2O(l) \rightarrow 2OH^-(aq)$, net ionic

         base: (H^+ ion acceptor) $O^{2-}(aq)$; acid: (H^+ ion donor) $H_2O(aq)$; spectator: K^+

Oxidation-Reduction Reactions

4.46 Oxidation and reduction can only occur together, not separately. When a metal reacts with oxygen, the metal atoms lose electrons and the oxygen atoms gain electrons. Free electrons do not exist under normal conditions. If electrons are lost by one substance they must be gained by another, and vice versa.

4.48 Elements (metals) from Table 4.5 in region A include Na, Mg, K, and Ca; those from region C are Hg and Au. Let's consider K and Au. Since metals from region A are more readily oxidized than those from region C, K will be oxidized to K^+ and Au^{3+} will be reduced to Au in the redox reaction. (Choose Au^{3+} because it is the Au ion shown in Table 4.5.)

In a balanced redox reaction, the number of electrons lost must equal the number of electrons gained. Since K loses 1 electron in forming K^+, while Au^{3+} gains 3 electrons when forming Au, 3 K atoms must be oxidized for every 1 Au^{3+} that is reduced. This relationship dictates the coefficients in the balanced redox reaction.

$$3K(s) + Au^{3+}(aq) \rightleftharpoons Au(s) + 3K^+(aq)$$

4.50 (a) +4 (b) +2 (c) +3 (d) –2 (e) +3 (f) +6

4.52 (a) acid-base reaction

    (b) oxidation-reduction reaction; Fe is reduced, C is oxidized

    (c) precipitation reaction

    (d) oxidation-reduction reaction; Zn is oxidized, N is reduced

4.54 (a) $2HCl(aq) + Ni(s) \rightarrow NiCl_2(aq) + H_2(g)$; $Ni(s) + 2H^+(aq) \rightarrow Ni^{2+}(aq) + H_2(g)$

    (b) $H_2SO_4(aq) + Fe(s) \rightarrow FeSO_4(aq) + H_2(g)$; $Fe(s) + 2H^+(aq) \rightarrow Fe^{2+}(aq) + H_2(g)$

    Products with the metal in a higher oxidation state are possible, depending on reaction conditions and acid concentration.

    (c) $2HBr(aq) + Mg(s) \rightarrow MgBr_2(aq) + H_2(g)$; $Mg(s) + 2H^+(aq) \rightarrow Mg^{2+}(aq) + H_2(g)$

    (d) $2CH_3COOH(aq) + Zn(s) \rightarrow ZnCH_3COO(aq) + H_2(g)$;

      $Zn(s) + 2CH_3COOH(aq) \rightarrow Zn^{2+}(aq) + 2CH_3COO^-(aq) + H_2(g)$

4.56 (a) $Mn(s) + NiCl_2(aq) \rightarrow MnCl_2(aq) + Ni(s)$

    (b) $Cu(s) + Cr(C_2H_3O_2)(aq) \rightarrow NR$

    (c) $2Cr(s) + 3NiSO_4(aq) \rightarrow Cr_2(SO_4)_3(aq) + 3Ni(s)$

    (d) $Pt(s) + HBr(aq) \rightarrow NR$

    (e) $H_2(g) + CuCl_2(aq) \rightarrow Cu(s) + 2HCl(aq)$

4.58 (a) $Br_2 + 2NaI \rightarrow 2NaBr + I_2$ indicates that Br_2 is more easily reduced than I_2.

      $Cl_2 + 2NaBr \rightarrow 2NaCl + Br_2$ shows that Cl_2 is more easily reduced than Br_2.

The order for ease of reduction is $Cl_2 > Br_2 > I_2$. Conversely, the order for ease of oxidation is $I^- > Br^- > Cl^-$.

(b)   Since the halogens are nonmetals, they tend to form anions when they react chemically. Nonmetallic character decreases going down a family and so does the tendency to gain electrons during a chemical reaction. Thus, the ease of reduction of the halogen, X_2, decreases going down the family and the ease of oxidation of the halide, X^-, increases going down the family.

(c) $Cl_2 + 2KI \rightarrow 2KCl + I_2$; $Br_2 + LiCl \rightarrow$ no reaction

Solution Composition; Molarity

4.60   (a)   The concentration of the remaining solution is unchanged, assuming the original solution was thoroughly mixed. Molar concentration is a **ratio** of moles solute to liters solution. Although there are fewer moles solute remaining in the flask, there is also less solution volume, so the ratio of moles solute/solution volume remains the same.

(b)   The concentration of the remaining solution is increased. The moles solute remaining in the flask are unchanged, but the solution volume decreases after evaporation. The ratio of moles solute/solution volume increases.

(c)   The second solution is five times as concentrated as the first. An equal volume of the more concentrated solution will contain five times as much solute (five times the number of moles and also five times the mass) as the 0.50 M solution. Thus, the mass of solute in the 2.50 M solution is 5×4.5 g $= 22.5$ g.

Mathematically:

$$\frac{\dfrac{2.50 \text{ mol solute}}{1 \text{ L solution}}}{\dfrac{0.50 \text{ mol solute}}{1 \text{ L solution}}} = \frac{x \text{ grams solute}}{4.5 \text{ g solute}}$$

$$\frac{2.50 \text{ mol solute}}{0.50 \text{ mol solute}} = \frac{x \text{ g solute}}{4.5 \text{ g solute}}; \ 5.0(4.5 \text{ g solute}) = 23 \text{ g solute}$$

The result has 2 sig figs; 22.5 rounds to 23 g solute.

4.62   (a) $M = \dfrac{\text{mol solute}}{\text{L solution}}; \ \dfrac{0.750 \text{ g Na}_2\text{SO}_4}{0.850 \text{ L}} \times \dfrac{1 \text{ mol Na}_2\text{SO}_4}{142.04 \text{ g Na}_2\text{SO}_4} = 6.21 \times 10^{-3} \ M \text{ Na}_2\text{SO}_4$

(b) $\text{mol} = M \times L; \ \dfrac{0.0475 \text{ mol KMnO}_4}{1 \text{ L}} \times 0.250 \text{ L} = 1.19 \times 10^{-2} \text{ mol KMnO}_4$

(c) $L = \dfrac{\text{mol}}{M}; \ \dfrac{0.250 \text{ mol HCl}}{11.6 \text{ mol HCl/L}} = 2.16 \times 10^{-2} \text{ L or } 21.6 \text{ mL}$

4.64   Calculate the mol of Na^+ at the two concentrations; the difference is the mol NaCl required to increase the Na^+ concentration to the desired level.

$$\frac{0.118 \text{ mol}}{L} \times 4.6 \text{ L} = 0.5428 = 0.54 \text{ mol Na}^+$$

$$\frac{0.138\,mol}{L} \times 4.6\,L = 0.6348 = 0.63\,mol\,Na^+$$

$(0.6348 - 0.5428) = 0.092 = 0.09\,mol\,NaCl$ (2 decimal places and 1 sig fig)

$$0.092\,mol\,NaCl \times \frac{58.5\,g\,NaCl}{mol} = 5.38 = 5\,g\,NaCl$$

4.66   *Analyze.* Given: BAC (definition from Exercise 4.65), vol of blood. Find: mass alcohol in bloodstream.

Plan. Change BAC (g/100 mL) to (g/L), then times vol of blood in L.

Solve. BAC = 0.10 g/100 mL

$$\frac{0.10\,g\,alcohol}{100\,mL\,blood} \times \frac{1000\,mL}{1\,L} \times 5.0\,L\,blood = 5.0\,g\,alcohol$$

4.68 $M = \dfrac{mol}{L}$; $mol = \dfrac{g}{MM}$ (MM is the symbol for molar mass in this manual.)

(a) $$\frac{0.488\,mol\,K_2Cr_2O_7}{1\,L} \times 50.0\,mL \times \frac{1\,L}{1000\,mL} \times \frac{294.2\,g\,K_2Cr_2O_7}{1\,mol\,K_2Cr_2O_7} = 7.18\,g\,K_2Cr_2O_7$$

(b) $$4.00\,g\,(NH_4)_2SO_4 \times \frac{1\,mol\,(NH_4)_2SO_4}{132.2\,g\,(NH_4)_2SO_4} \times \frac{1}{400.\,mL} \times \frac{1000\,mL}{1\,L}$$

$$= 0.0756\,M\,(NH_4)_2SO_4$$

(c) $$1.75\,g\,CuSO_4 \times \frac{1\,mol\,CuSO_4}{159.6\,g\,CuSO_4} \times \frac{1\,L}{0.0250\,mol\,CuSO_4} \times \frac{1000\,mL}{1\,L} = 439\,mL\,solution$$

4.70   (a) $0.1\,M\,CaCl_2 = 0.2\,M\,Cl^-$; $0.15\,M\,KCl = 0.15\,M\,Cl^-$

$0.1\,M\,CaCl_2$ has the higher Cl^- concentration.

(b) $0.1\,M\,KCl$ has a higher Cl^- concentration than $0.080\,M\,LiCl$. Total volume does not affect concentration.

(c) $0.050\,M\,HCl = 0.050\,M\,Cl^-$; $0.020\,M\,CdCl_2 = 0.040\,M\,Cl^-$

4.72   (a)   *Plan.* These two solutions have common ions. Find the ion concentration resulting from each solution, then add.

Solve. total volume = 42.0 mL + 37.6 mL = 79.6 mL

$$\frac{0.170\,M\,NaOH \times 42.0\,mL}{79.6\,mL} = 0.08970 = 0.0897\,M\,NaOH;$$

$0.0897\,M\,Na^+$, $0.0897\,M\,OH^-$

$$\frac{0.400\,M\,NaOH \times 37.6\,mL}{79.6\,mL} = 0.18894 = 0.189\,M\,NaOH;$$

0.189 M Na$^+$, 0.189 M OH$^-$

M Na$^+$ = 0.08970 M + 0.18894 M = 0.27864 = 0.2786 M Na$^+$

M OH$^-$ = M Na$^+$ = 0.2786 M OH$^-$

(b)   *Plan.* No common ions; just dilution.

Solve. 44.0 mL + 25.0 mL = 69.0 mL

$$\frac{0.100\,M\,Na_2SO_4 \times 44.0\,mL}{69.0\,mL} = 0.06377 = 0.0638\,M\,Na_2SO_4$$

2 × (0.06377 M) = 0.1275 = 0.128 M Na$^+$; 0.0638 M SO$_4^{2-}$

$$\frac{0.150\,M\,KCl \times 25.0\,mL}{69.0\,mL} = 0.054348 = 0.0543\,M\,KCl$$

0.0543 M K$^+$, 0.0543 M Cl$^-$

(c)   *Plan.* Calculate concentration of K$^+$ and Cl$^-$ due to the added solid. Then sum to get total concentration of Cl$^-$.

Solve. $\dfrac{3.60\,g\,KCl}{75.0\,mL\,soln} \times \dfrac{1\,mol\,KCl}{74.55\,g\,KCl} \times \dfrac{1000\,mL}{1\,L} = 0.6439 = 0.644\,M\,KCl$

0.250 M CaCl$_2$; 2(0.250 M) = 0.500 M Cl$^-$

total Cl$^-$ = 0.644 M + 0.500 M = 1.144 M Cl$^-$, 0.644 M K$^+$, 0.250 M Ca^{2+}

4.74   (a) $V_1 = M_2V_2/M_1$; $\dfrac{0.500\,M\,HNO_3 \times 0.450\,mL}{10.0\,M\,HNO_3} = 0.02250\,L = 22.5\,mL\,conc.\,HNO_3$

(b) $M_2 = M_1V_1/V_2$; $\dfrac{10.0\,M\,HNO_3 \times 25.0\,mL}{500\,mL} = 0.500\,M\,HNO_3$

4.76   (a)   The amount of AgNO$_3$ needed is:

0.150 M × 0.1750 L = 0.02625 = 0.263 mol AgNO$_3$

0.02625 mol AgNO$_3$ × $\dfrac{169.88\,g\,AgNO_3}{1\,mol\,AgNO_3}$ = 4.4594 = 4.46 g AgNO$_3$

Add this amount of solid to a 175 mL volumetric container, dissolve in a small amount of water, bring the total volume to exactly 175 mL, and agitate well.

(b)   Dilute the 3.6 M HNO$_3$ to prepare 100 mL of 0.50 M HNO$_3$. To determine the volume of 3.6 M HNO$_3$ needed, calculate the moles HNO$_3$ present in 100 mL of 0.50 M HNO$_3$ and then the volume of 6.0 M solution that contains this number of moles.

0.100 L × 0.50 M = 0.050 mol HNO$_3$ needed;

$$L = \frac{mol}{M}; \quad L\, 3.6\,M\,HNO_3 = \frac{0.050\,mol\ needed}{3.6\,M} = 0.01389\,L = 14\,mL$$

Thoroughly clean, rinse, and fill a buret with the 3.6 M HNO_3, taking precautions appropriate for working with a relatively concentrated acid. Dispense 14 mL of the 3.6 M acid into a 100 mL volumetric flask, add water to the mark, and mix thoroughly.

4.78 $50.000\,mL\ glycerol \times \dfrac{1.2656\,g\ glycerol}{1\,mL\ glycerol} = 63.280\,g\ glycerol$

$63.280\,g\ C_3H_8O_3 \times \dfrac{1\,mol\ C_3H_8O_3}{92.094\,g\ C_3H_8O_3} = 0.687124 = 0.68712\,mol\ C_3H_8O_3$

$M = \dfrac{0.687124\,mol\ C_3H_8O_3}{0.25000\,L\ solution} = 2.7485\,M\ C_3H_8O_3$

Solution Stoichiometry; Titrations

4.80   *Plan.* $M \times L$ = mol $Cd(NO_3)_2$; balanced equation → mol ratio → mol NaOH → g NaOH

Solve. $\dfrac{0.500\,mol\ Cd(NO_3)_2}{1\,L} \times 0.0350\,L = 0.0175\,mol\ Cd(NO_3)_2$

$Cd(NO_3)_2(aq) + 2NaOH(aq) \rightarrow Cd(OH)_2(s) + 2NaNO_3(aq)$

$0.0175\,mol\ Cd(NO_3)_2 \times \dfrac{2\,mol\ NaOH}{1\,mol\ Cd(NO_3)_2} \times \dfrac{40.00\,g\ NaOH}{1\,mol\ NaOH} = 1.40\,g\ NaOH$

4.82   (a) $2HCl(aq) + Ba(OH)_2(aq) \rightarrow BaCl_2(aq) + 2H_2O(l)$

$\dfrac{0.101\,mol\ Ba(OH)_2}{1\,L\ Ba(OH)_2} \times 0.0500\,L\ Ba(OH)_2 \times \dfrac{2\,mol\ HCl}{1\,mol\ Ba(OH)_2}$

$\times \dfrac{1\,L\ HCl}{0.120\,mol\ HCl} = 0.0842\,L\ or\ 84.2\,mL\ HCl\ soln$

(b) $H_2SO_4(aq) + 2NaOH(aq) \rightarrow Na_2SO_4(aq) + 2H_2O(l)$

$0.200\,g\ NaOH \times \dfrac{1\,mol\ NaOH}{40.00\,g\ NaOH} \times \dfrac{1\,mol\ H_2SO_4}{2\,mol\ NaOH} \times \dfrac{1\,L\ H_2SO_4}{0.125\,mol\ H_2SO_4}$

$= 0.0200\,L\ or\ 20.0\,mL\ H_2SO_4\ soln$

(c) $BaCl_2(aq) + Na_2SO_4(aq) \rightarrow BaSO_4(s) + 2NaCl(aq)$

$752\,mg = 0.752\,g\ Na_2SO_4 \times \dfrac{1\,mol\ Na_2SO_4}{142.1\,g\ Na_2SO_4} \times \dfrac{1\,mol\ BaCl_2}{1\,mol\ Na_2SO_4} \times \dfrac{1}{0.0558\,L}$

$= 0.0948\,M\ BaCl_2$

(d) $2HCl(aq) + Ca(OH)_2(aq) \rightarrow CaCl_2(aq) + 2H_2O(l)$

$0.0427\,L\ HCl \times \dfrac{0.208\,mol\ HCl}{1\,L\ HCl} \times \dfrac{1\,mol\ Ca(OH)_2}{2\,mol\ HCl} \times \dfrac{74.10\,g\ Ca(OH)_2}{1\,mol\ Ca(OH)_2} = 0.329\,g\ Ca(OH)_2$

4.84   See Exercise 4.81(a) for a more detailed approach.

$$\frac{0.115 \text{ mol NaOH}}{1 \text{ L}} \times 0.0425 \text{ L} \times \frac{1 \text{ mol CH}_3\text{COOH}}{1 \text{ mol NaOH}} \times \frac{60.05 \text{ g CH}_3\text{COOH}}{1 \text{ mol CH}_3\text{COOH}}$$

$$= 0.29349 = 0.293 \text{ g CH}_3\text{COOH in } 3.45 \text{ mL}$$

$$1.00 \text{ qt vinegar} \times \frac{1 \text{ L}}{1.057 \text{ qt}} \times \frac{1000 \text{ mL}}{1 \text{ L}} \times \frac{0.29349 \text{ g CH}_3\text{COOH}}{3.45 \text{ mL vinegar}} = 80.5 \text{ g CH}_3\text{COOH/qt}$$

4.86   The balanced equation for the titration is:

$$Sr(NO_3)_2(aq) + Na_2CrO_4(aq) \rightarrow SrCrO_4(s) + 2NaNO_3(aq)$$

Beginning with a 0.100 L sample, we can do the following conversions:

$$\text{volume soln} \rightarrow \text{g Sr(NO}_3)_2 \rightarrow \text{mol Sr(NO}_3)_2 \rightarrow \text{mol Na}_2\text{CrO}_4 \rightarrow \text{vol Na}_2\text{CrO}_4 \text{ soln}$$

$$0.100 \text{ L soln} \times \frac{6.82 \text{ g Sr(NO}_3)_2}{0.500 \text{ L soln}} \times \frac{1 \text{ mol Sr(NO}_3)_2}{211.6 \text{ g Sr(NO}_3)_2} \times \frac{1 \text{ mol Na}_2\text{CrO}_4}{1 \text{ mol Sr(NO}_3)_2}$$

$$\times \frac{1 \text{ L soln}}{0.0245 \text{ mol Na}_2\text{CrO}_4} = 0.263 \text{ L Na}_2\text{CrO}_4 \text{ soln}$$

4.88   (a) $HNO_3(aq) + NaOH(s) \rightarrow NaNO_3(aq) + H_2O(l)$

       (b)   Determine the limiting reactant, then the identity and concentration of ions remaining in solution. Assume that the $H_2O(l)$ produced by the reaction does **not** increase the total solution volume.

$$12.0 \text{ g NaOH} \times \frac{1 \text{ mol NaOH}}{40.00 \text{ g NaOH}} = 0.300 \text{ mol NaOH}$$

$$0.200 \text{ M HNO}_3 \times 0.0750 \text{ L HNO}_3 = 0.0150 \text{ mol HNO}_3.$$

The mol ratio is 1:1, so HNO_3 is the limiting reactant. No excess H^+ remains in solution. The remaining ions are OH^- (excess reactant), Na^+, and NO_3^- (spectators).

OH^-:   0.300 mol OH^- initial – 0.0150 mol OH^- react = 0.285 mol OH^- remain

          0.285 mol OH^-/0.0750 L soln = 3.80 M OH^-(aq)

Na^+:   0.300 mol Na^+/0.0750 L soln = 4.00 M Na^+(aq)

NO_3^-:   0.0150 mol NO_3^-/0.0750 L = 0.200 M NO_3^-(aq)

       (c)   The resulting solution is **basic** because of the large excess of OH^-(aq).

Additional Exercises

4.91   If we know the identity of a solute and need to measure its concentration in solution, titration is a good way to do this if three conditions are met. The solute must be a reactant in a chemical reaction of known stoichiometry, the second reactant must be available as a standard solution of well-known concentration, and the equivalence point of the titration must be detectable, either visually or instrumentally. If these conditions are met, a known quantity of solution can be precisely delivered by pipette or buret to a titration vessel. The unknown concentration is then determined by its stoichiometric relationship to the quantity of standard solution required to reach the equivalence point of the titration. Multiple titrations can be performed to ensure precision of the measurement. In summary, titration works well for measuring the unknown concentration of a solute because a well-known quantity of the solute can be measured volumetrically and titrations are repeatable.

4.93   The two precipitates formed are due to $AgCl(s)$ and $SrSO_4(s)$. Since no precipitate forms on addition of hydroxide ion to the remaining solution, the other two possibilities, Ni^{2+} and Mn^{2+}, are absent.

4.95   (a) $Al(OH)_3(s) + 3H^+(aq) \rightarrow Al^{3+}(aq) + 3H_2O(l)$

             (b) $Mg(OH)_2(s) + 2H^+(aq) \rightarrow Mg^{2+}(aq) + 2H_2O(l)$

             (c) $MgCO_3(s) + 2H^+(aq) \rightarrow Mg^{2+}(aq) + H_2O(l) + CO_2(g)$

             (d) $NaAl(CO_3)(OH)_2(s) + 4H^+(aq) \rightarrow Na^+(aq) + Al^{3+}(aq) + 3H_2O(l) + CO_2(g)$

             (e) $CaCO_3(s) + 2H^+(aq) \rightarrow Ca^{2+}(aq) + H_2O(l) + CO_2(g)$

     [In (c), (d) and (e), one could also write the equation for formation of bicarbonate, e.g., $MgCO_3(s) + H^+(aq) \rightarrow Mg^{2+} + HCO_3^-(aq).$]

4.96   (a) $2H^+(aq) + SO_3^{2-}(aq) \rightarrow H_2SO_3(aq)$; sulfurous acid

             (b) $H_2SO_3(aq) \rightarrow H_2O(l) + SO_2(g)$; sulfur dioxide

             (c)   The boiling point of $SO_2(g)$ is –10°C. It is a gas at room temperature (23°C) and pressure (1 atm).

             (d)   (i) $Na_2SO_3(aq) + 2HCl(aq) \rightarrow 2NaCl(aq) + H_2O(l) + SO_2(g)$

                         $SO_3^{2-}(aq) + 2H^+(aq) \rightarrow H_2O(l) + SO_2(g)$

                 (ii) $Ag_2SO_3(s) + 2HCl(aq) \rightarrow 2AgCl(s) + H_2O(l) + SO_2(g)$

                         $Ag_2SO_3(s) + 2H^+(aq) + 2Cl^-(aq) \rightarrow 2AgCl(s) + H_2O(l) + SO_2(g)$

                 (iii) $KHSO_3(s) + HCl(aq) \rightarrow KCl(aq) + H_2O(l) + SO_2(g)$

                         $KHSO_3(s) + H^+(aq) \rightarrow K^+(aq) + H_2O(l) + SO_2(g)$

                 (iv) $ZnSO_3(aq) + 2HCl(aq) \rightarrow ZnCl_2(aq) + H_2O(l) + SO_2(g)$

                         $SO_3^{2-}(aq) + 2H^+(aq) \rightarrow H_2O(l) + SO_2(g)$

4.98   A metal on Table 4.5 is able to displace the metal cations below it from their compounds. That is, zinc will reduce the cations below it to their metals.

     (a) $Zn(s) + Na^+(aq) \rightarrow$ no reaction

     (b) $Zn(s) + Pb^{2+}(aq) \rightarrow Zn^{2+}(aq) + Pb(s)$

     (c) $Zn(s) + Mg^{2+}(aq) \rightarrow$ no reaction

     (d) $Zn(s) + Fe^{2+}(aq) \rightarrow Zn^{2+}(aq) + Fe(s)$

     (e) $Zn(s) + Cu^{2+}(aq) \rightarrow Zn^{2+}(aq) + Cu(s)$

     (f) $Zn(s) + Al^{3+}(aq) \rightarrow$ no reaction

4.99   (a)   A : La_2O_3 Metals often react with the oxygen in air to produce metal oxides.

B : $La(OH)_3$ When metals react with water (HOH) to form H_2, OH^- remains.

C : $LaCl_3$ Most chlorides are soluble.

D : $La_2(SO_4)_3$ Sulfuric acid provides SO_4^{2-} ions.

(b) $4La(s) + 3O_2(g) \rightarrow 2La_2O_3(s)$

$2La(s) + 6HOH(l) \rightarrow 2La(OH)_3(s) + 3H_2(g)$

(There are no spectator ions in either of these reactions.)

molecular: $La_2O_3(s) + 6HCl(aq) \rightarrow 2LaCl_3(aq) + 3H_2O(l)$

net ionic: $La_2O_3(s) + 6H^+(aq) \rightarrow 2La^{3+}(aq) + 3H_2O(l)$

molecular: $La(OH)_3(s) + 3HCl(aq) \rightarrow LaCl_3(aq) + 3H_2O(l)$

net ionic: $La(OH)_3(s) + 3H^+(aq) \rightarrow La^{3+}(aq) + 3H_2O(l)$

molecular: $2LaCl_3(aq) + 3H_2SO_4(aq) \rightarrow La_2(SO_4)_3(s) + 6HCl(aq)$

net ionic: $2La^{3+}(aq) + 3SO_4^{2-}(aq) \rightarrow La_2(SO_4)_3(s)$

(c)   La metal is oxidized by water to produce $H_2(g)$, so La is definitely above H on the activity series. In fact, since an acid is not required to oxidize La, it is probably one of the more active metals.

4.102 Na^+ must replace the total positive (+) charge due to Ca^{2+} and Mg^{2+}. Think of this as moles of charge rather than moles of particles.

$$\frac{0.020 \text{ mol } Ca^{2+}}{1 \text{ L water}} \times 1.5 \times 10^3 \text{ L} \times \frac{2 \text{ mol + change}}{1 \text{ mol } Ca^{2+}} = 60 \text{ mol of + charge}$$

$$\frac{0.0040 \text{ mol } Mg^{2+}}{1 \text{ L water}} \times 1.5 \times 10^3 \text{ L} \times \frac{2 \text{ mol + charge}}{1 \text{ mol } Mg^{2+}} = 12 \text{ mol of + charge}$$

72 moles of + charge must be replaced; 72 mol Na^+ are needed.

4.103 $H_2C_4H_4O_6 + 2OH^-(aq) \rightarrow C_4H_4O_6^{2-}(aq) + 2H_2O(l)$

$$0.02465 \text{ L NaOH soln} \times \frac{0.2500 \text{ mol NaOH}}{1 \text{ L}} \times \frac{1 \text{ mol } H_2C_4H_4O_6}{2 \text{ mol NaOH}} \times \frac{1}{0.0500 \text{ L } H_2C_4H_4O_6}$$

$$= 0.06163 \; M \; H_2C_4H_4O_6 \text{ soln}$$

4.105   mol OH^- from NaOH(aq) + mol OH^- from $Zn(OH)_2(s)$ = mol H^+ from HBr

mol H^+ = M HBr $\times$ L HBr = 0.500 M HBr $\times$ 0.350 L HBr = 0.175 mol H^+

mol OH^- from NaOH = M NaOH $\times$ L NaOH = 0.500 M NaOH $\times$ 0.0885 L NaOH

$$= 0.04425 = 0.0443 \text{ mol } OH^-$$

mol OH^- from $Zn(OH)_2(s)$ = 0.175 mol H^+ – 0.04425 mol OH^- from NaOH = 0.13075

$$= 0.131 \text{ mol } OH^- \text{ from } Zn(OH)_2$$

$$0.13075 \text{ mol OH}^- \times \frac{1 \text{ mol Zn(OH)}_2}{2 \text{ mol OH}^-} \times \frac{99.41 \text{ g Zn(OH)}_2}{1 \text{ mol Zn(OH)}_2} = 6.50 \text{ g Zn(OH)}_2$$

Integrative Exercises

4.107 $Ba^{2+}(aq) + SO_4^{2-}(aq) \rightarrow BaSO_4(s)$

$$0.2815 \text{ g BaSO}_4 \times \frac{137.3 \text{ g Ba}}{233.4 \text{ g BaSO}_4} = 0.16560 = 0.1656 \text{ g Ba}$$

$$\text{mass \%} = \frac{\text{g Ba}}{\text{g sample}} \times 100 = \frac{0.16560 \text{ g Ba}}{3.455 \text{ g sample}} \times 100 = 4.793 \text{ \% Ba}$$

4.108   *Plan.* Write balanced equation.

$$\text{mass H}_2\text{SO}_4 \text{ soln} \xrightarrow{\text{mass \%}} \text{mass H}_2\text{SO}_4 \rightarrow \text{mol H}_2\text{SO}_4 \rightarrow \text{mol Na}_2\text{CO}_3 \rightarrow \text{mass Na}_2\text{CO}_3$$

Solve. $H_2SO_4(aq) + Na_2CO_3(s) \rightarrow Na_2SO_4(aq) + H_2O(l) + CO_2(g)$

$$5.0 \times 10^3 \text{ kg conc. H}_2\text{SO}_4 \times \frac{0.950 \text{ kg H}_2\text{SO}_4}{1.00 \text{ kg conc. H}_2\text{SO}_4} = 4.75 \times 10^3 = 4.8 \times 10^3 \text{ kg H}_2\text{SO}_4$$

$$4.75 \times 10^3 \text{ kg H}_2\text{SO}_4 \times \frac{1 \times 10^3 \text{ g}}{1 \text{ kg}} \times \frac{1 \text{ mol H}_2\text{SO}_4}{98.08 \text{ g H}_2\text{SO}_4} \times \frac{1 \text{ mol Na}_2\text{CO}_3}{1 \text{ mol H}_2\text{SO}_4}$$

$$\times \frac{105.99 \text{ g NaHCO}_3}{1 \text{ mol NaHCO}_3} \times \frac{1 \text{ kg}}{1 \times 10^3 \text{ g}} = 5.133 \times 10^3 = 5.1 \times 10^3 \text{ kg Na}_2\text{CO}_3$$

4.110   (a) $Na_2SO_4(aq) + Pb(NO_3)_2(s) \rightarrow PbSO_4(s) + 2NaNO_3(aq)$

(b)   Calculate mol of each reactant and compare.

$$1.50 \text{ g Pb(NO}_3)_2 \times \frac{1 \text{ mol Pb(NO}_3)_2}{331.2 \text{ g Pb(NO}_3)_2} = 0.004529 = 4.53 \times 10^{-3} \text{ mol Pb(NO}_3)_2$$

$0.100 \; M \text{ Na}_2\text{SO}_4 \times 0.125 \text{ L} = 0.0125 \text{ mol Na}_2\text{SO}_4$

Since the reactants combine in a 1:1 mol ratio, $Pb(NO_3)_2$ is the limiting reactant.

(c) $Pb(NO_3)_2$ is the limiting reactant, so no Pb^{2+} remains in solution. The remaining ions are: SO_4^{2-} (excess reactant), Na^+ and NO_3^- (spectators).

SO_4^{2-}:   0.0125 mol SO_4^{2-} initial – 0.00453 mol SO_4^{2-} reacted

$$= 0.00797 = 0.0080 \text{ mol SO}_4^{2-} \text{ remain}$$

0.00797 mol SO_4^{2-} / 0.125 L soln = 0.064 M SO_4^{2-}

Na^+:   Since the total volume of solution is the volume of $Na_2SO_4(aq)$ added, the concentration of Na^+ is unchanged.

0.100 M $Na_2SO_4 \times$ (2 mol Na^+ / 1 mol Na_2SO_4) = 0.200 M Na^+

NO_3^-: 4.53×10^{-3} mol $Pb(NO_3)_2 \times 2$ mol $NO_3^- \,/\, 1$ mol $Pb(NO_3)_2$

$$= 9.06 \times 10^{-3} \text{ mol } NO_3^-$$

9.06×10^{-3} mol $NO_3^- \,/\, 0.125$ L $= 0.0725\ M\ NO_3^-$

4.111   *Plan.* Cl^- is present in NaCl and $MgCl_2$; using mass %, calculate mass NaCl and $MgCl_2$ in mixture, mol Cl^- in each, then molarity of Cl^- in 0.500 L solution.   *Solve.*

$$7.50 \text{ mixture} \times \frac{0.765 \text{ g NaCl}}{1.00 \text{ g mixture}} \times \frac{1 \text{ mol NaCl}}{58.44 \text{ g NaCl}} \times \frac{1 \text{ mol Cl}^-}{1 \text{ mol NaCl}} = 0.09818 = 0.0982 \text{ mol Cl}^-$$

$$7.50 \text{ mixture} \times \frac{0.065 \text{ g MgCl}_2}{1.00 \text{ g mixture}} \times \frac{1 \text{ mol MgCl}_2}{95.21 \text{ g MgCl}_2} \times \frac{2 \text{ mol Cl}^-}{1 \text{ mol MgCl}} = 0.01024 = 0.010 \text{ mol Cl}^-$$

mol $Cl^- = 0.09818 + 0.01024 = 0.10842 = 0.108$ mol Cl^-; $M = \dfrac{0.10842 \text{ mol Cl}^-}{0.5000 \text{ L}} = 0.217\ M\ Cl^-$

4.112   *Plan.* $M = \dfrac{\text{mol Br}^-}{\text{L seawater}}$; mg $Br^- \rightarrow$ g $Br^- \rightarrow$ mol Br^-;

1 kg seawater $\rightarrow$ g $\xrightarrow{\text{density}}$ mL water $\rightarrow$ L seawater

Solve. 65 mg $Br^- \times \dfrac{1 \text{ g Br}^-}{1000 \text{ mg Br}^-} \times \dfrac{1 \text{ mol Br}^-}{79.90 \text{ g Br}^-} = 8.135 \times 10^{-4} = 8.1 \times 10^{-4}$ mol Br^-

1 kg seawater $\times \dfrac{1000 \text{ g}}{1 \text{ kg}} \times \dfrac{1 \text{ mL water}}{1.025 \text{ g water}} \times \dfrac{1 \text{ L}}{1000 \text{ mL}} = 0.9756$ L

$M\ Br^- = \dfrac{8.135 \times 10^{-4} \text{ mol Br}^-}{0.9756 \text{ L seawater}} = 8.3 \times 10^{-4}\ M\ Br^-$

4.114   (a) AsO_4^{3-}; +5

   (b) Ag_3PO_4 is silver phosphate; Ag_3AsO_4 is silver arsenate

   (c) $0.0250 \text{ L soln} \times \dfrac{0.102 \text{ mol Ag}^+}{1 \text{ L soln}} \times \dfrac{1 \text{ mol Ag}_3AsO_4}{3 \text{ mol Ag}^+} \times \dfrac{1 \text{ mol As}}{1 \text{ mol Ag}_3AsO_4} \times \dfrac{74.92 \text{ g As}}{1 \text{ mol As}}$

$$= 0.06368 = 0.0637 \text{ g As}$$

   mass percent $= \dfrac{0.06368 \text{ g As}}{1.22 \text{ g sample}} \times 100 = 5.22\% \text{ As}$

4.116   (a)   *Analyze.* Given: 55 gal of 50 ppb AsO_4^{3-}. Find. g Na_3AsO_4.

   Plan. Use the definition of ppb to calculate g AsO_4^{3-} in 55 gal of water. Then change AsO_4^{3-} to g Na_3SO_4 using molar masses. Assume the density of H_2O is 1.00 g/mL.

   Solve. 1 billion $= 1 \times 10^9$; 1 ppb $= \dfrac{1 \text{ g solute}}{1 \times 10^9 \text{ g solution}}$

   $\dfrac{1 \text{ g solute}}{1 \times 10^9 \text{ g solution}} \times \dfrac{1 \text{ g solution}}{1 \text{ mL solution}} \times \dfrac{1 \times 10^3 \text{ mL}}{1 \text{ L solution}} = \dfrac{\text{g AsO}_4^{3-}}{1 \times 10^6 \text{ L H}_2O}$

50 ppb AsO_4^{3-} = 50 g AsO_4^{3-} / 1×10^6 L H_2O

$$\frac{50 \text{ g } AsO_4^{3-}}{1 \times 10^6 \text{ L } H_2O} \times \frac{3.7854 \text{ L}}{1 \text{ gal}} \times 55 \text{ gal} \times \frac{207.89 \text{ g } Na_3AsO_4}{138.92 \text{ g } AsO_4^{3-}} = 0.015578$$

$$= 0.016 \text{ g } Na_3AsO_4$$

(b)   Parts per billion, ppb, is a mass ratio. If 90% of the arsenic, As, in the form of arsenate, AsO_4^{3-}, is retained in the bucket, then 90% of the mass of As (or AsO_4^{3-}) is in the bucket and 10% of the As (or AsO_4^{3-}) passes thru. Assuming 1×10^6 L of drinking water, 10% of the 500 g AsO_4^{3-} in the water is 50 g AsO_4^{3-}. Thus, only one pass through the bucket is required to reduce the AsO_4^{3-} concentration in drinking water from 500 ppb to 50 ppb, the acceptable standard.

4.117   (a)   mol HCl initial – mol NH_3 from air = mol HCl remaining

$$= \text{mol NaOH required for titration}$$

mol NaOH = 0.0588 M × 0.0131 L = 7.703×10^{-4} = 7.70×10^{-4} mol NaOH

$$= 7.70 \times 10^{-4} \text{ mol HCl remain}$$

mol HCl initial – mol HCl remaining = mol NH_3 from air

(0.0105 M HCl × 0.100 L) – 7.703×10^{-4} mol HCl = mol NH_3

10.5×10^{-4} mol HCl – 7.703×10^{-4} mol HCl = 2.80×10^{-4} = 2.8×10^{-4} mol NH_3

$$2.8 \times 10^{-4} \text{ mol } NH_3 \times \frac{17.03 \text{ g } NH_3}{1 \text{ mol } NH_3} = 4.77 \times 10^{-3} = 4.8 \times 10^{-3} \text{ g } NH_3$$

(b)   ppm is defined as molecules of NH_3/1×10^6 molecules in air.

Calculate molecules NH_3 from mol NH_3.

$$2.80 \times 10^{-4} \text{ mol } NH_3 \times \frac{0.022 \times 20^{23} \text{ molecules}}{1 \text{ mol}} = 1.686 \times 10^{20}$$

$$= 1.7 \times 10^{20} NH_3 \text{ molecules}$$

Calculate total volume of air processed, then g air using density, then molecules air using molar mass.

$$\frac{10.0 \text{ L}}{1 \text{ min}} \times 10.0 \text{ min} \times \frac{1.20 \text{ g air}}{1 \text{ L air}} \times \frac{1 \text{ mol air}}{29.0 \text{ g air}} \times \frac{6.022 \times 10^{23} \text{ molecules}}{1 \text{ mol}}$$

$$= 2.492 \times 10^{24} = 2.5 \times 10^{24} \text{ air molecules}$$

$$\text{ppm } NH_3 = \frac{1.686 \times 10^{20} NH_3 \text{ molecules}}{2.492 \times 10^{24} \text{ air molecules}} \times 1 \times 10^6 = 68 \text{ ppm } NH_3$$

(c)   68 ppm > 50 ppm. The manufacturer is **not** in compliance.

5 Thermochemistry

Visualizing Concepts

5.2   (a)   The internal energy, E, of the products is greater than that of the reactants, so the diagram represents an increase in the internal energy of the system.

         (b) ΔE for this process is positive, +.

         (c)   If no work is associated with the process, it is endothermic.

5.3   (a)   For an endothermic process, the sign of q is positive; the system gains heat. This is true only for system (iii).

         (b)   In order for ΔE to be less than 0, there is a net transfer of heat or work from the system to the surroundings. The magnitude of the quantity leaving the system is greater than the magnitude of the quantity entering the system. In system (i), the magnitude of the heat leaving the system is less than the magnitude of the work done on the system. In system (iii), the magnitude of the work done by the system is less than the magnitude of the heat entering the system. None of the systems has $\Delta E < 0$.

         (c)   In order for ΔE to be greater than 0, there is a net transfer of work or heat to the system from the surroundings. In system (i), the magnitude of the work done on the system is greater than the magnitude of the heat leaving the system. In system (ii), work is done on the system with no change in heat. In system (iii) the magnitude of the heat gained by the system is greater than the magnitude of the work done on the surroundings. $\Delta E > 0$ for all three systems.

5.5   (a) $w = -P\Delta V$. Since ΔV for the process is (−), the sign of w is (+).

         (b) $\Delta E = q + w$. At constant pressure, $\Delta H = q$. If the reaction is endothermic, the signs of ΔH and q are (+). From (a), the sign of w is (+), so the sign of ΔE is (+). The internal energy of the system increases during the change. (This situation is described by the diagram (ii) in Exercise 5.3.)

5.7   Cooling 1 kg H_2O releases more heat than cooling 1 kg Al . Equal masses of H_2O and Al are cooled by the same number of °C. Under these circumstances, the amount of heat released on cooling is proportional to the specific heat capacities of the substances. Specific heat capacity is expressed as J/g-K or energy per mass and temperature change. Since mass and temperature change are equal for the two substances, the one with the larger specific heat capacity loses the most heat. From Table 5.2, the specific heat capacity of $H_2O(l)$ is 4.18 J/g-K and of Al(s) is 0.90 J/g-K, so 1 kg H_2O releases more heat than 1 kg Al.

5.8　(a)　$N_2(g) + O_2(g) \rightarrow 2NO(g)$. Since $\Delta V = 0$, $w = 0$.

　　(b)　$\Delta H = 90.37$ kJ for production of 1 mol of NO(g). The definition of a formation reaction is one where elements combine to form one mole of a single product. The enthalpy change for such a reaction is the enthalpy of formation.

5.10　No. The standard enthalpy of formation of a compound is the change in enthalpy for the reaction that forms one mole of the compound from elements in their standard states. The equation

$$C(graphite) + 4H(g) + O(g) \rightarrow CH_3OH(l)$$

shows elemental carbon (graphite) in its standard state, but elemental hydrogen and oxygen exist as diatomic gases $H_2(g)$ and $O_2(g)$ at standard conditions. ΔH for this equation does not equal ΔH_f° for $CH_3OH(l)$.

The Nature of Energy

5.12　(a)　The kinetic energy of the ball **decreases** as it moves higher. As the ball moves higher and opposes gravity, kinetic energy is changed into potential energy.

　　(b)　The potential energy of the ball **increases** as it moves higher.

　　(c)　The heavier ball would go **half as high** as the tennis ball. At the apex of the trajectory, all initial kinetic energy has been changed into potential energy. The magnitude of the change in potential energy is $m\,g\,\Delta h$, which is equal to the energy initially imparted to the ball. If the same amount of energy is imparted to a ball with twice the mass, m doubles so Δh is half as large.

5.14　(a)　*Plan.* Convert lb → kg, mi/hr → m/s.

　　　Solve. $850\,\text{lb} \times \dfrac{1\,\text{kg}}{2.205\,\text{lb}} = 385.49 = 385\,\text{kg}$

　　　$\dfrac{66\,\text{mi}}{1\,\text{hr}} \times \dfrac{1.6093\,\text{km}}{1\,\text{mi}} \times \dfrac{1000\,\text{m}}{1\,\text{km}} \times \dfrac{1\,\text{hr}}{60\,\text{min}} \times \dfrac{1\,\text{min}}{60\,\text{sec}} = 29.504 = 30\,\text{m/s}$

　　　$E_k = 1/2\,mv^2 = 1/2 \times 385.49\,\text{kg} \times (29.504)^2\,\text{m}^2/\text{s}^2 = 1.7 \times 10^5\,\text{J}$

　　(b)　E_k is proportional to v^2, so if speed decreases by a factor of 2, kinetic energy decreases by a factor of 4.

　　(c)　Brakes stop a moving vehicle, so the kinetic energy of the motorcycle is primarily transferred to friction between brakes and wheels, and somewhat to deformation of the tire and friction between the tire and road.

5.16　(a)　*Analyze.* Given: 1 kwh; 1 watt = 1 J/s; 1 watt $\bullet$ s = 1 J.

　　　　　Find: conversion factor for joules and kwh.

　　　Plan. kwh → wh → ws → J

　　　Solve. $1\,\text{kwh} \times \dfrac{1000\,\text{w}}{1\,\text{kw}} \times \dfrac{60\,\text{min}}{\text{h}} \times \dfrac{60\,\text{s}}{\text{min}} \times \dfrac{1\,\text{J}}{1\,\text{w-s}} = 3.6 \times 10^6\,\text{J}$

　　　$1\,\text{kwh} = 3.6 \times 10^6\,\text{J}$

(b)   *Analyze.* Given: 100 watt bulb. Find: heat in kcal radiated by bulb or person in 24 hr.

Plan. 1 watt = 1 J/s; 1 kcal = 4.184×10^3 J; watt $\rightarrow$ J/s $\rightarrow$ J $\rightarrow$ kcal.   *Solve.*

$$100 \text{ watt} = \frac{100 \text{ J}}{1 \text{ s}} \times \frac{60 \text{ sec}}{\min} \times \frac{60 \min}{\text{hr}} \times 24 \text{ hr} \times \frac{1 \text{ kcal}}{4.184 \times 10^3 \text{ J}} = 2065 = 2.1 \times 10^3 \text{ kcal}$$

24 hr has 2 sig figs, but 100 watt is ambiguous. The answer to 1 sig fig would be

2×10^3 kcal.

Check. $(1 \times 10^2 \times 6 \times 10^1 \times 6/10^3) \approx 6^3 \times 10 \approx 2000$ kcal

5.18   (a)   The system is *open*, because it exchanges both mass and energy with the surroundings. Mass exchange occurs when solution flows into and out of the apparatus. The apparatus is not insulated, so energy exchange also occurs. Closed systems exchange energy but not mass, while isolated systems exchange neither.

         (b)   If the system is defined as shown, it can be closed by blocking the flow in and out, but leaving the flask full of solution.

5.20   (a)   Heat is the energy transferred from a hotter object to a colder object.

         (b)   Heat is transferred from one object (system) to another until the two objects (systems) are at the same temperature.

5.22   (a)   Electrostatic attraction; no work is done because the particles are held stationary.

         (b)   Magnetic attraction; work is done because the nail is moved a distance.

The First Law of Thermodynamics

5.24   (a) $\Delta E = q + w$

         (b)   The quantities q and w are negative when the system loses heat to the surroundings (it cools), or does work on the surroundings.

5.26   In each case, evaluate q and w in the expression $\Delta E = q + w$. For an exothermic process, q is negative; for an endothermic process, q is positive.

         (a)   q is positive and w is negative. $\Delta E = 850$ J – 382 J = 468 J. The process is endothermic.

         (b)   q is negative and w is essentially zero. $\Delta E = -3140$ J. The process is exothermic.

         (c)   q is negative and w is zero. $\Delta E = -6.47$ kJ. The process is exothermic.

5.28 $E_{el} = \dfrac{\kappa Q_1 Q_2}{r^2}$ For two oppositely charged particles, the sign of E_{el} is negative; the closer the particles, the greater the magnitude of E_{el}.

         (a)   The potential energy becomes less negative as the particles are separated (r increases).

(b) ΔE for the process is positive; the internal energy of the system increases as the oppositely charged particles are separated.

(c)   Work is done on the system to separate the particles so w is positive. We have no direct knowledge of the change in q, except that it cannot be large and negative, because overall $\Delta E = q + w$ is positive.

5.30   (a)   Independent. Potential energy is a state function.

(b)   Dependent. Some of the energy released could be employed in performing work, as is done in the body when sugar is metabolized; heat is not a state function.

(c)   Dependent. The work accomplished depends on whether the gasoline is used in an engine, burned in an open flame, or in some other manner. Work is not a state function.

Enthalpy

5.32   (a)   When a process occurs under constant external pressure, the enthalpy change (ΔH) equals the amount of heat transferred. $\Delta H = q_p$.

(b) $\Delta H = q_p$. If the system absorbs heat, q and ΔH are positive and the enthalpy of the system increases.

5.34   At constant volume ($\Delta V = 0$), $\Delta E = q_v$. According to the definition of enthalpy, $H = E + PV$, so $\Delta H = \Delta E + \Delta(PV)$. For an ideal gas at constant temperature and volume, $\Delta PV = V\Delta P = RT\Delta n$. For this reaction, there are 2 mol of gaseous product and 3 mol of gaseous reactants, so $\Delta n = -1$. Thus $V\Delta P$ or $\Delta(PV)$ is negative. Since $\Delta H = \Delta E + \Delta(PV)$, the negative $\Delta(PV)$ term means that ΔH will be smaller or more negative than ΔE.

5.36   The gas is the system. If 378 J of heat is added, $q = +378$ J. Work done by the system decreases the overall energy of the system, so $w = -56$ J.

5.38   (a) $ZnCO_3(s) \rightarrow ZnO(s) + CO_2(g)$

$\Delta H = 71.5$ kJ

(b)

5.40   *Plan.* Consider the sign of an enthalpy change that would convert one of the substances into the other.   *Solve.*

(a) $CO_2(s) \rightarrow CO_2(g)$. This change is sublimation, which is endothermic, $+\Delta H$. $CO_2(g)$ has the higher enthalpy.

(b) $H_2 \rightarrow 2H$. Breaking the H–H bond requires energy, so the process is endothermic, $+\Delta H$. Two moles of H atoms have higher enthalpy.

(c) $H_2O(g) \rightarrow H_2(g) + 1/2\ O_2(g)$. Decomposing H_2O into its elements requires energy and is endothermic, $+\Delta H$. One mole of $H_2(g)$ and 0.5 mol $O_2(g)$ at 25°C have the higher enthalpy.

(d) $N_2(g)$ at 100° → $N_2(g)$ at 300°. An increase in the temperature of the sample requires that heat is added to the system, $+q$ and $+\Delta H$. $N_2(g)$ at 300° has the higher enthalpy.

5.42   (a)   The reaction is endothermic, so heat is absorbed by the system during the course of reaction.

     (b) $45.0 \text{ g CH}_3\text{OH} \times \dfrac{1 \text{ mol CH}_3\text{OH}}{32.04 \text{ g CH}_3\text{OH}} \times \dfrac{90.7 \text{ kJ}}{1 \text{ mol CH}_3\text{OH}} = 127 \text{ kJ heat transferred (absorbed)}$

     (c) $25.8 \text{ kJ} \times \dfrac{2 \text{ mol H}_2}{90.7 \text{ kJ}} \times \dfrac{2.016 \text{ g H}_2}{1 \text{ mol H}_2} = 1.15 \text{ g H}_2 \text{ produced}$

The sign of ΔH is reversed for the reverse reaction: $\Delta H = -90.7 \text{ kJ}$

     (d) $50.9 \text{ g CO} \times \dfrac{1 \text{ mol CO}}{28.01 \text{ g CO}} \times \dfrac{-90.7 \text{ kJ}}{1 \text{ mol CO}} = -165 \text{ kJ heat transferred (released)}$

5.44   (a) $0.632 \text{ mol O}_2 \times \dfrac{-89.4 \text{ kJ}}{3 \text{ mol O}_2} = -18.83 = -18.8 \text{ kJ}$

     (b) $8.57 \text{ g KCl} \times \dfrac{1 \text{ mol KCl}}{74.55 \text{ g KCl}} \times \dfrac{-89.4 \text{ kJ}}{2 \text{ mol KCl}} = -5.1386 = -5.14 \text{ kJ}$

     (c)   Since the sign of ΔH is reversed for the reverse reaction, it seems reasonable that other characteristics would be reversed, as well. If the forward reaction proceeds spontaneously, the reverse reaction is probably not spontaneous. Also, we know from experience that KCl(s) does not spontaneously react with atmospheric $O_2(g)$, even at elevated temperature.

5.46   (a) $3C_2H_2(g) \rightarrow C_6H_6(l)$ $\Delta H = -630 \text{ kJ}$

     (b) $C_6H_6(l) \rightarrow 3C_2H_2(g)$ $\Delta H = 630 \text{ kJ}$
ΔH for the formation of 3 mol of acetylene is 630 kJ. ΔH for the formation of 1 mol of C_2H_2 is then 630 kJ/3 = 210 kJ.

     (c)   The exothermic reverse reaction is more likely to be thermodynamically favored.

     (d)

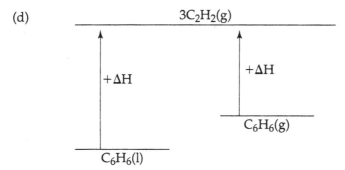

If the reactant is in the higher enthalpy gas phase, the overall ΔH for the reaction has a smaller positive value.

Calorimetry

The specific heat of water to four significant figures, **4.184 J/g·K**, will be used in many of the following exercises; temperature units of K and °C will be used interchangeably.

5.48   *Analyze.* Both objects are heated to 100°C. The two hot objects are placed in the same amount of cold water at the same temperature. Object A raises the water temperature more than object B. *Plan.* Apply the definition of heat capacity to heating the water and heating the objects to determine which object has the greater heat capacity.   *Solve.*

(a)   Both beakers of water contain the same mass of water, so they both have the same heat capacity. Object A raises the temperature of its water more than object B, so more heat was transferred from object A than from object B. Since both objects were heated to the same temperature initially, object A must have absorbed more heat to reach the 100° temperature. The greater the heat capacity of an object, the greater the heat required to produce a given rise in temperature. Thus, object A has the greater heat capacity.

(b)   Since no information about the masses of the objects is given, we cannot compare or determine the specific heats of the objects.

5.50   (a)   In Table 5.2, Hg(l) has the smallest specific heat, so it will require the smallest amount of energy to heat 50.0 g of the substance 10 k.

(b) $50.0 \, g \, Hg(l) \times 10 \, K \times \dfrac{0.14 \, J}{g \cdot K} = 70 \, J$

5.52 $62.0 \, g \text{ ethylene glycol} \times \dfrac{2.42 \, J}{g \cdot K} \times (40.5°C - 13.1°C) = 4.11 \times 10^3 \, J$

5.54   (a)   Following the logic in Solution 5.53, the dissolving process is endothermic, ΔH is positive. The total mass of the solution is (60.0 g H_2O + 3.88 g NH_4NO_3) = 63.88 = 63.9 g. The temperature change of the solution is 23.0 – 18.4 = 4.6°C. The heat lost by the surroundings is

$63.88 \, g \text{ solution} \times \dfrac{4.184 \, J}{1 g \cdot °C} \times 4.6°C \times \dfrac{1 \, kJ}{1000 \, J} = 1.229 = 1.2 \, kJ$

Thus, 1.2 kJ is absorbed when 3.88 g NH_4NO_3(s) dissolves.

$\dfrac{+1.229 \, kJ}{3.88 \, NH_4NO_3} \times \dfrac{80.05 \, g \, NH_4NO_3}{1 \, mol \, NH_4NO_3} = +25.36 = +25 \, kJ/mol \, NH_4NO_3$

(b)   This process is endothermic, because the temperature of the surroundings decreases, indicating that heat is absorbed by the system.

5.56   (a) $C_6H_5OH(s) + 7O_2(g) \rightarrow 6CO_2(g) + 3H_2O(l)$

(b) $q_{bomb} = -q_{rxn}; \Delta T = 26.37°C - 21.36°C = 5.01°C$

$q_{bomb} = \dfrac{11.66 \, kJ}{1°C} \times 5.01°C = 58.417 = 58.4 \, kJ$

At constant volume, $q_v = \Delta E$. ΔE and ΔH are very similar.

$$\Delta H_{rxn} \approx \Delta E_{rxn} = q_{rxn} = -q_{bomb} = \frac{-58.417\,kJ}{1.800\,g\,C_6H_5OH} = -32.454 = -32.5\,kJ/g\,C_6H_5OH$$

$$\Delta H_{rxn} = \frac{-32.454\,kJ}{1\,g\,C_6H_5OH} \times \frac{94.11\,g\,C_6H_5OH}{1\,mol\,C_6H_5OH} = \frac{-3.054 \times 10^3\,kJ}{mol\,C_6H_5OH}$$

$$= -3.05 \times 10^3\,kJ/mol\,C_6H_5OH$$

5.58   (a) $C = 1.640\,g\,C_6H_5COOH \times \dfrac{26.38\,kJ}{1\,g\,C_6H_5COOH} \times \dfrac{1}{4.95°C} = 8.740 = 8.74\,kJ/°C$

    (b) $\dfrac{8.740\,kJ}{°C} \times 4.68°C \times \dfrac{1}{1.320\,g\,sample} = 30.99 = 31.0\,kJ/g\,sample$

    (c)   If water is lost from the calorimeter, there is less water to heat, so the same amount of heat (kJ) from a reaction would cause a larger increase in the calorimeter temperature. The calorimeter constant, kJ/°C, would decrease, because °C is in the denominator of the expression.

Hess's Law

5.60   (a)   *Analyze/Plan.* Arrange the reactions so that in the overall sum, B, appears in both reactants and products and can be canceled. This is a general technique for using Hess's Law.   *Solve.*

$$
\begin{array}{ll}
A \rightarrow B & \Delta H = +30\,kJ \\
\underline{B \rightarrow C} & \underline{\Delta H = +60\,kJ} \\
A \rightarrow C & \Delta H = +90\,kJ
\end{array}
$$

    (b)

    Check. The process of A forming C can be described as A forming B and B forming C.

5.62

$$
\begin{array}{ll}
3H_2(g) + 3/2\,O_2(g) \rightarrow 3H_2O(g) & \Delta H = 3/2(-483.6\,kJ) \\
\underline{O_3(g) \rightarrow 3/2\,O_2(g)} & \underline{\Delta H = 1/2(-284.6\,kJ)} \\
3H_2(g) + O_3(g) \rightarrow 3H_2O(g) & \Delta H = -867.7\,kJ
\end{array}
$$

5.64

$$
\begin{array}{ll}
N_2O(g) \rightarrow N_2(g) + 1/2\,O_2(g) & \Delta H = 1/2\,(-163.2\,kJ) \\
NO_2(g) \rightarrow NO(g) + 1/2\,O_2(g) & \Delta H = 1/2(113.1\,kJ) \\
\underline{N_2(g) + O_2(g) \rightarrow 2NO(g)} & \underline{\Delta H = 180.7\,kJ} \\
N_2O(g) + NO_2(g) \rightarrow 3NO(g) & \Delta H = 155.7\,kJ
\end{array}
$$

Enthalpies of Formation

5.66 (a) Tables of ΔH_f° are useful because, according to Hess's law, the standard enthalpy of any reaction can be calculated from the standard enthalpies of formation for the reactants and products.

$$\Delta H_{rxn}^{\circ} = \Sigma \Delta H_f^{\circ} \text{ (products)} - \Sigma \Delta H_f^{\circ} \text{ (reactants)}$$

(b) The standard enthalpy of formation for any element in its standard state is zero. Elements in their standard states are the reference point for the enthalpy of formation scale.

(c) $6C(s) + 6H_2(g) + 3O_2(g) \rightarrow C_6H_{12}O_6(s)$

5.68 (a) $1/2\,H_2(g) + 1/2\,Br_2(l) \rightarrow HBr(g)$ $\qquad$ $\Delta H_f^{\circ} = -36.23 \text{ kJ}$

(b) $Ag(s) + 1/2\,N_2(g) + 3/2\,O_2(g) \rightarrow AgNO_3(s)$ $\qquad$ $\Delta H_f^{\circ} = -124.4 \text{ kJ}$

(c) $2Fe(s) + 3/2\,O_2(g) \rightarrow Fe_2O_3(s)$ $\qquad$ $\Delta H_f^{\circ} = -822.16 \text{ kJ}$

(d) $2C(s) + 2H_2(g) + O_2(g) \rightarrow CH_3COOH(l)$ $\qquad$ $\Delta H_f^{\circ} = -487.0 \text{ kJ}$

5.70 Use heats of formation to calculate ΔH° for the combustion of butane.

$C_4H_{10}(l) + 13/2\,O_2(g) \rightarrow 4CO_2(g) + 5H_2O(l)$

$\Delta H_{rxn}^{\circ} = 4\Delta H_f^{\circ}\,CO_2(g) + 5\Delta H_f^{\circ}\,H_2O(l) - \Delta H_f^{\circ}\,C_4H_{10}(l) - 13/2\,\Delta H_f^{\circ}\,O_2(g)$

$\Delta H_{rxn}^{\circ} = 4(-393.5 \text{ kJ}) + 5(-285.83 \text{ kJ}) - (-147.6 \text{ kJ}) - 13/2(0) = -2855.6 = -2856 \text{ kJ/mol } C_4H_{10}$

$5.00\text{ g } C_4H_{10} \times \dfrac{1\text{ mol } C_4H_{10}}{58.123\text{ g } C_4H_{10}} \times \dfrac{-2855.6 \text{ kJ}}{1\text{ mol } C_4H_{10}} = -246 \text{ kJ}$

5.72 (a) $\Delta H_{rxn}^{\circ} = 2\Delta H_f^{\circ}\,Br_2(l) + 2\Delta H_f^{\circ}\,H_2O(l) - 4\Delta H_f^{\circ}\,HBr(g) - \Delta H_f^{\circ}\,O_2(g)$

$\qquad = 2(0) + 2(-285.83 \text{ kJ}) - 4(-36.23 \text{ kJ}) - 0 = -426.74 \text{ kJ}$

(b) $\Delta H_{rxn}^{\circ} = \Delta H_f^{\circ}\,Na_2SO_4(s) + \Delta H_f^{\circ}\,H_2O(g) - 2\Delta H_f^{\circ}\,NaOH(s) - \Delta H_f^{\circ}\,SO_3(g)$

$\qquad = -1387.1 \text{ kJ} + (-241.82 \text{ kJ}) - 2(-425.6 \text{ kJ}) - (-395.2 \text{ kJ}) = -382.5 \text{ kJ}$

(c) $\Delta H_{rxn}^{\circ} = \Delta H_f^{\circ}\,CCl_4(l) + 4\Delta H_f^{\circ}\,HCl(g) - \Delta H_f^{\circ}\,CH_4(g) - 4\Delta H_f^{\circ}\,Cl_2(g)$

$\qquad = -139.3 \text{ kJ} + 4(-92.30 \text{ kJ}) - (-74.8 \text{ kJ}) - 4(0) = -433.7 \text{ kJ}$

(d) $\Delta H_{rxn}^{\circ} = 2\Delta H_f^{\circ}\,FeCl_3(s) + 3\Delta H_f^{\circ}\,H_2O(g) - \Delta H_f^{\circ}\,Fe_2O_3(g) - 6\Delta H_f^{\circ}\,HCl(g)$

$\qquad = 2(-400 \text{ kJ}) + 3(-241.82 \text{ kJ}) - (-822.16 \text{ kJ}) - 6(-92.30 \text{ kJ}) = -149.5 = -150 \text{ kJ}$

5.74 $\Delta H_{rxn}^{\circ} = \Delta H_f^{\circ}\,Ca(OH)_2(s) + \Delta H_f^{\circ}\,C_2H_2(g) - 2\Delta H_f^{\circ}\,H_2O(l) - \Delta H_f^{\circ}\,CaC_2(s)$

$-127.2 \text{ kJ} = -986.2 \text{ kJ} + 226.7 \text{ kJ} - 2(-285.83 \text{ kJ}) - \Delta H_f^{\circ}\,CaC_2(s)$

ΔH_f° for $CaC_2(s) = -60.6 \text{ kJ}$

5.76 (a) $10C(s) + 4H_2(g) \rightarrow C_{10}H_8(s)$ $\qquad$ formation

$\qquad C_{10}H_8(s) + 12O_2(g) \rightarrow 10CO_2(g) + 4H_2O(l)$ $\qquad$ combustion, $\Delta H^{\circ} = -5154 \text{ kJ}$

(b) $\Delta H_{rxn}^{\circ} = 10\Delta H_f^{\circ}\,CO_2(g) + 4\Delta H_f^{\circ}\,H_2O(l) - \Delta H_f^{\circ}\,C_{10}H_8(s) - 12\Delta H_f^{\circ}\,O_2(g)$

$$-5154 = 10(-393.5 \text{ kJ}) + 4(-285.83 \text{ kJ}) - \Delta H_f^\circ \, C_{10}H_8(s) - 12(0)$$

$$\Delta H_f^\circ \, C_{10}H_8(s) = 10(-393.5 \text{ kJ}) + 4(-285.83 \text{ kJ}) + 5154 \text{ kJ} = 76 \text{ kJ}$$

Check. The result has 0 decimal places because the heat of combustion has 0 decimal places.

5.78   (a) $CH_3OH(l) + 3/2 \, O_2(g) \rightarrow CO_2(g) + 2H_2O(g)$

      (b) $\Delta H_{rxn}^\circ = \Delta H_f^\circ \, CO_2(g) + 2\Delta H_f^\circ \, H_2O(g) - \Delta H_f^\circ \, CH_3OH(l) - 3/2 \, \Delta H_f^\circ \, O_2(g)$

$$= -393.5 \text{ kJ} + 2(-241.82 \text{ kJ}) - (-238.6 \text{ kJ}) - 3/2(0) = -638.54 = -638.5 \text{ kJ}$$

      (c) $\dfrac{-638.54 \text{ kJ}}{\text{mol CH}_3\text{OH}} \times \dfrac{1 \text{ mol CH}_3\text{OH}}{32.04 \text{ g}} \times \dfrac{0.791 \text{ g}}{\text{mL}} \times \dfrac{1000 \text{ mL}}{\text{L}}$

$$= 1.58 \times 10^4 \text{ kJ/L produced}$$

      (d) $\dfrac{1 \text{ mol CO}_2}{-638.54 \text{ kJ}} \times \dfrac{44.0095 \text{ g CO}_2}{\text{mol}} = 0.06892 \text{ g CO}_2/\text{kJ emitted}$

Foods and Fuels

5.80   (a)   Fats are appropriate for fuel storage because they are insoluble in water (and body fluids) and have a high fuel value.

      (b)   For convenience, assume 100 g of chips.

$$12 \text{ g protein} \times \frac{17 \text{ kJ}}{1 \text{ g protein}} \times \frac{1 \text{ Cal}}{4.184 \text{ kJ}} = 48.76 = 49 \text{ Cal}$$

$$14 \text{ g fat} \times \frac{38 \text{ kJ}}{1 \text{ g fat}} \times \frac{1 \text{ Cal}}{4.184 \text{ kJ}} = 127.15 = 130 \text{ Cal}$$

$$74 \text{ g carbohydrates} \times \frac{17 \text{ kJ}}{1 \text{ g carbohydrates}} \times \frac{1 \text{ Cal}}{4.184 \text{ kJ}} = 300.67 = 301 \text{ Cal}$$

$$\text{total Cal} = (48.76 + 127.15 + 300.67) = 476.58 = 480 \text{ Cal}$$

$$\% \text{ Cal from fat} = \frac{127.15 \text{ Cal fat}}{476.58 \text{ total Cal}} \times 100 = 26.68 = 27\%$$

(Since the conversion from kJ to Cal was common to all three components, we would have determined the same percentage by using kJ.)

      (c) $25 \text{ g fat} \times \dfrac{38 \text{ kJ}}{\text{g fat}} = x \text{ g protein} \times \dfrac{17 \text{ kJ}}{\text{g protein}}; \quad x = 56 \text{ g protein}$

5.82   Calculate the fuel value in a pound of M&M® candies.

$$96 \text{ fat} \times \frac{38 \text{ kJ}}{1 \text{ g fat}} = 3648 \text{ kJ} = 3.6 \times 10^3 \text{ kJ}$$

$$320 \text{ g carbohydrate} \times \frac{17 \text{ kJ}}{1 \text{ g carbohydrate}} = 5440 \text{ kJ} = 5.4 \times 10^3 \text{ kJ}$$

$$21 \text{ g protein} \times \frac{17 \text{ kJ}}{1 \text{ g protein}} = 357 \text{ kJ} = 3.6 \times 10^2 \text{ kJ}$$

$$\text{total fuel value} = 3648 \text{ kJ} + 5440 \text{ kJ} + 357 \text{ kJ} = 9445 \text{ kJ} = 9.4 \times 10^3 \text{ kJ/lb}$$

$$\frac{9445 \text{ kJ}}{\text{lb}} \times \frac{1 \text{ lb}}{453.6 \text{ g}} \times \frac{42 \text{ g}}{\text{serving}} = 874.5 \text{ kJ} = 8.7 \times 10^2 \text{ kJ/serving}$$

$$\frac{874.5 \text{ kJ}}{\text{serving}} \times \frac{1 \text{ kcal}}{4.184 \text{ kJ}} \times \frac{1 \text{ Cal}}{1 \text{ kcal}} = 209.0 \text{ Cal} = 2.1 \times 10^2 \text{ Cal/serving}$$

Check. 210 Cal is the approximate food value of a candy bar, so the result is reasonable.

5.84 $$177 \text{ mL} \times \frac{1.0 \text{ g wine}}{1 \text{ mL}} \times \frac{0.106 \text{ g ethanol}}{1 \text{ g wine}} \times \frac{1 \text{ mol ethanol}}{46.1 \text{ g ethanol}} \times \frac{1367 \text{ kJ}}{1 \text{ mol ethanol}} \times \frac{1 \text{ Cal}}{4.184 \text{ kJ}}$$

$$= 133 = 1.3 \times 10^2 \text{ Cal}$$

Check. A "typical" 6 oz. glass of wine has 150–250 Cal, so this is a reasonable result.

Note that alcohol is responsible for most of the food value of wine.

5.86 $$\Delta H^{\circ}_{rxn} = \Delta H^{\circ}_f \text{ CO}_2(g) + 2\Delta H^{\circ}_f \text{ H}_2\text{O}(g) - \Delta H^{\circ}_f \text{ CH}_4(g) - 2\Delta H^{\circ}_f \text{ O}_2(g)$$

$$= -393.5 \text{ kJ} + 2(-241.82 \text{ kJ}) - (-74.8 \text{ kJ}) - 2(0) \text{ kJ} = -802.3 \text{ kJ}$$

$$\Delta H^{\circ}_{rxn} = \Delta H^{\circ}_f \text{ CF}_4(g) + 4\Delta H^{\circ}_f \text{ HF}(g) - \Delta H^{\circ}_f \text{ CH}_4(g) - 4\Delta H^{\circ}_f \text{ F}_2(g)$$

$$= -679.9 \text{ kJ} + 4(-268.61 \text{ kJ}) - (-74.8 \text{ kJ}) - 4(0) \text{ kJ} = -1679.5 \text{ kJ}$$

The second reaction is twice as exothermic as the first. The "fuel values" of hydrocarbons in a fluorine atmosphere are approximately twice those in an oxygen atmosphere. Note that the difference in ΔH° values for the two reactions is in the ΔH°_f for the products, since the ΔH°_f for the reactants is identical.

Additional Exercises

5.88   (a) $E_p = mgh = 52.0 \text{ kg} \times 9.81 \text{ m/s}^2 \times 10.8 \text{ m} = 5509.3 \text{ J} = 5.51 \text{ kJ}$

(b) $E_k = 1/2 \, mv^2$; $v = (2E_k/m)^{1/2} = \left(\dfrac{2 \times 5509 \text{ kg} \cdot \text{m}^2/\text{s}^2}{52.0 \text{ kg}}\right)^{1/2} = 14.6 \text{ m/s}$

(c)   Yes, the diver does work on entering (pushing back) the water in the pool.

5.89   In the process described, one mole of solid CO_2 is converted to one mole of gaseous CO_2. The volume of the gas is much greater than the volume of the solid. Thus the system (that is, the mole of CO_2) must work against atmospheric pressure when it expands. To accomplish this work while maintaining a constant temperature requires the absorption of additional heat beyond that required to increase the internal energy of the CO_2. The remaining energy is turned into work.

5.91   Freezing is an exothermic process (the opposite of melting, which is clearly endothermic). When the system, the soft drink, freezes, it releases energy to the surroundings, the can. Some of this energy does the work of splitting the can.

5.92   (a) $q = 0$, $w > 0$ (work done to system), $\Delta E > 0$

(b)   Since the system (the gas) is losing heat, the sign of q is negative. The changes in state described in cases (a) and (b) are identical and ΔE is the same in both cases. The distribution of energy transferred as either work or heat is different in the two scenarios. In case (b), more work is required to compress the gas because some heat is lost to the surroundings. (The moral of this story is that the more energy lost by the system as heat, the greater the work on the system required to accomplish the desired change.)

5.94   If a function sometimes depends on path, then it is simply not a state function. Enthalpy is a state function, so ΔH for the two pathways leading to the same change of state pictured in Figure 5.9 must be the same. However, q is not the same for the both. Our conclusion must be that $\Delta H \neq q$ for these pathways. The condition for $\Delta H = q_p$ (other than constant pressure) is that the only possible work on or by the system is pressure-volume work. Clearly, the work being done in this scenario is not pressure-volume work, so $\Delta H \neq q$, even though the two changes occur at constant pressure.

5.95   *Plan.* Change mass of the iceberg to moles, then apply the enthalpy change for melting. 1 million = 1×10^6. *Solve.*

$$1.25 \times 10^6 \text{ metric tons ice} \times \frac{1000 \text{ kg}}{\text{metric ton}} \times \frac{1000 \text{ g}}{\text{kg}} \times \frac{1 \text{ mol H}_2\text{O}}{18.02 \text{ g H}_2\text{O}} \times \frac{6.01 \text{ kJ}}{\text{mol}} = 4.17 \times 10^{11} \text{ kJ}$$

5.97   Find the heat capacity of 1.7×10^3 gal H_2O.

$$C_{H_2O} = 1.7 \times 10^3 \text{ gal H}_2\text{O} \times \frac{4 \text{ qt}}{1 \text{ gal}} \times \frac{1 \text{ L}}{1.057 \text{ qt}} \times \frac{1 \times 10^3 \text{ cm}^3}{1 \text{ L}} \times \frac{1 \text{ g}}{1 \text{ cm}^3} \times \frac{4.184 \text{ J}}{1 \text{ g} \cdot {}^\circ\text{C}}$$

$$= 2.692 \times 10^7 \text{ J}/{}^\circ\text{C} = 2.7 \times 10^4 \text{ kJ}/{}^\circ\text{C}; \text{ then,}$$

$$\frac{2.692 \times 10^7 \text{ J}}{1 {}^\circ\text{C}} \times \frac{1 \text{ g} \cdot {}^\circ\text{C}}{0.85 \text{ J}} \times \frac{1 \text{ kg}}{1 \times 10^3 \text{ g}} \times \frac{1 \text{ brick}}{1.8 \text{ kg}} = 1.8 \times 10^4 \text{ or 18,000 bricks}$$

Check. $(1.7 \times {\sim}16 \times 10^6)/({\sim}1.6 \times 10^3) \approx 17 \times 10^3$ bricks; the units are correct.

5.98   (a) $q_{Cu} = \dfrac{0.385 \text{ J}}{\text{g} \cdot \text{K}} \times 121.0 \text{ g Cu} \times (30.1^\circ\text{C} - 100.4^\circ\text{C}) = -3274.9 = -3.27 \times 10^3 \text{ J}$

The negative sign indicates the 3.27×10^3 J are lost by the Cu block.

(b) $q_{H_2O} = \dfrac{4.184 \text{ J}}{\text{g} \cdot \text{K}} \times 150.0 \text{ g H}_2\text{O} \times (30.1^\circ\text{C} - 25.1^\circ\text{C}) = 3138 = 3.1 \times 10^3 \text{ J}$

The positive sign indicates that 3.14×10^3 J are gained by the H_2O.

(c)   The difference in the heat lost by the Cu and the heat gained by the water is 3.275×10^3 J – 3.138×10^3 J = 0.137×10^3 J = 1×10^2 J. The temperature change of the calorimeter is 5.0°C. The heat capacity of the calorimeter in J/K is

$$0.137 \times 10^3 \text{ J} \times \frac{1}{5.0^\circ\text{C}} = 27.4 = 3 \times 10 \text{ J/K}.$$

Since q_{H_2O} is known to one decimal place, the difference has one decimal place and the result has 1 sig fig.

If the rounded results from (a) and (b) are used,

$$C_{\text{calorimeter}} = \frac{0.2 \times 10^3 \text{ J}}{5.0 {}^\circ\text{C}} = 4 \times 10 \text{ J/K}.$$

(d) $q_{H_2O} = 3.275 \times 10^3 \, J = \dfrac{4.184 \, J}{g \cdot K} \times 150.0 \, g \times (\Delta T)$

$\Delta T = 5.22°C$; $T_f = 25.1°C + 5.22°C = 30.3°C$

5.99   (a)   From the mass of benzoic acid that produces a certain temperature change, we can calculate the heat capacity of the calorimeter.

$\dfrac{0.235 \, g \text{ benzoic acid}}{1.642°C \text{ change observed}} \times \dfrac{26.38 \, kJ}{1 \, g \text{ benzoic acid}} = 3.7755 = 3.78 \, kJ/°C$

Now we can use this experimentally determined heat capacity with the data for caffeine.

$\dfrac{1.525°C \text{ rise}}{0.265 \, g \text{ caffeine}} \times \dfrac{3.7755 \, kJ}{1°C} \times \dfrac{194.2 \, g \text{ caffeine}}{1 \, mol \text{ caffeine}} = 4.22 \times 10^3 \, kJ/mol \text{ caffeine}$

(b)   The overall uncertainty is approximately equal to the sum of the uncertainties due to each effect. The uncertainty in the mass measurement is 0.001/0.235 or 0.001/0.265, about 1 part in 235 or 1 part in 265. The uncertainty in the temperature measurements is 0.002/1.642 or 0.002/1.525, about 1 part in 820 or 1 part in 760. Thus the uncertainty in heat of combustion from each measurement is

$\dfrac{4220}{235} = 18 \, kJ$; $\dfrac{4220}{265} = 16 \, kJ$; $\dfrac{4220}{820} = 5 \, kJ$; $\dfrac{4220}{760} = 6 \, kJ$

The sum of these uncertainties is 45 kJ. In fact, the overall uncertainty is less than this because independent errors in measurement do tend to partially cancel.

5.101   (a) $Mg(s) + 2H_2O(l) \rightarrow Mg(OH)_2(s) + H_2(g)$

$\Delta H^°_{rxn} = \Delta H^°_f \, Mg(OH)_2(s) + \Delta H^°_f \, H_2(g) - 2\Delta H^°_f \, H_2O(l) - \Delta H^°_f \, Mg(s)$

$= -924.7 \, kJ + 0 - 2(-285.83 \, kJ) - 0 = -353.04 = -353.0 \, kJ$

(b)   Use the specific heat of water, 4.184 J/g-°C, to calculate the energy required to heat the water. Use the density of water at 25°C to calculate the mass of H_2O to be heated. (The change in density of H_2O going from 15°C to 85°C does not substantially affect the strategy of the exercise.) Then use the 'heat stoichiometry' in (a) to calculate mass of Mg(s) needed.

$25 \, mL \times \dfrac{0.997 \, g \, H_2O}{mL} \times \dfrac{4.184 \, J}{g \cdot °C} \times 70°C \times \dfrac{1 \, kJ}{1000 \, J} = 7.300 \, kJ = 7.3 \, kJ \text{ required}$

$7.300 \, kJ \times \dfrac{1 \, mol \, Mg}{353.04 \, kJ} \times \dfrac{24.305 \, g \, Mg}{1 \, mol \, Mg} = 0.5026 \, g = 0.50 \, g \, Mg \text{ needed}$

5.102   (a)   For comparison, balance the equations so that 1 mole of CH_4 is burned in each.

$CH_4(g) + O_2(g) \rightarrow C(s) + 2H_2O(l)$ $\Delta H° = -496.9 \, kJ$

$CH_4(g) + 3/2 \, O_2(g) \rightarrow CO(g) + 2H_2O(l)$ $\Delta H° = -607.4 \, kJ$

$CH_4(g) + 2O_2(g) \rightarrow CO_2(g) + 2H_2O(l)$ $\Delta H° = -890.4 \, kJ$

(b) $\Delta H_{rxn}^{\circ} = \Delta H_f^{\circ} \, C(s) + 2\Delta H_f^{\circ} \, H_2O(l) - \Delta H_f^{\circ} \, CH_4(g) - \Delta H_f^{\circ} \, O_2(g)$

$= 0 + 2(-285.83 \text{ kJ}) - (-74.8) - 0 = -496.9 \text{ kJ}$

$\Delta H_{rxn}^{\circ} = \Delta H_f^{\circ} \, CO(g) + 2\Delta H_f^{\circ} \, H_2O(l) - \Delta H_f^{\circ} \, CH_4(g) - 3/2 \, \Delta H_f^{\circ} \, O_2(g)$

$= (-110.5 \text{ kJ}) + 2(-285.83 \text{ kJ}) - (-74.8 \text{ kJ}) - 3/2(0) = -607.4 \text{ kJ}$

$\Delta H_{rxn}^{\circ} = \Delta H_f^{\circ} \, CO_2(g) + 2\Delta H_f^{\circ} \, H_2O(l) - \Delta H_f^{\circ} \, CH_4(g) - 2\Delta H_f^{\circ} \, O_2(g)$

$= -393.5 \text{ kJ} + 2(-285.83 \text{ kJ}) - (-74.8 \text{ kJ}) - 2(0) = -890.4 \text{ kJ}$

(c)   Assuming that $O_2(g)$ is present in excess, the reaction that produces $CO_2(g)$ represents the most negative ΔH per mole of CH_4 burned. More of the potential energy of the reactants is released as heat during the reaction to give products of lower potential energy. The reaction that produces $CO_2(g)$ is the most "down-hill" in enthalpy.

5.104   For nitroethane:

$$\frac{1368 \text{ kJ}}{1 \text{ mol } C_2H_5NO_2} \times \frac{1 \text{ mol } C_2H_5NO_2}{75.072 \text{ g } C_2H_5NO_2} \times \frac{1.052 \text{ g } C_2H_5NO_2}{1 \text{ cm}^3} = 19.17 \text{ kJ/cm}^3$$

For ethanol:

$$\frac{1367 \text{ kJ}}{1 \text{ mol } C_2H_5OH} \times \frac{1 \text{ mol } C_2H_5OH}{46.069 \text{ g } C_2H_5OH} \times \frac{0.789 \text{ g } C_2H_5OH}{1 \text{ cm}^3} = 23.4 \text{ kJ/cm}^3$$

For methylhydrazine:

$$\frac{1305 \text{ kJ}}{1 \text{ mol } CH_6N_2} \times \frac{1 \text{ mol } CH_6N_2}{46.072 \text{ g } CH_6N_2} \times \frac{0.874 \text{ g } CH_6N_2}{1 \text{ cm}^3} = 24.8 \text{ kJ/cm}^3$$

Thus, **methylhydrazine** would provide the most energy per unit volume, with ethanol a close second.

5.106   The reaction for which we want ΔH is:

$4NH_3(l) + 3O_2(g) \rightarrow 2N_2(g) + 6H_2O(g)$

Before we can calculate ΔH for this reaction, we must calculate ΔH_f for $NH_3(l)$.

We know that ΔH_f for $NH_3(g)$ is -46.2 kJ, and that for $NH_3(l) \rightarrow NH_3(g)$, $\Delta H = 23.2$ kJ

Thus, $\Delta H_{vap} = \Delta H_f \, NH_3(g) - \Delta H_f \, NH_3(l)$.

$23.2 \text{ kJ} = -46.2 \text{ kJ} - \Delta H_f \, NH_3(l); \; \Delta H_f \, NH_3(l) = -69.4 \text{ kJ/mol}$

Then for the overall reaction, the enthalpy change is:

$\Delta H_{rxn} = 6\Delta H_f \, H_2O(g) + 2\Delta H_f \, N_2(g) - 4\Delta H_f \, NH_3(l) - 3\Delta H_f \, O_2$

$= 6(-241.82 \text{ kJ}) + 2(0) - 4(-69.4 \text{ kJ}) - 3(0) = -1173.3 \text{ kJ}$

$$\frac{-1173.3 \text{ kJ}}{4 \text{ mol } NH_3} \times \frac{1 \text{ mol } NH_3}{17.0 \text{ g } NH_3} = \frac{0.81 \text{ g } NH_3}{1 \text{ cm}^3} \times \frac{1000 \text{ cm}^3}{1 \text{ L}} = \frac{1.4 \times 10^4 \text{ kJ}}{\text{L } NH_3}$$

(This result has two significant figures because the density is expressed to two figures.)

$$2CH_3OH(l) + 3O_2(g) \rightarrow 2CO_2(g) + 4H_2O(g)$$

$$\Delta H = 2(-393.5 \text{ kJ}) + 4(-241.82 \text{ kJ}) - 2(-239 \text{ kJ}) - 3(0) = -1276 \text{ kJ}$$

$$\frac{-1276 \text{ kJ}}{2 \text{ mol } CH_3OH} \times \frac{1 \text{ mol } CH_3OH}{32.04 \text{ g } CH_3OH} \times \frac{0.792 \text{ g } CH_3OH}{1 \text{ cm}^3} \times \frac{1000 \text{ cm}^3}{1 \text{ L}} = \frac{1.58 \times 10^4 \text{ kJ}}{\text{L } CH_3OH}$$

In terms of heat obtained per unit volume of fuel, methanol is a slightly better fuel than liquid ammonia.

5.107   **1,3-butadiene,** C_4H_6, MM = 54.092 g/mol

     (a) $C_4H_6(g) + 11/2\, O_2(g) \rightarrow 4CO_2(g) + 3H_2O(l)$

            $\Delta H^{\circ}_{rxn} = 4\Delta H^{\circ}_f\, CO_2(g) + 3\Delta H^{\circ}_f\, H_2O(l) - \Delta H^{\circ}_f\, C_4H_6(g) - 11/2\,\Delta H^{\circ}_f\, O_2(g)$

                   $= 4(-393.5 \text{ kJ}) + 3(-285.83 \text{ kJ}) - 111.9 \text{ kJ} + 11/2\,(0) = -2543.4 \text{ kJ/mol } C_4H_6$

     (b) $\dfrac{-2543.4 \text{ kJ}}{1 \text{ mol } C_4H_6} \times \dfrac{1 \text{ mol } C_4H_6}{54.092 \text{ g}} = 47.020 \rightarrow 47 \text{ kJ/g}$

     (c) $\% \text{ H} = \dfrac{6(1.008)}{54.092} \times 100 = 11.18\% \text{ H}$

1-butene, C_4H_8, MM = 56.108 g/mol

     (a) $C_4H_8(g) + 6O_2(g) \rightarrow 4CO_2(g) + 4H_2O(l)$

            $\Delta H^{\circ}_{rxn} = 4\Delta H^{\circ}_f\, CO_2(g) + 4\Delta H^{\circ}_f\, H_2O(l) - \Delta H^{\circ}_f\, C_4H_8(g) - 6\Delta H^{\circ}_f\, O_2(g)$

                 $= 4(-393.5 \text{ kJ}) + 4(-285.83 \text{ kJ}) - 1.2 \text{ kJ} - 6(0) = -2718.5 \text{ kJ/mol } C_4H_8$

     (b) $\dfrac{-2718.5 \text{ kJ}}{1 \text{ mol } C_4H_8} \times \dfrac{1 \text{ mol } C_4H_8}{56.108 \text{ g } C_4H_8} = 48.451 \rightarrow 48 \text{ kJ/g}$

     (c) $\% \text{ H} = \dfrac{8(1.008)}{56.108} \times 100 = 14.37\% \text{ H}$

n-butane, $C_4H_{10}(g)$, MM = 58.124 g/mol

     (a) $C_4H_{10}(g) + 13/2\, O_2(g) \rightarrow 4CO_2(g) + 5H_2O(l)$

            $\Delta H^{\circ}_{rxn} = 4\Delta H^{\circ}_f\, CO_2(g) + 5\Delta H^{\circ}_f\, H_2O(l) - \Delta H^{\circ}_f\, C_4H_{10}(g) - 13/2\,\Delta H^{\circ}_f\, O_2(g)$

                 $= 4(-393.5 \text{ kJ}) + 5(-285.83 \text{ kJ}) - (-124.7 \text{ kJ}) - 13/2(0)$

                 $= -2878.5 \text{ kJ/mol } C_4H_{10}$

     (b) $\dfrac{-2878.5 \text{ kJ}}{1 \text{ mol } C_4H_{10}} \times \dfrac{1 \text{ mol } C_4H_{10}}{58.124 \text{ g } C_4H_{10}} = 49.523 \rightarrow 50 \text{ kJ/g}$

     (c) $\% \text{ H} = \dfrac{10(1.008)}{58.124} \times 100 = 17.34\% \text{ H}$

     (d)   It is certainly true that as the mass % H increases, the fuel value (kJ/g) of the hydrocarbon increases, given the same number of C atoms. A graph of the data in parts (b) and (c) (see below) suggests that mass % H and fuel value are directly proportional when the number of C atoms is constant.

5.108   (a) $C_6H_{12}O_6(s) + 6O_2(g) \rightarrow 6CO_2(g) + 6H_2O(l)$

$\Delta H^\circ_{rxn} = 6\Delta H^\circ_f\ CO_2(g) + 6\Delta H^\circ_f\ H_2O(l) - \Delta H^\circ_f\ C_6H_{12}O_6(s) - 6\Delta H^\circ_f\ O_2(g)$

$= 6(-393.5\ kJ) + 6(-285.83\ kJ) - (-1273\ kJ) - 6(0)$

$= -2803\ kJ/mol\ C_6H_{12}O_6$

$C_{12}H_{22}O_{11}(s) + 12O_2(g) \rightarrow 12CO_2(g) + 11H_2O(l)$

$\Delta H^\circ_{rxn} = 12\Delta H^\circ_f\ CO_2(g) + 11\Delta H^\circ_f\ H_2O(l) - \Delta H^\circ_f\ C_{12}H_{22}O_{11}(s) - 12\Delta H^\circ_f\ O_2(g)$

$= 12(-393.5\ kJ) + 11(-285.83\ kJ) - (-2221\ kJ) - 12(0)$

$= -5645\ kJ/mol\ C_{12}H_{22}O_{11}$

(b) $\dfrac{-2803\ kJ}{1\ mol\ C_6H_{12}O_6} \times \dfrac{1\ mol\ C_6H_{12}O_6}{180.2\ g\ C_6H_{12}O_6} = -\dfrac{15.55\ kJ}{1\ g\ C_6H_{12}O_6} \rightarrow 16\ kJ/g\ C_6H_{12}O_6$ (fuel value)

$\dfrac{-5645\ kJ}{1\ mol\ C_{12}H_{22}O_{11}} \times \dfrac{1\ mol\ C_{12}H_{22}O_{11}}{342.3\ g\ C_{12}H_{22}O_{11}} = -\dfrac{16.49\ kJ}{1\ g\ C_{12}H_{22}O_{11}} \rightarrow 16\ kJ/g\ C_{12}H_{22}O_{11}$

(fuel value)

(c)   The average fuel value of carbohydrates (Section 5.8) is 17 kJ/g. These two carbohydrates have fuel values (16 kJ/g), slightly lower but in line with this average. (More complex carbohydrates supply more energy and raise the average value.)

5.110   *Plan.* Use dimensional analysis to calculate the amount of solar energy supplied per m^2 in 1 hr. Use stoichiometry to calculate the amount of plant energy used to produce sucrose per m^2 in 1 hr. Calculate the ratio of energy for sucrose to total solar energy, per m^2 per hr.

Solve. 1 W = 1 J/s, 1 kW = 1 kJ/s

$\dfrac{1.0\ kW}{m^2} = \dfrac{1.0\ kJ/s}{m^2} = \dfrac{1.0\ kJ}{m^2 \cdot s} \times \dfrac{60\ s}{1\ min} \times \dfrac{60\ min}{1\ hr} = \dfrac{3.6 \times 10^3\ kJ}{m^2 \cdot hr}$

$\dfrac{5645\ kJ}{mol\ sucrose} \times \dfrac{1\ mol\ sucrose}{342.3\ g\ sucrose} \times \dfrac{0.20\ g\ sucrose}{m^2 \cdot hr} = 3.298 = 3.3\ kJ/m^2 \cdot hr$

for sucrose production

$\dfrac{3.298\ kJ\ for\ sucrose}{3.6 \times 10^3\ kJ\ total\ solar} \times 100 = 0.092\%$ sunlight used to produce sucrose

5.111   (a) $6CO_2(g) + 6H_2O(l) \rightarrow C_6H_{12}O_6(s) + 6O_2(g)$, $\Delta H° = 2803$ kJ

This is the reverse of the combustion of glucose (Section 5.8 and Solution 5.108), so $\Delta H° = -(-2803)$ kJ = +2803 kJ.

$$\frac{5.5 \times 10^{16} \text{ g } CO_2}{\text{yr}} \times \frac{1 \text{ mol } CO_2}{44.01 \text{ g } CO_2} \times \frac{2803 \text{ kJ}}{6 \text{ mol } CO_2} = 5.838 \times 10^{17} = 5.8 \times 10^{17} \text{ kJ}$$

(b) $1 \text{ W} = 1 \text{ J/s}; 1 \text{ W-s} = 1 \text{ J}$

$$\frac{5.838 \times 10^{17} \text{ kJ}}{\text{yr}} \times \frac{1000 \text{ J}}{\text{kJ}} \times \frac{1 \text{ yr}}{365 \text{ d}} \times \frac{1 \text{ d}}{24 \text{ hr}} \times \frac{1 \text{ hr}}{60 \text{ min}} \times \frac{1 \text{ min}}{60 \text{ s}} \times \frac{1 \text{ W-s}}{\text{J}}$$

$$\times \frac{1 \text{ MW}}{1 \times 10^6 \text{ W}} = 1.851 \times 10^7 \text{ MW} = 1.9 \times 10^7 \text{ MW}$$

$$1.9 \times 10^7 \text{ MW} \times \frac{1 \text{ plant}}{10^3 \text{ MW}} = 1.9 \times 10^4 = 19,000 \text{ nuclear power plants}$$

Integrative Exercises

5.113   The situation in Figure 4.3 is a more complex version of that pictured in Exercise 5.28 and discussed in Solution 5.28. An ionic solid is an orderly arrangement of closely spaced ions, oppositely charged particles. The potential energy of any pair of these ions is described as $E_{el} = \kappa \, Q_1 Q_2 / r^2$. Separating these oppositely charged particles leads to an increase in the energy of the system, $+ \Delta E$. Work is done to the NaCl by the water molecules. Since both NaCl and water are part of the system, the net amount of work is zero. Since $\Delta E = q + w$, $\Delta E = q$ and both are positive. The dissolving process typically takes place at constant atmospheric pressure, so $\Delta H = q$ and ΔH is also positive.

To verify this conclusion, carry out the dissolution of NaCl in a constant pressure calorimeter. Begin with 1 L of H_2O, record the temperature, add 0.1 mol NaCl, and dissolve completely; record the final temperature. If ΔH is positive, the temperature will decrease.

5.114   (a),(b) $Ag^+(aq) + Li(s) \rightarrow Ag(s) + Li^+(aq)$

$$\Delta H° = \Delta H_f° \, Li^+(aq) - \Delta H_f° \, Ag^+(aq)$$

$$= -278.5 \text{ kJ} - 105.90 \text{ kJ} = -384.4 \text{ kJ}$$

$Fe(s) + 2Na^+(aq) \rightarrow Fe^{2+}(aq) + 2Na(s)$

$$\Delta H° = \Delta H_f° \, Fe^{2+}(aq) - 2\Delta H_f° \, Na^+(aq)$$

$$= -87.86 \text{ kJ} - 2(-240.1 \text{ kJ}) = +392.3 \text{ kJ}$$

$2K(s) + 2H_2O(l) \rightarrow 2KOH(aq) + H_2(g)$

$$\Delta H° = 2\Delta H_f° \, KOH(aq) - 2\Delta H_f° \, H_2O(l)$$

$$= 2(-482.4 \text{ kJ}) - 2(-285.83 \text{ kJ}) = -393.1 \text{ kJ}$$

(c)   Exothermic reactions are more likely to be favored, so we expect the first and third reactions be spontaneous and the second reaction to be nonspontaneous.

(d) In the activity series of metals, Table 4.5, any metal can be oxidized by the cation of a metal below it on the table.

Ag^+ is below Li, so the first reaction will occur.

Na^+ is above Fe, so the second reaction will not occur.

H^+ (formally in H_2O) is below K, so the third reaction will occur.

These predictions agree with those in part (c).

5.116   (a)   mol Cu = $M \times L$ = 1.00 $M \times$ 0.0500 L = 0.0500 mol

g = mol $\times$ / = 0.0500 $\times$ 63.546 = 3.1773 = 3.18 g Cu

(b)   The precipitate is copper(II) hydroxide, $Cu(OH)_2$.

(c) $CuSO_4(aq) + 2KOH(aq) \rightarrow Cu(OH)_2(s) + K_2SO_4(aq)$, complete

$Cu^{2+}(aq) + 2OH^-(aq) \rightarrow Cu(OH)_2(s)$, net ionic

(d)   The temperature of the calorimeter rises, so the reaction is exothermic and the sign of q is negative.

$$q = -6.2°C \times 100\,g \times \frac{4.184\,J}{1\,g\text{-}°C} = -2.6 \times 10^3\,J = -2.6\,kJ$$

The reaction as carried out involves only 0.050 mol of $CuSO_4$ and the stoichiometrically equivalent amount of KOH. On a molar basis,

$$\Delta H = \frac{-2.6\,kJ}{0.050\,mol} = -52\,kJ \text{ for the reaction as written in part (c)}$$

5.118   (a) $21.83\,g\,CO_2 \times \dfrac{1\,mol\,CO_2}{44.01\,g\,CO_2} \times \dfrac{1\,mol\,C}{1\,mol\,CO_2} \times \dfrac{12.01\,g\,C}{1\,mol\,C} = 5.9572 = 5.957\,g\,C$

$4.47\,g\,H_2O \times \dfrac{1\,mol\,H_2O}{18.02\,g\,H_2O} \times \dfrac{2\,mol\,H}{1\,mol\,H_2O} \times \dfrac{1.008\,g\,H}{mol\,H} = 0.5001 = 0.500\,g\,H$

The sample mass is (5.9572 + 0.5001) = 6.457 g

(b) $5.957\,g\,C \times \dfrac{1\,mol\,C}{12.01\,g\,C} = 0.4960\,mol\,C;\ \ 0.4960/0.496 = 1$

$0.500\,g\,H \times \dfrac{1\,mol\,H}{1.008\,g\,H} = 0.496\,mol\,H;\ \ 0.496/0.496 = 1$

The empirical formula of the hydrocarbon is CH.

(c)   Calculate $\Delta H°$ for 6.457 g of the sample.

$6.457\,g\,sample + O_2(g) \rightarrow 21.83\,g\,CO_2(g) + 4.47\,g\,H_2O(g),\ \Delta H°_{comb} = -311\,kJ$

$\Delta H°_{comb} = \Delta H°\,CO_2(g) + \Delta H°\,H_2O(g) - \Delta H°\,sample - \Delta H°\,O_2(g)$

$\Delta H°\,sample = \Delta H°\,CO_2(g) + \Delta H°\,H_2O(g) - \Delta H°_{comb}$

$\Delta H°\,CO_2(g) = 21.83\,g\,CO_2 \times \dfrac{1\,mol\,CO_2}{44.01\,g\,CO_2} \times \dfrac{-393.5\,kJ}{mol\,CO_2} = -195.185 = -195.2\,kJ$

$\Delta H°\,H_2O(g) = 4.47\,g\,H_2O \times \dfrac{1\,mol\,H_2O}{18.02\,g\,H_2O} \times \dfrac{-241.82\,kJ}{mol\,H_2O} = 59.985 = -60.0\,kJ$

ΔH° sample = -195.185 kJ $- 59.985$ kJ $- (-311\,\text{kJ}) = 55.83 = 56$ kJ

$$\frac{55.83\,\text{kJ}}{6.457\,\text{g sample}} \times \frac{13.02\,\text{g}}{\text{CH unit}} = 112.6 = 1.1 \times 10^2 \text{ kJ/CH unit}$$

(d)   The hydrocarbons in Appendix C with empirical formula CH are C_2H_2 and C_6H_6.

substance	ΔH_f°/mol	ΔH_f°/CH unit
$C_2H_2(g)$	226.7 kJ	113.4 kJ
$C_6H_6(g)$	82.9 kJ	13.8 kJ
$C_6H_6(l)$	49.0 kJ	8.17 kJ
sample		1.1×10^2 kJ

The calculated value of ΔH_f°/ CH unit for the sample is a good match with acetylene, $C_2H_2(g)$.

5.119   (a) $CH_4(g) \rightarrow C(g) + 4H(g)$ (i) reaction given

$CH_4(g) \rightarrow C(s) + 2H_2(g)$ (ii) reverse of formation

The differences are: the state of C in the products; the chemical form, atoms, or diatomic molecules, of H in the products.

(b)   i. $\Delta H^{\circ} = \Delta H_f^{\circ}\, C(g) + 4\Delta H_f^{\circ}\, H(g) - \Delta H_f^{\circ}\, CH_4(g)$

= 718.4 kJ + 4(217.94) kJ $-$ ($-$74.8) kJ = 1665.0 kJ

ii. $\Delta H^{\circ} = \Delta H_f^{\circ}\, CH_4 = -(-74.8)\,\text{kJ} = 74.8\,\text{kJ}$

The rather large difference in ΔH° values is due to the enthalpy difference between isolated gaseous C atoms and the orderly, bonded array of C atoms in graphite, C(s), as well as the enthalpy difference between isolated H atoms and H_2 molecules. In other words, it is due to the difference in the enthalpy stored in chemical bonds in C(s) and $H_2(g)$ versus the corresponding isolated atoms.

(c) $CH_4(g) + 4F_2(g) \rightarrow CF_4(g) + 4HF(g)$ $\Delta H^{\circ} = -1679.5$ kJ

The ΔH° value for this reaction was calculated in Solution 5.86.

$$3.45\,\text{g CH}_4 \times \frac{1\,\text{mol CH}_4}{16.04\,\text{g CH}_4} \times 0.21509 = 0.215\,\text{mol CH}_4$$

$$1.22\,\text{g F}_2 \times \frac{1\,\text{mol F}_2}{38.00\,\text{g F}_2} = 0.03211 = 0.0321\,\text{mol F}_2$$

There are fewer mol F_2 than CH_4, but 4 mol F_2 are required for every 1 mol of CH_4 reacted, so clearly F_2 is the limiting reactant.

$$0.03211\,\text{mol F}_2 \times \frac{-1679.5\,\text{kJ}}{4\,\text{mol F}_2} = -13.48 = -13.5\,\text{kJ heat evolved}$$

6 Electronic Structure of Atoms

Visualizing Concepts

6.1 (a) Speed is distance traveled per unit time. Measure the distance between the center point and a second reference point, possibly the edge of the container. Using a stop watch, measure the elapsed time between when a wave forms at the center and when it reaches the second reference point. Find the ratio of distance to time.

(b) Measure the distance between two wave crests (or troughs or any analogous points on two adjacent waves). Better yet, measure the distance between two crests (or analogous points) that are several waves apart and divide by the number of waves that separate them.

(c) Since speed is distance/time, and wavelength is distance, we can calculate frequency by dividing speed by wavelength, $\nu = c/\lambda$.

(d) We can measure frequency of the wave by dropping an object such as a cork in the water and counting the number of times per second it moves through a complete cycle of motion.

6.3 (a) The glowing stove burner is an example of black body radiation, the observational basis for Planck's quantum theory. The wavelengths emitted are related to temperature, with cooler temperatures emitting longer wavelengths and hotter temperatures emitting shorter wavelengths. At the hottest setting, the burner emits orange visible light. At the cooler low setting, the burner emits longer wavelengths out of the visible region, and the burner appears black.

(b) If the burner had a super high setting, the emitted wavelengths would be shorter than those of orange light and the glow color would be more blue. (See Figure 6.4 for color variation with wavelength.)

6.4 (a) Increase. The rainbow has shorter wavelength blue light on the inside and longer wavelength red light on the outside. (See Figure 6.4.)

(b) Decrease. Wavelength and frequency are inversely related. Wavelength increases so frequency decreases going from the inside to the outside of the rainbow.

(c) The light from the hydrogen discharge tube is not a continuous spectrum, so not all visible wavelengths will be in our "hydrogen discharge rainbow." Starting with the shortest wavelengths, it will be violet followed by blue-violet and blue-green on the inside. Then there will be a gap, and finally a red band. (See the H spectrum in Figure 6.13.)

6.6 (a) $\psi^2(x)$ will be positive or zero at all values of x, and have two maxima with larger magnitudes than the maximum in $\psi(x)$.

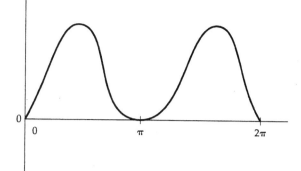

(b) The greatest probability of finding the electron is at the two maxima in $\psi^2(x)$ at $x = \pi/2$ and $3\pi/2$.

(c) There is zero probability of finding the electron at $x = \pi$. This value is called a node.

6.7   (a)   1

       (b)   p (dumbbell shape, node at the nucleus)

       (c)   The lobes in the contour representation would extend farther along the y axis. A larger principle quantum number (4p vs. 3p) implies a greater average distance from the nucleus for electrons occupying the orbital.

The Wave Nature of Light

6.10   (a)   Wavelength (λ) and frequency (ν) are inversely proportional; the proportionality constant is the speed of light (c). $\nu = c/\lambda$.

        (b)   Light in the 210–230 nm range is in the ultraviolet region of the spectrum. These wavelengths are slightly shorter than the 400 nm short-wavelength boundary of the visible region.

6.12   (a)   False. Electromagnetic radiation passes through water. The fact that you can see objects through a glass of water should make this clear.

        (b)   True.

        (c)   False. Infrared light has lower frequencies than visible light.

        (d)   False. A foghorn blast is a form of sound waves, which are not accompanied by oscillating electric and magnetic fields.

6.14   Wavelength of (a) gamma rays < (d) yellow (visible) light < (e) red (visible) light < (b) 93.1 MHz FM (radio) waves < (c) 680 kHz or 0.680 MHz AM (radio) waves

6.16   (a) $\nu = c/\lambda$; $\dfrac{2.998 \times 10^8 \text{ m}}{\text{s}} \times \dfrac{1}{10.0 \text{ Å}} \times \dfrac{1 \text{ Å}}{1 \times 10^{-10} \text{ m}} = 3.00 \times 10^{17} \text{ s}^{-1}$

        (b) $\lambda = c/\nu$; $\dfrac{2.998 \times 10^8 \text{ m}}{\text{s}} \times \dfrac{1 \text{ s}}{7.6 \times 10^{10}} = 3.94 \times 10^{-3} \text{ m}$

        (c)   The 1×10^{-9} m radiation in (a) is X-rays and can be observed by an X-ray detector. Radiation (b) is microwave.

(d) $25.5 \text{ fs} \times \dfrac{1 \times 10^{-15} \text{ s}}{1 \text{ fs}} \times \dfrac{2.998 \times 10^{8} \text{ m}}{\text{s}} = 7.64 \times 10^{-6} \text{ m } (7.64 \text{ μm})$

6.18   According to Figure 6.4, ultraviolet radiation has both higher frequency and shorter wavelength than infrared radiation. Looking forward to section 6.2, the energy of a photon is directly proportional to frequency ($E = h\nu$), so ultraviolet radiation yields more energy from a photovoltaic device.

Quantized Energy and Photons

6.20   Planck's original hypothesis was that energy could only be gained or lost in discreet amounts (quanta) with a certain minimum size. The size of the minimum energy change is related to the frequency of the radiation absorbed or emitted, $\Delta E = h\nu$, and energy changes occur only in multiples of $h\nu$.

Einstein postulated that light itself is quantized, that the minimum energy of a photon (a quantum of light) is directly proportional to its frequency, $E = h\nu$. If a photon that strikes a metal surface has less than the threshold energy, no electron is emitted from the surface. If the photon has energy equal to or greater than the threshold energy, an electron is emitted and any excess energy becomes the kinetic energy of the electron.

6.22   (a) $\nu = c/\lambda = \dfrac{2.998 \times 10^{8} \text{ m}}{\text{s}} \times \dfrac{1}{589 \text{ nm}} \times \dfrac{1 \text{ nm}}{1 \times 10^{-9} \text{ m}} = 5.0900 \times 10^{14}$

$= 5.09 \times 10^{14} \text{ s}^{-1}$

(b) $E = h\nu = 6.626 \times 10^{-34} \text{ J·s} \times 5.0900 \times 10^{14} \text{ s}^{-1} \times \dfrac{6.022 \times 10^{23} \text{ photons}}{\text{mol}}$

$\times 0.1 \text{ mol} = 2.03 \times 10^{4} \text{ J} = 20.3 \text{ kJ}$

(c) $\Delta E = h\nu = \dfrac{hc}{\lambda} = \dfrac{6.626 \times 10^{-34} \text{ J·s} \times 2.998 \times 10^{8} \text{ m/s}}{589 \text{ nm}} \times \dfrac{1 \text{ nm}}{1 \times 10^{-9} \text{ m}}$

$= 3.37 \times 10^{-19} \text{ J}$

(d)   The 589 nm light emission is characteristic of Na^{+}. If the pickle is soaked in a different salt long enough to remove all Na^{+}, the 589 nm light would not be observed. Emission at a different wavelength, characteristic of the new salt, would be observed.

6.24 $E = h\nu$

$AM : 6.626 \times 10^{-34} \text{ J·s} \times \dfrac{1010 \times 10^{3}}{1 \text{ s}} = 6.69 \times 10^{-28} \text{ J}$

$FM : 6.626 \times 10^{-34} \text{ J·s} \times \dfrac{98.3 \times 10^{6}}{1 \text{ s}} = 6.51 \times 10^{-26} \text{ J}$

The FM photon has about 100 times more energy than the AM photon.

6.26 $\dfrac{941 \times 10^{3} \text{ J}}{\text{mol } N_2} \times \dfrac{1 \text{ mol}}{6.022 \times 10^{23} \text{ photons}} = 1.563 \times 10^{-18} = 1.56 \times 10^{-18} \text{ J/photon}$

$$\lambda = hc/E = \frac{6.626 \times 10^{-34} \text{ J} \cdot \text{s}}{1.563 \times 10^{-18} \text{ J}} \times \frac{2.998 \times 10^8 \text{ m}}{1 \text{ s}} = 1.27 \times 10^{-7} \text{ m} = 127 \text{ nm}$$

According to Figure 6.4, this is ultraviolet radiation.

6.28   (a)   The radiation is microwave.

(b) $E_{photon} = hc/\lambda = \dfrac{6.626 \times 10^{-34} \text{ J} \cdot \text{s}}{3.55 \times 10^{-3} \text{ m}} \times \dfrac{2.998 \times 10^8 \text{ m}}{1 \text{ s}} = 5.5957 \times 10^{-23}$

$$= 5.60 \times 10^{-23} \text{ J/photon}$$

$$\frac{5.5957 \times 10^{-23} \text{ J}}{1 \text{ photon}} \times \frac{3.2 \times 10^8 \text{ photons}}{1 \text{ s}} \times \frac{60 \text{ s}}{1 \text{ min}} \times \frac{60 \text{ min}}{1 \text{ hr}} = 6.4463 \times 10^{-11}$$

$$= 6.4 \times 10^{-11} \text{ J/hr}$$

6.30   (a) $v = E/h = \dfrac{4.41 \times 10^{-19} \text{ J}}{6.626 \times 10^{-34} \text{ J} \cdot \text{s}} = 6.6556 \times 10^{14} = 6.66 \times 10^{14} \text{ s}^{-1}$

(b) $\lambda = hc/E = \dfrac{6.626 \times 10^{-34} \text{ J} \cdot \text{s}}{4.41 \times 10^{-19} \text{ J}} \times \dfrac{2.998 \times 10^8 \text{ m}}{\text{s}} = 4.50 \times 10^{-7} \text{ m} = 450 \text{ nm}$

(c) $E_{439} = hc/\lambda = \dfrac{6.626 \times 10^{-34} \text{ J} \cdot \text{s}}{439 \times 10^{-9} \text{ m}} \times \dfrac{2.998 \times 10^8 \text{ m}}{\text{s}} = 4.525 \times 10^{-19} = 4.53 \times 10^{-19} \text{ J}$

$E_K = E_{439} - E_{min} = 4.525 \times 10^{-19} \text{ J} - 4.41 \times 10^{-19} \text{ J} = 0.115 \times 10^{-19} = 1.1 \times 10^{-20} \text{ J}$

(d)   One electron is emitted per photon. Calculate the number of 439 nm photons in 1.00 μJ. The excess energy in each photon will become the kinetic energy of the electron; it cannot be "pooled" to emit additional electrons.

$$1.00 \text{ μJ} \times \frac{1 \times 10^{-6} \text{ J}}{\text{μJ}} \times \frac{1 \text{ photon}}{4.525 \times 10^{-19} \text{ J}} \times \frac{1 \text{ e}^-}{1 \text{ photon}} = 2.21 \times 10^{12} \text{ electrons}$$

Bohr's Model; Matter Waves

6.32   (a)   According to Bohr theory, when hydrogen emits radiant energy, electrons are moving from a higher allowed energy state to a lower one. Since only certain energy states are allowed, only certain energy changes can occur. These allowed energy changes correspond ($\lambda = hc/\Delta E$) to the wavelengths of the lines in the emission spectrum of hydrogen.

(b)   When a hydrogen atom changes from the ground state to an excited state, the single electron moves further away from the nucleus, so the atom "expands".

6.34   (a)   Absorbed.   (b)   Emitted.   (c)   Absorbed.

6.36   (a) $v = E/h = \dfrac{2.044 \times 10^{-18} \text{ J}}{6.626 \times 10^{-34} \text{ J} \cdot \text{s}} = 3.084 \times 10^{15} = 3.08 \times 10^{15} \text{ s}^{-1}$

$\lambda = c/v = \dfrac{2.998 \times 10^8 \text{ m}}{1 \text{ s}} \times \dfrac{1 \text{ s}}{3.084 \times 10^{15}} = 9.72 \times 10^{-8} \text{ m}$

Since the sign of ΔE is negative, radiation is emitted.

(b) $\Delta E = -2.18 \times 10^{-18} \, J (1/4 - 1/25) = -4.578 \times 10^{-19} = -4.58 \times 10^{-19} \, J$

$$\nu = \frac{4.578 \times 10^{-19} \, J}{6.626 \times 10^{-34} \, J\text{-}s} = 6.909 \times 10^{14} = 6.91 \times 10^{14} \, s^{-1}; \lambda = \frac{2.998 \times 10^8 \, m/s}{6.909 \times 10^{14}/s}$$

$\lambda = 4.34 \times 10^{-7} \, m$. Visible radiation is emitted.

(c) $\Delta E = -2.18 \times 10^{-18} \, J (1/36 - 1/9) = 1.817 \times 10^{-19} = 1.82 \times 10^{-19} \, J$

$$\nu = \frac{1.817 \times 10^{-19} \, J}{6.626 \times 10^{-34} \, J\text{-}s} = 2.742 \times 10^{14} = 2.74 \times 10^{14} \, s^{-1}; \lambda = \frac{2.998 \times 10^8 \, m/s}{2.742 \times 10^{14}/s}$$

$\lambda = 1.09 \times 10^{-6} \, m$. Radiation is absorbed.

6.38   (a)   Transitions with $n_f = 1$ have larger ΔE values and shorter wavelengths than those with $n_f = 2$. These transitions will lie in the ultraviolet region.

(b) $n_i = 2, n_f = 1;$ $\lambda = hc/E = \dfrac{6.626 \times 10^{-34} \, J\text{-}s \times 2.998 \times 10^8 \, m/s}{-2.18 \times 10^{-18} \, J \,(1/1 - 1/4)} = 1.21 \times 10^{-7} \, m$

$n_i = 3, n_f = 1;$ $\lambda = hc/E = \dfrac{6.626 \times 10^{-34} \, J\text{-}s \times 2.998 \times 10^8 \, m/s}{-2.18 \times 10^{-18} \, J \,(1/1 - 1/9)} = 1.03 \times 10^{-7} \, m$

$n_i = 4, n_f = 1;$ $\lambda = hc/E = \dfrac{6.626 \times 10^{-34} \, J\text{-}s \times 2.998 \times 10^8 \, m/s}{-2.18 \times 10^{-18} \, J \,(1/1 - 1/16)} = 0.972 \times 10^{-7} \, m$

6.40   (a) $2626 \, nm \times \dfrac{1 \times 10^{-9} \, m}{1 \, nm} = 2.626 \times 10^{-6} \, m$;   this line is in the infrared.

(b)   Absorption lines with $n_i = 1$ are in the ultraviolet and with $n_i = 2$ are in the visible. Thus, $n_i \geq 3$, but we do not know the exact value of n_i. Calculate the longest wavelength with $n_i = 3$ ($n_f = 4$). If this is less than 2626 nm, $n_i > 3$.

$$\lambda = hc/E = \frac{6.626 \times 10^{-34} \, J\text{-}s \times 2.998 \times 10^8 \, m/s}{-2.18 \times 10^{-18} \, J \,(1/16 - 1/9)} = 1.875 \times 10^{-6} \, m$$

This wavelength is shorter than 2.626×10^{-6} m, so $n_i > 3$; try $n_i = 4$ and solve for n_f as in Solution 6.39.

$$n_f = \left(\frac{1}{n_i^2} - \frac{hc}{\lambda(2.18 \times 10^{-18} \, J)} \right)^{-1/2} = \left(1/16 - \frac{6.626 \times 10^{-34} \, J\text{-}s \times 2.998 \times 10^8 \, m/s}{2.626 \times 10^{-6} \, m \times 2.18 \times 10^{-18} \, J} \right)^{-1/2} = 6$$

$n_f = 6, \, n_i = 4$

6.42 $\lambda = h/mv$; change mass to kg and velocity to m/s

mass of muon $= 206.8 \times 9.1094 \times 10^{-28} \, g \times \dfrac{1 \, kg}{1000 \, g} = 1.8838 \times 10^{-28} = 1.88 \times 10^{-28} \, kg$

$$\lambda = \frac{6.626 \times 10^{-34} \, kg\text{-}m^2\text{-}s}{1 \, s^2} \times \frac{1}{1.8838 \times 10^{-28} \, kg} \times \frac{1 \, s}{8.85 \times 10^3 \, m/s} = 3.97 \times 10^{-10} \, m$$

$$= 3.97 \, \text{Å}$$

6.44 $m_e = 9.1094 \times 10^{-31}$ kg (back inside cover of text)

$$\lambda = \frac{6.626 \times 10^{-34} \text{ kg-m}^2\text{-s}}{1 \text{ s}^2} \times \frac{1}{9.1094 \times 10^{-31} \text{ kg}} \times \frac{1 \text{ s}}{9.38 \times 10^6 \text{ m}} = 7.75 \times 10^{-11} \text{ m}$$

$$7.75 \times 10^{-11} \text{ m} \times \frac{1 \text{ Å}}{1 \times 10^{-10} \text{ m}} = 0.775 \text{ Å}$$

Since atomic radii and interatomic distances are on the order of 1–5 Å (Section 2.3), the wavelength of this electron is comparable to the size of atoms.

6.46 $\Delta x \geq = h/4\pi m \Delta v$; use masses in kg, Δv in m/s.

(a) $$\frac{6.626 \times 10^{-34} \text{ J-s}}{4\pi (9.109 \times 10^{-31} \text{ kg})(0.01 \times 10^5 \text{ m/s})} = 6 \times 10^{-8} \text{ m}$$

(b) $$\frac{6.626 \times 10^{-34} \text{ J-s}}{4\pi (1.675 \times 10^{-27} \text{ kg})(0.01 \times 10^5 \text{ m/s})} = 3 \times 10^{-11} \text{ m}$$

(c)   For particles moving with the same uncertainty in velocity, the more massive neutron has a much smaller uncertainty in position than the lighter electron. In our model of the atom, we know where the massive particles in the nucleus are located, but we cannot know the location of the electrons with any certainty, if we know their speed.

Quantum Mechanics and Atomic Orbitals

6.48   (a)   The Bohr model states with 100% certainty that the electron in hydrogen can be found 0.53 Å from the nucleus. The quantum mechanical model, taking the wave nature of the electron and the uncertainty principle into account, is a statistical model that states the probability of finding the electron in certain regions around the nucleus. While 0.53 Å might be the radius with highest probability, that probability would always be less than 100%.

    (b)   The equations of classical physics predict the instantaneous position, direction of motion, and speed of a macroscopic particle; they do not take quantum theory or the wave nature of matter into account. For macroscopic particles, these are not significant, but for microscopic particles like electrons, they are crucial. Schrödinger's equation takes these important theories into account to produce a statistical model of electron location given a specific energy.

    (c)   The square of the wave function has the physical significance of an amplitude, or probability. The quantity ψ^2 at a given point in space is the probability of locating the electron within a small volume element around that point at any given instant. The total probability, that is, the sum of ψ^2 over all the space around the nucleus, must equal 1.

6.50   (a)   For $n = 3$, there are three l values (2, 1, 0) and nine m_l values ($l = 2$; $m_l = -2, -1, 0, 1, 2$; $l = 1$, $m_l = -1, 0, 1$; $l = 0$, $m_l = 0$).

(b)   For $n = 5$, there are five l values (4, 3, 2, 1, 0) and twenty-five m_l values

($l = 4$, $m_l = -4$ to $+4$; $l = 3$, $m_l = -3$ to $+3$; $l = 2$, $m_l = -2$ to $+2$; $l = 1$, $m_l = -1$ to $+1$; $l = 0$, $= 0$).

In general, for each principal quantum number n there are n l-values and n^2 m_l-values. For each shell, there are n kinds of orbitals and n^2 total orbitals.

6.52   (a)   2, 1, 1; 2, 1, 0; 2, 1 –1   (b) 5, 2, 2; 5, 2, 1; 5, 2, 0; 5, 2, –1; 5, 2, –2

6.54

n	l	m_l	orbital
2	1	–1	2p (example)
1	0	0	1s
3	–3	2	not allowed (l < n and + only)
3	2	–2	3d
2	0	–1	not allowed ($m_l = -l$ to $+l$)
0	0	0	not allowed (n ≠ 0)
4	2	1	4d
5	3	0	5f

6.56

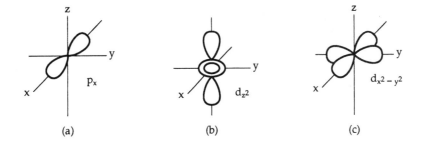

(a)   (b)   (c)

6.58   (a)   In an s orbital, there are ($n - 1$) nodes.

(b)   The $2p_x$ orbital has one node (the yz plane passing through the nucleus of the atom). The 3s orbital has two nodes.

(c)   Probability density, $\psi^2(r)$, is the probability of finding an electron at a single point, r. The radial probability function, P(r), is the probability of finding an electron at any point that is distance r from the nucleus. Figure 6.19 contains plots of P(r) vs. r for 1s, 2s, and 3s orbitals. The most obvious features of these plots are the radii of maximum probability for the three orbitals, and the number and location of nodes for the three orbitals.

By comparing plots for the three orbitals, we see that as n increases, the number of nodes increases and the radius of maximum probability (orbital size) increases.

(d)   2s = 2p < 3s < 4d < 5s. In the hydrogen atom, orbitals with the same n value are degenerate and energy increases with increasing n value.

Many-Electron Atoms and Electron Configurations

6.60   (a)   The electron with the greater average distance from the nucleus feels a smaller attraction for the nucleus and is higher in energy. Thus the 3p is higher in energy than 3s.

       (b)   Because it has a larger n value, a 3s electron has a greater average distance from the chlorine nucleus than a 2p electron. The 3s electron experiences a smaller attraction for the nucleus and requires less energy to remove from the chlorine atom.

6.62   (a)   The Pauli exclusion principle states that no two electrons can have the same four quantum numbers.

       (b)   An alternate statement of the Pauli exclusion principle is that a single orbital can hold a maximum of two electrons. Thus, the Pauli principle limits the maximum number of electrons in a main shell and its subshells, which determines when a new row of the periodic table begins.

6.64   (a)   4   (b)   14   (c)   2   (d)   2

6.66

Element	(a) C	(b) P	(c) Ne
Electron Configuration	$[He]2s^22p^2$	$[Ne]3s^23p^3$	$[He]2s^22p^6$
Core electrons	2	10	2
Valence electrons	4	5	8
Unpaired electrons	2	3	0

[The concept of "valence electrons" for noble gas elements is problematic, since they are mostly unreactive. We could list the core for neon as [Ne], with no valence or unpaired electrons.]

6.68   (a)   Ga: $[Ar]4s^23d^{10}4p^1$, 1 unpaired electron

       (b)   Ca: $[Ar]4s^2$, 0 unpaired electrons

       (c)   V: $[Ar]4s^23d^3$, 3 unpaired electrons

       (d)   I: $[Kr]5s^24d^{10}5p^5$, 1 unpaired electron

       (e)   Y: $[Kr]5s^24d^1$, 1 unpaired electron

       (f)   Pt: $[Xe]6s^14f^{14}5d^9$, 2 unpaired electrons

       (g)   Lu: $[Xe]6s^24f^{14}5d^1$, 1 unpaired electron

6.70   *Plan.* Write the electron configuration of the neutral transition metal atom, then remove electrons, ns first, then $(n-1)d$, to achieve the cationic charge. *Solve.*

       (a)   Zn: $[Ar]4s^23^{10}$; Zn^{2+} (remove two 4s electrons): $[Ar]3d^{10}$

       (b)   Pt: $[Xe]6s^14f^{14}5d^9$; Pt^{2+} (remove 6s and one 5d electron): $[Xe]4f^{14}5d^8$

       (c)   Cr: $[Ar]4s^13d^5$; Cr^{3+} (remove 4s and two 3d electrons): $[Ar]3d^3$

       (d)   Ti: $[Ar]4s^23d^2$; Ti^{4+} (remove all four valence electrons): $[Ar]$

6.72 (a) 7A (halogens)   (b) 4B   (c) 3A (row 4 and below)

   (d) the f-block elements Sm and Pu

6.74 Count the total number of electrons to assign the element.

   (a) N: $[He]2s^2 2p^3$ (b) Se: $[Ar]4s^2 3d^{10}4p^4$ (c) Rh: $[Kr]5s^2 4d^7$

Additional Exercises

6.76 (a) Elements that emit in the visible: Ba (dark blue), Ca (dark blue), K (dark blue), Na (yellow/orange). (The other wavelengths are in the ultraviolet.)

   (b) Au: shortest wavelength, highest energy

     Na: longest wavelength, lowest energy

   (c) $\lambda = c/v = \dfrac{2.998 \times 10^8 \text{ m/s}}{6.59 \times 10^{14}/\text{s}} \times \dfrac{1 \text{ nm}}{1 \times 10^{-9} \text{ m}} = 455 \text{ nm}, \quad \text{Ba}$

6.78 (a) $v = c/\lambda = \dfrac{2.998 \times 10^8 \text{ m/s}}{320 \text{ nm}} \times \dfrac{1 \text{ nm}}{1 \times 10^{-9} \text{ m}} = 9.37 \times 10^{14} \text{ s}^{-1}$

   (b) $E = hc/\lambda = \dfrac{6.626 \times 10^{-34} \text{ J-s} \times 2.998 \times 10^8 \text{ m/s}}{3.20 \times 10^{-7} \text{ m}} \times \dfrac{1 \text{ kJ}}{1000 \text{ J}} \times \dfrac{6.022 \times 10^{23} \text{ photons}}{\text{mole}}$

     $= 374 \text{ kJ/mol}$

   (c) UV-B photons have shorter wavelength and higher energy.

   (d) Yes. The higher energy UV-B photons would be more likely to cause sunburn.

6.79 $E = hc/\lambda \rightarrow$ J/photon; total energy = power × time; photons = total energy / J / photon

   $E = \dfrac{6.626 \times 10^{-34} \text{ J-s} \times 2.998 \times 10^8 \text{ m/s}}{780 \times 10^{-9} \text{ m}} = 2.5468 \times 10^{-19} = 2.55 \times 10^{-19} \text{ J/photon}$

   $0.10 \text{ mW} = \dfrac{0.10 \times 10^{-3} \text{ J}}{1 \text{ s}} \times 69 \text{ min} \times \dfrac{60 \text{ s}}{1 \text{ min}} = 0.4140 = 0.41 \text{ J}$

   $0.4140 \text{ J} \times \dfrac{1 \text{ photon}}{2.5468 \times 10^{-19} \text{ J}} = 1.626 \times 10^{18} = 1.6 \times 10^{18} \text{ photons}$

6.81 $\dfrac{2.6 \times 10^{-12} \text{ C}}{1 \text{ s}} \times \dfrac{1 e^-}{1.602 \times 10^{-19} \text{ C}} \times \dfrac{1 \text{ photon}}{1 e^-} = 1.623 \times 10^7 = 1.6 \times 10^7 \text{ photons/s}$

   $\dfrac{E}{\text{photon}} = hc/\lambda = \dfrac{6.626 \times 10^{-34} \text{ J-s}}{630 \text{ nm}} \times \dfrac{2.998 \times 10^8 \text{ m}}{1 \text{ s}} \times \dfrac{1 \text{ nm}}{1 \times 10^{-9} \text{ m}} \times \dfrac{1.623 \times 10^7 \text{ photon}}{\text{s}}$

     $= 5.1 \times 10^{-12} \text{ J/s}$

6.82 (a) $\dfrac{2.00 \times 10^5 \text{ J}}{\text{mol}} \times \dfrac{1 \text{ mol photons}}{6.022 \times 10^{23} \text{ photons}} = 3.321 \times 10^{-19} = 3.32 \times 10^{-19} \text{ J/photon}$

(b) $\lambda = \dfrac{hc}{E} = \dfrac{6.626 \times 10^{-34} \text{ J-s} \times 2.998 \times 10^8 \text{ m/s}}{3.321 \times 10^{-19} \text{ J}} = 5.98 \times 10^{-7} \text{ m}$

(c) 5.98×10^{-7} m = 598 nm is well within the visible portion of the electromagnetic spectrum and corresponds to yellow or yellow-orange light. Red light, with wavelengths near or greater than 700 nm, does not have sufficient energy to initiate electron transfer and darken the film.

6.83 (a) $\nu = c/\lambda$; $\dfrac{2.998 \times 10^8 \text{ m}}{\text{s}} \times \dfrac{1}{680 \text{ nm}} \times \dfrac{1 \text{ nm}}{1 \times 10^{-9} \text{ m}} = 4.4088 \times 10^{14} = 4.41 \times 10^{14} \text{ s}^{-1}$

(b) Calculate J/photon using $E = hc/\lambda$; change to kJ/mol.

$E_{photon} = \dfrac{6.626 \times 10^{-34} \text{ J-s}}{680 \times 10^{-9} \text{ m}} \times \dfrac{2.998 \times 10^8 \text{ m}}{\text{s}} = 2.9213 \times 10^{-19} = 2.92 \times 10^{-19} \text{ J/photon}$

$\dfrac{2.9213 \times 10^{-19} \text{ J}}{\text{photon}} \times \dfrac{6.022 \times 10^{23} \text{ photons}}{\text{mol}} \times \dfrac{1 \text{ kJ}}{1000 \text{ J}} = 175.92 = 176 \text{ kJ/mol}$

(c) Nothing. The incoming (incident) radiation does not transfer sufficient energy to an electron to overcome the attractive forces holding the electron in the metal.

(d) For frequencies greater than ν_o, any "extra" energy not needed to remove the electron from the metal becomes the kinetic energy of the ejected electron. The kinetic energy of the electron is directly proportional to this extra energy.

(e) Let E_{total} be the total energy of an incident photon, E_{min} be the minimum energy required to eject an electron, and E_k be the "extra" energy that becomes the kinetic energy of the ejected electron.

$E_{total} = E_{min} + E_k$, $E_k = E_{total} - E_{min} = h\nu - h\nu_o$, $E_k = h(\nu - \nu_o)$. The slope of the line is the value of h, Planck's constant.

6.85 (a) Gaseous atoms of various elements in the sun's atmosphere typically have ground state electron configurations. When these atoms are exposed to radiation from the sun, the electrons change from the ground state to one of several allowed excited states. Atoms absorb the wavelengths of light which correspond to these allowed energy changes. All other wavelengths of solar radiation pass through the atmosphere unchanged. Thus, the dark lines are the wavelengths that correspond to allowed energy changes in atoms of the solar atmosphere. The continuous background is all other wavelengths of solar radiation.

(b) The scientist should record the absorption spectrum of pure neon or other elements of interest. The black lines should appear at the same wavelengths regardless of the source of neon.

6.86 (a) He^+ is hydrogen-like because it is a one-electron particle. A He atom has two electrons. The Bohr model is based on the interaction of a single electron with the nucleus, but does not accurately account for additional interactions when two or more electrons are present.

(b)   Divide each energy by the smallest value to find the integer relationship.

H: $2.18 \times 10^{-18}/2.18 \times 10^{-18} = 1$; $Z = 1$

He$^+$: $8.72 \times 10^{-18}/2.18 \times 10^{-18} = 4$; $Z = 2$

Li^{2+}: $1.96 \times 10^{-17}/2.18 \times 10^{-18} = 9$; $Z = 3$

The ground-state energies are in the ratio of 1:4:9, which is also the ratio Z^2, the square of the nuclear charge for each particle.

The ground state energy for hydrogen-like particles is:

$E = R_H Z^2$. (By definition, n = 1 for the ground state of a one-electron particle.)

(c)   Z = 6 for C. $E = -2.18 \times 10^{-18} J (6)^2 = -7.85 \times 10^{-17} J$

6.88   *Plan.* Change keV to J/electron. Calculate v from kinetic energy. $\lambda = h/mv$.   *Solve.*

$$18.6 \text{ keV} \times \frac{1000 \text{ eV}}{\text{keV}} \times \frac{96.485 \text{ kJ}}{1 \text{ eV-mol}} \times \frac{1000 \text{ J}}{1 \text{ kJ}} \times \frac{1 \text{ mol}}{6.022 \times 10^{23} \text{ electrons}}$$

$$= 2.980 \times 10^{-15} = 2.98 \times 10^{-15} \text{ J/electron}$$

$$E_k = mv^2/2; \; v^2 = 2E_k/m; \; v = \sqrt{2E_k/m}$$

$$v = \left(\frac{2 \times 2.980 \times 10^{-15} \text{ kg-m}^2/\text{s}^2}{9.1094 \times 10^{-31} \text{ kg}}\right)^{1/2} = 8.089 \times 10^7 = 8.09 \times 10^7 \text{ m/s}$$

$$\lambda = h/mv = \frac{6.626 \times 10^{-34} \text{ J-s}}{9.1094 \times 10^{-31} \text{ kg} \times 8.089 \times 10^7 \text{ m/s}} \times \frac{1 \text{ kg-m}^2/\text{s}^2}{1 \text{ J}} = 8.99 \times 10^{-12} \text{ m} = 8.99 \text{ pm}$$

6.89   Heisenberg postulated that the dual nature of matter places a limitation on how precisely we can know both the position and momentum of an object. This limitation is significant at the subatomic particle level. The *Star Trek* transporter (presumably) dissembles humans into their protons, neutrons and electrons, moves the particles at high speed (possibly the speed of light) to a new location, and reassembles the particles into the human. Heisenberg's uncertainty principle indicates that if we know the momentum (*mv*) of the moving particles, we can't precisely know their position (*x*). If a few of the subatomic particles don't arrive in exactly the correct location, the human would not be reassembled in their originall form. So, the "Heisenberg compensator" is necessary to make sure that the transported human arrives at the new location intact.

6.91   (a)   Probability density, $[\psi(r)]^2$, is the probability of finding an electron at a single point at distance r from the nucleus. The radial probability function, $4\pi r^2$, is the probability of finding an electron at any point on the sphere defined by radius r. $P(r) = 4\pi r^2 [\psi(r)]^2$ (Figure 6.21).

(b)   The term $4\pi r^2$ explains the differences in plots of the two functions. Plots of the probability density, $[\psi(r)^2]$ for s orbitals shown in Figure 6.23 each have their maximum value at r = 0, with (n – 1) smaller maxima at greater values of r. The plots of radial probability, P(r), for the same s orbitals shown in Figure 6.19 have values of zero at r = 0 and the size of the maxima increases. P(r) is the product of $[\psi(r)]^2$ and $4\pi r^2$. At r = 0, the value of $[\psi(r)]^2$ is finite and large, but the value of

$4\pi r^2$ is zero, so the value of $P(r)$ is zero. As r increases, the values of $[\psi(r)]^2$ vary as shown in Figure 6.23, but the values of $4\pi r^2$ increase continuously, leading to the increasing size of $P(r)$ maxima as r increases.

(c)

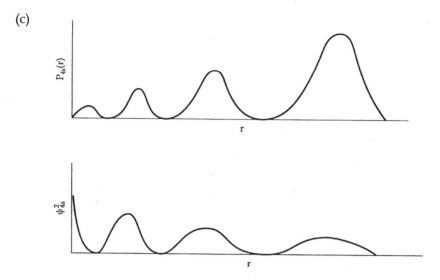

6.92   What the noble gas elements have in common are completed ns and np subshells. Since the Pauli principle limits the number of electrons per orbital to two, this leads to the first three magic numbers, $2(1s^2)$, $10(1s^22s^22p^6)$, and $18(1s^22s^22p^63s^23p^6)$. In the fourth row, $(n-1)$ d orbitals begin to fill as their energy falls below that of the np orbitals. This leads to the next two magic numbers, $36(1s^22s^22p^63s^23p^64s^23d^{10}4p^6)$ and $54(1s^22s^22p^63s^23p^64s^23d^{10}4p^65s^24d^{10}5p^6)$. In the sixth row, the energy of the 4f orbitals falls below that of the $(n-1)$d and np subshells, and it fills. This explains the final magic number, $86(1s^22s^22p^63s^23p^64s^23d^{10}4p^65s^24d^{10}5p^66s^24f^{14}5d^{10}6p^6)$.

6.93   (a)   The p_z orbital has a nodal plane where $z = 0$. This is the xy plane.

        (b)   The d_{xy} orbital has four lobes and two nodal planes, the two planes where $x = 0$ and $y = 0$. These are the yz and xz planes.

        (c)   The $d_{x^2-y^2}$ has four lobes and two nodal planes, the planes where $x^2 - y^2 = 0$. These are the planes that bisect the x and y axes and contain the z axis.

6.94   (a)   In the absence of a magnetic field, electrons with opposite m_s values have the same energy. Because electrons with opposite spins will have oppositely oriented magnetic fields, only the interaction of the magnetic fields of the electrons with an external magnetic field will cause the energies of the electrons to be different and observable.

        (b)   According to Figure 6.28, the particle with its magnetic field parallel to the external field will have the lower energy. The left electron has its magnetic field oriented parallel to the described magnetic orientation, so it will be lower in energy.

        (c)   Microwave photons used to excite unpaired electrons in the ESR experiment have higher energy than radio wave photons used to excite nuclei in NMR.

6.95   (a)   This is the frequency of radiowaves that excite the nuclei from one spin state to the other.

(b) $\Delta E = h\nu = 6.626 \times 10^{-34} \text{ J-s} \times \dfrac{450 \times 10^6}{\text{s}} = 2.98 \times 10^{-25} \text{ J}$

(c)   Since $\Delta E = 0$ in the absence of a magnetic field, it is reasonable to assume that the stronger the external field, the greater ΔE. (In fact, ΔE is directly proportional to field strength). Because ΔE is relatively small [see part (b)], the two spin states are almost equally populated, with a very slight excess in the lower energy state. The stronger the magnetic field, the larger ΔE, the greater number of nuclei in the lower energy spin state. With more nuclei in the lower energy state, more are able to absorb the appropriate radio wave photons and reach the higher energy state. This increases the intensity of the NMR signal, which provides more information and more reliable information than a weak absorption signal.

6.97   (a)   Se: $[Ar]4s^2 3d^{10}4p^4$

(b)   Rh: $[Kr]5s^2 4d^7$

(c)   Si: $[Ne]3s^2 3p^2$

(d)   Hg: $[Xe]6s^2 4f^{14}5d^{10}$

(e)   Hf: $[Xe]6s^2 4f^{14}5d^2$

6.98   The core would be the electron configuration of element 118. If no new subshell begins to fill, the condensed electron configuration of element 126 would be similar to those of elements vertically above it on the periodic chart, Pu and Sm. The condensed configuration would be $[118]8s^2 6f^6$. On the other hand, the 5g subshell could begin to fill after 8s, resulting in the condensed configuration $[118]8s^2 5g^6$. Exceptions are also possible (likely).

Integrative Exercises

6.100 $\Delta H^\circ_{rxn} = \Delta H^\circ_f O_2(g) + \Delta H^\circ_f O(g) - \Delta H^\circ_f O_3(g)$

$\Delta H^\circ_{rxn} = 0 + 247.5 \text{ kJ} - 142.3 \text{ kJ} = +105.2 \text{ kJ}$

$\dfrac{105.2 \text{ kJ}}{\text{mol O}_3} \times \dfrac{1 \text{ mol O}_3}{6.022 \times 10^{23} \text{ molecules}} \times \dfrac{1000 \text{ J}}{1 \text{ kJ}} = \dfrac{1.747 \times 10^{-19} \text{ J}}{\text{O}_3 \text{ molecule}}$

$\Delta E = hc/\lambda; \lambda = \dfrac{hc}{\Delta E} = \dfrac{6.626 \times 10^{-34} \text{ J-s} \times 2.998 \times 10^8 \text{ m/s}}{1.747 \times 10^{-19} \text{ J}} = 1.137 \times 10^{-6} \text{ m}$

Radiation with this wavelength is in the infrared portion of the spectrum. (Clearly, processes other than simple photodissociation cause O_3 to absorb ultraviolet radiation.)

6.101   (a)   The electron configuration of Zr is $[Kr]5s^2 4d^2$ and that of Hf is $[Xe]6s^2 4f^{14}5d^2$. Although Hf has electrons in f orbitals as the rare earth elements do, the 4f subshell in Hf is filled, and the 5d electrons primarily determine the chemical properties of the element. Thus, Hf should be chemically similar to Zr rather than the rare earth elements.

(b) $ZrCl_4(s) + 4Na(l) \rightarrow Zr(s) + 4NaCl(s)$

This is an oxidation-reduction reaction; Na is oxidized and Zr is reduced.

(c) $2ZrO_2(s) + 4Cl_2(g) + 3C(s) \rightarrow 2ZrCl_4\ (s) + CO_2(g) + 2CO(g)$

$$55.4\,g\,ZrO_2 \times \frac{1\,mol\,ZrO_2}{123.2\,g\,ZrO_2} \times \frac{2\,mol\,ZrCl_4}{2\,mol\,ZrO_2} \times \frac{233.0\,g\,ZrCl_4}{1\,mol\,ZrCl_4} = 105\,g\,ZrCl_4$$

(d)   In ionic compounds of the type MCl_4 and MO_2, the metal ions have a 4+ charge, indicating that the neutral atoms have lost four electrons. Zr, $[Kr]5s^2 4d^2$, loses the four electrons beyond its Kr core configuration. Hf, $[Xe]6s^2 4f^{14} 5d^2$, similarly loses its four 6s and 5d electrons, but not electrons from the "complete" 4f subshell.

6.102   (a)   Each oxide ion, O^{2-}, carries a 2- charge. Each metal oxide is a neutral compound, so the metal ion or ions must adopt a total positive charge equal to the total negative charge of the oxide ions in the compound. The table below lists the electron configuration of the neutral metal atom, the positive charge of each metal ion in the oxide, and the corresponding electron configuration of the metal ion.

i.   K: $[Ar]\,4s^1$ 1+ $[Ar]$

ii.   Ca: $[Ar]\,4s^2$ 2+ $[Ar]$

iii.   Sc: $[Ar]\,4s^2 3d^1$ 3+ $[Ar]$

iv.   Ti: $[Ar]\,4s^2 3d^2$ 4+ $[Ar]$

v.   V: $[Ar]\,4s^2 3d^3$ 5+ $[Ar]$

vi.   Cr: $[Ar]\,4s^1 3d^5$ 6+ $[Ar]$

Each metal atom loses all (valence) electrons beyond the Ar core configuration. In K_2O, Sc_2O_3 and V_2O_5, where the metal ions have odd charges, two metal ions are required to produce a neutral oxide.

(b)   i.   potassium oxide

ii.   calcium oxide

iii.   scandium(III) oxide

iv.   titanium (IV) oxide

v.   vanadium (V) oxide

vi.   chromium (VI) oxide

(Roman numerals are required to specify the charges on the transition metal ions, because more than one stable ion may exist.)

(c)   Recall that $\Delta H_f^\circ = 0$ for elements in their standard states. In these reactions, M(s) and $H_2(g)$ are elements in their standard states.

i. $K_2O(s) + H_2(g) \rightarrow 2K(s) + H_2O(g)$

$$\Delta H^\circ = \Delta H_f^\circ\,H_2O(g) + 2\Delta H_f^\circ\,K(s) - \Delta H\,K_2O(s) - \Delta H_f^\circ\,H_2(g)$$

$$\Delta H^\circ = -241.82\,kJ + 2(0) - (-363.2\,kJ) - 0 = 121.4\,kJ$$

ii. $CaO(s) + H_2(g) \rightarrow Ca(s) + H_2O(g)$

$\Delta H° = \Delta H_f° \, H_2O(g) + \Delta H_f° \, Ca(s) - \Delta H_f° \, CaO(s) - \Delta H_f° \, H_2(g)$

$\Delta H° = -241.82 \, kJ + 0 - (-635.1 \, kJ) - 0 = 393.3 \, kJ$

iii. $TiO_2(s) + 2H_2(g) \rightarrow Ti(s) + 2H_2O(g)$

$\Delta H° = 2\Delta H_f° \, H_2O(g) + \Delta H_f° \, Ti(s) - \Delta H_f° \, TiO_2(s) - 2\Delta H_f° \, H_2(g)$

$= 2(-241.82) + 0 - (-938.7) - 2(0) = 455.1 \, kJ$

iv. $V_2O_5(s) + 5H_2(g) \rightarrow 2V(s) + 5H_2O(g)$

$\Delta H° = 5\Delta H_f° \, H_2O(g) + 2\Delta H_f° \, V(s) - \Delta H_f° \, V_2O_5(s) - 5\Delta H_f° \, H_2(g)$

$= 5(-241.82) + 2(0) - (-1550.6) - 5(0) = 341.5 \, kJ$

(d) $\Delta H_f°$ becomes more negative moving from left to right across this row of the periodic chart. Since Sc lies between Ca and Ti, the median of the two $\Delta H_f°$ values is approximately –785 kJ/mol. However, the trend is clearly not linear. Dividing the $\Delta H_f°$ values by the positive charge on the pertinent metal ion produces the values –363, –318, –235, and –310. The value between Ca^{2+} (–318) and Ti^{4+} (–235) is Sc^{3+} (–277). Multiplying (–277) by 3, a value of approximately –830 kJ results. A reasonable range of values for $\Delta H_f°$ of $Sc_2O_3(s)$ is then –785 to –830 kJ/mol.

6.104   (a) ^{238}U: 92 p, 146 n, 92 e; ^{235}U: 92 p, 143 n, 92 e

In keeping with the definition isotopes, only the number of neutrons is different in the two nuclides. Since the two isotopes have the same number of electrons, they will have the same electron configuration.

(b)   U: $[Rn]7s^25f^4$

(c)   From Figure 6.31, the actual electron configuration is $[Rn]7s^25f^36d^1$. The energies of the 6d and 5f orbitals are very close, and electron configurations of many actinides include 6d electrons.

(d) $^{238}_{92}U \rightarrow ^{234}_{90}Th + ^4_2He$ ^{234}Th has 90 p, 144 n, 90 e. ^{238}U has lost 2 p, 2 n, 2 e.

These are organized into ^4_2He shown in the nuclear reaction above.

(e)   From Figure 6.31, the electron configuration of Th is $[Rn]7s^26d^2$. This is not really surprising because there are so many rare earth electron configurations that are exceptions to the expected orbital filling order. However, Th is the only rare earth that has two d valence electrons. Furthermore, the configuration of Th is different than that of Ce, the element above it on the periodic chart, so the electron configuration is at least interesting.

6.105   In each case, radiant energy from the sun in shown as hν.

inert black material   photovoltaic cell

green leaf

Energy from the sun irradiates the earth without external intervention. In the absence of an appropriate receiver, the energy is dissipated as heat, as the diagram for the inert black material shows. When sunlight hits a material that can convert the energy to a useable form, such as a leaf or a photovoltaic cell, the sun becomes a sustainable energy source.

7 Periodic Properties of the Elements

Visualizing Concepts

7.1   (a)   The light bulb itself represents the nucleus of the atom. The brighter the bulb, the more nuclear charge the electron "sees." A frosted glass lampshade between the bulb and our eyes reduces the brightness of the bulb. The shade is analogous to core electrons in the atom shielding outer electrons (our eyes) from the full nuclear charge (the bare light bulb).

       (b)   Increasing the wattage of the light bulb mimics moving right along a row of the periodic table. The brighter bulb inside the same shade is analogous to having more protons in the nucleus while the core electron configuration doesn't change.

       (c)   Moving down a family, both the nuclear charge and the core electron configuration changes. To simulate the addition of core electrons farther from the nucleus, we would add larger frosted glass shades as well as increase the wattage of the bulb to show the increase in Z. The effect of the shade should dominate the increase in wattage, so that the brightness of the light decreases moving down a column.

7.3   (a)   The bonding atomic radius of A, r_A, is $d_1/2$. The distance d_2 is the sum of the bonding atomic radii of A and X, $r_A + r_X$. Since we know that $r_A = d_1/2$, $d_2 = r_X + d_1/2$, $r_X = d_2 - d_1/2$.

       (b)   The length of the X-X bond is $2r_X$.

            $2r_X = 2(d_2 - d_1/2) = 2d_2 - d_1$.

7.4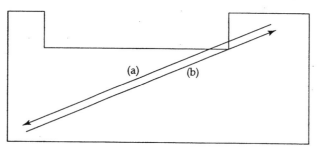

           Lines (a) and (b) coincide, but their directions are opposite. Line (a) goes from upper right to lower left, and line (b) from lower left to upper right.

       (c)   From the diagram, we observe that the trends in bonding atomic radius (size) and ionization energy are opposite each other. As bonding atomic radius increases increases ionization energy decreases, and vice versa.

7.6   (a) $X + 2F_2 \rightarrow XF_4$

        (b)   If X is a nonmetal, XF_4 is a molecular compound. If X is a metal, XF_4 is ionic. For an ionic compound with this formula, X would have a charge of 4+, and a much smaller bonding atomic radius than F^-. X in the diagram has about the same bonding radius as F, so it is likely to be a nonmetal.

Periodic Table; Effective Nuclear Charge

7.8   Assuming *eka-* means one place below or under, *eka-manganese* on Table 7.2 is technetium, Tc.

7.10   (a)   The verification of the existence of many new elements by accurately measuring their atomic weights spurred interest in a classification scheme. Mendeleev (and Meyer) noted that certain chemical and physical properties recur periodically when the elements are arranged by increasing atomic weight. The accurate atomic weights provided a common property on which to base a classification scheme of the elements.

        (b)   Moseley realized that the characteristic X-ray frequencies emitted by each element were related to a unique integer that he assigned to each element. We now know this integer as the atomic number, the number of protons in the nucleus of an atom. In general, atomic weight increases as atomic number increases, but there are a few exceptions. If elements are arranged by increasing atomic number, a few seeming contradictions in the Mendeleev table (the positions of Ar and K or Te and I) are eliminated.

        (c)   The main determining factor of physical and especially chemical properties is electron configuration. For electrically neutral elements, the number of electrons equals the number of protons, which in turn is the atomic number of an element. Atomic weight is related to mass number, protons plus neutrons. The number of neutrons in its nucleus does influence the mass of an atom, but mass is a minor or non-factor in determining properties.

7.12   (a)   Electrostatic attraction for the nucleus lowers the energy of an electron, while electron-electron repulsions increase this energy. The concept of effective nuclear charge allows us to model this increase in the energy of an electron as a smaller net attraction to a nucleus with a smaller positive charge, Z_{eff}.

        (b)   In Be (or any element), the 1s electrons are not shielded by any core electrons, so they experience a much greater Z_{eff} than the 2s electrons.

7.14   Follow the method in the preceding question to calculate Z_{eff} values.
        (a)   Si: $Z = 14$; $[Ne]3s^23p^2$. 10 electrons in the Ne core. $Z_{eff} = 14 - 10 = 4$
               Cl: $Z = 17$; $[Ne]3s^23p^5$. 10 electrons in the Ne core. $Z_{eff} = 17 - 10 = 7$

        (b)   Si: $1s^22s^22p^63s^23p^2$. $S = 0.35(3) + 0.85(8) + 1(2) = 9.85$. $Z_{eff} = 14 - 9.85 = 4.15$
               Cl: $1s^22s^22p^63s^23p^5$. $S = 0.35(6) + 0.85(8) + 1(2) = 10.9$. $Z_{eff} = 17 - 10.9 = 6.10$

(c)   The Slater values of 4.15 (Si) and 6.10 (Cl) are closer to the results of detailed calculations, 4.29 (Si) and 6.12 (Cl).

(d)   The Slater method of approximation more closely approximates the gradual increase in Z_{eff} moving across a row. The 'core 100%-effective' approximation underestimates Z_{eff} for Si but overestimates it for Cl. Slater values are closer to detailed calculations, and a better indication of the change in Z_{eff} moving from Si to Cl.

7.16   Mg < P < K < Ti < Rh. The shielding of electrons in the $n = 3$ shell by 1s, 2s and 2p core electrons in these elements is approximately equal, so the effective nuclear charge increases as Z increases.

Atomic and Ionic Radii

7.18   (a)   Since the quantum mechanical description of the atom does not specify the exact location of electrons, there is no specific distance from the nucleus where the last electron can be found. Rather, the electron density decreases gradually as the distance from the nucleus increases. There is no quantum mechanical "edge" of an atom.

(b)   When nonbonded atoms touch, it is their electron clouds that interact. These interactions are primarily repulsive because of the negative charges of electrons. Thus, the size of the electron clouds determines the nuclear approach distance of nonbonded atoms.

7.20   The distance between Si atoms in solid silicon is two times the bonding atomic radius from Figure 7.7. The Si–Si distance is 2×1.11 Å = 2.22 Å.

7.22   Bi–I = 2.81 Å = $r_{Bi} + r_{I}$. From Figure 7.7, $r_{I} = 1.33$ Å.

r_{Bi} = [Bi–I] – r_{I} = 2.81 Å – 1.33 Å = 1.48 Å.

7.24   (a)   The vertical difference in radius is due to a change in principal quantum number of the outer electrons. The horizontal difference in radius is due to the change in electrostatic attraction between the outer electron and a nucleus with one more or one fewer proton. Adding or subtracting a proton has a much smaller radius effect than moving from one principal quantum level to the next.

(b)   Si < Al < Ge < Ga. This order is predicted by the trends in increasing atomic radius moving to the left in a row and down a column of the periodic chart, assuming that changes moving down a column are larger [see part (a)]. That is, the order above assumes that the change from Si to Ge is larger than the change from Si to Al. This order is confirmed by the values in Figure 7.7.

7.26   (a)   Na < Ca < Ba   (b)   As < Sb < Sn

(c)   Be < Si < Al. This order assumes the increase in radius from the second to the third row is greater than the decrease moving right in the third row. Radii in Figure 7.7 confirm this assumption.

7.28   (a)   As Z stays constant and the number of electrons increases, the electron-electron repulsions increase, the electrons spread apart, and the ions become larger.

$I^- > I > I^+$

(b)   Going down a column, the increasing average distance of the outer electrons from the nucleus causes the size of particles with like charge to increase.

$Ca^{2+} > Mg^{2+} > Be^{2+}$

(c)   Fe: $[Ar]4s^2 3d^6$; Fe^{2+}: $[Ar]3d^6$; Fe^{3+}: $[Ar]3d^5$. The 4s valence electrons in Fe are on average farther from the nucleus than the 3d electrons, so Fe is larger than Fe^{2+}. Since there are five 3d orbitals, in Fe^{2+} at least one orbital must contain a pair of electrons. Removing one electron to form Fe^{3+} significantly reduces repulsion, increasing the nuclear charge experienced by each of the other d electrons and decreasing the size of the ion. $Fe > Fe^{2+} > Fe^{3+}$

7.30   The order of radii is $Ca > Ca^{2+} > Mg^{2+}$, so the largest sphere is Ca, the intermediate one is Ca^{2+}, and the smallest is Mg^{2+}.

7.32   (a) Cl^-: Ar   (b) Sc^{3+}: Ar

(c) Fe^{2+}: $[Ar]3d^6$. Fe^{2+} has 24 electrons. Neutral Cr has 24 electrons, $[Ar]4s^1 3d^5$. Because transition metals fill the s subshell first but also lose s electrons first when they form ions, many transition metal ions do not have isolectronic neutral atoms.

(d) Zn^{2+}: $[Ar]3d^{10}$; no isoelectronic neutral atom [same reason as (c)].

(e) Sn^{4+}: $[Kr]4d^{10}$; no isoelectronic neutral atom [same reason as (c)], but Sn^{4+} is isoelectronic with Cd^{2+}.

7.34   (a) K^+ (larger Z) is smaller.

(b) Cl^- and K^+: $[Ne]3s^2 3p^6$. 10 core electrons
Cl^-, $Z = 17$. $Z_{eff} = 17 - 10 = 7$
K^+, $Z = 19$. $Z_{eff} = 19 - 10 = 9$

(c)   Valence electron, $n = 3$; 7 other $n = 3$ electrons; eight $n = 2$ electrons; two $n = 1$ electrons. $S = 0.35(7) + 0.85(8) + 1(2) = 11.25$
Cl^-: $Z_{eff} = 17 - 11.25 = 5.75$. K^+: $Z_{eff} = 19 - 11.25 = 7.75$

(d)   For isoelectronic ions, the electron configurations and therefore shielding values (S) are the same. Only the nuclear charge changes. So, as nuclear charge (Z) increases, effective nuclear charge (Z_{eff}) increases and ionic radius decreases.

7.36   (a) $Se < Se^{2-} < Te^{2-}$ (b) $Co^{3+} < Fe^{3+} < Fe^{2+}$ (c) $Ti^{4+} < Sc^{3+} < Ca$ (d) $Be^{2+} < Na^+ < Ne$

7.38   Make a table of d(measured), d(ionic radii) and d(covalent radii). Use these values to make comparisons for (b) and (c). The estimated distances are just the sum of the various radii from Figure 7.8.

(a)

	d(measured), Å	d(ionic radii), Å	d(covalent radii), Å
Li–F	2.01	2.09	2.05
Na–Cl	2.82	2.83	2.53
K–Br	3.30	3.34	3.10
Rb–I	3.67	3.72	3.44

(b)   The agreement between measured distances in specific ionic compounds, and predicted distances based on ionic radii is not perfect. Ionic radii are averages compiled from distances in many ionic compounds containing the ion in question. The sum of these average radii may not give an exact match for the distance in any specific compound, but it will give good distance estimates for many ionic compounds. Also, there is uncertainty in all measured data. Note that all estimates from ionic radii are within 0.08 Å of the measured distances.

(c)   Distance estimates from bonding atomic radii are not as accurate as those from ionic radii. This indicates that the bonding in these four compounds is more accurately described as ionic, rather than covalent. The details of these two models will be discussed in Chapter 8.

Ionization Energies; Electron Affinities

7.40   (a) $Sn(g) \rightarrow Sn^+(g) + 1e^-$; $Sn^+(g) \rightarrow Sn^{2+}(g) + 1e^-$

(b) $Ti^{3+}(g) \rightarrow Ti^{4+}(g) + 1e^-$

7.42   (a)   The effective nuclear charges of Li and Na are similar, but the outer electron in Li has a smaller n-value and is closer to the nucleus than the outer electron in Na. More energy is needed to overcome the greater attraction of the Li electron for the nucleus.

(b)   Sc: $[Ar] 4s^2 3d^1$; Ti: $[Ar] 4s^2 3d^2$. The fourth ionization of titanium involves removing a 3d valence electron, while the fourth ionization of Sc requires removing a 3p electron from the [Ar] core. The effective nuclear charges experienced by the two 3d electrons in Ti are much more similar than the effective nuclear charges of a 3d valence electron and a 3p core electron in Sc. Thus, the difference between the third and fourth ionization energies of Sc is much larger.

(c)   The electron configuration of Li^+ is $1s^2$ or [He] and that of Be^+ is $[He]2s^1$. Be^+ has one more valence electron to lose while Li^+ has the stable noble gas configuration of He. It requires much more energy to remove a 1s core electron close to the nucleus of Li^+ than a 2s valence electron farther from the nucleus of Be^+.

7.44   (a)   Moving from F to I in group 7A, first ionization energies decrease and atomic radii increase. The greater the atomic radius, the smaller the electrostatic attraction of an outer electron for the nucleus and the smaller the ionization energy of the element.

(b) First ionization energies increase slightly going from K to Kr and atomic sizes decrease. As valence electrons are drawn closer to the nucleus (atom size decreases), it requires more energy to completely remove them from the atom (first ionization energy increases). Each trend has a discontinuity at Ga, owing to the increased shielding of the 4p electrons by the filled 3d subshell.

7.46 (a) Ti. Effective nuclear charge increases moving both right across a row and up a family. The 4s valence electrons in Ti experience the greater Z_{eff} and have greater first ionization energy than 6s electrons of Ba. Recall that transition metals like Ti lose ns electrons first when forming ions.

(b) Cu. The 4s electrons of Cu are closer to the nucleus and shielded mainly by an [Ar] core, while the 5s electrons of Ag are further from the nucleus and shielded by a [Kr] core. Recall that transition elements lose ns electrons first when forming ions.

(c) Cl. Effective nuclear charge increases moving both right across a row and up a family. Valence electrons in Cl, which is to the right and above Ge, experience the greater Z_{eff} and have the larger first ionization energy.

(d) Sb. Z_{eff} and first ionization energy increase moving up a family and right across a row. Even the excess nuclear charge (Z) associated with filling the 4f subshell between Sb and Pb does not totally offset these trends.

7.48 (a) Cr^{3+}: $[Ar]3d^3$

(b) N^{3-}: $[He]2s^22p^6 = [Ne]$, noble-gas configuration

(c) Sc^{3+}: [Ar], noble-gas configuration   (d) Cu^{2+}: $[Ar]3d^9$

(e) Tl^+: $[Xe]6s^24f^{14}5d^{10}$ (f) Au^+: $[Xe]4f^{14}5d^{10}$

7.50 (a) Cu^{2+}, 1 unpaired electron   (b) Tl^+, 0 unpaired electrons

7.52   Li $+1e^- \rightarrow$ Li$^-$;   Be $+ 1e^- \rightarrow$ Be$^-$
[He]2s^1 [He]2s^2 [He]2s^2 [He]2s^{2}2p^1

Adding an electron to Li completes the 2s subshell. The added electron experiences essentially the same effective nuclear charge as the other valence electron, except for the repulsion of pairing electrons in an orbital. There is an overall stabilization; ΔE is negative.

An extra electron in Be would occupy the higher energy 2p subshell. This electron is shielded from the full nuclear charge by the 2s electrons and does not experience a stabilization in energy; ΔE is positive.

7.54   Ionization energy of F$^-$: $F^-(g) \rightarrow F(g) + 1e^-$

Electron affinity of F: $F(g) + 1e^- \rightarrow F^-(g)$

The two processes are the reverse of each other. The energies are equal in magnitude but opposite in sign. $I_1 (F^-) = -E (F)$

7.56 $\quad$ $Mg^+(g)$ $\quad + \quad 1e^-$ $\quad \rightarrow \quad$ $Mg(g)$

$\quad\quad$ [Ne] $3s^1$ $\quad\quad\quad\quad\quad\quad\quad\quad\quad$ [Ne] $3s^2$

This process is the reverse of the first ionization of Mg. The magnitude of the energy change for this process is the same as the magnitude of the first ionization energy of Mg, 738 kJ/mol.

Properties of Metals and Nonmetals

7.58 $\quad$ S < Si < Ge < Ca. S is a nonmetal, Si and Ge are metalloids, and Ca is a metal. We expect that electrical conductivity increases as metallic character increases. Since metallic character increases going down a column and to the left in a row, the order of increasing electrical conductivity is as shown above.

7.60 $\quad$ Metallic character increases moving down a family and to the left in a period. Use these trends to select the element with greater metallic character.

$\quad\quad$ (a) $\quad$ Li $\quad\quad\quad\quad$ (b) $\quad$ Na $\quad\quad$ (c) $\quad$ Sn $\quad\quad$ (d) $\quad$ Al

7.62 $\quad$ Follow the logic in Sample Exercise 7.8. Scandium is a metal, so we expect Sc_2O_3 to be ionic. Metal oxides are usually basic and react with acid to form a salt and water. We choose $HNO_3(aq)$ as the acid for our equation.

$\quad\quad$ $Sc_2O_3(s) + 6HNO_3(aq) \rightarrow 2Sc(NO_3)_3(aq) + 3H_2O(l)$.

The net ionic equation is:

$\quad\quad$ $Sc_2O_3(s) + 6H^+(aq) \rightarrow 2Sc^{3+}(aq) + 3H_2O(l)$

7.64 $\quad$ The more nonmetallic the central atom, the more acidic the oxide. In order of increasing acidity: $CaO < Al_2O_3 < SiO_2 < CO_2 < P_2O_5 < SO_3$

7.66 $\quad$ (a) $\quad$ $XCl_4(l) + 2H_2O(l) \rightarrow XO_2(s) + 4HCl(g)$. The second product is $HCl(g)$.

$\quad\quad$ (b) $\quad$ If X were a metal, both the oxide and the chloride would be high melting solids. If X were a nonmetal, XO_2 would be a nonmetallic, molecular oxide and probably gaseous, like CO_2, NO_2, and SO_2. Neither of these statements describes the properties of XO_2 and XCl_4, so X is probably a metalloid.

$\quad\quad$ (c) $\quad$ Use the *Handbook of Chemistry* to find formulas and melting points of oxides, and formulas and boiling points of chlorides of selected metalloids.

metalloid	formula of oxide	m.p. of oxide	formula of chloride	b.p. of chloride
boron	B_2O_3	460°C	BCl_3	12°C
silicon	SiO_2	~1700°C	$SiCl_4$	58°C
germanium	GeO	710°C	$GeCl_2$	decomposes
	GeO_2	~1100°C	$GeCl_4$	84°C
arsenic	As_2O_3	315°C	$AsCl_3$	132°C
	As_2O_5	315°C		

Boron, arsenic, and, by analogy, antimony, do not fit the description of X, because the formulas of their oxides and chlorides are wrong. Silicon and germanium, in the same family, have oxides and chlorides with appropriate formulas. Both SiO_2 and GeO_2 melt above 1000°C, but the boiling point of $SiCl_4$ is much closer to that of XCl_4. Element X is silicon.

7.68   (a) $K_2O(s) + H_2O(l) \rightarrow 2KOH(aq)$

(b) $P_2O_3(l) + 3H_2O(l) \rightarrow 2H_3PO_3(aq)$

(c) $Cr_2O_3(s) + 6HCl(aq) \rightarrow 2CrCl_3(aq) + 3H_2O(l)$

(d) $SeO_2(s) + 2KOH(aq) \rightarrow K_2SeO_3(aq) + H_2O(l)$

Group Trends in Metals and Nonmetals

7.70   (a)   Rb: $[Kr]5s^1$, r = 2.11 Å   Ag: $[Kr]5s^14d^{10}$, r = 1.53 Å

The electron configurations both have a [Kr] core and a single 5s electron; Ag has a completed 4d subshell as well. The radii are very different because the 5s electron in Ag experiences a much greater effective nuclear charge. Ag has a much larger Z (47 vs. 37), and although the 4d electrons in Ag shield the 5s electron somewhat, the increased shielding does not compensate for the large increase in Z.

(b)   Ag is much less reactive (less likely to lose an electron) because its 5s electron experiences a much larger effective nuclear charge and is more difficult to remove.

7.72   (a)   Cs is much more reactive than Li toward H_2O because its valence electron is less tightly held (greater n value), and Cs is more easily oxidized.

(b)   The purple flame indicates that the metal is potassium (see Figure 7.26).

(c) $K_2O_2(s)$ + $H_2O(l)$ $\rightarrow$ $H_2O_2(aq)$ + $K_2O(aq)$
potassium peroxide   hydrogen peroxide

7.74   (a) $2Cs(s) + 2H_2O(l) \rightarrow 2CsOH(aq) + H_2(g)$

(b) $Sr(s) + 2H_2O(l) \rightarrow Sr(OH)_2(aq) + H_2(g)$

(c) $2Na(s) + O_2(g) \rightarrow Na_2O_2(s)$ (See Equation [7.21].)

(d) $Ca(s) + I_2(s) \rightarrow CaI_2(s)$

7.76   (a)   The reactions of the alkali metals with hydrogen and with a halogen are redox reactions. In both classes of reaction, the alkali metal loses electrons and is oxidized. Both hydrogen and the halogen gain electrons and are reduced. The product is an ionic solid, where either hydride ion, H^-, or a halide ion, X^-, is the anion and the alkali metal is the cation.

(b) $Ca(s) + F_2(g) \rightarrow CaF_2(s)$ $Ca(s) + H_2(g) \rightarrow CaH_2(s)$

Both products are ionic solids containing Ca^{2+} and the corresponding anion in a 1:2 ratio.

7.78   *Plan.* Predict the physical and chemical properties of At based on the trends in properties in the halogen (7A) family.   *Solve.*

(a)   F, at the top of the column, is a gas; I, immediately above At, is a solid; the melting points of the halogens increase going down the column. At is likely to be a solid at room temperature.

(b)   All halogens form ionic compounds with Na; they have the generic formula NaX. The compound formed by At will have the formula NaAt.

7.80   Xe has a lower ionization energy than Ne. The valence electrons in Xe are much farther from the nucleus than those of Ne ($n = 5$ vs $n = 2$) and much less tightly held by the nucleus; they are more "willing" to be shared than those in Ne. Also, Xe has empty 5d orbitals that can help to accommodate the bonding pairs of electrons, while Ne has all its valence orbitals filled.

7.82   (a) $Cl_2(g) + H_2O(l) \rightarrow HCl(aq) + HOCl(aq)$

(b) $Ba(s) + H_2(g) \rightarrow BaH_2(s)$

(c) $2Li(s) + S(s) \rightarrow Li_2S(s)$

(d) $Mg(s) + F_2(g) \rightarrow MgF_2(s)$

Additional Exercises

7.84   (a)   4s

(b)   To a first approximation, s and p valence electrons do not shield each other, so we expect the 4s and 4p electrons in As to experience a similar Z_{eff}. However, since s electrons have a finite probability of being very close to the nucleus (Figure 7.4), they experience less shielding than p electrons with the same n-value. Since $Z_{eff} = Z - S$ and Z is the same for all electrons in As, if S is smaller for 4s than 4p, Z_{eff} will be greater for 4s electrons and they will have a lower energy.

7.86   Close approach by two positively charged nuclei is impossible because of the large electrostatic repulsion between like-charged particles at small distances. The additional space between the nuclei in a molecule like F_2 is occupied by bonding electrons, which are electrostatically stabilized by attraction to both nuclei. The electrons also provide a buffer between the two nuclei.

7.87   (a) $Z_{eff} = Z - S$. According to our simple model, moving from C to N, causes Z to increase by 1 and S to remain the same, so Z_{eff} is greater by 1 for N than for C.

(b)   Using Slater's rules, Z increases by 1 moving from C to N, but S also increases by 0.35, the value for an electron with the same n value as the one of interest. The increase in Z_{eff} from C to N should be $(1 - 0.35) = 0.65$.

(c)   In this case, Slater's rules predict the change in Z_{eff} more accurately than the simple model in (a).

(d) In O, $[He]2s^2 2p^4$, one of the p orbitals is doubly occupied, which significantly increases electron-electron repulsion. This leads to an overall higher energy for the 2p electrons. In our model, this appears as a larger S and a smaller Z_{eff} for O. The change from N to O is then smaller.

7.89 (a) The estimated distances in the table below are the sum of the radii of the group 5A elements and H from Figure 7.7.

bonded atoms	estimated distance	measured distance
P – H	1.43	1.419
As – H	1.56	1.519
Sb – H	1.75	1.707

In general, the estimated distances are a bit longer than the measured distances. This probably shows a systematic bias in either the estimated radii or in the method of obtaining the measured values.

(b) The principal quantum number of the outer electrons and thus the average distance of these electrons from the nucleus increases from P (n = 3) to As (n = 4) to Sb (n = 5). This causes the systematic increase in M – H distance.

7.90 Ge – H distance = $r_{Ge} + r_H$ = 1.22 + 0.37 = 1.59Å

Ge – Cl distance = $r_{Ge} + r_{Cl}$ = 1.22 + 0.99 = 2.21 Å

7.92 (a) Hg^{2+}

(b) No. According to Figure 2.24, Hg is not essential for life, even in trace amounts.

(c) Zn^{2+}: $[Ar]3d^{10}$, r = 0.88 Å; Cd^{2+}: $[Kr]4d^{10}$, r = 1.09 Å; Hg^{2+}: $[Xe]4f^{14}5d^{10}$, r ≈ 1.19 Å.

If there were no 4f electrons in Hg^{2+}, we would expect an ionic radius of around 1.29 Å, an increase of ~0.20 Å due to the increase in principal quantum number of the valence electrons. However, the increase in Z due to filling of the 4f orbitals is not completely offset by shielding. The 5d valence electrons in Hg^{2+} experience a greater than expected Z_{eff} which largely offsets the increase in principal quantum number; ionic radius of Hg^{2+} is smaller than expected. This phenomenon is known as the *lanthanide contraction* and affects the physical properties of elements in the sixth period and beyond. See also Solution 7.91.

(d) Since the ionic radius of Hg^{2+} is similar to that of Cd^{2+}, Hg^{2+} will be physiologically more similar to Cd^{2+}.

(e) By common knowledge, and verified with WebElements.com™, both Hg and Hg^{2+} are extremely toxic to humans.

7.93 (a) $2Sr(s) + O_2(g) \rightarrow 2SrO(s)$

(b) Assume that the corners of the cube are at the centers of the outermost O^{2-} ions, and that the edges pass through the centers of perimeter Sr^{2+} ions. The length of an edge is then $r(O^{2-}) + 2r(Sr^{2+}) + r(O^{2-}) = 2r(O^{2-}) + 2r(Sr^{2+}) = 2(1.32 \text{ Å}) + 2(1.26 \text{ Å}) = 5.16 \text{ Å}$.

(c) Density is the ratio of mass to volume.

$$d = \frac{\text{mass SrO in cube}}{\text{vol cube}} = \frac{\text{\# SrO units} \times \text{mass of SrO}}{\text{vol cube}}$$

Calculate the mass of 1 SrO unit in grams and the volume of the cube in cm^3; solve for number of SrO units.

$$\frac{103.62 \text{ g SrO}}{\text{mol}} \times \frac{1 \text{ mol SrO}}{6.022 \times 10^{23} \text{ SrO units}} = 1.7207 \times 10^{-22} = 1.721 \times 10^{-22} \text{ g/SrO unit}$$

$$V = (5.16)^3 \text{ Å}^3 \times \frac{(1 \times 10^{-8})^3 \text{ cm}^3}{\text{Å}^3} = 1.3739 \times 10^{-22} = 1.37 \times 10^{-22} \text{ cm}^3$$

$$d = \frac{\text{\# of SrO units} \times 1.7207 \times 10^{-22} \text{ g/SrO unit}}{1.3739 \times 10^{-22} \text{ cm}^3} = 5.10 \text{ g/cm}^3$$

$$\text{\# of SrO units} = 5.10 \text{ g/cm}^3 \times \frac{1.3739 \times 10^{-22} \text{ cm}^3}{1.7207 \times 10^{-22} \text{ g/SrO unit}} = 4.07 \text{ units}$$

Since the number of formula units must be an integer, there are four SrO formula units in the cube. Using average values for ionic radii to estimate the edge length probably leads to the small discrepancy.

7.95 The statement is somewhat true, but more accurate if changed to read: "A negative value for the electron affinity of an atom occurs when the outermost electrons only incompletely shield the added electron from the nucleus." This new statement totally explains the negative electron affinity of Br and the positive value for Kr. For Br^-, the electron is added to the 4p subshell and is incompletely shielded by the "other" 4s and 4p electrons. For Kr^-, the electron is added to the 5s subshell, which is effectively shielded by the spherical Kr core.

7.97 (a) P: [Ne] $3s^2 3p^3$; S: [Ne] $3s^2 3p^4$. In P, each 3p orbital contains a single electron, while in S one 3p orbital contains a pair of electrons. Removing an electron from S eliminates the need for electron pairing and reduces electrostatic repulsion, so the overall energy required to remove the electron is smaller than in P, even though Z is greater.

(b) C: [He] $2s^2 2p^2$; N: [He] $2s^2 2p^3$; O: [He] $2s^2 2p^4$. An electron added to a N atom must be paired in a relatively small 2p orbital, so the additional electron-electron repulsion more than compensates for the increase in Z and the electron affinity is smaller (less exothermic) than that of C. In an O atom, one 2p orbital already contains a pair of electrons, so the additional repulsion from an extra electron is offset by the increase in Z and the electron affinity is greater (more exothermic). Note from Figure 7.14 that the electron affinity of O is only slightly more exothermic than that of C, although the value of Z has increased by 2.

(c) O^+: [He] $2s^2 2p^3$; O^{2+}: [He] $2s^2 2p^2$; F: [He] $2s^2 2p^5$; F^+: [He] $2s^2 2p^4$. Both 'core-only' [Z_{eff} (F) = 7; Z_{eff} (O^+) = 6] and Slater [Z_{eff} (F) = 5.2; Z_{eff} (O^+) = 4.9] predict that F has a greater Z_{eff} than O^+. Variation in Z_{eff} does not offer a satisfactory explanation. The decrease in electron-electron repulsion going from F to F^+ energetically favors ionization and causes it to be less endothermic than the second ionization of O, where there is no significant decrease in repulsion.

(d) Mn^{2+}: [Ar]$3d^5$; Mn^{3+}; [Ar] $3d^4$; Cr^{2+}: [Ar] $3d^4$; Cr^{3+}: [Ar] $3d^3$; Fe^{2+}: [Ar] $3d^6$; Fe^{3+}: [Ar] $3d^5$. The third ionization energy of Mn is expected to be larger than that of Cr because of the larger Z value of Mn. The third ionization energy of Fe is less than that of Mn because going from $3d^6$ to $3d^5$ reduces electron repulsions, making the process less endothermic than predicted by nuclear charge arguments.

7.99   (a)   For both H and the alkali metals, the added electron will complete an *ns* subshell (1s for H and *ns* for the alkali metals) so shielding and repulsion effects will be similar. For the halogens, the electron is added to an *np* subshell, so the energy change is likely to be quite different.

         (b)   True. Only He has a smaller estimated "bonding" atomic radius, and no known compounds of He exist. The electron configuration of H is $1s^1$. The single 1s electron experiences no repulsion from other electrons and feels the full unshielded nuclear charge. It is held very close to the nucleus. The outer electrons of all other elements that form compounds are shielded by a spherical inner core of electrons and are less strongly attracted to the nucleus, resulting in larger atomic radii.

         (c)   Ionization is the process of removing an electron from an atom. For the alkali metals, the *ns* electron being removed is effectively shielded by the core electrons, so ionization energies are low. For the halogens, a significant increase in nuclear charge occurs as the *np* orbitals fill, and this is not offset by an increase in shielding. The relatively large effective nuclear charge experienced by *np* electrons of the halogens is similar to the unshielded nuclear charge experienced by the H 1s electron. Both H and the halogens have large ionization energies.

7.100   Since Xe reacts with F_2, and O_2 has approximately the same ionization energy as Xe, O_2 will probably react with F_2. Possible products would be O_2F_2, analogous to XeF_2, or OF_2.

$$O_2(g) + F_2(g) \rightarrow O_2F_2(g)$$
$$O_2(g) + 2F_2(g) \rightarrow 2OF_2(g)$$

7.102   Sr: [Kr]$5s^2$, Z = 38. Ca: [Ar]$4s^2$, Z = 20. Sr and Ca are in the same family. They have the same valence electron configuration so we expect their chemical properties to be similar. Both are metals that form cations with +2 charge. From trends, we expect Sr to have a larger covalent radius than Ca, and Sr^{2+} to have a larger ionic radius than Ca^{2+}; Figure 7.8 corroborates this relationship. However, the size difference is not as great as we might expect. The [Kr] core of Sr includes a full 3d subshell, which increases the Z and Z_{eff} for Sr and Sr^{2+} relative to Ca and Ca^{2+}.

Ca and Sr both react with water at room temperature to form $Ca^{2+}(aq)$ and $Sr^{2+}(aq)$. The predominant species in water supplies (very dilute aqueous solutions) and in the form for uptake by plants are $Ca^{2+}(aq)$ and $Sr^{2+}(aq)$. The main factors that determine transport properties are size and charge. Since Ca^{2+} and Sr^{2+} have the same charge and similar size, it is not surprising that they are transported together and assimilated by organisms similarly. Since they have similar properties, they are likely to interact with bio-molecules similarly. So, "normal" (nonradioactive) Sr acts like Ca, which is not only safe, but essential to organisms. This is why "normal" Sr is not very dangerous, but radioactive Sr is lethal. Radioactive Sr replaces Ca and is readily incorporated into organisms.

7.103   Moving one place to the right in a horizontal row of the table, for example, from Li to Be, there is an increase in ionization energy. Moving downward in a given family, for example from Be to Mg, there is usually a decrease in ionization energy. Similarly, atomic size decreases in moving one place to the right and increases in moving downward. Thus, two elements such as Li and Mg that are diagonally related tend to have similar ionization energies and atomic sizes. This in turn gives rise to some similarities in chemical behavior. Note, however, that the valences expected for the elements are not the same. That is, lithium still appears as Li^+, magnesium as Mg^{2+}.

7.104   (a)   *Plan.* Use qualitative physical (bulk) properties to narrow the range of choices, then match melting point and density to identify the specific element.   *Solve.*

         Hardness varies widely in metals and nonmetals, so this information is not too useful. The relatively high density, appearance, and ductility indicate that the element is probably less metallic than copper. Focus on the block of nine main group elements centered around Sn. Pb is not a possibility because it was used as a comparison standard. The melting point of the five elements closest to Pb are:

            Tl, 303.5°C; In, 156.1°C; Sn, 232°C; Sb, 630.5°C; Bi, 271.3°C

         The best match is In. To confirm this identification, the density of In is 7.3 g/cm^3, also a good match to properties of the unknown element.

        (b)   In order to write the correct balanced equation, determine the formula of the oxide product from the mass data, assuming the unknown is In.

         5.08 g oxide – 4.20 g In = 0.88 g O

         4.20 g In/114.82 g/mol = 0.0366 mol In; 0.0366/0.0366 = 1

         0.88 g O/16.00 g/mol = 0.0550 mol O; 0.0550/0.0366 = 1.5

         Multiplying by 2 produces an integer ratio of 2 In: 3 O and a formula of In_2O_3. The balanced equation is: $4 In(s) + 3 O_2(g) \rightarrow 2 In_2O_3(s)$

        (c)   According to Figure 7.2, the element In was discovered between 1843–1886. The investigator who first recorded this data in 1822 could have been the first to discover In.

Integrative Exercises

7.105   (a) $\nu = c/\lambda$; 1 Hz $= 1$ s^{-1}

Ne: $\nu = \dfrac{2.998 \times 10^8 \text{ m/s}}{14.610 \text{ Å}} \times \dfrac{1 \text{ Å}}{1 \times 10^{-10} \text{ m}} = 2.052 \times 10^{17} \text{ s}^{-1} = 2.052 \times 10^{17}$ Hz

Ca : $\nu = \dfrac{2.998 \times 10^8 \text{ m/s}}{3.358 \times 10^{-10} \text{ m}} = 8.928 \times 10^{17}$ Hz

Zn : $\nu = \dfrac{2.998 \times 10^8 \text{ m/s}}{1.435 \times 10^{-10} \text{ m}} = 20.89 \times 10^{17}$ Hz

Zr : $\nu = \dfrac{2.998 \times 10^8 \text{ m/s}}{0.786 \times 10^{-10} \text{ m}} = 38.14 \times 10^{17} = 38.1 \times 10^{17}$ Hz

Sn : $\nu = \dfrac{2.998 \times 10^8 \text{ m/s}}{0.491 \times 10^{-10} \text{ m}} = 61.06 \times 10^{17} = 61.1 \times 10^{17}$ Hz

(b)

Element	Z	ν	$\nu^{1/2}$
Ne	10	2.052×10^{17}	4.530×10^8
Ca	20	8.928×10^{17}	9.449×10^8
Zn	30	20.89×10^{17}	14.45×10^8
Zr	40	38.14×10^{17}	19.5×10^8
Sn	50	61.06×10^{17}	24.7×10^8

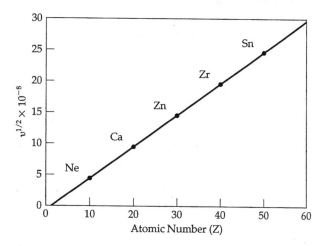

(c)   The plot in part (b) indicates that there is a linear relationship between atomic
number and the square root of the frequency of the X-rays emitted by an element.
Thus, elements with each integer atomic number should exist. This relationship
allowed Moseley to predict the existence of elements that filled "holes" or gaps in
the periodic table.

(d)   For Fe, Z = 26. From the graph, $v^{1/2} = 12.5 \times 10^8$, $v = 1.56 \times 10^{18}$ Hz.

$$\lambda = c/v = \frac{2.998 \times 10^8 \text{ m/s}}{1.56 \times 10^{18} \text{ s}^{-1}} \times \frac{1 \text{ Å}}{1 \times 10^{-10} \text{ m}} = 1.92 \text{ Å}$$

(e) $\lambda = 0.980 \text{ Å} = 0.980 \times 10^{-10}$ m

$$v = c/\lambda = \frac{2.998 \times 10^8 \text{ m/s}}{0.980 \times 10^{-10} \text{ m}} = 30.6 \times 10^{17} \text{ Hz}; \; v^{1/2} = 17.5 \times 10^8$$

From the graph, $v^{1/2} = 17.5 \times 10^8$, Z = 36. The element is krypton, Kr.

7.107   (a) $E = hc/\lambda$; 1 nm = 1×10^{-9} m; 58.4 nm = 58.4×10^{-9} m;

1 eV = 96.485 kJ/mol, 1 eV - mol = 96.485 kJ

$$E = \frac{6.626 \times 10^{-34} \text{ J-s} \times 2.998 \times 10^8 \text{ m/s}}{58.4 \times 10^{-9} \text{ m}} = 3.4015 \times 10^{-18} = 3.40 \times 10^{-18} \text{ J/photon}$$

$$\frac{3.4015 \times 10^{-18} \text{ J}}{\text{photon}} \times \frac{1 \text{ kJ}}{1000 \text{ J}} \times \frac{6.022 \times 10^{23} \text{ photons}}{\text{mol}} \times \frac{1 \text{ eV - mol}}{96.485 \text{ kJ}} = 21.230 = 21.2 \text{ eV}$$

(b) $Hg(g) \rightarrow Hg^+(g) + 1e^-$

(c) $I_1 = E_{58.4} - E_K = 21.23 \text{ eV} - 10.75 \text{ eV} = 10.48 = 10.5 \text{ eV}$

$$10.48 \text{ eV} \times \frac{96.485 \text{ kJ}}{1 \text{ eV - mol}} = 1.01 \times 10^3 \text{ kJ/mol}$$

(d)   From Figure 7.12, iodine (I) appears to have the ionization energy closest to that of Hg, approximately 1000 kJ/mol.

7.108   (a) $Na(g) \rightarrow Na^+(g) + 1e^-$ (ionization energy of Na)

               $Cl(g) + 1e^- \rightarrow Cl^-(g)$ (electron affinity of Cl)

               $Na(g) + C1(g) \rightarrow Na^+(g) + Cl^-(g)$

(b) $\Delta H = I_1 (Na) + E_1 (Cl) = +496 \text{ kJ} - 349 \text{ kJ} = +147 \text{ kJ}$, endothermic

(c)   The reaction $2Na(s) + Cl_2(g) \rightarrow 2NaCl(s)$ involves many more steps than the reaction in part (a). One important difference is the production of NaCl(s) versus NaCl(g). The condensation NaCl(g) $\rightarrow$ NaCl(s) is very exothermic and is the step that causes the reaction of the elements in their standard states to be exothermic, while the gas phase reaction is endothermic.

7.110   (a) $r_{Bi} = r_{BiBr_3} - r_{Br} = 2.63 \text{ Å} - 1.14 \text{ Å} = 1.49 \text{ Å}$

(b) $Bi_2O_3(s) + 6HBr(aq) \rightarrow 2BiBr_3(aq) + 3H_2O(l)$

(c) Bi_2O_3 is soluble in acid solutions because it act as a base and undergoes acid-base reactions like the one in part (b). It is insoluble in base because it cannot acts as an acid. Thus, Bi_2O_3 is a basic oxide, the oxide of a metal. Based on the properties of its oxide, Bi is characterized as a metal.

(d)   Bi: $[Xe]6s^2 4f^{14} 5d^{10} 6p^3$. Bi has five outer electrons in the 6p and 6s subshells. If all five electrons participate in bonding, compounds such as BiF_5 are possible. Also, Bi has a large enough atomic radius (1.49 Å) and low-energy orbitals available to accommodate more than four pairs of bonding electrons.

(e)   The high ionization energy and relatively large negative electron affinity of F, coupled with its small atomic radius, make it the most electron withdrawing of the halogens. BiF_5 forms because F has the greatest tendency to attract electrons from Bi. Also, the small atomic radius of F reduces repulsions between neighboring bonded F atoms. The strong electron withdrawing properties of F are also the reason that only F compounds of Xe are known.

7.111   (a) $4KO_2(s) + 2CO_2(g) \rightarrow 2K_2CO_3(s) + 3O_2(g)$

(b)   K, +1; O, –1/2 (O_2^- is superoxide ion); C, +4; O, –2 → K, +1; C, +4; O, –2; O, 0

(c) $18.0 \text{ g CO}_2 \times \dfrac{1 \text{ mol CO}_2}{44.01 \text{ g CO}_2} \times \dfrac{4 \text{ mol KO}_2}{2 \text{ mol CO}_2} \times \dfrac{71.10 \text{ g KO}_2}{1 \text{ mol KO}_2} = 58.2 \text{ g KO}_2$

$18.0 \text{ g CO}_2 \times \dfrac{1 \text{ mol CO}_2}{44.01 \text{ g CO}_2} \times \dfrac{3 \text{ mol O}_2}{2 \text{ mol CO}_2} \times \dfrac{32.00 \text{ g O}_2}{1 \text{ mol O}_2} = 19.6 \text{ g O}_2$

8 Basic Concepts of Chemical Bonding

Visualizing Concepts

8.1 *Analyze/Plan.* Count the number of electrons in the Lewis symbol. This corresponds to the 'A'-group number of the family. *Solve.*

(a) Group 14 or 4A

(b) Group 2 or 2A

(c) Group 15 or 5A

(These are the appropriate groups in the s and p blocks, where Lewis symbols are most useful.)

8.3 *Analyze/Plan.* Count the valence electrons in the orbital diagram, take ion charge into account, and find the element with this orbital electron count on the periodic chart. Write the complete electron configuration for the ion. *Solve.*

(a) This ion has seven 3d electrons. Transition metals, or d-block elements, have valence electrons in d-orbitals. Transition metal ions first lose electrons from the 4s orbital, then from 3d if required by the charge. This 2+ ion has lost two electrons from 4s, none from 3d. The transition metal with seven 3d-electrons is cobalt, Co.

(b) The electron configuration of Co is $[Ar]4s^2 3d^7$. (The configuration of Co^{2+} is $[Ar]3d^7$).

8.4 *Analyze/Plan.* This question is a "reverse" Lewis structure. Count the valence electrons shown in the Lewis structure. For each atom, assume zero formal charge and determine the number of valence electrons an unbound atom has. Name the element. *Solve.*

A: 1 shared e^- pair = 1 valence electron + 3 unshared pairs = 7 valence electrons, F

E: 2 shared pairs = 2 valence electrons + 2 unshared pairs = 6 valence electrons, O

D: 4 shared pairs = 4 valence electrons, C

Q: 3 shared pairs = 3 valence electrons + 1 unshared pair = 5 valence electrons, N

X: 1 shared pair = 1 valence electron, no unshared pairs, H

Z: same as X, H

Check. Count the valence electrons in the Lewis structure. Does the number correspond to the molecular formula CH_2ONF? 12 e^- pair in the Lewis structure. $CH_2ONF = 4 + 2 + 6 + 5 + 7 = 24\ e^-$, 12 e^- pair. The molecular formula we derived matches the Lewis structure.

8.6   *Analyze/Plan.* Given an oxyanion of the type XO_4^{n-}, find the identity of X from elements in the third period. Use the generic Lewis structure to determine the identity of X, and to draw the ion-specific Lewis structures. Use the definition of formal charge, [# of valence electrons – # of nonbonding electrons – (# bonding electrons/2)], to draw Lewis structures where X has a formal charge of zero. *Solve.*

(a)   According to the generic Lewis structure, each anion has 12 nonbonding and 4 bonding electron pairs, for a total of 32 electrons. Of these 32 electrons, the 4 O atoms contribute $(4 \times 6) = 24$, and the overall negative charges contribute 1, 2 or 3. # X electrons = $32 - 24 - n$.

For $n = 1-$, X has $(32 - 24 - 1) = 7$ valence electrons. X is Cl, and the ion is ClO_4^-.

For $n = 2-$, X has $(32 - 24 - 2) = 6$ valence electrons. X is S, and the ion is SO_4^{2-}.

For $n = 3-$, X has $(32 - 24 - 3) = 5$ valence electrons. X is P, and the ion is PO_4^{3-}.

Check. The identity of the ions is confirmed in Figure 2.27.

(b)   In the generic Lewis structure, X has 0 nonbonding electrons and $(8/2) = 4$ bonding electrons. Differences in formal charge are due to difference in the number of valence electrons on X.

For PO_4^{3-}, formal charge of P is $(5 - 4) = +1$.

For SO_4^{2-}, formal charge of S is $(6 - 4) = +2$.

For ClO_4^-, formal charge of Cl is $(7 - 4) = +3$.

(c)   In order to reduce the formal charge of X to zero, X must have more bonding electrons. This is accomplished by changing the appropriate number of lone pairs on O to multiple bonds between X and O.

(d)   In part (c) the Lewis structures that cause the formal charge on X to be zero all violate the octet rule. According to Section 8.7, the best single Lewis structure for an anion is the one that obeys the octet rule.

Lewis Symbols

8.8   (a)   Atoms will gain, lose or share electrons to achieve the nearest noble-gas electron configuration. Except for H and He, this corresponds to eight electrons in the valence shell, thus the term octet rule.

(b)   S: $[Ne]3s^2 3p^4$ A sulfur atom has six valence electrons, so it must gain two electrons to achieve an octet.

(c) $1s^2 2s^2 2p^3 = [He]2s^2 2p^3$ The atom (N) has five valence electrons and must gain three electrons to achieve an octet.

8.10　(a)　Ti: $[Ar]4s^23d^2$. Ti has four (4) valence electrons. These valence electrons are available for chemical bonding, while core electrons do not participate in chemical bonding.

(b)　Hf: $[Xe]6s^24f^{14}5d^2$

(c)　If Hf and Ti both behave as if they have four (4) valence electrons, the 6s and 5d orbitals in Hf behave as valence orbitals and the 4f behaves as a core orbital. This is reasonable because 4f is complete and 4f electrons are, on average, closer to the nucleus than 5d or 6s electrons.

8.12　(a)　$\overset{\cdot}{Ca}\cdot$　　(b)　$\cdot\overset{\cdot\cdot}{\underset{\cdot}{P}}\cdot$　　(c)　$\left[:\overset{\cdot\cdot}{\underset{\cdot\cdot}{Mg}}:\right]^{2+}$ or Mg^{2+}　　(d)　$\left[:\overset{\cdot\cdot}{\underset{\cdot\cdot}{S}}:\right]^{2-}$ or S^{2-}

Ionic Bonding

8.14　$Ca\cdot\overset{\cdot}{\ } + \overset{\cdot\cdot}{\cdot}\overset{\cdot\cdot}{\underset{\cdot\cdot}{F}}: + \overset{\cdot\cdot}{\cdot}\overset{\cdot\cdot}{\underset{\cdot\cdot}{F}}: \longrightarrow Ca^{2+} + 2\left[:\overset{\cdot\cdot}{\underset{\cdot\cdot}{F}}:\right]$

8.16　(a)　BaF_2　　　(b)　CsCl　　　(c)　Li_3N　　　(d)　Al_2O_3

8.18　(a)　Zn^{2+}: $[Ar]3d^{10}$

(b)　Te^{2-}: $[Kr]5s^24d^{10}5p^6 = [Xe]$, noble-gas configuration

(c)　Sc^{3+}: $[Ar]$, noble-gas configuration

(d)　Rh^{3+}: $[Kr]4d^6$

(e)　Tl^+: $[Xe]6s^24f^{14}5d^{10}$

(f)　Bi^{3+}: $[Xe]6s^24f^{14}5d^{10}$

8.20　(a)　NaCl, 788 kJ/mol; KF, 808 kJ/mol

The two factors that affect lattice energies are ionic charge and radius. The ionic charges, 1+ and 1–, are the same in the two compounds. Since lattice energy is inversely proportional to the ion separation (d), we expect the compound with the smaller lattice energy, NaCl, to have the larger ion separation. That is, the K–F distance should be shorter than the Na–Cl distance.

(b)　Na–Cl, 1.16 Å + 1.67 Å = 2.83 Å

K–F, 1.52 Å + 1.19 Å = 2.71 Å

This estimate of the relative ion separations agrees with the estimate from lattice energies. Ionic radii indicate that the K–F distance is shorter than the Na–Cl distance.

8.22　(a)　According to Equation 8.4, electrostatic attraction increases with increasing charges of the ions and decreases with increasing radius of the ions. Thus, lattice energy (i) increases as the charges of the ions increase and (ii) decreases as the sizes of the ions increase.

(b)   KBr < NaF < MgO < ScN. This order is confirmed by the lattice energies given in Table 8.2. ScN has the highest lattice energy, because its ions have 3+ and 3– charges. Na^+ is smaller than K^+, and F^- is smaller than Br^-. The ion separation is smaller in NaF, so it has the larger lattice energy.

8.24   (a)   The ion charges in CaF_2 and BaF_2 are the same, 2+ for the cations and 1– for the anions. Ba^{2+} is a larger cation than Ca^{2+}, so the ion separation, d, is greater in BaF_2 and the lattice energy is smaller than that of CaF_2.

         (b)   The ions have 1+ and 1– charges in all three compounds. In NaCl the cationic and anionic radii are smaller than in the other two compounds, so it has the largest lattice energy. In RbBr and CsBr, the anion is the same, but the Cs cation is larger, so CsBr has the smaller lattice energy.

         (c)   In BaO, the magnitude of the charges of both ions is 2; in KF, the magnitudes are 1. Charge considerations alone predict that BaO will have the higher lattice energy. The distance effect is less clear; O^{2-} and F^- are isoelectronic, so F^-, with the larger Z, has a slightly smaller radius. Ba^{2+} is two rows lower on the periodic chart than K^+, but it has a greater positive charge, so the radii are probably similar. In any case, the ionic separations in the two compounds are not very different, and the charge effect dominates.

8.26   By analogy to the Born-Haber cycle for NaCl(s), Figure 8.4, the enthalpy of formation for $NaCl_2(s)$ is

$$\Delta H_f^\circ\, NaCl_2(s) = -\Delta H_{latt} NaCl_2 + \Delta H_f^\circ\, Na(g) + 2\,\Delta H_f^\circ\, Cl(g) + I_1(Na) + I_2(Na) + 2E(Cl)$$

         (a) $\Delta H_f^\circ\, NaCl_2(s) = -\Delta H_{latt} NaCl_2 + 107.7\ kJ + 2(121.7\ kJ) + 496\ kJ + 4562\ kJ$

                                                       $+\ 2(-349\ kJ)$

             $\Delta H_f^\circ\, NaCl_2(s) = -\Delta H_{latt} NaCl_2 + 4711\ kJ$

             The collective energy of the "other" steps in the cycle (vaporization and ionization of Na^{2+}, dissociation of Cl_2 and electron affinity of Cl) is +4711 kJ. In order for the sign of $\Delta H_f^\circ\, NaCl_2$ to be negative, the lattice energy would have to be greater than 4711 kJ.

         (b) $\Delta H_f^\circ\, NaCl_2(s) = -(2326\ kJ) + 4711\ kJ = 2385\ kJ$

             This value is large and positive.

8.28   (a) $MgCl_2$, 2326 kJ; $SrCl_2$, 2127 kJ. Since the ionic radius of Ca^{2+} is greater than that of Mg^{2+}, but less than that of Sr^{2+}, the ion separation (d) in $CaCl_2$ will be intermediate as well. We expect the lattice energy of $CaCl_2$ to be in the range 2200–2250 kJ.

         (b)   By analogy to Figure 8.4:

$$\Delta H_{latt} = -\Delta H_f^\circ\, CaCl_2 + \Delta H_f^\circ\, Ca(g) + 2\Delta H_f^\circ\, Cl(g) + I_1(Ca) + I_2(Ca) + 2E(Cl)$$
$$= -(-795.8\ kJ) + 179.3\ kJ + 2(121.7\ kJ) + 590\ kJ + 1145\ kJ + 2(-349\ kJ) = +2256\ kJ$$

             This value is near the range predicted in part (a).

Covalent Bonding, Electronegativity, and Bond Polarity

8.30   K and Ar. K is an active metal with one valence electron. It is most likely to achieve an octet by losing this single electron and to participate in ionic bonding. Ar has a stable octet of valence electrons; it is not likely to form chemical bonds of any type.

8.32

8.34   (a)   The H atoms must be terminal because H can form only one bond.

14 e⁻, 7 e⁻ pairs

$$H-\ddot{\underset{..}{O}}-\ddot{\underset{..}{O}}-H$$

(b)   From Solution 8.33, O_2 has a double bond. The O–O bond in H_2O_2 is a single bond, and thus longer than the O–O bond in O_2.

8.36   (a)   The electronegativity of the elements increases going from left to right across a row of the periodic chart.

(b)   Electronegativity decreases going down a family of the periodic chart.

(c)   Generally, the trends in electronegativity are the same as those in ionization energy and opposite those in electron affinity. That is, the more positive the ionization energy and the more negative the electron affinity (ignoring a few exceptions), the greater the electronegativity of an element.

8.38   Electronegativity increases going up and to the right in the periodic table.

(a)   O   (b)   Al   (c)   Cl   (d)   F

8.40   The more different the electronegativity values of the two elements, the more polar the bond.

(a)   O–F < C–F < Be–F. This order is clear from the periodic trend.

(b)   S–Br < C–P < O–Cl. Refer to the electronegativity values in Figure 8.6 to confirm the order of bond polarity. The 3 pairs of elements all have the same positional relationship on the periodic chart. The more electronegative element is one row above and one column to the left of the less electronegative element. This leads us to conclude that ΔEN is similar for the 3 bonds, which is confirmed by values in Figure 8.6. The most polar bond, O–Cl, involves the most electronegative element, O. Generally, the largest electronegativity differences tend to be between row 2 and row 3 elements. The 2 bonds in this exercise involving elements in row 2 and row 3 do have slightly greater ΔEN than the S–Br bond, between elements in rows 3 and 4.

(c)   C–S < N–O < B–F. You might predict that N–O is least polar since the elements are adjacent on the table. However, the big decrease going from the second row to the third means that the electronegativity of S is not only less than that of O, but essentially the same as that of C. C–S is the least polar.

8.42   (a)   The more electronegative element, Br, will have a stronger attraction for the shared electrons and adopt a partial negative charge.

       (b)   Q is the charge at either end of the dipole.

$$Q = \frac{\mu}{r} = \frac{1.21\,D}{2.49\,\text{Å}} \times \frac{1\,\text{Å}}{1 \times 10^{-10}\,m} \times \frac{3.34 \times 10^{-30}\,C\cdot m}{1\,D} \times \frac{1\,e}{1.60 \times 10^{-19}\,C}$$
$$= 0.1014 = 0.101\,e$$

The charges on I and Br are 0.101 e.

8.44   Generally, compounds formed by a metal and a nonmetal are described as ionic, while compounds formed from two or more nonmetals are covalent. However, substances with metals in a high oxidation states often have properties of molecular compounds.

       (a) $TiCl_4$, metal and nonmetal, Ti(IV) is a relatively high oxidation state, molecular (by contrast with CaF_2, which is definitely ionic), titanium tetrachloride

                CaF_2, metal and nonmetal, ionic, calcium fluoride

       (b) ClF_3, two nonmetals, molecular, chlorine trifluoride

                VF_3, metal and nonmetal, ionic, vanadium(III) fluoride

       (c) $SbCl_5$, metalloid and nonmetal, molecular, antimony pentachloride

                AlF_3, metal and nonmetal, ionic, aluminum fluoride

Lewis Structures; Resonance Structures

8.46   (a)   12 e⁻, 6 e⁻ pairs   (b)   14 valence e⁻, 7 e⁻ pairs

            H—C=Ö   H—Ö—Ö—H
                 |
                 H

       (c)   50 valence e⁻, 25 e⁻ pairs   (d)   26 valence e⁻, 13 e⁻ pairs

               :F: :F:
                | |
         :F—C—C—F:
                | |
               :F: :F:

           (Choose the Lewis structure that
            obeys the octet rule, Section 8.7)

       (e)   26 valence e⁻, 13 e⁻ pairs   (f)   10 e⁻, 5 e⁻ pairs

            H—Ö—S—Ö—H   H—C≡C—H
                   |
                :O:

           (Choose the Lewis structure that
            obeys they octet rule, Section 8.7.)

8.48   (a)   26 e⁻, 13 e⁻ pairs

The octet rule is satisfied for all atoms in the structure.

(b)   F is more electronegative than P. Assuming F atoms hold all shared electrons, the oxidation number of each F is –1. The oxidation number of P is +3.

(c)   Assuming perfect sharing, the formal charges on all F and P atoms are 0.

(d)   The oxidation number on P is +3; the formal charge is 0. These represent extremes in the possible electron distribution, not the best picture. By virtue of their greater electronegativity, the F atoms carry a partial negative charge, and the P atom a partial positive charge.

8.50   Formal charges are given near the atoms, oxidation numbers are listed below the structures.

(a)   18 e⁻, 9 e⁻ pairs

$$:\!\overset{..}{\underset{..}{O}}\!\!-\!\!\overset{..}{S}\!=\!\overset{..}{\underset{..}{O}}$$
$$\quad -1 \quad +1 \quad\;\; 0$$

ox. #: S, +4; O, –2

(b)   24 e⁻, 12 e⁻ pairs

$$-1:\!\overset{..}{\underset{..}{O}}\!\!-\!\!\overset{+2}{S}\!=\!\overset{..}{\underset{..}{O}}\;0$$
$$\qquad\quad\; |$$
$$\qquad\quad :\!\overset{}{\underset{..}{O}}\!:$$
$$\qquad\quad\;\; -1$$

ox. #: S, +6; O, –2

(c)   26 e⁻, 13 e⁻ pairs

$$\left[\; -1:\!\overset{..}{\underset{..}{O}}\!\!-\!\!\overset{+1}{\overset{..}{S}}\!\!-\!\!\overset{..}{\underset{..}{O}}\!:-1 \;\right]^{2-}$$
$$\qquad\qquad\quad |$$
$$\qquad\qquad :\!\overset{}{\underset{..}{O}}\!:$$
$$\qquad\qquad\;\; -1$$

ox. #: S, +4; O, –2

(d) $SO_2 < SO_3 < SO_3^{2-}$

Double bonds are shorter than single bonds. SO_2 has two resonance structures with alternating single and double bonds, for an approximate average "one-and-a-half" bond. SO_3 has three resonance structures with one double and two single bonds, for an approximately, "one-and-a-third" bond. SO_3^{2-} has all single bonds. The order of increasing bond length is the order of decreasing bond type.

SO_2 (1.5) < SO_3 (1.3) < SO_3^{2-} (1.0).

8.52   (a)   16 e⁻, 8 e⁻ pairs

$$\left[\overset{..}{\underset{..}{O}}\!=\!N\!=\!\overset{..}{\underset{..}{O}}\right]^{+} \longleftrightarrow \left[:\!O\!\equiv\!N\!-\!\overset{..}{\underset{..}{O}}\!:\right]^{+} \longleftrightarrow \left[:\!\overset{..}{\underset{..}{O}}\!-\!N\!\equiv\!O\!:\right]^{+}$$

(b)   More than one correct Lewis structure can be drawn, so resonance structures are needed to accurately describe the structure.

(c) NO_2^+ has 16 valence electrons. Consider other triatomic molecules involving second-row nonmetallic elements. O_3^{2+} or C_3^{4-} are not "common" (or stable). CO_2 is common and matches the description (as does N_3^-, azide ion).

8.54   The Lewis structures are as follows:

5 e⁻ pairs   9 e⁻ pairs

12 e⁻ pairs

The average number of electron pairs in the N–O bond is 3.0 for NO^+, 1.5 for NO_2^-, and 1.33 for NO_3^-. The more electron pairs shared between two atoms, the shorter the bond. Thus the N–O bond lengths vary in the order $NO^+ < NO_2^- < NO_3^-$.

8.56   (a)

(b)   The resonance model of this molecule has bonds that are neither single nor double, but somewhere in between. This results in bond lengths that are intermediate between C–C single and C=C double bond lengths.

(c)

Exceptions to the Octet Rule

8.58   Carbon, in group 14, needs to form four single bonds to achieve an octet, as in CH_4. Nitrogen, in group 15, needs to form three, as in NH_3. If G = group number and n = the number of single bonds, G + n = 18 is a general relationship for the representative non-metals.

Check: O as in H_2O (G = 16) + (n = 2 bonds) = 18

8.60   In the third period, atoms have the space and available orbitals to accommodate extra electrons. Since atomic radius increases going down a family, elements in the third period and beyond are less subject to destabilization from additional electron-electron repulsions. Also, the third shell contains d orbitals that are relatively close in energy to 3s and 3p orbitals (the ones that accommodate the octet) and provide an allowed energy state for the extra electrons.

8.62   (a)   8 e⁻, 4 e⁻ pairs

$$\left[\begin{array}{c} H \\ | \\ H-N-H \\ | \\ H \end{array}\right]^{+}$$

(b)   16 e⁻, 8 e⁻ pairs

$$\left[\:\ddot{S}-C\equiv N\:\right]^{-} \longleftrightarrow \left[\ddot{S}=C=\ddot{N}\:\right]^{-} \longleftrightarrow \left[\:S\equiv C-\ddot{N}\:\right]^{-}$$

Three resonance structures, all obey the octet rule. The middle structure is probably the largest contributor to the actual structure. (See Sample Exercise 8.9.)

(c)   26 e⁻, 13 e⁻ pairs

$$:\ddot{Cl}-\overset{\textstyle P}{\underset{\textstyle :\ddot{Cl}:}{|}}-\ddot{Cl}:$$

(d)   34 e⁻, 17 e⁻ pairs

$$:\ddot{F}-\overset{\textstyle :\ddot{F}:}{\underset{\textstyle :\ddot{F}:}{Te}}-\ddot{F}:$$

Does not obey the octet rule.

(e)   22 e⁻, 11 e⁻ pair

$$:\ddot{F}-\ddot{Xe}-\ddot{F}:$$

Does not obey the octet rule.

8.64   (a)   19 e⁻, 9.5 e⁻ pairs, odd electron molecule

$$:\ddot{O}-\ddot{Cl}-\ddot{O}: \longleftrightarrow :\ddot{O}-\ddot{Cl}-\dot{\ddot{O}}\cdot \longleftrightarrow \cdot\dot{\ddot{O}}-\ddot{Cl}-\ddot{O}:$$

(b)   None of the structures satisfies the octet rule. In each structure, one atom has only 7 e⁻ around it. If a molecule has an odd number of electrons in the valence shell, no Lewis structure can satisfy the octet rule.

(c) $$:\ddot{O}-\ddot{Cl}-\ddot{O}: \longleftrightarrow :\ddot{O}-\ddot{Cl}-\ddot{O}\cdot \longleftrightarrow \cdot\ddot{O}-\ddot{Cl}-\ddot{O}:$$
        $\,-1\quad +2\quad -1 \qquad\quad -1\quad +1\quad\ 0 \qquad\quad\ 0\quad +1\quad -1$

Formal charge arguments predict that the two resonance structures with the odd electron on O are most important. This contradicts electronegativity arguments, which would predict that the less electronegative atom, Cl, would be more likely to have fewer than 8 e⁻ around it.

Bond Enthalpies

8.66 (a) $\Delta H = 3D(\text{C–Br}) + D(\text{C–H}) + D(\text{Cl–Cl}) - 3D(\text{C–Br}) - (\text{C–Cl}) - D(\text{H–Cl})$

    $= D(\text{C–H}) + D(\text{Cl–Cl}) - D(\text{C–Cl}) - D(\text{H–Cl})$

    $\Delta H = 413 + 242 - 328 - 431 = -104 \text{ kJ}$

(b) $\Delta H = 4D(\text{C–H}) + 2D(\text{C–S}) + 2D(\text{S–H}) + D(\text{C–C}) + 2D(\text{H–Br})$

    $-4D(\text{S–H}) - D(\text{C–C}) - 2D(\text{C–Br}) - 4D(\text{C–H})$

    $= 2D(\text{C–S}) + 2D(\text{H–Br}) - 2D(\text{S–H}) - 2D(\text{C–Br})$

    $\Delta H = 2(259) + 2(366) - 2(339) - 2(276) = 20 \text{ kJ}$

(c) $\Delta H = 4D(\text{N–H}) + D(\text{N–N}) + D(\text{Cl–Cl}) - 4D(\text{N–H}) - 2D(\text{N–Cl})$ (

    $= D(\text{N–N}) + D(\text{Cl–Cl}) - 2D(\text{N–Cl})$

    $\Delta H = 163 + 242 - 2(200) = 5 \text{ kJ}$

8.68 *Plan.* Draw structural formulas so bonds can be visualized. *Solve.*

(a)

    $\Delta H = 2D(\text{C–C}) + 8D(\text{C–H}) + 5D(\text{O=O}) - 6D(\text{C=O}) - 8D(\text{O–H})$

    $= 2(348) + 8(413) + 5(495) - 6(799) - 8(463) = -2023 \text{ kJ}$

(b)

    $\Delta H = D(\text{C–C}) + 5D(\text{C–H}) + D(\text{C–O}) + D(\text{O–H}) + 3D(\text{O=O}) - 4D(\text{C=O}) - 6D(\text{O–H})$

    $= 348 + 5(413) + 358 + 3(495) - 4(799) - 5(463) = -1255 \text{ KJ}$

(c) *Plan.* Use bond enthalpies to calculate ΔH for the reaction with $S_8(g)$ as a product. Then,

    $8H_2S(g) \rightarrow 8H_2(g) + S_8(g)$ ΔH

    $\underline{S_8(g) \rightarrow S_8(s)}$ $\underline{-\Delta H_f^\circ \text{ for } S_8(g)}$

    $8H_2S(g) \rightarrow 8H_2(g) + S_8(s)$ $\Delta H_{rxn} = [\Delta H - \Delta H_f^\circ S_8(g)]$

    $\Delta H = 16D(\text{S–H}) - 8(\text{H–H}) - 8(\text{S–S})$

    $= 16(339) - 8(436) - 8(266) = -192 \text{ kJ}$

    $\Delta H_{rxn} = \Delta H - \Delta H_f^\circ S_8(g) = -192 \text{ kJ} - 102.3 \text{ kJ} = -294.3 = -294 \text{ kJ}$

8.70   (a)

$$\Delta H = 4D(C-H) + D(C=C) + D(H-H) - 6D(C-H) - D(C-C)$$

$$= D(C=C) + D(H-H) - 2D(C-H) - D(C-C)$$

$$\Delta H = 614 + 436 - 2(413) - 348 = -124 \text{ kJ}$$

(b) $\Delta H° = \Delta H_f° \ C_2H_6(g) - \Delta H_f° \ C_2H_4(g) - \Delta H_f° \ H_2(g)$

$$= -84.68 - 52.30 - 0 = -136.98 \text{ kJ}$$

The values of ΔH for the reaction differ because the bond enthalpies used in part (a) are average values that can differ from one compound to another. For example, the exact enthalpy of a C–H bond in C_2H_4 is probably not equal to the enthalpy of a C–H bond in C_2H_6. Thus, reaction enthalpies calculated from average bond enthalpies are estimates. On the other hand, standard enthalpies of formation are measured quantities and should lead to accurate reaction enthalpies. The advantage of average bond enthalpies is that they can be used for reactions where no measured enthalpies of formation are available.

8.72   (a)   (i)

$$\Delta H = 2D(F-F) - 4D(C-F) = 2(155) - 4(485) = -1630 \text{ kJ}$$

(ii)

$$\Delta H = D(C≡O) + 3D(F-F) - 4D(C-F) - 2D(O-F)$$

$$= 1072 + 3(155) - 4(485) - 2(190) = -783 \text{ kJ}$$

(iii)

$$\Delta H = 2D(C=O) + 4D(F-F) - 4D(C-F) - 4D(O-F)$$

$$= 2(799) + 4(155) - 4(485) - 4(190) = -482 \text{ kJ}$$

Reaction (i) is most exothermic.

(b)   The more oxygen atoms bound to carbon, the less exothermic the reaction in this series.

Additional Exercises

8.74   (a)   Lattice energy is proportional to Q_1Q_2/d. For each of these compounds, Q_1Q_2 is the same. The anion H^- is present in each compound, but the ionic radius of the cation increases going from Be to Ba. Thus, the value of d (the cation-anion separation) increases and the ratio Q_1Q_2/d decreases. This is reflected in the decrease in lattice energy going from BeH_2 to BaH_2.

        (b)   Again, Q_1Q_2 for ZnH_2 is the same as that for the other compounds in the series and the anion is H^-. The lattice energy of ZnH_2, 2870 kJ, is closest to that of MgH_2, 2791 kJ. The ionic radius of Zn^{2+} is similar to that of Mg^{2+}.

8.76   AlN and ScN have the same charges on cations (3+) and anions (3–), so differences in lattice energy are due to differences in ion separation (d). With the same anion in both compounds, any separation difference depends on the difference in ionic radii of Al^{3+} and Sc^{3+}. The Al^{3+} cation is isoelectronic with Ne in the second row, while Sc^{3+} is isoelectronic with Ar in the third row. Size increases going down a family, so the ionic radius of Al^{3+} is smaller than that of Sc^{3+}. Lattice energy is inversely proportional to ion separation, so AlN, with the smaller separation, has the larger lattice energy.

8.77 $$E = \frac{-8.99 \times 10^9 \text{ J-m}}{C^2} \times \frac{4(1.60 \times 10^{-19} \text{ C})^2}{(1.14+1.26) \times 10^{-10} \text{ m}} = -3.836 \times 10^{-18} = -3.84 \times 10^{-18} \text{ J}$$

On a molar basis: $(-3.836 \times 10^{-18} \text{ J})(6.022 \times 10^{23}) = -2.310 \times 10^6 \text{ J} = -2310 \text{ kJ}$

Note that the absolute value of this potential energy is less than the lattice energy of CaO, 3414 kJ/mol. The difference represents the added energy of putting all the $Ca^{2+}O^{2-}$ ion pairs together in a three-dimensional array, similar to the one in Figure 8.3.

8.78 $E = Q_1Q_2/d$; $k = 8.99 \times 10^9$ J-m/coul^2

        (a) Na^+, Br^-: $$E = \frac{-8.99 \times 10^9 \text{ J-m}}{C^2} \times \frac{(1 \times 1.60 \times 10^{-19} \text{ C})^2}{(1.16+1.82) \times 10^{-10} \text{ m}} = -7.7230 \times 10^{-19}$$
$$= -7.72 \times 10^{-19} \text{ J}$$

            The sign of E is negative because one of the interacting ions is an anion; this is an attractive interaction.

            On a molar basis: $-7.723 \times 10^{-19} \times 6.022 \times 10^{23} = -4.65 \times 10^5 \text{ J} = -465 \text{ kJ}$

        (b) Rb^+, Br^-: $$E = \frac{-8.99 \times 10^9 \text{ J-m}}{C^2} \times \frac{(1 \times 1.60 \times 10^{-19} \text{ C})^2}{(1.66+1.82) \times 10^{-10} \text{ m}} = -6.61 \times 10^{-19} \text{ J}$$

            On a molar basis: $-3.98 \times 10^5 \text{ J} = -398 \text{ kJ}$

        (c) Sr^{2+}, S^{2-}: $$E = \frac{-8.99 \times 10^9 \text{ J-m}}{C^2} \times \frac{(2 \times 1.60 \times 10^{-19} \text{ C})^2}{(1.32+1.70) \times 10^{-10} \text{ m}} = -3.05 \times 10^{-18} \text{ J}$$

            On a molar basis: $-1.84 \times 10^6 \text{ J} = -1.84 \times 10^3 \text{ kJ}$

8.80   Molecule (b) H_2S and ion (c) NO_2^- contain polar bonds. The atoms that form the bonds (H–S) and N–O) have different electronegativity values.

8.81   (a) $2NaAlH_4(s) \rightarrow 2NaH(s) + 2Al(s) + 3H_2(g)$

(b)   Hydrogen is the only nonmetal in $NaAlH_4$, so we expect it to be most electronegative. (The position of H on the periodic table is problematic. Its electronegativity does not fit the typical trend for Gp 1A elements.) For the two metals, Na and Al, electronegativity increases moving up and to the right on the periodic table, so Al is more electronegative. The least electronegative element in the compound is Na.

(c)   Covalent bonds hold polyatomic anions together; elements involved in covalent bonding have smaller electronegativity differences than those that are involved in ionic bonds. Possible covalent bonds in $NaAlH_4$ are Na–H and Al–H. Al and H have a smaller electronegativity difference than Na and H and are more likely to form covalent bonds. The anion has an overall 1– charge, so it can be thought of as four hydride ions and one Al^{3+} ion. The formula is AlH_4^-. For the purpose of counting valence electrons, assume neutral atoms.

$8\,e^-$, $4\,e^-$ pairs

$$
\left[
\begin{array}{c}
H \\
| \\
H - Al - H \\
| \\
H
\end{array}
\right]^{-}
$$

8.82   (a)   B–O. The most polar bond will be formed by the two elements with the greatest difference in electronegativity. Since electronegativity increases moving right and up on the periodic chart, the possibilities are B–O and Te–O. These two bonds are likely to have similar electronegativitiy differences (3 columns apart vs. 3 rows apart). Values from Figure 8.6 confirm the similarity, and show that B–O is slightly more polar.

(b)   Te–I. Both are in the fifth row of the periodic chart and have the two largest covalent radii among this group of elements.

(c) TeI_2. Te needs to participate in two covalent bonds to satisfy the octet rule, and each I atom needs to participate in one bond, so by forming a TeI_2 molecule, the octet rule can be satisfied for all three atoms.

$$
\ddot{\underset{\cdot\cdot}{I}} - \ddot{\underset{\cdot\cdot}{Te}} - \ddot{\underset{\cdot\cdot}{I}}
$$

(d) B_2O_3. Although this is probably not a purely ionic compound, it can be understood in terms of gaining and losing electrons to achieve a noble-gas configuration. If each B atom were to lose $3\,e^-$ and each O atom were to gain $2\,e^-$, charge balance and the octet rule would be satisfied.

P_2O_3. Each P atom needs to share $3\,e^-$ and each O atom $2\,e^-$ to achieve an octet. Although the correct number of electrons seem to be available, a correct Lewis structure is difficult to imagine. In fact, phosphorus (III) oxide exists as P_4O_6 rather than P_2O_3 (Chapter 22).

8.84   To calculate empirical formulas, assume 100 g of sample.

(a) $\dfrac{47.7 \text{ g Cr}}{51.9961 \text{g / mol}} = 0.9174$ mol Cr; $0.9174/0.9174 = 1$ Cr

$$\frac{52.3 \text{ g F}}{18.9984 \text{ g/mol}} = 2.753 \text{ mol F}; \; 2.753/0.9174 = 3 \text{ F}$$

The empirical formula of compound 1 is CrF_3.

(b) $\dfrac{45.7 \text{ g Mo}}{95.94 \text{ g/mol}} = 0.4763 \text{ mol Mo}; \; 0.4763/0.4763 = 1 \text{ Mo}$

$\dfrac{54.3 \text{ g F}}{18.9984 \text{ g/mol}} = 2.858 \text{ mol F}; \; 2858/0.4763 = 6 \text{ F}$

The empirical formula of Compound 2 is MoF_6.

(c) The high-melting green powder is ionic and contains the group 6B metal in the lower oxidation state, Cr(III). The green powder is CrF_3, chromium(III) fluoride.

The colorless liquid is molecular and contains the group 6B metal in the higher oxidation state, Mo(VI). The colorless liquid is MoF_6, molybdenum hexafluoride.

8.86 Use the method detailed in Section 8.5, *A Closer Look*, to estimate partial charges from electronegativity values. From Figure 8.6, the electronegativity of Br is 2.8 and of Cl is 3.0.

Br has $2.8/(3.0 + 2.8) = 0.48$ of the charge of the bonding e^- pair.

Cl has $3.0/(3.0 + 2.8) = 0.52$ of the charge of the bonding e^- pair.

This amounts to $0.52 \times 2e = 1.04e$ on Cl or $0.04e$ more than a neutral Cl atom. This implies a -0.04 charge on Cl and $+0.04$ charge on Br.

From Figure 7.7, the covalent radius of Br is 1.14 Å and of Cl is 0.99 Å. The Br–Cl separation is 2.13 Å.

$$\mu = Qr = 0.04e \times \frac{1.60 \times 10^{-19} \text{ C}}{e} \times 2.13 \text{ Å} \times \frac{1 \times 10^{-10} \text{ m}}{\text{Å}} \times \frac{1 D}{3.34 \times 10^{-30} \text{ C-m}} = 0.41 \text{ D}$$

Clearly, this method is approximate. The estimated dipole moment of 0.41 D is within 28% of the measured value of 0.57 D.

8.87 I_3^- has a Lewis structure with an expanded octet of electrons around the central I.

$$:\ddot{\underset{..}{I}} - \ddot{\underset{..}{I}} - \ddot{\underset{..}{I}}:$$

F cannot accommodate an expanded octet because it is too small and has no available d orbitals in its valence shell.

8.89 (a) 14e^-, 7 e^- pairs   32 e^-, 16 e^- pairs

$$\left[:\ddot{\underset{..}{Cl}} - \ddot{\underset{..}{O}}:\right]^-$$

FC on Cl = 7 – [6 + 1/2(2)] = 0

$$\left[\begin{array}{c} :\ddot{O}: \\ | \\ :\ddot{\underset{..}{O}} - \underset{|}{Cl} - \ddot{\underset{..}{O}}: \\ :\ddot{\underset{..}{O}}: \end{array} \right]^-$$

FC on Cl = 7 – [0 + 1/2(8)] = +3

(b) The oxidation number of Cl is +1 in ClO^- and +7 in ClO_4^-.

(c)   The definition of formal charge assumes that all bonding pairs of electrons are equally shared by the two bonded atoms, that all bonds are purely covalent. The definition of oxidation number assumes that the more electronegative element in the bond gets all of the bonding electrons, that the bonds are purely ionic. These two definitions represent the two extremes of how electron density is distributed between bonded atoms.

In ClO^- and ClO_4^-, Cl is the less electronegative element, so the oxidation numbers have a higher positive value than the formal charges. The true description of the electron density distribution is somewhere between the extremes indicated by formal charge and oxidation number.

(d)   Oxidizing power is the tendency of a substance to be reduced, to gain electrons. Oxidation numbers show the maximum electron deficiency (or excess) of a substance. The higher the oxidation number of the central atom in an oxyanion, the greater its electron deficiency and oxidizing power. Formal charges can also be used to show oxidizing (or reducing) power, but trends are less obvious because the magnitudes are smaller.

8.90   (a)

$$:N\equiv N-\ddot{\underset{..}{O}}: \longleftrightarrow \underset{..}{:}\ddot{N}-N\equiv O: \longleftrightarrow :\ddot{N}=N=\ddot{O}:$$

       0   +1   −1   −2 +1 −1   −1 +1   0

In the leftmost structure, the more electronegative O atom has the negative formal charge, so this structure is likely to be most important.

(b)   In general, the more shared pairs of electrons between two atoms, the shorter the bond, and vice versa. That the N–N bond length in N_2O is slightly longer than the typical N≡N indicates that the middle and right resonance structures where the N atoms share less than three electron pairs are contributors to the true structure. That the N–O bond length is slightly shorter than a typical N=O indicates that the middle structure, where N and O share more than two electron pairs, does contribute to the true structure. This physical data indicates that while formal charge can be used to predict which resonance form will be more important to the observed structure, the influence of minor contributors on the true structure cannot be ignored.

8.92   (a) $\Delta H = 5D(C–H) + D(C–C) + D(C–O) + D(O–H) – 6D(C–H) – 2D(C–O)$

          $= D(C–C) + D(O–H) – D(C–H) – D(C–O)$

          $= 348 \text{ kJ} + 463 \text{ kJ} – 413 \text{ kJ} – 358 \text{ kJ}$

    $\Delta H = +40 \text{ kJ}$; ethanol has the lower enthalpy

(b) $\Delta H = 4D(C–H) + D(C–C) + 2D(C–O) – 4D(C–H) – D(C–C) – D(C=O)$

          $= 2D(C–O) – D(C=O)$

          $= 2(358 \text{ kJ}) – 799 \text{ kJ}$

    $\Delta H = –83 \text{ kJ}$; acetaldehyde has the lower enthalpy

(c) $\Delta H = 8D(C–H) + 4D(C–C) + D(C=C) - 8D(C–H) - 2D(C–C) - 2D(C=C)$

      $= 2D(C–C) - D(C=C)$

      $= 2(348 \text{ kJ}) - 614 \text{ kJ}$

   $\Delta H = +82 \text{ kJ}$; cyclopentene has the lower enthalpy

(d) $\Delta H = 3D(C–H) + D(C–N) + D(C \equiv N) - 3D(C–H) - D(C–C) - D(C \equiv N)$

      $= D(C–N) - D(C–C)$

      $= 293 \text{ kJ} - 348 \text{ kJ}$

   $\Delta H = -55 \text{ kJ}$; acetonitrile has the lower enthalpy

8.93   (a)

$\Delta H = 20D(C–H) + 8D(C–C) + 12D(C–O) + 24D(O–N) + 12D(N=O)$

     $- [6D(N \equiv N) + 24D(C=O) + 20D(H–O) + D(O=O)]$

$\Delta H = 20(413) + 8(348) + 12(358) + 24(201) + 12(607)$

     $- [6(941) + 24(799) + 20(463) + 495]$

   $= -7129 \text{ kJ}$

$$1.00 \text{ g } C_3H_5N_3O_9 \times \frac{1 \text{ mol } C_3H_5N_3O_9}{227.1 \text{ g } C_3H_5N_3O_9} \times \frac{-7129 \text{ kJ}}{4 \text{ mol } C_3H_5N_3O_9} = 7.85 \text{ kJ/g } C_3H_5N_3O_9$$

(b) $4C_7H_5N_3O_6(s) \rightarrow 6N_2(g) + 7CO_2(g) + 10H_2O(g) + 21C(s)$

8.95

When comparing the same pair of bonded atoms (C–N vs. C=N vs. C≡N), the shorter the bond the greater the bond energy, but the two quantities are not necessarily directly proportional. The plot clearly shows that there are no simple length/strength correlations for single bonds alone, double bonds alone, triple bonds alone, or among different pairs of bonded atoms (all C–C bonds vs. all C–N bonds, etc.).

Integrative Exercises

8.98   (a) $Sr(s) \rightarrow Sr(g)$ ΔH_f° Sr(g) [ΔH_{sub}° Sr(s)]

                $Sr(g) \rightarrow Sr^+(g) + 1\,e^-$ I_1 Sr

                $Sr^+(g) \rightarrow Sr^{2+}(g) + 1\,e^-$ I_2 Sr

                $Cl_2(g) \rightarrow 2Cl(g)$ $2\,\Delta H_f^{\circ}$ Cl(g) [$D(Cl_2)$]

                $2Cl(g) + 2\,e^- \rightarrow 2Cl^-(g)$ $2E_1$ Cl

                $\underline{SrCl_2(s) \rightarrow Sr(s) + Cl_2(g)}$ $\underline{-\Delta H_f^{\circ} \; SrCl_2}$

                $SrCl_2(s) \rightarrow Sr^{2+}(g) + 2Cl^-(g)$ ΔH_{latt}

  (b) $\Delta H_f^{\circ}\, SrCl_2(s) = \Delta H_f^{\circ}\, Sr(g) + I_1(Sr) + I_2(Sr) + 2\Delta H_f^{\circ}\, Cl(g) + 2E(Cl) - \Delta H_{latt}\, SrCl_2$

          $\Delta H_f^{\circ}\, SrCl_2(s) = 164.4\,\text{kJ} + 549\,\text{kJ} + 1064\,\text{kJ} + 2(121.7)\,\text{kJ} + 2(-349)\,\text{kJ} - 2127\,\text{kJ}$

                    $= -804\,\text{kJ}$

8.99   The pathway to the formation of K_2O can be written:

        $2K(s) \rightarrow 2K(g)$ $2\Delta H_f^{\circ}$ K(g)

        $2K(g) \rightarrow 2K^+(g) + 2\,e^-$ $2\,I_1(K)$

        $1/2\,O_2(g) \rightarrow O(g)$ ΔH_f° O(g)

        $O(g) + 1\,e^- \rightarrow O^-(g)$ $E_1(O)$

        $O^-(g) + 1\,e^- \rightarrow O^{2-}(g)$ $E_2(O)$

        $\underline{2K^+(g) + O^{2-}(g) \rightarrow K_2O(s)}$ $\underline{-\Delta H_{latt}\, K_2O(s)}$

        $2K(s) + 1/2\,O_2(g) \rightarrow K_2O(s)$ ΔH_f° $K_2O(s)$

        $\Delta H_f^{\circ}\, K_2O(s) = 2\Delta H_f^{\circ}\, K(g) + 2\,I_1(K) + \Delta H_f^{\circ}\, O(g) + E_1(O) + E_2(O) - \Delta H_{latt}\, K_2O(s)$

        $E_2(O) = \Delta H_f^{\circ}\, K_2O(s) + \Delta H_{latt}\, K_2O(s) - 2\Delta H_f^{\circ}\, K(g) - 2\,I_t(K) - \Delta H_f^{\circ}\, O(g) - E_1(O)$

        $E_2(O) = -363.2\,\text{kJ} + 2238\,\text{kJ} - 2(89.99)\,\text{kJ} - 2(419)\,\text{kJ} - 247.5\,\text{kJ} - (-141)\,\text{kJ}$

             $= +750\,\text{kJ}$

8.101   (a)   Even though Cl has the greater (more negative) electron affinity, F has a much larger ionization energy, so the electronegativity of F is greater.

           F: k(IE–EA) = k(1681 – (–328)) = k(2009)

           Cl: k(IE–EA) = k(1251 – (–349)) = k(1600)

(b) Electronegativiy is the ability of an atom in a molecule to attract electrons to itself. It can be thought of as the ability to hold its own electrons (as measured by ionization energy) and the capacity to attract the electrons of other atoms (as measured by electron affinity). Thus, both properties are relevant to the concept of electronegativity.

(c) EN = k(IE - EA). For F: 4.0 = k(2009), k = 4.0/2009 = 2.0×10^{-3}

(d) Cl: EN = 2.0×10^{-3} (1600) = 3.2

O: EN = 2.0×10^{-3} (1314 – (–141)) = 2.9

These values do not follow the trend on Figure 8.6. The Pauling scale on the figure shows O to be second only to F in electronegativity, more electronegative than Cl. The simple definition EN = k(IE – EA) that employs thermochemical properties of isolated, gas phase atoms does not take into account the complex bonding environment of molecules.

8.102   (a)   Assume 100 g.

$$14.52 \text{ g C } \times \frac{1 \text{ mol}}{12.011 \text{ g C}} = 1.209 \text{ mol C}; 1.209 / 1.209 = 1$$

$$1.83 \text{ g H } \times \frac{1 \text{ mol}}{1.008 \text{ g H}} = 1.816 \text{ mol H}; 1.816 / 1.209 = 1.5$$

$$64.30 \text{ g Cl } \times \frac{1 \text{ mol}}{35.453 \text{ g Cl}} = 1.814 \text{ mol Cl}; 1.814 / 1.209 = 1.5$$

$$19.35 \text{ g O } \times \frac{1 \text{ mol}}{15.9994 \text{ g O}} = 1.209 \text{ mol O}; 1.209 / 1.209 = 1.0$$

Multiplying by 2 to obtain an integer ratio, the empirical formula is $C_2H_3Cl_3O_2$.

(b) The empirical formula weight is 2(12.0) + 3(1.0) + 3(35.5) + 2(16) = 165.5. The empirical formula is the molecular formula.

(c) 44 e^-, 22 e^- pairs

8.104   (a) C_2H_2: 10 e^-, 5 e^- pair N_2: 10 e^-, 5 e^- pair

H—C≡C—H   :N≡N:

(b) N_2 is an extremely stable, unreactive compound. Under appropriate conditions, it can be either oxidized (Section 22.7) or reduced (Sections 14.7 and 15.2). C_2H_2 is a reactive gas, used in combination with O_2 for welding and as starting material for organic synthesis (Section 25.3).

(c) $2N_2(g) + 5O_2(g) \rightarrow 2N_2O_5(g)$

          $2C_2H_2(g) + 5O_2(g) \rightarrow 4CO_2(g) + 2H_2O(g)$

(d) $\Delta H^{\circ}_{rxn} (N_2) = 2\Delta H^{\circ}_f N_2O_5(g) - 2\Delta H^{\circ}_f N_2(g) - 5\Delta H^{\circ}_f O_2(g)$

              $= 2(11.30) - 2(0) - 5(0) = 22.60 \text{ kJ}$

$\Delta H^{\circ}_{ox} = 11.30 \text{ kJ/mol } N_2$

$\Delta H^{\circ}_{rxn} (C_2H_2) = 4\Delta H^{\circ}_f CO_2(g) + 2\Delta H^{\circ}_f H_2O(g) - 2\Delta H^{\circ}_f C_2H_2(g) - 5\Delta H^{\circ}_f O_2(g)$

               $= 4(-393.5 \text{ kJ}) + 2(-241.82 \text{ kJ}) - 2(226.7 \text{ kJ}) - 5(0)$

               $= -2511.0 \text{ kJ}$

$\Delta H^{\circ}_{ox} (C_2H_2) = -1255.5 \text{ kJ/mol } C_2H_2$

The oxidation of C_2H_2 is highly exothermic, which means that the energy state of the combined products is much lower than that of the reactants. The reaction is "downhill" in an energy sense, and occurs readily. The oxidation of N_2 is mildly endothermic (energy of products higher than reactants) and the reaction does not readily occur. This is in agreement with the general reactivities from part (b).

Referring to bond enthalpies in Table 8.4, when the C–H bonds are taken into account, even more energy is required for bond breaking in the oxidation of C_2H_2 than in the oxidation of N_2. The difference seems to be in the enthalpies of formation of the products. $CO_2(g)$ and $H_2O(g)$ have extremely exothermic ΔH°_f values, which cause the oxidation of C_2H_2 to be energetically favorable. $N_2O_5(g)$ has an endothermic ΔH°_f value, which causes the oxidation of N_2 to be energetically unfavorable.

8.105   (a)   Assume 100 g of compound

$$69.6 \text{ g S} \times \frac{1 \text{ mol S}}{32.07 \text{ g}} = 2.17 \text{ mol S}$$

$$30.4 \text{ g N} \times \frac{1 \text{ mol N}}{14.01 \text{ g}} = 2.17 \text{ mol N}$$

S and N are present in a 1:1 mol ratio, so the empirical formula is SN. The empirical formula weight is 46. MM/FW = 184.3/46 = 4 The molecular formula is S_4N_4.

(b)   44 e⁻, 22 e⁻ pairs. Because of its small radius, N is unlikely to have an expanded octet. Begin with alternating S and N atoms in the ring. Try to satisfy the octet rule with single bonds and lone pairs. At least two double bonds somewhere in the ring are required.

These structures carry formal charges on S and N atoms as shown. Other possibilities include:

These structures have zero formal charges on all atoms and are likely to contribute to the true structure. Note that the S atoms that are shown with two double bonds are not necessarily linear, because S has an expanded octet. Other resonance structures with four double bonds are.

In either resonance structure, the two 'extra' electron pairs can be placed on any pair of S atoms in ring, leading to a total of 10 resonance structures. The sulfur atoms alternately carry formal charges of +1 and –1. Without further structural information, it is not possible to eliminate any of the above structures. Clearly, the S_4N_4 molecule stretches the limits of the Lewis model of chemical bonding.

(c) Each resonance structure has 8 total bonds and more than 8 but less than 16 bonding e$^-$ pairs, so an "average" bond will be intermediate between a S-N single and double bond. We estimate an average S-N single bond length to be 1.77 Å (sum of bonding atomic radii from Figure 7.7). We do not have a direct value for a S-N double bond length. Comparing double and single bond lengths for C–C (1.34 Å, 1.54 Å), N–N (1.24 Å, 1.47 Å) and O–O (1.21 Å, 1.48 Å) bonds from Table 8.5, we see that, on average, a double bond is approximately 0.23 Å shorter than a single bond. Applying this difference to the S–N single bond length, we estimate the S–N double bond length as 1.54 Å. Finally, the intermediate S–N bond length in S_4N_4 should be between these two values, approximately 1.60–165 Å. (The measured bond length is 1.62 Å.)

(d) $S_4N_4 \rightarrow 4S(g) + 4N(g)$

$\Delta H = 4\Delta H_f^\circ \, S(g) + 4\Delta H_f^\circ \, N(g) - \Delta H_f^\circ \, S_4N_4$

$\Delta H = 4(222.8 \text{ kJ}) + 4(472.7 \text{ kJ}) - 480 \text{ kJ} = 2302 \text{ kJ}$

This energy, 2302 kJ, represents the dissociation of 8 S–N bonds in the molecule; the average dissociation energy of one S–N bond in S_4N_4 is then 2302 kJ/8 bonds = 287.8 kJ.

8.106 (a) Yes. In the structure shown in the exercise, each P atom needs 1 unshared pair to complete its octet. This is confirmed by noting that only 6 of the 10 valence e$^-$ pairs are bonding pairs.

(b) There are six P–P bonds in P_4.

(c) $20\,e^-$, $10\,e^-$ pr

$$\ddot{P}\!=\!\!P\!=\!\!P\!=\!\ddot{P}$$

In this Lewis structure, the octet rule is satisfied for all atoms. However, it requires P=P, which are uncommon because P has a covalent radius that is too large to accommodate the side-to-side π overlap of parallel p orbitals required for double bond formation.

(d)   From left to right, the formal charges are on the P atoms in the linear structure are -1, $+1$, $+1$, -1. In the tetrahedral structure, all formal charges are zero. Clearly the linear structure does not minimize formal charge and is probably less stable than the tetrahedral structure, owing to the difficulty of P=P bond formation (see above).

8.107   (a) $C_6H_6(g) \rightarrow 6H(g) + 6C(g)$

$$\Delta H^\circ = 6\Delta H^\circ_f \ H(g) + 6\Delta H^\circ_f \ C(g) - \Delta H^\circ_f \ C_6H_6(g)$$

$\Delta H^\circ = 6(217.94)\ kJ + 6(718.4)\ kJ - 82.9\ kJ = 5535\ kJ$

(b) $C_6H_6(g) \rightarrow 6CH(g)$

(c)

$C_6H_6(g) \rightarrow 6H(g) + 6C(g)$	ΔH°	$5535\ kJ$
$6H(g) + 6C(g) \rightarrow 6CH(g)$	$-6D(C\!-\!H)$	$-6(413)\ kJ$
$C_6H_6(g) \rightarrow 6CH(g)$		$3057\ kJ$

3057 kJ is the energy required to break the six C–C bonds in $C_6H_6(g)$. The average bond dissociation energy for one carbon-carbon bond in $C_6H_6(g)$ is

$$\frac{3057\ kJ}{6\ C-C\ bonds} = 509.5\ kJ.$$

(d)   The value of 509.5 kJ is between the average value for a C–C single bond (348 kJ) and a C=C double bond (614 kJ). It is somewhat greater than the average of these two values, indicating that the carbon-carbon bond in benzene is a bit stronger than we might expect.

8.108   (a) $Br_2(l) \rightarrow 2Br(g)$ $\Delta H^\circ = 2\Delta H^\circ_f \ Br(g) = 2(111.8)\ kJ = 223.6\ kJ$

(b) $CCl_4(l) \rightarrow C(g) + 4Cl(g)$

$$\Delta H^\circ = \Delta H^\circ_f \ C(g) + 4\Delta H^\circ_f \ Cl(g) - \Delta H^\circ_f \ CCl_4(l)$$

$$= 718.4\ kJ + 4(121.7)\ kJ - (-139.3)\ kJ = 1344.5$$

$$\frac{1344.5\ kJ}{4\ C-Cl\ bonds} = 336.1\ kJ$$

(c)

$$H_2O_2(l) \rightarrow 2H(g) + 2O(g)$$
$$2H(g) + 2O(g) \rightarrow 2OH(g)$$
$$\overline{H_2O_2(l) \rightarrow 2OH(g)}$$

$$D(O-O)(l) = 2\Delta H_f^\circ H(g) + 2\Delta H_f^\circ O(g) - \Delta H_f^\circ H_2O_2(l) - 2D(O-H)(g)$$

$$= 2(217.94)\ kJ + 2(247.5)\ kJ - (-187.8)\ kJ - 2(463)\ kJ$$

$$= 193\ kJ$$

(d)   The data are listed below.

bond	D gas kJ/mol	D liquid kJ/mol
Br–Br	193	223.6
C–Cl	328	336.1
O–O	146	192.7

Breaking bonds in the liquid requires more energy than breaking bonds in the gas phase. For simple molecules, bond dissociation from the liquid phase can be thought of in two steps:

molecule (l) → molecule (g)

molecule (g) → atoms (g)

The first step is evaporation or vaporization of the liquid and the second is bond dissociation in the gas phase. Average bond enthalpy in the liquid phase is then the sum of the enthalpy of vaporization for the molecule and the gas phase bond dissociation enthalpies, divided by the number of bonds dissociated. This is greater than the gas phase bond dissociation enthalpy owing to the contribution from the enthalpy of vaporization.

9 Molecular Geometry and Bonding Theories

Visualizing Concepts

9.2   (a)   120°

       (b)   If the blue balloon expands, the angle between red and green balloons decreases.

       (c)   Nonbonding (lone) electron pairs exert greater repulsive forces than bonding pairs, resulting in compression of adjacent bond angles.

9.4   (a)   4 e^- domains

       (b)   The molecule has a non-zero dipole moment, because the C–H and C–F bond dipoles do not cancel each other.

       (c)   The dipole moment vector bisects the F–C–F and H–C–H angles, with the negative end of the vector toward the F atoms.

9.6   (a)   90° angles are characteristic of hybrids with a d atomic orbital contribution. This pair of orbitals could be sp^3d or sp^3d^2.

       (b)   Angles of 109.5° are characteristic of sp^3 hybrid orbitals only.

       (c)   Angles of 120° can be formed by sp^2 hybrids or sp^3d hybrids.

9.8   (a)   Recall that π bonds require p atomic orbitals, so the maximum hybridization of a C atom involved in a double bond is sp^2 and in a triple bond is sp. There are 6 C atoms in the molecule. Starting on the left, the hybridizations are: sp^2, sp^2, sp^3, sp, sp, sp^3.

       (b)   All single bonds are σ bonds. Double and triple bonds each contain 1 σ bond. This molecule has 8 C–H σ bonds and 5 C–C σ bonds, for a total of 13 σ bonds.

       (c)   Double bonds have 1 π bond and triple bonds have 2 π bonds. This molecule has a total of 3 π bonds.

9.10   (a)   The diagram has five electrons in MOs formed by 2p atomic orbitals. C has two 2p electrons, so X must have three 2p electrons. X is N.

       (b)   The molecule has an unpaired electron, so it is paramagnetic.

       (c)   Atom X is N, which is more electronegative than C. The atomic orbitals of the more electronegative N are slightly lower in energy than those of C. The lower energy π_{2p} bonding molecular orbitals will have a greater contribution from the lower energy N atomic orbitals. (Higher energy π_{2p}^* MOs will have a greater contribution from higher energy C atomic orbitals.)

Molecular Shapes; the VSEPR Model

9.12   (a)   In a symmetrical tetrahedron, the four bond angles are equal to each other, with values of 109.5°. The H–C–H angles in CH_4 and the O–Cl–O angles in ClO_4^- will have values close to 109.5°.

       (b)   'Planar' molecules are flat, so trigonal planar BF_3 is flat. In the trigonal pyramidal NH_3 molecule, the central N atom sits out of the plane of the three H atoms; this molecule is not flat.

9.14   (a)   The number of electron domains in a molecule or ion is the number of bonds (double and triple bonds count as one domain) **plus** the number of nonbonding (lone) electron pairs.

       (b)   A *bonding electron domain* is a region between two bonded atoms that contains one or more pairs of bonding electrons. A *nonbonding electron domain* is localized on a single atom and contains one pair of nonbonding electrons (a lone pair).

9.16   *Analyze/Plan.* See Table 9.1. *Solve.*

       (a)   trigonal planar   (b)   tetrahedral

       (c)   trigonal bipyramidal   (d)   octahedral

9.18   If the electron-domain geometry is trigonal bipyramidal, there are five total electron domains around the central atom. An AB_3 molecule has three bonding domains, so there must be two nonbonding domains on A.

9.20

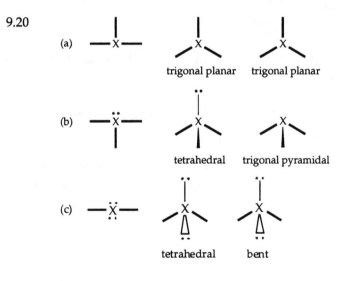

9.22   bent (b), linear (l), octahedral (oh), seesaw (ss) square pyramidal (sp), square planar (spl), tetrahedral (td), trigonal bipyramidal (tbp), trigonal planar (tr), trigonal pyramidal (tp), T-shaped (T)

	Molecule or ion	Valence electrons	Lewis structure		Electron-domain geometry	Molecular geometry
(a)	PF_3	26			td	tp
(b)	CH_3^+	6			tr	tr
(c)	BrF_3	28			tbp	T
(d)	ClO_4^-	32			td	td
(e)	XeF_2	22			tbp	l
(f)	BrO_2^-	20			td	b

*More than one resonance structure is possible. All equivalent resonance structures predict the same molecular geometry.

9.24   (a)   Electron-domain geometries: i, octahedral; ii, tedrahedral; iii, trigonal bipyramial

         (b)   nonbonding electron domains: i, 2; ii, 0; iii, 1

         (c)   S or Se. Shape iii has five electron domains, so A must be in or below the third row of the periodic table. This eliminates Be and C. Assuming each F atom has three nonbonding electron domains and forms only single bonds with A, A must have six valence electrons to produce these electron-domain and molecular geometries.

         (d)   Xe. (See Table 9.3) Assuming F behaves typically, A must be in or below the third row and have eight valence electrons. Only Xe fits this description. (Noble gas elements above Xe have not been shown to form molecules of the type AF_4. See Section 7.8.)

9.26   (a)   1 – 109°, 2 – 120°   (b)   3 – 109°, 4 – 120°

         (c)   5 – 109°, 6 – 109°   (d)   7 – 180°, 8 – 109°

9.28

$$\left[\text{H}-\overset{\cdot\cdot}{\underset{\cdot\cdot}{\text{N}}}-\text{H}\right]^{-} \quad \text{H}-\overset{\cdot\cdot}{\text{N}}-\text{H} \quad \left[\text{H}-\overset{\text{H}}{\underset{\text{H}}{\text{N}}}-\text{H}\right]^{+}$$
$$\qquad\qquad\qquad\qquad \text{H}$$

Each species has four electron domains around the N atom, but the number of nonbonding domains decreases from two to zero, going from NH_2^- to NH_4^+. Since nonbonding domains exert greater repulsive forces on adjacent domains, the bond angles expand as the number of nonbonding domains decreases.

9.30   (a) ClO_2^- 20 e$^-$, 10 e$^-$ pr

$$\left[\overset{\cdot\cdot}{\underset{\cdot\cdot}{\text{O}}}-\overset{\cdot\cdot}{\underset{\cdot\cdot}{\text{Cl}}}-\overset{\cdot\cdot}{\underset{\cdot\cdot}{\text{O}}}\right]^{-}$$

4 e$^-$ domains around Cl, tetrahedral e$^-$ domain geometry,

bent molecular geometry bond angle $\leq 109.5°$

NO_2^- 18 e$^-$, 9 e$^-$ pr

$$\left[\overset{\cdot\cdot}{\text{O}}=\overset{\cdot\cdot}{\text{N}}-\overset{\cdot\cdot}{\underset{\cdot\cdot}{\text{O}}}\right]^{-} \longleftrightarrow \left[\overset{\cdot\cdot}{\underset{\cdot\cdot}{\text{O}}}-\overset{\cdot\cdot}{\text{N}}=\overset{\cdot\cdot}{\text{O}}\right]^{-}$$

3 e$^-$ domains about N (both resonance structures), trigonal planar e$^-$ domain geometry bent molecular geometry bond angle $\leq 120°$

Both molecular geometries are described as "bent" because both molecules have two nonlinear bonding electron domains. The bond angles (the angle between the two bonding domains) in the two ions are different because the total number of electron domains, and thus the electron domain geometries are different.

(b) XeF_2 22 e$^-$, 11 e$^-$ pr

$$\overset{\cdot\cdot}{\underset{\cdot\cdot}{\text{F}}}-\overset{\cdot\cdot}{\text{Xe}}-\overset{\cdot\cdot}{\underset{\cdot\cdot}{\text{F}}}$$

5 e$^-$ domains around Xe, trigonal bipyramidal e$^-$ domain geometry, linear molecular geometry

The question here really is: why do the three nonbonding domains all occupy the equatorial plane of the trigonal bipyramid? In a tbp, there are several different kinds of repulsions, bonding domain-bonding domain (bd-bd), bonding domain-nonbonding domain (bd-nd), and nonbonding domain-nonbonding domain (nd-nd). Each of these can have 90°, 120°, or 180° geometry. Since nonbonding domains occupy more space, 90° nd-nd repulsions are most significant and least desirable. The various electron domains arrange themselves to minimize these 90° nd-nd interactions. The arrangement shown above has no 90° nd-nd repulsions. An arrangement with one or two nonbonding domains in axial positions would lead to at least two 90° nd-nd repulsions, a less stable situation. (To convince yourself, tabulate the number and kinds of repulsions for each possible tbp arrangement of 2bd's and 3nd's.)

Polarity of Polyatomic Molecules

9.32   (a)   If PH_3 were planar, the P–H bond dipoles would cancel, and the molecule would be nonpolar. Since PH_3 is polar, the 3 P–H bond dipoles do not cancel, and the molecule can't be planar.

       (b) O_3, 18 e⁻, 9 e⁻ pr;   :Ö=Ö—Ö: ⟷ :Ö—Ö=Ö:

           trigonal planar e⁻ domain geometry
           bent molecular geometry

           Since all atoms are the same, the individual bond dipoles are zero. However, the central O atom has a lone pair of electrons which cause an unequal electron (and charge) distribution in the molecule. The unopposed lone pair is the source of the dipole moment in O_3.

9.34   (a)   For a molecule with polar bonds to be nonpolar, the polar bonds must be (symmetrically) arranged so that the bond dipoles cancel. In most cases, nonbonding e⁻ domains must be absent from the central atom. Square planar structures may not meet the second condition.

       (b) AB_2: linear e⁻ domain geometry (edg), linear molecular geometry (mg), trigonal bipyramidal edg, linear mg

           AB_3: trigonal planar edg, trigonal planar mg

           AB_4: tetrahedral edg, tetrahedral mg; octahedral edg, square planar mg

9.36   (a)   Nonpolar, in a symmetrical tetrahedral structure (Figure 9.1) the bond dipoles cancel.

       (b)   Polar, there is an unequal charge distribution due to the nonbonded electron pair on N.

           H—N(H)—H (with lone pair on N)

       (c)   Polar, there is an unequal charge distribution due to the nonbonded electron pair on S.

           F, F, F, F on S:

       (d)   Nonpolar, the bond dipoles and the nonbonded electron pairs cancel.

           F, F, F, F on Xe

(e)   Polar, the C–H and C–Br bond dipoles are not equal and do not cancel.

(f)   Nonpolar, in a symmetrical trigonal planar structure, the bond dipoles cancel.

9.38   Each C–Cl bond is polar. The question is whether the vector sum of the C–Cl bond dipoles in each molecule will be nonzero. In the *ortho* and *meta* isomers, the C–Cl vectors are at 60° and 120° angles, respectively, and their resultant dipole moments are nonzero. In the *para* isomer, the C–Cl vectors are opposite, at an angle of 180°, with a resultant dipole moment of zero. The *ortho* and *meta* isomers are polar, the *para* isomer is nonpolar.

Orbital Overlap; Hybrid Orbitals

9.40   (a)

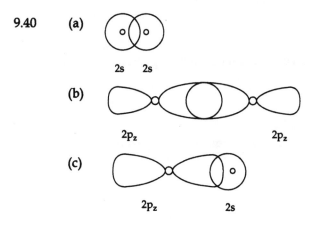

9.42   By analogy to the H_2 molecule shown in Figure 9.15, as the distance between the atoms decreases, the overlap between their valence orbitals increases. According to Figure 7.7, the bonding atomic radius for the halogens is on the order F < Cl < Br < I. The order of bond lengths in the molecules is I–F < I–Cl < I–Br < I–I. If the extent of orbital overlap increases as the distance between atoms decreases, I–F has the greatest overlap and I_2 the least. The order for extent of orbital overlap is I–I < I–Br < I–Cl < I–F.

9.44   In order for atomic orbitals to mix or hybridize, they must have the same principal quantum number. In each principal quantum level, there are a maximum of three *p* orbitals. Any hybrid orbital can have contribution from a maximum of three *p* orbitals. Hybrid orbitals designated sp⁴ or sp⁵ would require contribution from four or five *p* orbitals, which is not possible.

9.46   (a)   S: $[Ne]3s^2 3p^4$

3s   3p sp^3

(b)   The hybrid orbitals are called sp^3.

(c)

(d)   The hybrid orbitals formed in (a) would not be appropriate for SF_4. There are five electron domains in SF_4, four bonding and one nonbonding, so five hybrid orbitals are required. A set of four sp^3 hybrid orbitals could not accommodate all the electron pairs around S.

9.48   (a)   32 e^-, 16 e^- pairs

$$:\overset{..}{\underset{..}{Cl}}:$$
$$|$$
$$:\overset{..}{\underset{..}{Cl}}—Si—\overset{..}{\underset{..}{Cl}}:$$
$$|$$
$$:\overset{..}{\underset{..}{Cl}}:$$

4 e^- pairs around Si, tetrahedral e^- domain geometry, sp^3 hybridization

(b)   10 e^-, 5 e^- pairs

H—C≡N:

2 e^- domains around C, linear e^- domain geometry, sp hybridization

(c)   24 e^-, 12 e^- pairs

$$:\overset{..}{\underset{..}{O}}—S—\overset{..}{\underset{..}{O}}:$$
$$\|$$
$$:O:$$

(other resonance structures are possible)
3 e^- domains around S, trigonal planar e^- domain geometry, sp^2 hybridization

(d)   22 e^-, 11 e^- pairs

$$\left[\begin{array}{c} :\overset{..}{Cl}: \\ | \\ >I—: \\ | \\ :\overset{..}{\underset{..}{Cl}}: \end{array} \right]^-$$

5 e^- domains around I, trigonal bipyramidal e^- domain geometry, sp^3d hybridization   (In a trigonal bipyramid, placing nonbonding e^- pairs in the equatorial position minimizes repulsion.)

(e)   36 e⁻, 18 e⁻ pairs

6 e⁻ domains around Br, octahedral e⁻ domain geometry, sp^3d^2 hybridization

Multiple Bonds

9.50   (a)   Two unhybridized p orbitals remain, and the atom can form two pi bonds.

       (b)   It would be much easier to twist or rotate around a single sigma bond. Sigma bonds are formed by end-to-end overlap of orbitals and the bonding electron density is symmetric about the internuclear axis. Rotating (twisting) around a sigma bond can be done without disrupting either the orbital overlap or bonding electron density, without breaking the bond.

             The π part of a double bond is formed by side-to-side overlap of p atomic orbitals perpendicular to the internuclear axis. This π overlap locks the atoms into position and makes twisting difficult. Also, only a small twist (rotation) destroys overlap of the p orbitals and breaks the π bond.

9.52   (a)

$$H-\ddot{N}-\ddot{N}-H \qquad :N\equiv N:$$
$$\quad\; |\quad\; |$$
$$\quad\; H\quad H$$

       (b)   The N atoms in N_2H_4 are sp^3 hybridized; there are no unhybridized p orbitals available for π bonding. In N_2, the N atoms are sp hybridized, with two unhybridized p orbitals on each N atom available to form the two π bonds in the N≡N triple bond.

       (c)   The N–N triple bond is N_2 is significantly stronger than the N–N single bond in N_2H_4, because it consists of one σ and two π bonds, rather than a 'plain' sigma bond. Generally, bond strength increases as the extent of orbital overlap increases. The additional overlap from the two π bonds adds to the strength of the N–N bond in N_2.

9.54   (a)   The C bound to O has three electron domains and is sp^2 hybridized; the other three C atoms are sp^3 hybridized.

       (b) $C_4H_8O_2$ has 4(4) + 8(1) + 2(6) = 36 valence electrons.

       (c)   13 pairs or 26 total valence electrons form σ bonds

       (d)   1 pair or 2 total valence electrons form π bonds

       (e)   4 pairs or 8 total valence electrons are nonbonding

9.56   (a)   1, 120°; 2, 120°; 3, 109°

       (b)   1, sp^2; 2, sp^2; 3, sp^3

       (c)   21 σ bonds

9.58   (a)   24 e⁻, 12 e⁻ pairs   (b)

3 electron domains around S, trigonal planar electron-domain geometry, sp^2 hybrid orbitals

(c)   The multiple resonance structures indicate delocalized π bonding. All four atoms lie in the trigonal plane of the sp^2 hybrid orbitals. On each atom there is a p atomic orbital perpendicular to this plane in the correct orientation for π overlap. The resulting delocalized π electron cloud is Y-shaped (the shape of the molecule) and has electron density above and below the plane of the molecule.

Molecular Orbitals

9.60   (a)   An MO, since the AOs come from two different atoms.

(b)   A hybrid orbital, since the AOs are on the same atom.

(c)   Yes. The Pauli principle, that no two electrons can have the same four quantum numbers, means that an orbital can hold at most two electrons. (Since n, l, and m_1 are the same for a particular orbital and m_s has only two possible values, an orbital can hold at most two electrons). This is true for atomic and molecular orbitals.

9.62   (a)

(b) σ_{1s}^*

          σ_{1s}

(c)   Bond order = 1/2 (2-1) = ½

(d)   If one electron moves from σ_{1s} to σ^*_{1s}, the bond order becomes –1/2. There is a net increase in energy relative to isolated H atoms, so the ion will decompose.

$$H_2^- \xrightarrow{h\nu} H + H^-.$$

9.64   (a)   Zero

(b)   The two π_{2p} molecular orbitals are degenerate; they have the same energy, but they have different spatial orientations 90° apart.

(c)   In the bonding MO the electrons are stabilized by both nuclei. In an antibonding MO, the electrons are directed away from the nuclei, so π_{2p} is lower in energy than π^*_{2p}.

9.66   (a) O_2^{2-} has a bond order of 1.0, while O_2^- has a bond order of 1.5. For the same bonded atoms, the greater the bond order the shorter the bond, so O_2^- has the shorter bond.

       (b)   The two possible orbital energy level diagrams are:

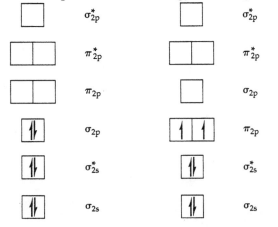

           The magnetic properties of a molecule reveal whether it has unpaired electrons. If the σ_{2p} MOs are lower in energy, B_2 has no unpaired electrons. If the π_{2p} MOs are lower in energy than the σ_{2p} MO, there are two unpaired electrons. The magnetic properties of B_2 must indicate that it has unpaired electrons.

       (c)   According to Figure 9.46, the two highest-energy electrons of O_2 are in antibonding π^*_{2p} MOs and O_2 has a bond order of 2.0 Removing these two electrons to form O_2^{2+} produces an ion with bond order 3.0. O_2^{2+} has a stronger O–O bond than O_2, because O_2^{2+} has a greater bond order.

9.68   (a)   Substances with unpaired electrons are attracted into a magnetic field. This property is called *paramagnetism*.

       (b)   Weigh the substance normally and in a magnetic field, as shown in Figure 9.47. Paramagnetic substances appear to have a larger mass when weighed in a magnetic field.

       (c)   See Figures 9.37 and 9.46. O_2^+, one unpaired electron; N_2^{2-}, two unpaired electrons; Li_2^+, one unpaired electron

9.70   Determine the number of "valence" (non-core) electrons in each molecule or ion. Use the homonuclear diatomic MO diagram from Figure 9.43 (shown below) to calculate bond order and magnetic properties of each species. The electronegativity difference between heteroatomics increases the energy difference between the 2s AO on one atom and the 2p AO on the other, rendering the "no interaction" MO diagram in Figure 9.43 appropriate.

σ_{2p}^*

π_{2p}^*

π_{2p}

σ_{2p}

σ_{2s}^*

σ_{2s}

(a) CO^+:   9 e$^-$, B.O. = (7 – 2) / 2 = 2.5, paramagnetic

(b) NO^-: 12 e$^-$, B.O. = (8 – 4) / 2 = 2.0, paramagnetic

(c) OF^+: 12 e$^-$, B.O. = (8 – 4) / 2 = 2.0, paramagnetic

(d) NeF^+: 14 e$^-$, B.O. = (8 – 6) / 2 = 1.0, diamagnetic

9.72   (a)   The bond order of NO is $[1/2\,(8 - 3)]$ = 2.5. The electron that is lost is in an antibonding molecular orbital, so the bond order in NO^+ is 3.0. The increase in bond order is the driving force for the formation of NO^+.

(b)   To form NO^-, an electron is added to an antibonding orbital, and the new bond order is $[1/2\,(8 - 4)]$ = 2. The order of increasing bond order and bond strength is: $NO^- < NO < NO^+$.

(c) NO^+ is isoelectronic with N_2, and NO^- is isoelectronic with O_2.

9.74   (a)   I: $5s, 5p_x, 5p_y, 5p_z$; Br: $4s, 4p_x, 4p_y, 4p_z$

(b)   By analogy to F_2, the BO of IBr will be 1.

(c)   I and Br have valence atomic orbitals with different principal quantum numbers. This means that the radial extensions (sizes) of the valence atomic orbital that contribute to the MO are different. The n = 5 valence AOs on I are larger than the n = 4 valence AOs on Br.

(d) σ_{np}^*

(e)   None

Additional Exercises

9.75   (a)   The physical basis of VSEPR is the electrostatic repulsion of like-charged particles, in this case groups or domains of electrons. That is, owing to electrostatic repulsion, electron domains will arrange themselves to be as far apart as possible.

(b)   The σ-bond electrons are localized in the region along the internuclear axes. The positions of the atoms and geometry of the molecule are thus closely tied to the locations of these electron pairs. Because the π-bond electrons are distributed above and below the plane that contains the σ bonds, these electron pairs do not, in effect, influence the geometry of the molecule. Thus, all σ- and π-bond electrons localized between two atoms are located in the same electron domain.

9.77   For any triangle, the law of cosines gives the length of side c as $c^2 = a^2 + b^2 - 2ab \cos\theta$.

Let the edge length of the cube (uy = vy = vz) = X

The length of the face diagonal (uv) is

$(uv)^2 = (uy)^2 + (vy)^2 - 2(uy)(vy) \cos 90$

$(uv)^2 = X^2 + X^2 - 2(X)(X) \cos 90$

$(uv)^2 = 2X^2; uv = \sqrt{2}X$

The length of the body diagonal (uz) is

$(uz)^2 = (vz^2) + (uv)^2 - 2(vz)(uv) \cos 90$

$(uz)^2 = X^2 + (\sqrt{2}X)^2 - 2(X)(\sqrt{2}X) \cos 90$

$(uz)^2 = 3X^2; uz = \sqrt{3}X$

For calculating the characteristic tetrahedral angle, the appropriate triangle has vertices u, v, and w. Theta, θ, is the angle formed by sides wu and wv and the hypotenuse is side uv.

$wu = wv = uz/2 = \sqrt{3}/2X; uv = \sqrt{2}X$

$(\sqrt{2}X)^2 = (\sqrt{3}/2X)^2 + (\sqrt{3}/2X)^2 - 2(\sqrt{3}/2X)(\sqrt{3}/2) \cos\theta$

$2X^2 = 3/4 X^2 + 3/4 X^2 - 3/2 X^2 \cos\theta$

$2X^2 = 3/2 X^2 - 3/2 X^2 \cos\theta$

$1/2 X^2 = -3/2 X^2 \cos\theta$

$\cos\theta = -(1/2 X^2) / (3/2 X^2) = -1/3 = -0.3333$

$\theta = 109.47°$

9.78   (a)   40 e$^-$, 20 e$^-$ pairs

             5 e$^-$ domains
             trigonal pyramidal electron domain geometry

   (b)   The greater the electronegativity of the terminal atom, the larger the negative charge centered on the atom, the greater the effective size of the P–X electron domain. A P–F bond will produce a larger electron domain than a P–Cl bond.

   (c)   The molecular geometry (shape) is also trigonal bipyramidal, because all five electron domains are bonding domains. Because we predicted the P–F electron domain to be larger, the maximum number of three P–F bonds will occupy the equatorial plane of the molecule, minimizing the number of 90° P–F to P–F repulsions. This is the same argument that places a "larger" nonbonding domain in the equatorial position of a molecule like SF_4. The P–Cl bond is then axial, as shown in the Lewis structure.

(d)   The molecular geometry is distorted from a perfect trigonal bipyramid because not all electron domains are alike. The 90° P–F to P–F repulsions will be greater than the 90° P–F to P–Cl repulsions, so the F(axial)–P–F angles will be slightly greater than 90°, and the Cl(axial)–P–F angles will be slightly less than 90°. The equatorial F–P–F angles of 120° are distorted little, if at all.

9.80   (a)

$$H-\underset{\underset{H}{|}}{\overset{\overset{H}{|}}{C}}-\underset{\underset{H}{|}}{\overset{:\ddot{O}:}{C}}-\overset{:\ddot{O}:}{C}-\ddot{\underset{..}{O}}-H$$

$3(4) + 3(6) + 6(1) = 36$ e⁻, 18 e⁻ pr

(b)   There are 11 σ and 1 π bonds.

(c)   The C=O on the right-hand C atom is shortest. For the same bonded atoms, in this case C and O, the greater the bond order, the shorter the bond.

(d, e) The right-most C has three e⁻ domains, so the hybridization is sp^2; bond angles about this C atom are approximately 120°. The middle and left-hand C atoms both have four e⁻ domains, are sp^3 hybridized, and have bond angles of approximately 109°.

9.82   (a)   The compound on the right has a dipole moment. In the square planar trans structure on the left, all equivalent bond dipoles can be oriented opposite each other, for a net dipole moment of zero.

(b)

The cis orientation of the Cl atoms in cisplatin means that when they leave, the Pt can bind two adjacent N sites on DNA. This "chelate" orientation (see Chapter 24) tightly binds cisplatin to DNA. Transplatin can bind only one DNA N atom at a time. Thus, to avoid bumping by transplatin NH₃ groups and DNA, the plane of transplatin must rotate away from the DNA backbone. This is a much looser bonding situation than for cisplatin.

9.83

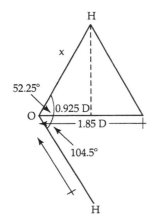

(a) The bond dipoles in H_2O lie along the O–H bonds with the positive end at H and the negative end at O. The dipole moment vector of the H_2O molecule is the resultant (vector sum) of the two bond dipoles. This vector bisects the H–O–H angle and has a magnitude of 1.85 D with the negative end pointing toward O.

(b) Since the dipole moment vector bisects the H–O–H bond angle, the angle between one H–O bond and the dipole moment vector is 1/2 the H–O–H bond angle, 52.25°. Dropping a perpendicular line from H to the dipole moment vector creates the right triangle pictured. If x = the magnitude of the O–H bond dipole, x cos (52.25) = 0.925 D. x = 1.51 D.

(c) The X–H bond dipoles (Table 8.3) and the electronegativity values of X (Figure 8.6) are

	Electronegativity	Bond dipole
F	4.0	1.82
O	3.5	1.51
Cl	3.0	1.08

Since the electronegativity of O is midway between the values for F and Cl, the O–H bond dipole should be approximately midway between the bond dipoles of HF and HCl. The value of the O–H bond dipole calculated in part (b) is consistent with this prediction.

9.84 (a) XeF_6 50 e⁻, 25 e⁻ pairs

(b) There are seven electron domains around Xe, and the maximum number of e⁻ domains in Table 9.3 is six.

(c) Tie seven balloons together and see what arrangement they adopt (seriously! see Figure 9.5). Alternatively, go to the chemical literature where VSEPR was first proposed and see if there is a preferred orientation for seven e⁻ domains.

(d)   Since the hybrid orbitals for five e⁻ domains involve one d orbital and for six pairs, two d orbitals, a reasonable suggestion would be sp^3d^3.

(e)   One of the seven e⁻ domains is a nonbonded domain. The question is whether it occupies an axial or equatorial position. The equatorial plane of a pentagonal bipyramid has F–Xe–F angles of 72°. Placing the nonbonded domain in the equatorial plane would create severe repulsions between it and the adjacent bonded domains. Thus, the nonbonded domain will reside in the axial position. The molecular structure is a pentagonal pyramid.

9.86   (a)   16 e⁻, 8 e⁻ pairs

$$\left[\overset{\bullet\bullet}{\underset{\bullet\bullet}{N}}=N=\overset{\bullet\bullet}{\underset{\bullet\bullet}{N}}\right]^- \longleftrightarrow \left[:N\equiv N-\overset{\bullet\bullet}{\underset{\bullet\bullet}{N}}:\right]^- \longleftrightarrow \left[:\overset{\bullet\bullet}{\underset{\bullet\bullet}{N}}-N\equiv N:\right]^-$$

(b)   The observed bond length of 1.16 Å is intermediate between the values for N=N, 1.24 Å, and N ≡ N, 1.10 Å. This is consistent with the resonance structures, which indicate contribution from formally double and triple bonds to the true bonding picture in N_3^-.

(c)   In each resonance structure, the central N has two electron domains, so it must be sp hybridized. It is difficult to predict the hybridization of terminal atoms in molecules where there are resonance structures because there are a different number of electron domains around the terminal atoms in each structure. Since the "true" electronic arrangement is a combination of all resonance structures, we will assume that the terminal N–N bonds have some triple bond character and that the terminal N atoms are sp hybridized. (There is no experimental measure of hybridization at terminal atoms, since there are no bond angles to observe.)

(d)   In each resonance structure, N–N σ bonds are formed by sp hybrids and π bonds are formed by unhybridized p orbitals. Nonbonding e⁻ pairs can reside in sp hybrids or p atomic orbitals.

(e)   Recall that electrons in 2s orbitals are on the average closer to the nucleus than electrons in 2p orbitals. Since sp hybrids have greater s orbital character, it is reasonable to expect the radial extension of sp orbitals to be smaller than that of sp^2 or sp^3 orbitals and σ bonds formed by sp orbitals to be slightly shorter than those formed by other hybrid orbitals, assuming the same bonded atoms.

There are no solitary σ bonds in N_3^-. That is, the two σ bonds in N_3^- are each accompanied by at least one π bond between the bonding pair of atoms. Sigma bonds that are part of a double or triple bond must be shorter so that the p orbitals can overlap enough for the π bond to form. Thus, the observation is not applicable to this molecule. (Comparison of C–H bond lengths in C_2H_2, C_2H_4, C_2H_6 and related molecules would confirm or deny the observation.)

9.88   (a)   Each C atom is surrounded by three electron domains (two single bonds and one double bond), so bond angles at each C atom will be approximately 120°.

Since there is free rotation around the central C–C single bond, other conformations are possible.

(b)   According to Table 8.5, the average C–C length is 1.54 Å, and the average C=C length is 1.34 Å. While the C=C bonds in butadiene appear "normal," the central C–C is significantly shorter than average. Examination of the bonding in butadiene reveals that each C atom is sp^2 hybridized and the π bonds are formed by the remaining unhybridized 2p orbital on each atom. If the central C–C bond is rotated so that all four C atoms are coplanar, the four 2p orbitals are parallel, and some delocalization of the π electrons occurs.

9.89   (a)   The diagram shows two s atomic orbitals with opposite phases. Because they are spherically symmetric, the interaction of s orbitals can only produce a σ molecular orbital. Because the two orbitals in the diagram have opposite phases, the interaction excludes electron density from the region between the nuclei. The resulting MO has a node between the two nuclei and is labeled σ_{2s}^*. The principal quantum number designation is arbitrary, because it defines only the size of the pertinent AOs and MOs. Shapes and phases of MOs depend only on these same characteristics of the interacting AOs.

(b)   The diagram shows two p atomic orbitals with oppositely phased lobes pointing at each other. End-to-end overlap produces a σ-type MO; opposite phases mean a node between the nuclei and an antibonding MO. The interaction results in a σ_{2p}^* MO.

(c)   The diagram shows parallel p atomic orbitals with like-phased lobes aligned. Side-to-side overlap produces a π-type MO; overlap of like-phased lobes concentrates electron density between the nuclei and a bonding MO. The interaction results in a π_{2p} MO.

9.90   (a) $C_5H_5^-$, 26 e$^-$, 13 e$^-$ pair

No. According to the single Lewis structure above, the four C atoms involved in double bonds would be sp^2 hybridized, but the C atom with the lone pair would be sp^3 hybridized. Not all C atoms would have the same hybridization.

(b)   Equivalent sp^2 hydridization at all C atoms is consistent with the planar structure of the ion. The VSEPR model applied to the single Lewis structure above does not predict uniform hybridization or planarity of the ion. The VSEPR/Lewis model is consistent with the known structural features of the ion only if the other resonance structures are considered.

(c)   If all C atoms are sp^2 hybridized, as required by the planar structure of the ion, the three hybrid orbitals on each C are used to form the σ framework of the molecule. The unshared pair would then reside in an unhybridized p orbital.

(d)   Yes, there are resonance structures equivalent to the Lewis structure above. It is possible to place the unshared pair on any of the five C atoms, resulting in five equivalent resonance structures (four plus the one above).

(e)   The five resonance structures indicate that there is delocalization of the π electron density in the molecule. The interior circle conveys uniform delocalization over the entire ring, as in benzene.

(f)   In a rigid planar structure with each C atom sp^2 hybridized, the unhybridized p orbitals are aligned parallel, facilitating overlap to form a delocalized π network above and below the plane of the molecule. Since the unshared pair occupies an unhybridized p orbital (part c), it is part of the delocalized π network, along with the four electrons involved in π bonds; the delocalized system contains a total of six electrons.

9.92   Paramagnetic materials appear to weigh more when the mass measurement is made in the presence of a magnetic field. Air contains $O_2(g)$, which is paramagnetic because of its two unpaired electrons in a π_{2p}^* molecular orbital (Figure 9.46). In the presence of a magnetic field, mass measurements made in air would be skewed by the paramagnetism of O_2, and lead to inaccurate results.

9.93   We will refer to azo benzene (on the left) as A and hydrazobenzene (on the right) as H.

(a)   A: sp^2; H: sp^3

(b)   A: Each N and C atom has one unhybridized p orbital. H: Each C atom has one unhybridized p orbital, but the N atoms have no unhybridized p orbitals.

(c)   A: 120°; H: 109°

(d)   Since all C and N atoms in A have unhybridized p orbitals, all can participate in delocalized π bonding. The delocalized π system extends over the entire molecule, including both benzene rings and the azo "bridge." In H, the N atoms have no unhybridized p orbitals, so they cannot participate in delocalized π bonding. Each of the benzene rings in H is delocalized, but the network cannot span the N atoms in the bridge.

(e)   This is consistent with the answer to (d). In order for the unhybridized p orbitals in A to overlap, they must be parallel. This requires a planar σ bond framework where all atoms in the molecule are coplanar.

(f)   For a molecule to be useful in a solar energy conversion device, it must absorb visible light. This requires a HOMO-LUMO energy gap in the visible region. For organic molecules, the size of the gap is related to the number of conjugated π bonds; the more conjugated π bonds, the smaller the gap and the more likely the molecule is to be colored. Azobenzene has seven conjugated π bonds (π network delocalized over the entire molecule) and appears red-orange. Hydrazobenzene has only three conjugated π bonds (π network on benzene rings only) and appears white. Thus, the smaller HOMO-LUMO energy gap in A causes it to be both intensely colored and a more useful molecule for solar energy conversion.

9.94   (a)   H: $1s^1$; F: $[He]2s^22p^5$

When molecular orbitals are formed from atomic orbitals, the total number of orbitals is conserved. Since H and F have a total of five valence AOs ($H_{1s} + F_{2s} + 3F_{2p}$), the MO diagram for HF has five MOs.

(b)   H and F have a total of eight valence electrons. Since each MO can hold a maximum of two electrons, four of the five MOs would be occupied.

(c)   No. The MO diagram for NO in Figure 9.49 has eight MOs, and the diagram for HF has only five. The number, type, and energy spacing of the MOs in HF will be different.

(d)

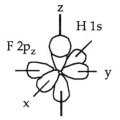

If H and F lie on the z axis, then the $2p_z$ orbital of F will overlap with the 1s orbital of H.

(e)   Since F is more electronegative than H, the valence orbitals on F are at lower energy than those on H.

The HF MO diagram has 6 nonbonding, 2 bonding and 0 antibonding electrons. The BO = [2 – 0]/2 = 1. (Nonbonding electrons do not "count" toward bond order.)

(f)   H—$\ddot{\text{F}}$:

In the Lewis structure for HF, the nonbonding electrons are on the (more electronegative) F atom, as they are in the MO diagram.

9.95   (a)   CO, 10 e^-, 5 e^- pair

:C≡O:

(b) The bond order for CO, as predicted by the MO diagram in Figure 9.49, is $1/2[8 - 2] = 3.0$. A bond order of 3.0 agrees with the triple bond in the Lewis structure.

(c) Applying the MO diagram in Figure 9.49 to the CO molecule, the highest energy electrons would occupy the π_{2p} MOs. That is, π_{2p} would be the HOMO, highest occupied molecular orbital. If the true HOMO of CO is a σ-type MO, the order of the π_{2p} and σ_{2p} orbitals must be reversed. Figure 9.45 shows how the interaction of the 2s orbitals on one atom and the 2p orbitals on the other atom can affect the relative energies of the resulting MOs. This 2s–2p interaction in CO is significant enough so that the σ_{2p} MO is higher in energy than the π_{2p} MOs, and the σ_{2p} is the HOMO.

(d) We expect the atomic orbitals of the more electronegative element to have lower energy than those of the less electronegative element. When atoms of the two elements combine, the lower energy atomic orbitals make a greater contribution to the bonding MOs and the higher energy atomic orbitals make a larger contribution to the antibonding orbitals. Thus, the π_{2p} bonding MOs will have a greater contribution from the more electronegative O atom.

Integrative Exercises

9.96 (a) Assume 100 g of compound

$$2.1\,g\,H \times \frac{1\,mol\,H}{1.008\,g\,H} = 2.1\,mol\,H; 2.1/2.1 = 1$$

$$29.8\,g\,N \times \frac{1\,mol\,N}{14.01\,g\,N} = 2.13\,mol\,N; 2.13/2.1 \approx 1$$

$$68.1\,g\,O \times \frac{1\,mol\,O}{16.00\,g\,O} = 4.26\,mol\,O; 4.26/2.1 \approx 2$$

The empirical formula is HNO_2; formula weight = 47. Since the approximate molar mass is 50, the molecular formula is HNO_2.

(b) Assume N is central, since it is unusual for O to be central, and part (d) indicates as much. HNO_2: 18 valence e⁻

$$\ddot{O}{=}\ddot{N}{-}\ddot{O}{-}H \longleftrightarrow :\ddot{O}{-}N{=}\ddot{O}{-}H$$
$$\phantom{:\ddot{O}{-}N{=}}{-1}0{+1}$$

The second resonance form is a minor contributor due to unfavorable formal charges.

(c) The electron domain geometry around N is trigonal planar with an O–N–O angle of approximately 120°. If the resonance structure on the right makes a significant contribution to the molecular structure, all four atoms would lie in a plane. If only the left structure contributes, the H could rotate in and out of the molecular plane. The relative contributions of the two resonance structures could be determined by measuring the O–N–O and N–O–H bond angles.

(d) 3 VSEPR e⁻ domains around N, sp^2 hybridization

(e) 3 σ, 1 π for both structures (or for H bound to N).

9.98   (a) PX_3, 26 valence e⁻, 13⁻ pairs

4 electron domains around P, tetrahedral e⁻ domain geometry,
bond angles ≤ 109°

(b)   As electronegativity increases (I < Br < Cl < F), the X–P–X angles decreases.

(c)   For P–I, ΔEN is (2.5 - 2.1) = 0.4 and the bond dipole is small. For P–F, ΔEN is (4.0 – 2.1) = 1.9 and the bond dipole is large. The greater the ΔEN and bond dipole, the larger the magnitude of negative charge centered on X. The more negative charge centered on X, the greater the repulsion between X and the nonbonding electron pair on P and the smaller the bond angle. (Since electrostatic attractions and repulsions vary as 1/r, the shorter P–F bond may also contribute to this effect.)

(d) $PBrCl_4$, 40 valence electrons, 20 e⁻ pairs. The molecule will have trigonal bipyramidal electron-domain geometry (similar to PCl_5 in Table 9.3.) Based on the argument in part (c), the P–Br bond will have smaller repulsions with P–Cl bonds than P–Cl bonds have with each other. Therefore, the Br will occupy an axial position in the trigonal bipyramid, so that the more unfavorable P–Cl to P–Cl repulsions can be situated at larger angles in the equatorial plane.

9.100   (a)   C ≡ C   839 kJ/mol   (1 σ, 2 π)

C=C   614 kJ/mol   (1 σ, 1 π)

C–C   348 kJ/mol   (1 σ)

The contribution from 1 π bond would be (614–348) 266 kJ/mol. From a second π bond, (839 – 614), 225 kJ/mol. An average π bond contribution would be (266 + 225)/2 = 246 kJ/mol.

This is $\dfrac{246 \text{ kJ/}\pi \text{ bond}}{348 \text{ kJ/}\sigma \text{ bond}} \times 100 = 71\%$ of the average enthalpy of a σ bond.

(b)   N ≡ N   941 kJ/mol

N=N   418 kJ/mol

N–N   163 kJ/mol

first π = (418 – 163) = 255 kJ/mol

second π = (941 – 418) = 523 kJ/mol

average π bond enthalpy = (255 + 523)/2 = 389 kJ/mol

This is $\dfrac{389 \text{ kJ}/\pi \text{ bond}}{163 \text{ kJ}/\sigma \text{ bond}} \times 100 = 240\%$ of the average enthalpy of a σ bond.

N–N σ bonds are weaker than C–C σ bonds, while N–N π bonds are stronger than C–C π bonds. The relative energies of C–C σ and π bonds are similar, while N–N π bonds are much stronger than N–N σ bonds.

(c) N_2H_4, 14 valence e^-, 7 e^- pairs

$$\text{H}-\overset{\displaystyle ..}{\underset{\displaystyle |}{\text{N}}}-\overset{\displaystyle ..}{\underset{\displaystyle |}{\text{N}}}-\text{H}$$
$$\text{H}\text{H}$$

4 electron domains around N, sp^3 hybridization

N_2H_2, 12 valence e^-, 6 e^- pairs

$$\text{H}-\overset{..}{\text{N}}=\overset{..}{\text{N}}-\text{H}$$

3 electron domains around N, sp^2 hybridization

N_2, 10 valence e^-, 5 e^- pairs

$$:\text{N}\equiv\text{N}:$$

2 electron domains around N, sp hybridization

(d)   In the three types of N–N bonds, each N atom has a nonbonding or lone pair of electrons. The lone pair to bond pair repulsions are minimized going from 109° to 120° to 180° bond angles, making the π bonds stronger relative to the σ bond. In the three types of C–C bonds, no lone-pair to bond-pair repulsions exist, and the σ and π bonds have more similar energies.

9.102   (a)   1 eV = 96.485 kJ/mol

$$\text{H}_2: 15.4 \text{ eV} \times \frac{96.485 \text{ kJ/mol}}{1 \text{ eV}} = 1486 = 1.49 \times 10^3 \text{ kJ/mol}$$

$$\text{N}_2: 15.6 \text{ eV} \times \frac{96.485 \text{ kJ/mol}}{1 \text{ eV}} = 1505 = 1.51 \times 10^3 \text{ kJ/mol}$$

$$\text{O}_2: 12.1 \text{ eV} \times \frac{96.485 \text{ kJ/mol}}{1 \text{ eV}} = 1167 = 1.17 \times 10^3 \text{ kJ/mol}$$

$$\text{F}_2: 15.7 \text{ eV} \times \frac{96.485 \text{ kJ/mol}}{1 \text{ eV}} = 1515 = 1.52 \times 10^3 \text{ kJ/mol}$$

(b)

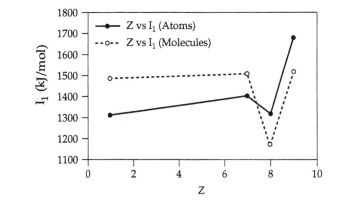

(c) In general, I_1 for atoms and molecules increases going across a row of the periodic chart. In both cases, there is a discontinuity at oxygen. The details of the trends are different. The deviation at O is larger for the molecules than the atoms, while the increase at F is much greater for the atoms than the molecules.

(d) According to Figures 9.35 and 9.46, H_2, N_2, and F_2 are diamagnetic and O_2 is paramagnetic. That is, ionization in H_2, N_2, and F_2 has to overcome spin-pairing energy, while ionization of O_2 removes an already unpaired electron. Thus, the ionization energy of O_2 is much less than I_1 for H_2, N_2, and F_2.

Despite differences in bond order, bond length, and the bonding or antibonding nature of the HOMO in H_2, N_2, and F_2, the ionization energies for these molecules are very similar.

9.104 (a) The molecular and empirical formulas of the four molecules are:

benzene: molecular, C_6H_6; empirical, CH

napthalene: molecular, $C_{10}H_8$; empirical, C_5H_4

anthracene: molecular, $C_{14}H_{10}$; empirical, C_7H_5

tetracene: molecular, $C_{18}H_{12}$, empirical, C_3H_2

(b) Yes. Since the compounds all have different empirical formulas, combustion analysis could in principle be used to distinguish them. In practice, the mass % of C in the four compounds is not very different, so the data would have to be precise to at least 3 decimal places and 4 would be better.

(c) $C_{10}H_8(s) + 12O_2(g) \rightarrow 10CO_2(g) + 4H_2O(g)$

(d) $\Delta H_{comb} = 5D(C=C) + 6D(C-C) + 8D(C-H) + 12D(O=O) - 20D(C=O) - 8D(O-H)$

$= 5(614) + 6(348) + 8(413) + 12(495) - 20(799) - 8(463)$

$= -5282 \text{ kJ/mol } C_{10}H_8$

(e) Yes. For example, the resonance structures of naphthalene are:

(f) Colored compounds absorb visible light and appear the color of the visible light that they reflect. Colorless compounds typically absorb shorter wavelength, higher energy light. The energy of light absorbed corresponds to the energy gap between the HOMO and LUMO of the molecule. That tetracene absorbs longer wavelength, lower energy visible light indicates that it has the smallest HOMO-LUMO energy gap of the four molecules. Tetracene also has the most conjugated double bonds of the four molecules. We might conclude that the more conjugated double bonds in an organic molecule, the smaller the HOMO-LUMO energy gap. More information about the absorption spectra of anthracene, naphthalene and benzene is needed to confirm this conclusion.

9.105   (a) – (e)

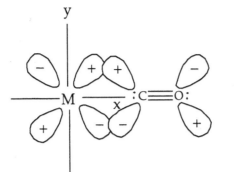

(c)   The two lobes of a p AO have opposite phases. These are shown on the diagram as + and –. An antibonding MO is formed when p AOs with opposite phases interact.

(d)   Note that the d_{xy} AO has lobes that lie between, not on, the x and y axes.

(e)   A π bond forms by overlap of orbitals on M and C. There is electron density above and below, but not along, the M–C axis.

(f)   According to Exercise 9.95, the HOMO of CO is a σ-type MO. So the appropriate MO diagram is shown on the left side of Figure 9.46. A lone CO molecule has 10 valence electrons, the HOMO is σ_{2p} and the bond order is 3.0. The LUMO is π_{2p}^*.

When M and CO interact as shown in the π_{2p}^* diagram, d–π back bonding causes the π_{2p}^* to become partially occupied. Electron density in the π_{2p}^* decreases electron density in the bonding molecular orbitals and decreases the BO of the bound CO. The strength of the C–O bond in a metal–CO complex decreases relative to the strength of the C–O bond in an isolated CO molecule.

9.106   (a) $:\overset{..}{\underset{..}{Cl}}-\overset{..}{\underset{..}{Cl}}:$

(b)   4 e⁻ domains on each Cl, sp³ hybridization at each Cl

(c)   Since both Cl atoms employ sp³ hybrid orbitals, the Cl–Cl bond is formed by end-to-end, σ overlap of two sp³ hybrid orbitals on different Cl atoms.

(d)   In the AO model, two of the lone pairs occupy mutually perpendicular p orbitals on the same Cl atom. The angle between these two lone pair electron domains is 90°. In the hybrid orbital model, all three lone pairs occupy sp³ hybrid orbitals on the same Cl atom. The angle between these lone pair domains is 109°. If we could measure the "positions" of the lone pairs, we could distinguish between models by the angles between electron domains.

(e)   According to the MO diagram for F_2, and by inference Cl_2, shown in Figure 9.46, the HOMO for Cl_2 is π_{3p}^* and it contains four electrons. If we ionize $Cl_2 \rightarrow Cl_2^+$, the number of antibonding electrons decreases and the BO increases. Cl_2^+ should have a stronger, shorter Cl–Cl bond than Cl_2 has. Conversely, adding an electron to form Cl_2^- adds an electron to the antibonding σ_{3p}^*. This reduces the BO in Cl_2^-, resulting in a weaker, longer Cl–Cl bond.

So, generate Cl_2^+ and Cl_2^- and measure the Cl–Cl distances. If the order of Cl–Cl bond lengths is $Cl_2^+ < Cl_2 < Cl_2^-$, the data supports an MO model.

10 Gases

Visualizing Concepts

10.1   It would be much easier to drink from a straw on Mars. When a straw is placed in a glass of liquid, the liquid level in the straw equals the liquid level in the glass. The atmosphere exerts equal pressure inside and outside the straw. When we drink through a straw, we withdraw air, thereby reducing the pressure on the liquid inside. If only 0.007 atm is exerted on the liquid in the glass, a very small reduction in pressure inside the straw will cause the liquid to rise.

Another approach is to consider the gravitational force on Mars. Since the pull of gravity causes atmospheric pressure, the gravity on Mars must be much smaller than that on Earth. With a very small Martian gravity holding liquid in a glass, it would be very easy to raise the liquid through a straw.

10.3   At constant temperature and volume, pressure depends on total number of particles (Charles' Law). In order to reduce the pressure by a factor of 2, the number of particles must be reduced by a factor of 2. At the lower pressure, the container would have half as many particles as at the higher pressure.

10.4   (a)   At constant pressure and temperature, the container volume is directly proportional to the number of particles present (Avogadro's Law). As the reaction proceeds, 3 gas molecules are converted to 2 gas molecules, so the container volume decreases. If the reaction goes to completion, the final volume would be 2/3 of the initial volume.

(b)   At constant volume, pressure is directly proportional to the number of particles (Charles' Law). Since the number of molecules decreases as the reaction proceeds, the pressure also decreases. At completion, the final pressure would be 2/3 the initial pressure.

10.6   (a)   Partial pressure depends on the number of particles of each gas present. Red has the fewest particles, then yellow, then blue. $P_{red} < P_{yellow} < P_{blue}$

(b) $P_{gas} = \chi_{gas} P_t$. Calculate the mole fraction, χ_{gas} = [mol gas / total moles] or [particles gas / total particles]. This is true because Avogadro's number is a counting number, and mole ratios are also particle ratios.

χ_{red} = 2 red atoms / 10 total atoms = 0.2; P_{red} = 0.2(0.90 atm) = 0.18 atm

χ_{yellow} = 3 yellow atoms / 10 total atoms = 0.3; P_{yellow} = 0.3(0.90 atm) = 0.27 atm

χ_{blue} = 5 blue atoms / 10 total atoms = 0.5; P_{blue} = 0.5(0.90 atm) = 0.45 atm

10.7

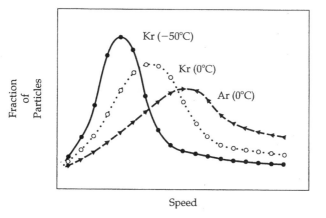

10.9   (a)   Total pressure is directly related to total number of particles (or total mol particles). $P(ii) < P(i) = P(iii)$

(b)   Partial pressure of He is directly related to number of He atoms (yellow) or mol He atoms. $P_{He}(iii) < P_{He}(ii) < P_{He}(i)$

(c)   Density is total mass of gas per unit volume. We can use the atomic or molar masses of He (4) and N_2(28), as relative masses of the particles.

$$mass(i) = 5(4) + 2(28) = 76$$

$$mass(ii) = 3(4) + 1(28) = 40$$

$$mass(iii) = 2(4) + 5(28) = 148$$

Since the container volumes are equal, $d(ii) < d(i) < d(iii)$.

(d)   At the same temperature, all gases have the same "avg" kinetic energy. The average kinetic energies of the particles in the three containers are equal.

10.10 CCl_4, Cl_2 and H_2O would deviate most, while He, N and H_2 would deviate least. At high pressure, gas molecules have relatively little free space in which to move. Both intermolecular attractive forces and the nonzero volume of molecules are important. At low temperature, molecules have less kinetic energy, are less able to overcome attractive forces, and deviations due to intermolecular attractive forces dominate (see Figure 10.24). In Table 10.3, values of a show corrections for intermolecular attraction. The gases CCl_4, Cl_2 and H_2O have the largest values of a, while He, Ne and H_2 have the smallest. The magnitude of a increases as the size and complexity of the molecule increases. CCl_4 and Cl_2 are relatively large gases; H_2O is polar and has hydrogen bonding. He, Ne and H_2 are small, monatomic or nonpolar diatomic. The magnitude of a is predictable from molecular structure.

Gas Characteristics; Pressure

10.12   (a)   Because gas molecules are far apart and in constant motion, the gas expands to fill the container. Attractive forces hold liquid molecules together and the volume of the liquid does not change.

(b) H_2O and CCl_4 molecules are too dissimilar to displace each other and mix in the liquid state. All mixtures of gases are homogeneous. (See Solution 10.11 (c)).

(c)　Because gas molecules are far apart, the mass present in 1 mL of a gas is very small. The mass of a gas present in 1 L is on the same order of magnitude as the mass of a liquid present in 1 mL.

10.14　$P = m \times a / A$; $1\,Pa = 1\,kg/m\text{-}s^2$; $A = 3.0\,cm \times 4.1\,cm \times 4 = 49.2 = 49\,cm^2$

$$\frac{262\,kg}{49.2\,cm^2} \times \frac{9.81\,m}{s^2} \times \frac{(100)^2\,cm^2}{1\,m^2} = 5.224 \times 10^5\,\frac{kg}{m\text{-}s^2} = 5.2 \times 10^5\,Pa$$

10.16　Using the relationship derived in Solution 10.15 for two liquids under the influence of gravity, $(d \times h)_{1id} = (d \times h)_{Hg}$. At 752 torr, the height of an Hg barometer is 752 mm.

$$\frac{1.20\,g}{1\,mL} \times h_{1id} = \frac{13.6\,g}{1\,mL} \times 760\,mm;\ h_{1id} = \frac{13.6\,g/mL \times 752\,mm}{1.20\,g/mL} = 8.52 \times 10^3\,mm = 8.52\,m$$

10.18　The mercury would fill the tube completely; there would be no vacuum at the closed end. This is because atmospheric pressure will support a mercury column higher than 70 cm, while our tube is only 50 cm. No mercury flows from the tube into the dish and no vacuum forms at the top of the tube.

10.20　(a)　$0.850\,atm \times \dfrac{760\,torr}{1\,atm} = 646\,torr$

　　　　(b)　$785\,torr \times \dfrac{101.325\,kPa}{760\,torr} = 105\,kPa$

　　　　(c)　$655\,mm\,Hg \times \dfrac{1\,atm}{760\,mm\,Hg} = 0.862\,atm$

　　　　(d)　$1.323 \times 10^5\,Pa \times \dfrac{1\,atm}{1.01325 \times 10^5\,Pa} = 1.3057 = 1.306\,atm$

　　　　(e)　$2.50\,atm \times \dfrac{1.01325 \times 10^5\,Pa}{1\,atm} \times \dfrac{1\,bar}{1 \times 10^5\,Pa} = 2.53\,bar$

10.22　(a)　$\dfrac{1.63105\,Pa}{1\,Titan\,atm} \times \dfrac{1\,Earth\,atm}{101,325\,Pa} = \dfrac{1.60972 \times 10^{-5}\,Earth\,atm}{1\,Titan\,atm}$

　　　　(b)　$\dfrac{90\,Earth\,atm}{1\,Venus\,atm} \times \dfrac{101.3\,kPa}{1\,Earth\,atm} = \dfrac{9.1 \times 10^3\,kPa}{1\,Venus\,atm}$

10.24　(a)　The atmosphere is exerting 15.4 cm = 154 mm Hg (torr) more pressure than the gas.

$$P_{gas} = P_{atm} - 154\,torr = \left(0.966\,atm \times \frac{760\,torr}{1\,atm}\right) - 154\,torr = 580\,torr$$

　　　　(b)　The gas is exerting 8.7 mm Hg (torr) more pressure than the atmosphere.

$$P_{gas} = P_{atm} + 8.7\,torr = \left(0.99\,atm \times \frac{760\,torr}{1\,atm}\right) + 8.7\,torr = 761.1\,torr = 7.6 \times 10^2\,torr$$

(Atmospheric pressure of 0.99 atm determines that the result has 2 sig figs.)

The Gas Laws

10.26 *Analyze.* Given: initial P, V, T. Find: final values of P, V, T for certain changes of condition. *Plan.* Select the appropriate gas law relationships from Section 10.3; solve for final conditions, paying attention to units.　*Solve.*

(a)　$P_1V_1 = P_2V_2$; the proportionality holds true for any pressure or volume units.

$P_1 = 752$ torr, $V_1 = 4.38$ L, $P_2 = 1.88$ atm

$$V_2 = \frac{P_1V_1}{P_2} = \frac{752 \text{ torr} \times 4.38 \text{ L}}{1.88 \text{ atm}} \times \frac{1 \text{ atm}}{760 \text{ torr}} = 2.31 \text{ L}$$

Check. As pressure increases, volume should decrease; our result agrees with this.

(b)　$V_1/T_1 = V_2/T_2$; T must be in Kelvins for the relationship to be true.

$V_1 = 4.38$ L, $T_1 = 21°C = 294$ K, $T_2 = 175°C = 448$ K

$$V_2 = \frac{V_1T_2}{T_1} = \frac{4.38 \text{ L} \times 448 \text{ K}}{294 \text{ K}} = 6.67 \text{ L}$$

Check. As temperature increases, volume should increase; our result is consistent with this.

10.28 According to Avogadro's hypothesis, the mole ratios in the chemical equation will be volume ratios for the gases if they are at the same temperature and pressure.

$$N_2(g) + 3H_2(g) \rightarrow 2NH_3(g)$$

The volumes of H_2 and N_2 are in a stoichiometric $\dfrac{3.6 \text{ L}}{1.2 \text{ L}}$ or $\dfrac{3 \text{ vol } H_2}{1 \text{ vol } N_2}$ ratio, so either can be used to determine the volume of $NH_3(g)$ produced.

$$1.2 \text{ L } N_2 \times \frac{2 \text{ mol } NH_3}{1 \text{ mol } N_2} = 2.4 \text{ L } NH_3(g) \text{ produced.}$$

The Ideal-Gas Equation

(In *Solutions to Exercises*, the symbol for molar mass is MM.)

10.30 (a)　STP stands for standard temperature, 0°C (or 273 K), and standard pressure, 1 atm.

(b)　$V = \dfrac{nRT}{P}$; $V = 1 \text{ mol} \times \dfrac{0.08206 \text{ L - atm}}{\text{K - mol}} \times \dfrac{273 \text{ K}}{1 \text{ atm}}$

$V = 22.4$ L for 1 mole of gas at STP

(c)　25°C + 273 = 298 K

$$V = \frac{nRT}{P}; V = 1 \text{ mol} \times \frac{0.08206 \text{ L - atm}}{\text{K - mol}} \times \frac{298 \text{ K}}{1 \text{ atm}}$$

$V = 24.5$ L for 1 mol of gas at 1 atm and 25°C

10.32 n = g/MM; PV = nRT = gRT/MM; MM = gRT/PV.

2-L flask: MM = 4.8 RT/2.0(X) = 2.4 RT/X

3-L flask: MM = 0.36 RT/3.0 (0.1 X) = 1.2 RT/X

$$\frac{4.8 \, RT}{2.0 X} = \frac{2.4 \, RT}{X}$$

$$\frac{0.36 \, RT}{3.0(0.1X)} = \frac{1.2 \, RT}{X}$$

The molar masses of the two gases are not equal. The gas in the 2-L flask has a molar mass that is twice as large as the gas in the 3-L flask.

10.34   *Analyze/Plan.* Follow the strategy for calculations involving many variables given in Section 10.4.   *Solve.*

(a) $n = 1.50$ mol, $P = 0.985$ atm, $T = -6°C = 267$ K

$$V = \frac{nRT}{P} = 1.50\,\text{mol} \times \frac{0.08206\,\text{L-atm}}{\text{K-mol}} \times \frac{267\,\text{K}}{0.985\,\text{atm}} = 33.4\,\text{L}$$

(b) $n = 3.33 \times 10^{-3}$ mol, $V = 325$ mL $= 0.325$ L

$$P = 750\,\text{torr} \times \frac{1\,\text{atm}}{760\,\text{torr}} = 0.9868 = 0.987\,\text{atm}$$

$$T = \frac{PV}{nR} = 0.9868\,\text{atm} \times \frac{0.325\,\text{L}}{3.33 \times 10^{-3}\,\text{mol}} \times \frac{1\,\text{K-mol}}{0.08206\,\text{L-atm}} = 1174 = 1.17 \times 10^3 \text{ K}$$

(c) $n = 4.67 \times 10^{-2}$ mol, $V = 413$ mL $= 0.413$ L, $T = 138°C = 411$ K

$$P = \frac{nRT}{V}; = 0.0467\,\text{mol} \times \frac{0.08206\,\text{L-atm}}{\text{K-mol}} \times \frac{411\,\text{K}}{0.413\,\text{L}} = 3.81\,\text{atm}$$

(d) $V = 55.7$ L, $T = 54°C = 327$ K,

$$P = 11.25\,\text{kPa} \times \frac{1\,\text{atm}}{101.325\,\text{kPa}} = 0.11103 = 0.1110 \text{ atm}$$

$$n = \frac{PV}{RT} = 0.11103\,\text{atm} \times \frac{\text{K-mol}}{0.08206\,\text{L-atm}} \times \frac{55.7\,\text{L}}{327\,\text{K}} = 0.230\,\text{mol}$$

10.36   Find the volume of the tube in cm^3; 1 cm^3 = 1 mL.

$r = d/2 = 2.5$ cm$/2 = 1.25 = 1.3$ cm; $h = 5.5$ m $= 5.5 \times 10^2$ cm

$V = \pi r^2 h = 3.14159 \times (1.25\,\text{cm})^2 \times (5.5 \times 10^2\,\text{cm}) = 2.700 \times 10^3 \text{ cm}^3 = 2.7$ L

$$PV = \frac{g}{MM}RT;\ g = \frac{MM \times PV}{RT};\ P = 1.78\,\text{torr} \times \frac{1\,\text{atm}}{760\,\text{torr}} = 2.342 \times 10^{-3} = 2.34 \times 10^{-3}\,\text{atm}$$

$$g = \frac{20.18\,\text{g Ne}}{1\,\text{mol Ne}} \times \frac{\text{K-mol}}{0.08206\,\text{L-atm}} \times \frac{2.342 \times 10^{-3}\,\text{atm} \times 2.700\,\text{L}}{308\,\text{K}} = 5.049 \times 10^{-3} = 5.0 \times 10^{-3}\,\text{g Ne}$$

10.38 $P_{O_3} = 3.0 \times 10^{-3}$ atm; $T = 250$ K; $V = 1$ L (exact)

$$\#\text{ of } O_3 \text{ molecules} = \frac{PV}{RT} \times 6.022 \times 10^{23}$$

$$\# = \frac{3.0 \times 10^{-3}\,\text{atm} \times 1\,\text{L}}{250\,\text{K}} \times \frac{\text{K-mol}}{0.08206\,\text{L-atm}} \times \frac{6.022 \times 10^{23}\,\text{molecules}}{\text{mol}}$$

$$= 8.8 \times 10^{19}\,O_3 \text{ molecules}$$

10.40   (a) $V = 0.250$ L, $T = 23°C = 296$ K, $n = 2.30\,\text{g } C_3H_8 \times \dfrac{1\,\text{mol } C_3H_8}{44.1\,\text{g } C_3H_8} = 0.052154$

$$= 0.0522\,\text{mol}$$

$$P = \frac{nRT}{V} = 0.052154\,\text{mol} \times \frac{0.08206\,\text{L-atm}}{\text{K-mol}} \times \frac{296\,\text{K}}{0.250\,\text{L}} = 5.07\,\text{atm}$$

10 Gases   **Solutions to Black Exercises**

(b)   STP = 1.00 atm, 273 K

$$V = \frac{nRT}{P} = 0.052154 \text{ mol} \times \frac{0.08206 \text{ L-atm}}{\text{K-mol}} \times \frac{273 \text{ K}}{1.00 \text{ atm}} = 1.1684 = 1.17 \text{ L}$$

(c)   °C = 5/9 (°F – 32°); K = °C + 273.15 = 5/9 (130°F – 32°) + 273.15 = 327.59 = 328 K

$$P = \frac{nRT}{V} = 0.052154 \text{ mol} \times \frac{0.08206 \text{ L-atm}}{\text{K-mol}} \times \frac{327.59 \text{ K}}{0.250 \text{ L}} = 5.608 = 5.61 \text{ atm}$$

10.42 $V = 65.0 \text{ L}, T = 23°C = 296 \text{ K}, P = 16,500 \text{ kPa} \times \dfrac{1 \text{ atm}}{101.325} = 162.84 = 163 \text{ kPa}$

(a) $g = \dfrac{MM \times PV}{RT}; g = \dfrac{32.0 \text{ g } O_2}{1 \text{ mol } O_2} \times \dfrac{K\text{-mol}}{0.08206 \text{ L-atm}} \times \dfrac{162.84 \text{ atm}}{296 \text{ K}} \times 65.0 \text{ L}$

$$= 1.39446 \times 10^4 \text{ g } O_2 = 13.9 \text{ kg } O_2$$

(b) $V_2 = \dfrac{P_1 V_1 T_2}{T_1 P_2} = \dfrac{16,500 \text{ kPa} \times 65.0 \text{ L} \times 273 \text{ K}}{296 \text{ K} \times 101.325 \text{ kPa}} = 9.76 \times 10^3 \text{ L}$

(c) $T_2 = \dfrac{P_2 T_1}{P_1} = \dfrac{150.0 \text{ atm} \times 296 \text{ K}}{16,500 \text{ kPa}} \times \dfrac{101.325 \text{ kPa}}{1 \text{ atm}} = 272.7 = 273 \text{ K}$

(d) $P_2 = \dfrac{P_1 V_1 T_2}{V_2 T_1} = \dfrac{16,500 \text{ kPa} \times 65.0 \text{ L} \times 297 \text{ K}}{55.0 \text{ L} \times 296 \text{ K}} = 19,566 = 1.96 \times 10^4 \text{ kPa}$

10.44   mass $= 1800 \times 10^{-9} \text{ g} = 1.8 \times 10^{-6} \text{ g}; V = 1 \text{ m}^3 = 1 \times 10^3 \text{ L}; T = 273 + 10° C = 283 \text{ K}$

(a) $P = \dfrac{gRT}{MM \times V}; P = \dfrac{1.8 \times 10^{-6} \text{ g Hg} \times 1 \text{ mol Hg}}{200.6 \text{ g Hg}} \times \dfrac{0.08206 \text{ L-atm}}{K\text{-mol}} \times \dfrac{283 \text{ K}}{1 \times 10^3 \text{ L}}$

$$= 2.1 \times 10^{-10} \text{ atm}$$

(b) $\dfrac{1.8 \times 10^{-6} \text{ g Hg}}{1 \text{ m}^3} \times \dfrac{1 \text{ mol Hg}}{200.6 \text{ g Hg}} \times \dfrac{6.022 \times 10^{23} \text{ Hg atoms}}{1 \text{ mol Hg}} = 5.4 \times 10^{15} \text{ Hg atoms/m}^3$

(c) $1600 \text{ km}^3 \times \dfrac{1000^3 \text{ m}^3}{1 \text{ km}^3} \times \dfrac{1.8 \times 10^{-6} \text{ g Hg}}{1 \text{ m}^3} = 2.9 \times 10^6 \text{ g Hg/day}$

Further Applications of the Ideal-Gas Equation

10.46 $CO_2 < SO_2 < HBr$. For gases at the same conditions, density is directly proportional to molar mass. The order of increasing molar mass is the order of increasing density. CO_2, 44 g/mol $< SO_2$, 64 g/mol $< HBr$, 81 g/mol.

10.48   (b) Xe atoms have a higher mass than N_2 molecules. Because both gases at STP have the same number of molecules per unit volume, the Xe gas must be denser.

10.50   (a) $d = \dfrac{MM \times P}{RT}$; MM = 146.1 g/mol, T = 21°C = 294 K, P = 707 torr

$$d = \dfrac{146.1 \text{ g}}{1 \text{ mol}} \times \dfrac{K\text{-mol}}{0.08206 \text{ L-atm}} \times \dfrac{707 \text{ torr}}{294 \text{ K}} \times \dfrac{1 \text{ atm}}{760 \text{ torr}} = 5.63 \text{ g/L}$$

159

(b) $MM = \dfrac{dRT}{P} = \dfrac{7.135\,g}{1\,L} \times \dfrac{0.08206\,L\text{-}atm}{K\text{-}mol} \times \dfrac{285\,K}{743\,torr} \times \dfrac{760\,torr}{1\,atm} = 171\,g/mol$

10.52 $MM = \dfrac{gRT}{PV} = \dfrac{0.846\,g}{0.354\,L} \times \dfrac{0.08206\,L\text{-}atm}{K\text{-}atm} \times \dfrac{373\,K}{752\,torr} \times \dfrac{760\,torr}{1\,atm} = 73.9\,g/mol$

10.54 $n_{H_2} = \dfrac{P_{H_2}V}{RT}$; $\quad P = 814\,torr \times \dfrac{1\,atm}{760\,torr} = 1.071 = 1.07\,atm$; $T = 273 + 21°C = 294\,K$

$n_{H_2} = 1.071\,atm \times \dfrac{K\text{-}mol}{0.08206\,L\text{-}atm} \times \dfrac{53.5\,L}{294\,K} = 2.3751 = 2.38\,mol\,H_2$

$2.3751\,mol\,H_2 \times \dfrac{1\,mol\,CaH_2}{2\,mol\,H_2} \times \dfrac{42.10\,g\,CaH_2}{1\,mol\,CaH_2} = 50.0\,g\,CaH_2$

10.56   Follow the logic in Sample Exercise 10.9. The $H_2(g)$ will be used in a balloon, which operates at atmospheric pressure. Since atmospheric pressure is not explicitly given, assume 1 atm (infinite sig figs).

$n = \dfrac{PV}{RT} = 1 \times \dfrac{3.1150 \times 10^4\,L}{295\,K} \times \dfrac{K\text{-}mol}{0.08206\,L\text{-}atm} = 1.28678 \times 10^3 = 1.29 \times 10^3\,mol\,H_2$

From the balanced equation, 1 mol of Fe produces 1 mol of H_2, so 1.29×10^3 mol Fe are required.

$1.28678 \times 10^3\,mol\,Fe \times \dfrac{55.845\,g\,Fe}{mol\,Fe} \times \dfrac{1\,kg}{1000\,g} = 71.86 = 71.9\,kg\,Fe$

10.58   The gas sample is a mixture of $C_2H_2(g)$ and $H_2O(g)$. Find the partial pressure of C_2H_2, then moles CaC_2 and C_2H_2.

$P_t = 745\,torr = P_{C_2H_2} + P_{H_2O}$. $\quad P_{H_2O}$ at $23°C = 21.07\,torr$

$P_{C_2H_2} = (745\,torr - 21.07\,torr) \times \dfrac{1\,atm}{760\,torr} = 0.95254 = 0.953\,atm$

$0.752\,g\,CaC_2 \times \dfrac{1\,mol\,CaC_2}{64.10\,g} \times \dfrac{1\,mol\,C_2H_2}{1\,mol\,CaC_2} = 0.01173 = 0.0117\,mol\,C_2H_2$

$V = 0.01173\,mol \times \dfrac{0.08206\,L\text{-}atm}{K\text{-}mol} \times \dfrac{296\,K}{0.95254\,atm} = 0.299\,L\,C_2H_2$

Partial Pressures

10.60   (a)   The partial pressure of gas A is **not affected** by the addition of gas C. The partial pressure of A depends only on moles of A, volume of container, and conditions; none of these factors changes when gas C is added.

       (b)   The total pressure in the vessel **increases** when gas C is added, because the total number of moles of gas increases.

       (c)   The mole fraction of gas B **decreases** when gas C is added. The moles of gas B stay the same, but the total moles increase, so the mole fraction of B (nB/nt) decreases.

10.62   Given mass, V and T of O_2 and He, find the partial pressure of each gas. Sum to find the total pressure in the tank.

$V = 10.0 \text{ L}; \quad T = 19°C; \quad 19 + 273 = 292 \text{ K}$

$n_{O_2} = 51.2 \text{ g O}_2 \times \dfrac{1 \text{ mol O}_2}{31.999 \text{ g O}_2} = 1.600 = 1.60 \text{ mol O}_2$

$n_{He} = 32.6 \text{ g He} \times \dfrac{1 \text{ mol He}}{4.0026 \text{ g He}} = 8.1447 = 8.14 \text{ mol He}$

$P_{O_2} = 1.600 \text{ mol} \times \dfrac{0.08206 \text{ L-atm}}{\text{K-mol}} \times \dfrac{292 \text{ K}}{10.0 \text{ L}} = 3.8338 = 3.84 \text{ atm}$

$P_{He} = 8.1447 \text{ mol} \times \dfrac{0.08206 \text{ L-atm}}{\text{K-mol}} \times \dfrac{292 \text{ K}}{10.0 \text{ L}} = 19.5159 = 19.5 \text{ atm}$

$P_t = 3.8338 + 19.5159 = 23.3497 = 23.3 \text{ atm}$

10.64 $V \text{ C}_4\text{H}_{10}\text{O(l)} \xrightarrow{\text{density}} \text{ mass C}_4\text{H}_{10}\text{O} \rightarrow \text{mol C}_4\text{H}_{10}\text{O} \rightarrow P_{C_4H_{10}O} \text{ at given V, T.}$

$P_t = P_{N_2} + P_{O_2} + P_{C_4H_{10}O}; \quad T = 273.15 + 35.0°C = 308.15 = 308.2 \text{ K}$

(a) $4.00 \text{ mL C}_4\text{H}_{10}\text{O} \times \dfrac{0.7134 \text{ g C}_4\text{H}_{10}\text{O}}{\text{mL}} \times \dfrac{1 \text{ mol C}_4\text{H}_{10}\text{O}}{74.12 \text{ g C}_4\text{H}_{10}\text{O}} = 0.0385 \text{ mol C}_4\text{H}_{10}\text{O}$

$P = \dfrac{nRT}{V} = 0.03850 \text{ mol} \times \dfrac{308.15}{5.00 \text{ L}} \times \dfrac{0.08206 \text{ L-atm}}{\text{K-mol}} = 0.1947 = 0.195 \text{ atm}$

(b) $P_t = P_{N_2} + P_{O_2} + P_{C_4H_{10}O} = 0.751 \text{ atm} + 0.208 \text{ atm} + 0.195 \text{ atm} = 1.154 \text{ atm}$

10.66 $n_{N_2} = 10.25 \text{ g N}_2 \times \dfrac{1 \text{ mol}}{28.02 \text{ g}} = 0.3658 \text{ mol}; \quad n_{H_2} = 1.83 \text{ g H}_2 \times \dfrac{1 \text{ mol}}{2.016 \text{ g}} = 0.9077 = 0.908 \text{ mol}$

$n_{NH_3} = 7.95 \text{ g NH}_3 \times \dfrac{1 \text{ mol}}{17.03 \text{ g}} = 0.4668 \text{ mol} = 0.467 \text{ mol}$

$n_t = 0.3658 + 0.9077 + 0.4668 = 1.7403 = 1.740 \text{ mol}$

$P_{N_2} = \dfrac{n_{N_2}}{n_t} \times P_t = \dfrac{0.3658}{1.7403} \times 1.85 \text{ atm} = 0.389 \text{ atm}$

$P_{H_2} = \dfrac{0.9077}{1.7403} \times 1.85 \text{ atm} = 0.965 \text{ atm}; \quad P_{NH_3} = \dfrac{0.4668}{1.7403} \times 1.85 \text{ atm} = 0.496 \text{ atm}$

10.68   (a) $n_{O_2} = 5.08 \text{ g O}_2 \times \dfrac{1 \text{ mol}}{32.00 \text{ g}} = 0.159 \text{ mol}; \quad n_{N_2} = 7.17 \text{ g N}_2 \times \dfrac{1 \text{ mol}}{28.02 \text{ g}} = 0.256 \text{ mol}$

$n_{H_2} = 1.32 \text{ g H}_2 \times \dfrac{1 \text{ mol}}{2.016 \text{ g}} = 0.655 \text{ mol}; \quad n_t = 0.159 + 0.256 + 0.655 = 1.070 \text{ mol}$

$\chi_{O_2} = \dfrac{n_{O_2}}{n_t} = \dfrac{0.159}{1.07} = 0.149; \quad \chi_{N_2} = \dfrac{n_{N_2}}{n_t} = \dfrac{0.256}{1.07} = 0.239$

$\chi_{H_2} = \dfrac{0.655}{1.07} = 0.612$

(b) $P_{O_2} = n \times \dfrac{RT}{V}; \quad P_{O_2} = 0.159 \text{ mol} \times \dfrac{0.08206 \text{ L-atm}}{\text{K-mol}} \times \dfrac{288 \text{ K}}{12.40 \text{ L}} = 0.303 \text{ atm}$

$$P_{N_2} = 0.256 \text{ mol} \times \frac{0.08206 \text{ L-atm}}{\text{K-mol}} \times \frac{288 \text{ K}}{12.40 \text{ L}} = 0.488 \text{ atm}$$

$$P_{H_2} = 0.655 \text{ mol} \times \frac{0.08206 \text{ L-atm}}{\text{K-mol}} \times \frac{288 \text{ K}}{12.40 \text{ L}} = 1.25 \text{ atm}$$

10.70   Calculate the pressure of the gas in the second vessel directly from mass and conditions using the ideal-gas equation.

(a) $P_{SO_2} = \dfrac{gRT}{MV} = \dfrac{3.00 \text{ g SO}_2}{64.07 \text{ g SO}_2/\text{mol}} \times \dfrac{0.08206 \text{ L-atm}}{\text{K-mol}} \times \dfrac{299 \text{ K}}{10.0 \text{ L}} = 0.11489 = 0.115 \text{ atm}$

(b) $P_{N_2} = \dfrac{gRT}{MV} = \dfrac{2.35 \text{ g N}_2}{28.01 \text{ g N}_2/\text{mol}} \times \dfrac{0.08206 \text{ L-atm}}{\text{K-mol}} \times \dfrac{299 \text{ K}}{10.0 \text{ L}} = 0.20585 = 0.206 \text{ atm}$

(c) $P_t = P_{SO_2} + P_{N_2} = 0.11489 \text{ atm} + 0.20585 \text{ atm} = 0.321 \text{ atm}$

Kinetic-Molecular Theory; Graham's Law

10.72   (a)   False. The average kinetic energy per molecule in a collection of gas molecules is the same for all gases at the same temperature.

         (b)   True.

         (c)   False. The molecules in a gas sample at a given temperature exhibit a distribution of kinetic energies.

         (d)   True.

10.74   Newton's model provides no explanation of the effect of a change in temperature on the pressure of a gas at constant volume or on the volume of a gas at constant pressure. On the other hand, the assumption that the average kinetic energy of gas molecules increases with increasing temperature explains Charles' Law, that an increase in temperature requires an increase in volume to maintain constant pressure.

10.76   (a)   They have the same number of molecules (equal volumes of gases at the same temperature and pressure contain equal numbers of molecules).

         (b) N_2 is more dense because it has the larger molar mass. Since the volumes of the samples and the number of molecules are equal, the gas with the larger molar mass will have the greater density.

         (c)   The average kinetic energies are equal (statement 5, section 10.7).

         (d) CH_4 will effuse faster. The lighter the gas molecules, the faster they will effuse (Graham's Law).

10.78   (a)   *Plan.* The greater the molecular (and molar) mass, the smaller the rms speed of the molecules. Calculate the molar mass of each gas, and place them in decreasing order of mass and increasing order of rms speed. *Solve.*

         $CO = 28 \text{ g/mol}; SF_6 = 146 \text{ g/mol}; H_2S = 34 \text{ g/mol}; Cl_2 = 71 \text{ g/mol};$

         $HBr = 81 \text{ g/mol}.$ In order of increasing speed (and decreasing molar mass):

         $SF_6 < HBr < Cl_2 < H_2S < CO$

(b)   *Plan.* Follow the logic of Sample Exercise 10.14. *Solve.*

$$u_{CO} = \sqrt{\frac{3RT}{M}} = \left(\frac{3 \times 8.314 \text{ kg-m}^2/\text{s}^2\text{-K-mol} \times 300 \text{ K}}{28.0 \times 10^{-3} \text{ kg/mol}} \right)^{1/2} = 5.17 \times 10^2 \text{ m/s}$$

$$u_{Cl_2} = \left(\frac{3 \times 8.314 \text{ kg-m}^2/\text{s}^2\text{-K-mol} \times 300 \text{ K}}{70.9 \times 10^{-3} \text{ kg/mol}} \right)^{1/2} = 3.25 \times 10^2 \text{ m/s}$$

As expected, the lighter molecule moves at the greater speed.

10.80 $\dfrac{\text{rate}^{235}U}{\text{rate}^{238}U} = \sqrt{\dfrac{238.05}{235.04}} = \sqrt{1.0128} = 1.0064$

There is a slightly greater rate enhancement for $^{235}U(g)$ atoms than $^{235}UF_6(g)$ molecules (1.0043), because ^{235}U is a greater percentage (100%) of the mass of the diffusing particles than in $^{235}UF_6$ molecules. The masses of the isotopes were taken from *The Handbook of Chemistry and Physics*.

10.82   The time required is proportional to the reciprocal of the effusion rate.

$$\frac{\text{rate (X)}}{\text{rate (O}_2)} = \frac{105 \text{ s}}{31 \text{ s}} = \left[\frac{32 \text{ g O}_2}{MM_x} \right]^{1/2}; \quad MM_x = 32 \text{ g O}_2 \times \left[\frac{105}{31} \right]^2 = 370 \text{ g/mol (two sig figs)}$$

Nonideal-Gas Behavior

10.84   Ideal-gas behavior is most likely to occur at high temperature and low pressure, so the atmosphere on Mercury is more likely to obey the ideal-gas law. The higher temperature on Mercury means that the kinetic energies of the molecules will be larger relative to intermolecular attractive forces. Further, the gravitational attractive forces on Mercury are lower because the planet has a much smaller mass. This means that for the same column mass of gas (Figure 10.1), atmospheric pressure on Mercury will be lower.

10.86   The constant *a* is a measure of the strength of intermolecular attractions among gas molecules; *b* is a measure of molecular volume. Both increase with increasing molecular mass and structural complexity.

10.88   *Analyze.* Conditions and amount of $CCl_4(g)$ are given. *Plan.* Use ideal-gas equation and van der Waals equation to calculate pressure of gas at these conditions.   *Solve.*

(a) $P = 1.00 \text{ mol} \times \dfrac{0.08206 \text{ L-atm}}{\text{K-mol}} \times \dfrac{313 \text{ K}}{28.0 \text{ L}} = 0.917 \text{ atm}$

(b) $P = \dfrac{nRT}{V-nb} - \dfrac{an^2}{V^2} = \dfrac{1.00 \times 0.08206 \times 313}{28.0 - (1.00 \times 0.1383)} - \dfrac{20.4(1.00)^2}{(28.0)^2} = 0.896 \text{ atm}$

Check. The van der Waals result indicates that the real pressure will be less than the ideal pressure. That is, intermolecular forces reduce the effective number of particles and the real pressure. This is reasonable for 1 mole of gas at relatively low temperature and pressure.

(c)   According to Table 10.3, CCl_4 has larger a and b values. That is, CCl_4 experiences stronger intermolecular attractions and has a larger molecular volume than Cl_2 does. CCl_4 will deviate more from ideal behavior at these conditions than Cl_2 will.

Additional Exercises

10.90   Only item (b) is satisfactory. Item (c) would not have supported a column of Hg because it is open at both ends. The atmosphere would exert pressure on the top of the column as well as on the reservoir; the column would only be as high as the reservoir and the height would not change with changing pressure. Item (d) is not tall enough to support a nearly 760 mm Hg column. Items (a) and (e) are inappropriate for the same reason: they don't have a uniform cross-sectional area. The height of the Hg column is a direct measure of atmospheric pressure only if the cross-sectional area is constant over the entire tube.

10.91 $P_1V_1 = P_2V_2$; $V_2 = P_1V_1/P_2$

$$V_2 = \frac{3.0\,\text{atm} \times 1.0\,\text{mm}^3}{695\,\text{torr}} \times \frac{760\,\text{torr}}{1\,\text{atm}} = 3.3\,\text{mm}^3$$

10.93 $P = \dfrac{nRT}{V}$; $n = 1.4 \times 10^{-5}$ mol, $V = 0.600$ L, $T = 23°C = 296$ K

$$P = 1.4 \times 10^{-5}\,\text{mol} \times \frac{0.08206\,\text{L-atm}}{\text{K-mol}} \times \frac{296\,\text{K}}{0.600\,\text{L}} = 5.7 \times 10^{-4}\,\text{atm} = 0.43\,\text{mm Hg}$$

10.94   (a)   Change mass CO_2 to mol CO_2. $P = 1.00$ atm, $T = 27°C = 300$ K.

$$6 \times 10^6\,\text{tons CO}_2 \times \frac{2000\,\text{lb}}{\text{ton}} \times \frac{453.6\,\text{g}}{\text{lb}} \times \frac{1\,\text{mol CO}_2}{44.01\,\text{g CO}_2} = 1.237 \times 10^{11} = 1 \times 10^{11}\,\text{mol}$$

$$V = \frac{nRT}{P} = 1.237 \times 10^{11}\,\text{mol} \times \frac{300\,\text{K}}{1.00\,\text{atm}} \times \frac{0.08206\,\text{L-atm}}{\text{K-mol}} =$$
$$3.045 \times 10^{12} = 3 \times 10^{12}\,\text{L}$$

(b) $1.237 \times 10^{11}\,\text{mol CO}_2 \times \dfrac{44.01\,\text{g CO}_2}{\text{mol CO}_2} \times \dfrac{1\,\text{cm}^3}{1.2\,\text{g}} \times \dfrac{1\,\text{L}}{1000\,\text{cm}^3} = 4.536 \times 10^9 = 5 \times 10^9\,\text{L}$

(c) $n = 1.237 \times 10^{11}$ mol, $P = 90$ atm, $T = 36°C = 309$ K

$$V = \frac{nRT}{P} = 1.237 \times 10^{11}\,\text{mol} \times \frac{309\,\text{K}}{90\,\text{atm}} \times \frac{0.08206\,\text{L-atm}}{\text{K-mol}} = \quad 3.485 \times 10^{10} = 3 \times 10^{10}\,\text{L}$$

10.96   Volume of laboratory $= 54\,\text{m}^2 \times 3.1\,\text{m} \times \dfrac{1000\,\text{L}}{1\,\text{m}^3} = 1.674 \times 10^5 = 1.7 \times 10^5\,\text{L}$

Calculate the **total** moles of gas in the laboratory at the conditions given.

$$n_t = \frac{PV}{RT} = 1.00\,\text{atm} \times \frac{\text{K-mol}}{0.08206\,\text{L-atm}} \times \frac{1.674 \times 10^5\,\text{L}}{297\,\text{K}} = 6.869 \times 10^3 = 6.9 \times 10^3\,\text{mol gas}$$

An $Ni(CO)_4$ concentration of 1 part in 10^9 means 1 mol $Ni(CO)_4$ in 1×10^9 total moles of gas.

$$\frac{x \text{ mol Ni(CO)}_4}{6.869 \times 10^3 \text{ mol gas}} = \frac{1}{10^9} = 6.869 \times 10^{-6} \text{ mol Ni(CO)}_4$$

$$6.869 \times 10^{-6} \text{ mol Ni(CO)}_4 \times \frac{170.74 \text{ g Ni(CO)}_4}{1 \text{ mol Ni(CO)}_4} = 1.2 \times 10^{-3} \text{ g Ni(CO)}_4$$

10.97   (a)   mol = g/MM; assume mol Ar = mol X;

$$\frac{\text{g Ar}}{39.948 \text{ g / mol}} = \frac{\text{g X}}{\text{MM X}}; \quad \frac{3.224 \text{ g Ar}}{39.948 \text{ g / mol}} = \frac{8.102 \text{ g X}}{\text{MM X}}$$

$$\text{MM X} = \frac{(8.102 \text{ g X})(39.948 \text{ g / mol})}{3.224 \text{ g Ar}} = 100.39 = 100.4 \text{ g / mol}$$

       (b)   Assume mol Ar = mol X. For gases, PV= nRT and n = PV/RT. For moles of the two gases to be equal, the implied assumption is that P, V, and T are constant. Since we use the same container for both gas samples, constant V is a good assumption. Constant P and T are not explicitly stated.

               We also assume that the gases behave ideally. At ambient conditions, this is a reasonable assumption.

10.99   (a) $n = \dfrac{PV}{RT} = 0.980 \text{ atm} \times \dfrac{\text{K - mol}}{0.08206 \text{ L - atm}} \times \dfrac{0.524 \text{ L}}{347 \text{ K}} = 0.018034 = 0.0180 \text{ mol air}$

             $\text{mol O}_2 = 0.018034 \text{ mol air} \times \dfrac{0.2095 \text{ mol O}_2}{1 \text{ mol air}} = 0.003778 = 0.00378 \text{ mol O}_2$

       (b) $C_8H_{18}(l) + 25/2 \, O_2(g) \rightarrow 8CO_2(g) + 9H_2O(g)$

             (The H_2O produced in an automobile engine is in the gaseous state.)

             $0.003778 \text{ mol O}_2 \times \dfrac{1 \text{ mol C}_8\text{H}_{18}}{12.5 \text{ mol O}_2} \times \dfrac{114.2 \text{ g C}_8\text{H}_{18}}{1 \text{ mol C}_8\text{H}_{18}} = 0.0345 \text{ g C}_8\text{H}_{18}$

10.100   (a)   Pressure percent = mol percent. Change pressure/mol percents to mol fraction. Partial pressure of each gas is mol fraction (χ) times total pressure. $P_x = \chi_x P_t$

             $P_{N_2} = 0.748(0.980 \text{ atm}) = 0.733 \text{ atm}; \quad P_{O_2} = 0.153(0.980 \text{ atm}) = 0.150 \text{ atm}$

             $P_{CO_2} = 0.037(0.980 \text{ atm}) = 0.03626 = 0.036 \text{ atm}$

             $P_{H_2O} = 0.062(0.980 \text{ atm}) = 0.06076 = 0.060 \text{ atm}$

       (b)   PV = nRT, n = PV/RT; P = 0.036 atm, V = 0.455 L, T = 37°C = 310 K

             $n = 0.03626 \text{ atm} \times \dfrac{0.455 \text{ L}}{310 \text{ K}} \times \dfrac{\text{K - mol}}{0.08206 \text{ L - atm}} = 6.486 \times 10^{-4} = 6.5 \times 10^{-4} \text{ mol}$

       (c) $C_6H_{12}O_6 + 6O_2 \rightarrow 6CO_2 + 6H_2O$

             $6.486 \times 10^{-4} \text{ mol CO}_2 \times \dfrac{1 \text{ mol C}_6\text{H}_{12}\text{O}_6}{6 \text{ mol CO}_2} \times \dfrac{180.15 \text{ g C}_6\text{H}_{12}\text{O}_6}{1 \text{ mol C}_6\text{H}_{12}\text{O}_6} = 0.01947$

                                               $= 0.019 \text{ g C}_6\text{H}_{12}\text{O}_6$

10.102 $MM_{avg} = \dfrac{dRT}{P} = \dfrac{1.104\ g}{1\ L} \times \dfrac{0.08206\ L\text{-}atm}{K\text{-}mol} \times \dfrac{300\ K}{435\ torr} \times \dfrac{760\ torr}{1\ atm} = 47.48 = 47.5\ g/mol$

χ = mole fraction O_2; $1 - \chi$ = mole fraction Kr

$47.48\ g = \chi(32.00) + (1 - \chi)(83.80)$

$36.3 = 51.8\ \chi;\ \chi = 0.701;\ 70.1\%\ O_2$

10.103   (a)   The quantity d/P = MM/RT should be a constant at all pressures for an ideal gas. It is not, however, because of nonideal behavior. If we graph d/P vs P, the ratio should approach ideal behavior at low P. At P = 0, d/P = 2.2525. Using this value in the formula MM = d/P × RT, MM = 50.46 g/mol.

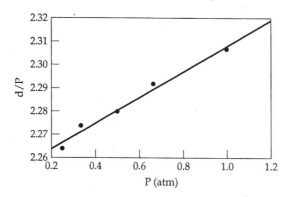

        (b)   The ratio d/P varies with pressure because of the finite volumes of gas molecules and attractive intermolecular forces.

10.105 $u = (3RT/MM)^{1/2}$; $u_2 = 1.100\ u_1$; $T_1 = -33° = 240\ K$

$u_1 = (3RT_1/MM)^{1/2}$; $u_1^2 = 3(240)R/MM = 720R/MM$

$u_1 = (720R/MM)^{1/2}$; $u_2 = 1.100u_1 = (1.100)(720R/MM)^{1/2}$

$(1.100)(720R/MM)^{1/2} = (3RT_2/MM)^{1/2}$

$(1.100)^2(720R/MM) = 3RT_2/MM;\ (1.100)^2(720) = 3T_2$

$T_2 = (1.100)(720)/3 = 290.4 = 290\ K = 17°C$

A 10.0% increase in rms speed (u) requires a 50 K (or 50°C) increase in temperature.

10.106   (a)   Assumption 3 states that attractive and repulsive forces between molecules are negligible. All gases in the list are nonpolar. The largest and most structurally complex molecule, SF_6, is most likely to depart from this assumption.

        (b)   The monatomic gas Ne is smallest and least structurally complex, so it will behave most like an ideal gas.

        (c)   Root-mean-square speed is directly related to molecular mass. The lightest gas, CH_4, has the highest rms speed.

        (d)   The heaviest and most structurally complex is SF_6. We guess that this one will occupy the greatest molecular volume relative to total volume. A quantitative measure is the b value in Table 10.3, with units of L/mol. Unfortunately, SF_6 does not appear in Table 10.3.

(e)   Average kinetic energy is only related to absolute (K) temperature. At the same temperature, they all have the same average kinetic molecular energy.

(f)   Rate of effusion is inversely related to molecular mass. The lighter the molecule, the faster it effuses. Ne and CH_4 have smaller molecular masses and effuse faster than N_2.

10.108   (a)   The initial drop in the value of PV/RT is due to attractive forces between molecules. These intermolecular forces cause the molecules to "stick" together and behave as if there are fewer net particles in the sample. At lower pressures, this is the dominant effect. At the same time, the real volume of gas molecules causes the amount of free space in the gas sample to be less than the container volume. Using the container volume to calculate PV/RT gives a value larger than that for an ideal gas (which assumes that the total container volume is free space). At higher pressures, this effect more than compensates for molecular attraction, and PV/RT is greater than 1.

(b)   As the temperature of a gas increases, the average kinetic energy of the particles increases. The increased kinetic energy overcomes the attractive forces between molecules and keeps them separate.

10.109   The larger and heavier the particle, in this case a single atom, the more likely it is to deviate from ideal behavior. Other than Rn, Xe is the largest (atomic radius = 1.45 Å), heaviest (molar mass = 131.3 g/mol) and most dense (9.73 g/L) noble gas. Its susceptibility to intermolecular interactions is also demonstrated by its ability to form compounds like XeF_4.

10.111   (a) $120.00 \text{ kg } N_2(g) \times \dfrac{1000 \text{ g}}{1 \text{ kg}} \times \dfrac{1 \text{ mol } N_2}{28.0135 \text{ g N}} = 4283.6 \text{ mol } N_2$

$P = \dfrac{nRT}{V} = 4283.6 \text{ mol} \times \dfrac{0.08206 \text{ L-atm}}{K\text{-mol}} \times \dfrac{553 \text{ K}}{1100.0 \text{ L}} = 176.72 = 177 \text{ atm}$

(b)   According to Equation [10.26], $P = \dfrac{nRT}{V - nb} - \dfrac{n^2 a}{V^2}$

$P = \dfrac{(4283.6 \text{ mol})(0.08206 \text{ L-atm/K-mol})(553 \text{ K})}{1100.0 \text{ L} - (4283.6 \text{ mol})(0.0391 \text{ L/mol})} - \dfrac{(4283.6 \text{ mol})^2 (1.39 \text{ L}^2\text{-atm/mol}^2)}{(1100.0 \text{ L})^2}$

$P = \dfrac{194{,}388 \text{ L-atm}}{1100.0 \text{ L} - 167.5 \text{ L}} - 21.1 \text{ atm} = 208.5 \text{ atm} - 21.1 \text{ atm} = 187.4 \text{ atm}$

(c)   The pressure corrected for the real volume of the N_2 molecules is 208.5 atm, 31.8 atm higher than the ideal pressure of 176.7 atm. The 21.1 atm correction for intermolecular forces reduces the calculated pressure somewhat, but the "real" pressure is still higher than the ideal pressure. The correction for the real volume of molecules dominates. Even though the value of b is small, the number of moles of N_2 is large enough so that the molecular volume correction is larger than the attractive forces correction.

Integrative Exercises

10.113   *Plan.* Write the balanced equation for the combustion of methanol. Since amounts of both reactants are given, determine the limiting reactant. Use mole ratios to calculate mol H_2O produced, based on the amount of limiting reactant.   *Solve.*

methanol = $CH_3OH(l)$. $2CH_3OH(l) + 3O_2(g) \rightarrow 2CO_2(g) + 4H_2O(g)$

$$25.0 \, mL \, CH_3OH \times \frac{0.850 \, g \, CH_3OH}{mL} \times \frac{1 \, mol \, CH_3OH}{32.04 \, g} = 0.6632 = 0.663 \, mol \, CH_3OH$$

$$mol \, O_2 = n = \frac{PV}{RT} = 1.00 \, atm \times \frac{12.5 \, L}{273 \, K} \times \frac{K \cdot mol}{0.08206 \, L \cdot atm} = 0.5580 = 0.558 \, mol \, O_2$$

$$0.558 \, mol \, O_2 \times \frac{2 \, mol \, CH_3OH}{3 \, mol \, O_2} = 0.372 \, mol \, CH_3OH$$

0.558 mol O_2 can react with only 0.372 mol CH_3OH, so O_2 is the limiting reactant. Note that a large volume of $O_2(g)$ is required to completely react with a relatively small volume of $CH_3OH(l)$.

$$0.558 \, mol \, O_2 \times \frac{4 \, mol \, H_2O}{3 \, mol \, O_2} = 0.744 \, mol \, H_2O$$

10.115   (a)   *Plan.* Use the ideal-gas law to calculate the moles CO_2 that react.

  Solve. P(reacted) = P(initial) – P(final), at constant V, T. Since both CaO and BaO react with CO_2 in a 1:1 mole ratio, mol CaO + mol BaO = mol CO_2. Use molar masses to calculate % CaO in sample.

$$P(reacted) = 730 \, torr - 150 \, torr = 580 \, torr; 580 \, torr \times \frac{1 \, atm}{760 \, torr} = 0.76316 = 0.763 \, atm$$

$$n = \frac{PV}{RT} = 0.76316 \, atm \times \frac{1.0 \, L}{298 \, K} \times \frac{K \cdot mol}{0.08206 \, L \cdot atm} = 0.03121 = 0.0312 \, mol \, CO_2$$

  (b)   *Plan.* Use the stoichiometry of the reaction and definition of moles to calculate the mass and Mass % of CaO.

  Solve. $CaO(s) + CO_2(s) \rightarrow CaCO_3(s)$. $BaO(s) + CO_2(g) \rightarrow BaCO_3(s)$

mol CO_2 reacted = mol CaO + mol BaO

Let x = g CaO, 4.00 – x = g BaO

$$0.03121 = \frac{x}{56.08} + \frac{4.00 - x}{153.3}$$

$0.03121(56.08)(153.3) = 153.3x + 56.08(4.00 - x)$

$268.3 = (153.3x - 56.08x) + 224.3$

$43.98 = 97.22x, x = 0.452 = 0.45$ g CaO

$$\frac{0.452 \, g \, CaO}{4.00 \, g \, sample} \times 100 = 11.3 = 11\% \, CaO$$

(By strict sig fig rules, the result has 2 sig figs, because 268 – 224 = 44 has 0 decimal places and 2 sig figs.)

10.116 (a) $5.00 \text{ g HCl} \times \dfrac{1 \text{ mol HCl}}{36.46 \text{ g HCl}} = 0.1371 = 0.137 \text{ mol HCl}$

$5.00 \text{ g NH}_3 \times \dfrac{1 \text{ mol NH}_3}{17.03 \text{ g NH}_3} = 0.2936 = 0.294 \text{ mol NH}_3$

The gases react in a 1:1 mole ratio, HCl is the limiting reactant and is completely consumed. (0.2936 mol − 0.1371 mol) = 0.1565 = 0.157 mol NH_3 remain in the system. $NH_3(g)$ is the only gas remaining after reaction. $V_t = 4.00$ L

(b) $P = \dfrac{nRT}{V} = 0.1565 \text{ mol} \times \dfrac{0.08206 \text{ L-atm}}{\text{K-mol}} \times \dfrac{298 \text{ K}}{4.00 \text{ L}} = 0.957 \text{ atm}$

10.118 (a) *Analyze/Plan.* $AgF + S_8 \xrightarrow{\Delta}$ unknown gas

The gas probably contains the elements S and F in an unknown mole ratio. (Most compounds containing Ag are ionic and therefore solids.) Calculate the molar mass of the gas from its density at the given conditions. Determine the relative amount of F from the data on the reaction of the gas with water to produce HF. *Solve.*

$MM = \dfrac{dRT}{P} = \dfrac{0.803 \text{ g}}{L} \times \dfrac{0.08206 \text{ L-atm}}{\text{K-mol}} \times \dfrac{305 \text{ K}}{150 \text{ mm}} \times \dfrac{760 \text{ mm}}{1 \text{ atm}} = 101.83$

$= 102 \text{ g/mol}$

mol F in 480 mL sample: $M \times L = \text{mol}$

0.081 *M* HF × 0.080 L = 0.00648 = 0.0065 mol HF = 0.0065 mol F in the sample total mol gas in 480 mL sample: n = PV/RT

$n = 126 \text{ mm} \times \dfrac{1 \text{ atm}}{760 \text{ mm}} \times \dfrac{0.480 \text{ L}}{301 \text{ K}} \times \dfrac{\text{K-mol}}{0.08206 \text{ L-atm}} = 3.222 \times 10^{-3}$

$= 3.22 \times 10^{-3} \text{ mol}$

mole ratio of S and F:

total g gas = 3.222×10^{-3} mol gas × 101.83 g/mol; = 0.32808 = 0.328 g gas

g F = 6.48×10^{-3} mol F × 18.998 g/mol F = 0.12311 = 0.123 g F

g S = 0.32808 − 0.12311 = 0.20497 = 0.205 g S

mol S = 0.205 g S/32.07 g/mol = 0.0639 mol S

0.00639 mol S/0.00648 mol F = 1:1 mole ratio of S:F

The empirical formula is SF; empirical FW = 32.07 + 19.00 = 51.07;

since MM = 102, the empirical formula is S_2F_2

Check. The empirical and molecular formula weights are in an integer ratio, so the result is reasonable.

(b) 26 valence e⁻, 13 e⁻ pairs

(c)   According to VSEPR the electron domain geometry about S in each of the molecules will be tetrahedral, with bond angles of 109° or less. The left molecule will have a structure similar to hydrogen peroxide, H_2O_2; the 4 atoms are not necessarily coplanar, and in fact we expect a dihedral angle of approximately 110°. Since the right molecule has a single central atom, we can describe the molecular geometry as trigonal pyramidal.

From the given bond distances, we expect the single-bond covalent radii to be S = 1.02 Å, F = 0.72 Å. The simple conclusion is that the S–S distances will be ~2.04 Å and the S–F distances ~1.74 Å. In fact, the S–S distance in the left compound is 1.89 Å and in the right it is 1.86 Å. Clearly each of these bonds has some double bond character, as indicated by one of the resonance structures for the right molecule. The actual S-F distances are 1.63 Å (left) and 1.60 Å (right), also shorter than the predicted S–F single bond distance. One possible conclusion is that each of the bonds has some double bond character, that some of the "nonbonding" electron density is incorporated into a delocalized π-bonding network in the molecules.

10.119   (a)   19 e⁻, 9.5 e⁻ pairs

$$\ddot{\text{O}}—\dot{\text{C}}\text{l}—\ddot{\text{O}}:$$

Resonance structures can be drawn with the odd electron on O, but electronegativity considerations predict that it will be on Cl for most of the time.

(b) ClO_2 is very reactive because it is an odd-electron molecule. Adding an electron (reduction) both pairs the odd electron and completes the octet of Cl. Thus, ClO_2 has a strong tendency to gain an electron and be reduced.

(c) ClO_2^-, 20 e⁻, 10 e⁻ pairs

$$\left[:\ddot{\text{O}}—\ddot{\text{C}}\text{l}—\ddot{\text{O}}:\right]$$

(d)   4 e⁻ domains around Cl, O–Cl–O bond angle ~109°

(e)   Calculate mol Cl_2 from ideal-gas equation; determine limiting reactant; mass ClO_2 via mol ratios.

$$\text{mol } Cl_2 = \frac{PV}{RT} = 1.50 \text{ atm} \times \frac{2.00 \text{ L}}{294 \text{ K}} \times \frac{\text{K-mol}}{0.08206 \text{ L-atm}} = 0.1243 = 0.124 \text{ mol } Cl_2$$

$$10.0 \text{ g NaClO}_2 \times \frac{1 \text{ mol NaClO}_2}{90.44 \text{ g}} = 0.1106 = 0.111 \text{ mol NaClO}_2$$

2 mol $NaClO_2$ are required for 1 mol Cl_2, so $NaClO_2$ is the limiting reactant. For every 2 mol $NaClO_2$ reacted, 2 mol ClO_2 are produced, so mol ClO_2 = mol $NaClO_2$.

$$0.1106 \text{ mol ClO}_2 \times \frac{67.45 \text{ g ClO}_2}{\text{mol}} = 7.46 \text{ g ClO}_2$$

10.121 After reaction, the flask contains $IF_5(g)$ and whichever reactant is in excess. Determine the limiting reactant, which regulates the moles of IF_5 produced and moles of excess reactant.

$$I_2(s) + 5F_2(g) \rightarrow 2\,IF_5(g)$$

$$10.0\,g\,I_2 \times \frac{1\,mol\,I_2}{253.8\,g\,I_2} \times \frac{5\,mol\,F_2}{1\,mol\,I_2} = 0.1970 = 0.197\,mol\,F_2$$

$$10.0\,g\,F_2 \times \frac{1\,mol\,F_2}{38.00\,g\,F_2} = 0.2632 = 0.263\,mol\,F_2\;\text{available}$$

I_2 is the limiting reactant; F_2 is in excess.

0.263 mol F_2 available – 0.197 mol F_2 reacted = 0.066 mol F_2 remain.

$$10.0\,g\,I_2 \times \frac{1\,mol\,I_2}{253.8\,g\,I_2} \times \frac{2\,mol\,IF_5}{1\,mol\,I_2} = 0.0788\,mol\,IF_5\;\text{produced}$$

(a) $P_{IF_5} = \dfrac{nRT}{V} = 0.0788\,mol \times \dfrac{0.08206\,L\text{-}atm}{K\text{-}mol} \times \dfrac{398\,K}{5.00\,L} = 0.515\,atm$

(b) $\chi_{IF_5} = \dfrac{mol\,IF_5}{mol\,IF_5 + mol\,F_2} = \dfrac{0.0788}{0.0788 + 0.066} = 0.544$

10.122 (a) $MgCO_3(s) + 2HCl(aq) \rightarrow MgCl_2(aq) + H_2O(l) + CO_2(g)$

$CaCO_3(s) + 2HCl(aq) \rightarrow CaCl_2(aq) + H_2O(l) + CO_2(g)$

(b) $n = \dfrac{PV}{RT} = 743\,torr \times \dfrac{1\,atm}{760\,torr} \times \dfrac{K\text{-}mol}{0.08206\,L\text{-}atm} \times \dfrac{1.72\,L}{301\,K}$

$$= 0.06808 = 0.0681\,mol\;CO_2$$

(c) $x = g\,MgCO_3,\; y = g\,CaCO_3,\; x + y = 6.53\,g$

mol $MgCO_3$ + mol $CaCO_3$ = mol CO_2 total

$$\frac{x}{84.32} + \frac{y}{100.09} = 0.06808;\; y = 6.53 - x$$

$$\frac{x}{84.32} + \frac{6.53 - x}{100.09} = 0.06808$$

$$100.09x - 84.32x + 84.32(6.53) = 0.06808\,(84.32)(100.09)$$

$$15.77x + 550.610 = 574.549;\; x = 1.52\,g\,MgCO_3$$

$$\text{mass \% } MgCO_3 = \frac{1.52\,g\,MgCO_3}{6.53\,g\,sample} \times 100 = 23.3\%$$

[By strict sig fig rules, the answer has 2 sig figs: 15.77x + 551 (3 digits from 6.53) = 575; 575 – 551 = 24 (no decimal places, 2 sig figs) leads to 1.5 g $MgCO_3$.]

11 Intermolecular Forces, Liquids, and Solids

Visualizing Concepts

11.2 **(a)(a)** Hydrogen bonding; H–F interactions qualify for this narrowly defined interaction.

(a)(b) London dispersion forces, the only intermolecular forces between nonpolar F_2 molecules.

(a)(c) Ion-dipole forces between Na^+ cation and the negative end of a polar covalent water molecule.

(a)(d) Dipole-dipole forces between oppositely charged portions of two polar covalent SO_2 molecules.

(b) Ion-dipole forces in **(a)(c)** and hydrogen bonding in **(a)(a)** are stronger than the other two.

11.3 The viscosity of glycerol will be greater than that of 1-propanol. Viscosity is the resistance of a substance to flow. The stronger the intermolecular forces in a liquid, the greater its viscosity. Hydrogen bonding is the predominant force for both molecules. Glycerol has three times as many O–H groups and many more H-bonding interactions than 1-propanol, so it experiences stronger intermolecular forces and greater viscosity. (Both molecules have the same carbon-chain length, so dispersion forces are similar.)

11.5 The stronger the intermolecular forces, the greater the average kinetic energy required to escape these forces, and the higher the boiling point. $CH_3CH_2CH_2OH$ has hydrogen bonding, by virtue of its –OH group, so it has the higher boiling point. Dispersion forces are similar because molar masses are the same for both molecules.

11.6 **(a)** 360 K, the normal boiling point; 260 K, normal freezing point. The left-most line is the freezing/melting curve, the right-most line is the condensation/boiling curve. The normal boiling and freezing points are the temperatures of boiling and freezing at 1 atm pressure.

(b) The material is solid in the green zone, liquid in the blue zone, and gas in the tan zone. (i) gas (ii) solid (iii) liquid

(c) The triple point, where all three phases are in equilibrium, is the point where the three lines on the phase diagram meet. For this substance, the triple point is approximately 185 K at 0.45 atm.

Molecular Comparisons of Gases, Liquids, and Solids

11.10   (a)   In solids, particles are in essentially fixed positions relative to each other, so the average energy of attraction is stronger than average kinetic energy. In liquids, particles are close together but moving relative to each other. The average attractive energy and average kinetic energy are approximately balanced. In gases, particles are far apart and in constant, random motion. Average kinetic energy is much greater than average energy of attraction.

       (b)   As the temperature of a substance is increased, the average kinetic energy of the particles increases. In a collection of particles (molecules), the state is determined by the strength of interparticle forces relative to the average kinetic energy of the particles. As the average kinetic energy increases, more particles are able to overcome intermolecular attractive forces and move to a less ordered state, from solid to liquid to gas.

       (c)   If a gas is placed under very high pressure, the particles undergo many collisions with the container and with each other. The large number of particle-particle collisions increases the likelihood that intermolecular attractions will cause the molecules to coalesce (liquefy).

11.12   (a)   The average distance between molecules is greater in the liquid state. Density is the ratio of the mass of a substance to the volume it occupies. For the same substance in different states, mass will be the same. The smaller the density, the greater the volume occupied, and the greater the distance between molecules. The liquid at 130° has the lower density (1.08 g/cm^3), so the average distance between molecules is greater.

       (b)   As the temperature of a substance increases, the average kinetic energy and speed of the molecules increases. At the melting point the molecules, on average, have enough kinetic energy to break away from the very orderly array that was present in the solid. As the translational motion of the molecules increases, the occupied volume increases and the density decreases. Thus, the solid density, 1.266 g/cm^3 at 15°C, is greater than the liquid density, 1.08 g/cm^3 at 130°C.

Intermolecular Forces

11.14   Intermolecular forces are based on charge attraction and repulsion. Because substances must be electrically neutral overall, there are opposite charges (full or partial) in every substance. While repulsions occur, net forces are attractive because attractions lower the overall energy of the sample, and matter tends to exist in the lowest possible energy state.

11.16   (a) CH_3OH experiences hydrogen bonding, but CH_3SH does not.

       (b)   Both gases are influenced by London dispersion forces. The heavier the gas particles, the stronger the London dispersion forces. The heavier Xe is a liquid at the specified conditions, while the lighter Ar is a gas.

       (c)   Both gases are influenced by London dispersion forces. The larger, diatomic Cl_2 molecules are more polarizable, experience stronger dispersion forces, and have the higher boiling point.

(d)   Acetone and 2-methylpropane are molecules with similar molar masses and London dispersion forces. Acetone also experiences dipole-dipole forces and has the higher boiling point.

11.18   (a)   True. A more polarizable molecule can develop a larger transient dipole, increasing the strength of electrostatic attractions and dispersion forces among molecules.

(b)   False. The noble gases are all monoatomic. Going down the column, the atomic radius and the size of the electron cloud increase. The larger the electron cloud, the more polarizable the atom, the stronger the London dispersion forces and the higher the boiling point.

(c)   False. The smaller the molecule, the more tightly its electron cloud is held. Tightly held electron clouds are not very polarizable and dispersion forces are weak.

(d)   True. It is usually true that as the number of electrons in a molecule increases, the size of the molecule increases. Larger molecules tend to have diffuse electron clouds, which are more polarizable. Thus, the statement that more electrons lead to increased dispersion forces (and greater polarizability) is correct.

11.20   For molecules with similar structures, the strength of dispersion forces increases with molecular size (molecular weight and number of electrons in the molecule).

(a) Br_2

(b) $CH_3CH_2CH_2CH_2CH_2SH$

(c) $CH_3CH_2CH_2Cl$. These two molecules have the same molecular formula and molecular weight (C_3H_7Cl, molecular weight = 78.5 amu), so the shapes of the molecules determine which has the stronger dispersion forces. According to Figure 11.6, the cylindrical (not branched) molecule will have stronger dispersion forces.

11.22   Both molecules experience hydrogen bonding through their –OH groups and dispersion forces between their hydrocarbon portions. The position of the –OH group in isopropyl alcohol shields it somewhat from approach by other molecules and slightly decreases the extent of hydrogen bonding. Also, isopropyl alcohol is less rod-like (it has a shorter chain) than propyl alcohol, so dispersion forces are weaker. Since hydrogen bonding and dispersion forces are weaker in isopropyl alcohol, it has the lower boiling point.

11.24   (a)   HF has the higher boiling point because hydrogen bonding is stronger than dipole-dipole forces.

(b) $CHBr_3$ has the higher boiling point because it has the higher molar mass, which leads to greater polarizability and stronger dispersion forces.

(c)   ICl has the higher boiling point because it is a polar molecule. For molecules with similar structures and molar masses, dipole-dipole forces are stronger than dispersion forces.

11.26   (a) C_6H_{14}, dispersion; C_8H_{18}, dispersion. C_8H_{18} has the higher boiling point due to greater molar mass and similar strength of forces.

(b) C_3H_8, dispersion; CH_3OCH_3, dipole-dipole, and dispersion. CH_3OCH_3 has the higher boiling point due to stronger intermolecular forces and similar molar mass.

(c)   HOOH, hydrogen bonding, dipole-dipole, and dispersion; HSSH, dipole-dipole, and dispersion. HOOH has the higher boiling point due to the influence of hydrogen bonding (Figure 11.7).

(d) NH_2NH_2, hydrogen bonding, dipole-dipole, and dispersion; CH_3CH_3, dispersion. NH_2NH_2 has the higher boiling point due to much stronger intermolecular forces.

11.28   (a)   In the solid state, NH_3 molecules are arranged so as to form the maximum number of hydrogen bonds. At the melting point, the average kinetic energy of the molecules is large enough so that they are free to move relative to each other. As they move, old hydrogen bonds break and new ones form, but the strict relative order required for maximum hydrogen bonding is no longer present.

(b)   In the liquid state, molecules are moving relative to one another while touching, which makes some hydrogen bonding possible. When molecules achieve enough kinetic energy to vaporize, the distance between them increases beyond the point where hydrogen bonds can form.

Viscosity and Surface Tension

11.30   (a)   *Cohesive* forces bind molecules to each other.

Adhesive forces bind molecules to surfaces.

(b)   The cohesive forces are hydrogen bonding among water molecules and also hydrogen bonding among cellulose molecules in the paper towel. Adhesive forces are any attractive forces between water and cellulose (the paper towel), likely also hydrogen bonding. If adhesive forces between cellulose and water weren't significant, paper towels wouldn't absorb water.

(c)   The shape of a meniscus depends on the strength of the cohesive forces within a liquid relative to the adhesive forces between the walls of the tube and the liquid. Adhesive forces between polar water molecules and silicates in glass (Figure 11.16) are even stronger than cohesive hydrogen-bonding forces among water molecules, so the meniscus is U-shaped (concave-upward).

11.32   (a) $H—\ddot{N}—\ddot{N}—H \qquad H—\ddot{\underset{..}{O}}—\ddot{\underset{..}{O}}—H \qquad H—\ddot{\underset{..}{O}}—H$
                    $\underset{H}{|} \quad \underset{H}{|}$

(b)   All have bonds (N–H or O–H, respectively) capable of forming hydrogen bonds. Hydrogen bonding is the strongest intermolecular interaction between neutral molecules and leads to very strong cohesive forces in liquids. The stronger the cohesive forces in a liquid, the greater the surface tension.

Phase Changes

11.34   (a)   condensation, exothermic

          (b)   sublimation, endothermic

          (c)   vaporization (evaporation), endothermic

          (d)   freezing, exothermic

11.36   (a)   Liquid ethyl chloride at room temperature is far above its boiling point. When the liquid contacts the metal surface, heat sufficient to vaporize the liquid is transferred from the metal to the ethyl chloride, and the heat content of the molecules increases. At constant atmospheric pressure, $\Delta H = q$, so the heat content and the enthalpy content of $C_2H_5Cl(g)$ is higher than that of $C_2H_5Cl(l)$. This indicates that the specific heat of the gas is less than that of the liquid, because the heat content of the gas is starts at a higher level.

          (b)   Liquid C_2H_5Cl is vaporized (boiled), C_2H_5Cl (g) is warmed to the final temperature, and the solid surface is cooled to the final temperature. The enthalpy of vaporization (ΔH_{vap}) of $C_2H_5Cl(l)$, the specific heat of $C_2H_5Cl(g)$, and the specific heat of the solid surface must be considered.

11.38   Energy released when 200 g of H_2O is cooled from 15°C to 0°C:

$$\frac{4.184\,J}{g\text{-}K} \times 200\,g\,H_2O \times 15°C = 12.55 \times 10^3\,J = 13\,kJ$$

Energy released when 200 g of H_2O is frozen (there is no change in temperature during a change of state):

$$\frac{334\,J}{g} \times 200\,g\,H_2O = 6.68 \times 10^4\,J = 66.8\,kJ$$

Total energy released = 12.55 kJ + 66.8 kJ = 79.35 = 79.4 kJ

Mass of freon that will absorb 40.9 kJ when vaporized:

$$79.35\,kJ \times \frac{1 \times 10^3\,J}{1\,kJ} \times \frac{1\,g\,CCl_2F_2}{289\,J} = 275\,g\,CCl_2F_2$$

11.40   Consider the process in steps, using the appropriate thermochemical constant.

Heat the liquid from 10.00°C to 47.6°C, $\Delta T = 37.6°C = 37.6\,K$, using the specific heat of the liquid.

$$50.0\,g\,C_2Cl_3F_3 \times \frac{0.91\,J}{g\text{-}K} \times 37.6\,K \times \frac{1\,kJ}{1000\,J} = 1.711 = 1.7\ kJ$$

Boil the liquid at 47.6°C (320.6 K), using the enthalpy of vaporization.

$$50.0\,g\,C_2Cl_3F_3 \times \frac{1\,mol\,C_2Cl_3F_3}{187.4\,g\,C_2Cl_3F_3} \times \frac{27.49\,kJ}{mol} = 7.3346 = 7.33\ kJ$$

Heat the gas from 47.6°C to 85.00°C, $\Delta T = 37.4°C = 37.4\,K$, using the specific heat of the gas.

$$50.0 \, g \, C_2Cl_3F_3 \times \frac{0.67 \, J}{g \cdot K} \times 37.4 \, K \times \frac{1 \, kJ}{1000 \, J} = 1.253 = 1.3 \, kJ$$

The total energy required is 1.711 kJ + 7.335 kJ + 1.253 kJ = 10.299 = 10.3 kJ.

11.42   (a) CCl_3F, CCl_2F_2 and $CClF_3$ are polar molecules that experience dipole-dipole and London dispersion forces with like molecules. CF_4 is a nonpolar compound that experiences only dispersion forces.

         (b)   According to Solution 11.41(b), the higher the critical temperature, the stronger the intermolecular attractive forces of a substance. Therefore, the strength of intermolecular attraction increases moving from right to left across the series and as molecular weight increases. $CF_4 < CClF_3 < CCl_2F_2 < CCl_3F$.

         (c)   The increasing intermolecular attraction with increasing molecular weight indicates that the critical temperature and pressure of CCl_4 will be greater than that of CCl_3F. Looking at the numerical values in the series, an increase of 88 K in critical temperature and 3.1 atm in critical pressure to the corresponding values for CCl_3F seem reasonable.

Physical Property	CCl_3F	CCl_4(predicted)	CCl_4 (CRC)
Critical Temperature (K)	47.1	557	556.6
Critical Pressure (atm)	43.5	46.6	44.6

            The predicted values for CCl_4 are in very good agreement with literature values. The key concept is that dispersion, not dipole-dipole, forces dominate the physical properties in the series.

Vapor Pressure and Boiling Point

11.44   (a)   The pressure difference on the manometer is 130 mm Hg and the gas in the vessel is essentially 100% molecules of the substance in the vapor phase. When the vessel is evacuated, virtually all air is removed. As the frozen liquid warms, it establishes a vapor pressure of 130 mm Hg. This is the pressure difference on the manometer.

         (b)   The pressure difference is 1 atm. The gas is 130 mm Hg of the molecular vapor and the rest is air. The liquid vaporizes in contact with the atmosphere, so atmospheric pressure is maintained above the liquid, but the equilibrium gas composition reflects the amount of vapor necessary to maintain 130 mm pressure, plus enough air to maintain a total pressure of 1 atm.

         (c)   The pressure difference is 890 mm Hg (1 atm + 130 mm Hg) and the gas is a mixture of 130 mm vapor and 1 atm air. The initial air pressure in the flask is 1 atm and no air is allowed to escape. The gas in the flask is not in equilibrium with the atmosphere and the final pressure in the flask does not equal atmospheric pressure. After a most of the liquid vaporizes, the total gas pressure is the result of 130 mm vapor and 1 atm air.

11.46   (a)   False. The heavier (and larger) CBr_4 has stronger dispersion forces, a higher boiling point, lower vapor pressure and is less volatile.

(b)   True.

(c)   False.

(d)   False

11.48   (a)   On a humid day, there are more gaseous water molecules in the air and more are recaptured by the surface of the liquid, making evaporation slower.

(b)   At high altitude, atmospheric pressure is lower and water boils at a lower temperature. Because the boiling water is cooler, it takes longer to brew tea at high altitude. (Water may actually boil faster at high altitude, because the boiling temperature is lower.)

11.50   (a)

A plot of vapor pressure vs. temperature data for H_2O from Appendix B is shown above. The vapor pressure of water at body temperature, 37°C, is approximately 50 torr.

(b)   The data point at 760.0 torr, 100°C is the normal boiling point of H_2O. This is the temperature at which the vapor pressure of H_2O is equal to a pressure of 1 atm or 760 torr.

(c)   At an external (atmospheric) pressure of 633 torr, the boiling point of H_2O is approximately 96°C.

(d)   At an external pressure of 774 torr, the boiling point of water is approximately 100.5°C.

(e)   Follow the logic in Sample Exercise 10.14 to calculate rms speeds at the two temperatures. The rms speed is one way to represent the "average" speed of a large collection of particles.

$u = (3RT/MM)^{1/2}$. $MM = 18.0$ g/mol $= 18.0 \times 10^{-3}$ kg/mol

$R = 8.314$ kg - m^2/s^2 - mol - K. At T = 96°C = 369 K,

$$u = \left(\frac{3(8.314 \text{ kg - m}^2/\text{s}^2 \text{ - mol - K}) \, 369 \text{ K}}{18.0 \times 10^{-3} \text{ kg}/\text{mol}} \right)^{1/2} = 715 \text{ m}/\text{s}$$

At T = 100.5 °C = 373.6 K,

$$u = \left(\frac{3(8.314 \text{ kg - m}^2/\text{s}^2 \text{ - mol - K}) \, 373.6 \text{ K}}{18.0 \times 10^{-3} \text{ kg}/\text{mol}} \right)^{1/2} = 719.50 = 720 \text{ m}/\text{s}$$

The difference in the two rms speeds is less than 1% of the average of the two values. Given the precision of estimating boiling temperatures from the plot, the two rms speeds are essentially equal.

Phase Diagrams

11.52   (a)   The *triple point* on a phase diagram represents the temperature and pressure at which the gas, liquid, and solid phases are in equilibrium.

(b)   No. A phase diagram represents a closed system, one where no matter can escape and no substance other than the one under consideration is present; air cannot be present in the system. Even if air is excluded, at 1 atm of external pressure, the triple point of water is inaccessible, regardless of temperature [see Sample Exercise 11.6(b)].

11.54   (a)   Solid CO_2 sublimes to form $CO_2(g)$ at a temperature of about –60°C.

(b)   Solid CO_2 melts to form $CO_2(l)$ at a temperature of about –50°C. The $CO_2(l)$ boils when the temperature reaches approximately –40°C.

11.56   (a)

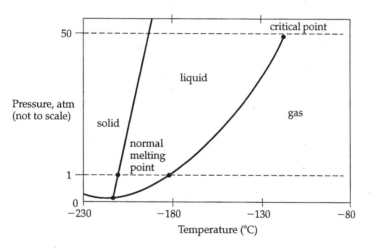

On the diagram above, the triple point is at the dot closest to the x-axis. The normal melting point is at the intersection of the solid-liquid line and the 1-atm dashed line. The normal boiling point is at the intersection of the liquid-gas line and the 1-atm line. The critical point is clearly marked.

(b) $O_2(s)$ will not float on $O_2(l)$. $O_2(s)$ is denser than $O_2(l)$ because the solid-liquid line on the phase diagram is normal. That is, as pressure increases, the melting temperature increases. [Note that the solid-liquid line for O_2 is nearly vertical, indicating a small difference in the densities of $O_2(s)$ and $O_2(l)$].

(c) $O_2(s)$ will melt when heated at a pressure of 1 atm, since this is a much greater pressure than the pressure at the triple point.

Structures of Solids

11.58   In amorphous silica (SiO_2) the regular structure of quartz is disrupted; the loose, disordered structure, Figure 11.30(b), has many vacant "pockets" throughout. There are fewer SiO_2 groups per volume in the amorphous solid; the packing is less efficient and less dense.

11.60   (a)   Ti: 8 corners × 1/8 sphere/corner + [1 center × 1 sphere/center] = 2 Ti atoms

              O: 4 faces × 1/2 sphere/face + [2 interior × 1 sphere/interior] = 4 O atoms

              Formula: TiO_2

        (b)   Rutile is an ionic solid; ion-ion forces among Ti^{4+} cations and O^{2-} anions are quite strong, owing to the magnitudes of the charges, and lead to the ordered structure.

        (c)   From inspection of the unit cell diagram, the Ti at the center of the cell is bound to six (6) O atoms. Each totally interior O atom is bound to three titanium atoms. The coordination number of Ti is 6; the coordination number of O is 3.

11.62   (a)   8 corners × 1/8 atom/corner + 6 faces × ½ atom/face = 4 atoms

        (b)   Each aluminum atom is in contact with 12 nearest neighbors, 6 in one plane, 3 above that plane, and 3 below. Its coordination number is thus 12.

        (c)   The length of the face diagonal of a face-centered cubic unit cell is four times the radius of the metal and $\sqrt{2}$ times the unit cell dimension (usually designated a for cubic cells).

$$4 \times 1.43 \text{ Å} = \sqrt{2} \times a; \; a = \frac{4 \times 1.43 \text{ Å}}{\sqrt{2}} = 4.0447 = 4.04 \text{ Å} = 4.04 \times 10^{-8} \text{ cm}$$

        (d)   The density of the metal is the mass of the unit cell contents divided by the volume of the unit cell.

$$\text{density} = \frac{4 \text{ Al atoms}}{(4.0447 \times 10^{-8} \text{ cm})^3} \times \frac{26.98 \text{ g Al}}{6.022 \times 10^{23} \text{ Al atoms}} = 2.71 \text{ g/cm}^3$$

11.64   Avogadro's number is the number of KCl formula units in 74.55 g of KCl.

$$74.55 \text{ g KCl} \times \frac{1 \text{ cm}^3}{1.984 \text{ g}} \times \frac{(1 \times 10^{10} \text{ pm})^3}{1 \text{ cm}^3} \times \frac{4 \text{ KCl units}}{628^3 \text{ pm}^3} = 6.07 \times 10^{23} \text{ KCl formula units}$$

11.66   (a) Na^+, 6   (b) Zn^{2+}, 4   (c) Ca^{2+}, 8

11.68   (a)   According to Figure 11.42(b), there are 4 HgS units in a unit cell with the zinc blende structure. [4 complete Hg^{2+} ions, 6(1/2) + 8(1/8) S^{2-} ions]

$$\text{density} = \frac{4 \text{ HgS units}}{(5.852 \text{ Å})^3} \times \frac{232.655 \text{ g}}{6.022 \times 10^{23} \text{ HgS units}} \times \left(\frac{1 \text{ Å}}{1 \times 10^{-8} \text{ cm}} \right)^3 = 7.711 \text{ g/cm}^3$$

(b) We expect Se^{2-} to have a larger ionic radius than S^{2-}, since Se is below S in the chalcogen family and both ions have the same charge. Thus, HgSe will occupy a larger volume and the unit cell edge will be longer.

(c) For HgSe:

$$\text{density} = \frac{4 \, \text{HgSe units}}{(6.085 \, \text{Å})^3} \times \frac{279.55 \, \text{g HgSe}}{6.022 \times 10^{23} \, \text{HgSe units}} \times \left(\frac{1 \, \text{Å}}{1 \times 10^{-8} \, \text{cm}} \right)^3 = 8.241 \, \text{g/cm}^3$$

Even though HgSe has a larger unit cell volume than HgS, it also has a larger molar mass. The mass of Se is more than twice that of S, while the radius of Se^{2-} is only slightly larger than that of S^{2-} (Figure 7.8). The greater mass of Se accounts for the greater density of HgSe.

Bonding in Solids

11.70 (a) According to Figure 11.41(a), diamond is a covalent-network solid. Carbon atoms are bound into a three-dimensional network by strong covalent (single) bonds. Silicon, with the same structure, is also a covalent-network solid.

(b) Si is a metalloid and O is a nonmetal; we expect their bonding to be covalent. Silica, $SiO_2(s)$, will be either molecular or covalent network. By analogy with Si(s), each Si atom will be tetrahedral and participate in four covalent bonds. If the stoichiometry of the solid is SiO_2, a network structure is required for Si to form four single covalent bonds with O. Discreet linear SiO_2 molecules analogous to CO_2 are unlikely, because the covalent radius of Si is too large for effective π-overlap. To confirm this hypothesis, the *Handbook of Chemistry and Physics* lists the melting points of the various forms of $SiO_2(s)$ to be greater than 1600°C. Very high melting points are characteristic of covalent-network solids.

11.72 (a) ionic　　　　　　(b) metallic

(c) covalent-network (Based on melting point, it is definitely a network solid. Transition metals in high oxidation states often form bonds with nonmetals that have significant covalent character. It could also be characterized as ionic with same covalent character to the bonds.)

(d) molecular　　　　　(e) molecular　　　　　(f) molecular

11.74 (a) metallic

(b) molecular or metallic (physical properties of metals vary widely)

(c) covalent-network or ionic　　　(d) covalent-network　　　(e) ionic

11.76 According to Table 11.7, the solid could be either ionic with low water solubility or network covalent. Due to the extremely high sublimation temperature, it is probably covalent-network.

11.78 (a) HF, hydrogen bonding versus dipole-dipole for HCl

(b) C(graphite), covalent-network bonding versus London dispersion forces for CH_4

(c) KCl, ionic versus dispersion forces for nonpolar Cl_2

(d) MgF_2, higher charge on Mg^{2+} than Li^+

Additional Exercises

11.80　(a)　Correct.

　　　(b)　The lower boiling liquid must experience less total intermolecular forces.

　　　(c)　If both liquids are structurally similar nonpolar molecules, the lower boiling liquid has a lower molecular weight than the higher boiling liquid.

　　　(d)　Correct.

　　　(e)　At their boiling points, both liquids have vapor pressures of 760 mm Hg.

11.81　(a)　The cis isomer has stronger dipole-dipole forces; the trans isomer is nonpolar. The higher boiling point of the cis isomer supports this conclusion.

　　　(b)　While boiling points are primarily a measure of strength of intermolecular forces, melting points are influenced by crystal packing efficiency as well as intermolecular forces. Since the nonpolar trans isomer with weaker inter-molecular forces has the higher melting point, it must pack more efficiently.

11.82　(a)　In dibromomethane, CH_2Br_2, the dispersion force contribution will be larger than for CH_2Cl_2, because bromine is more polarizable than the lighter element chlorine. At the same time, the dipole-dipole contribution for CH_2Cl_2 is greater than for CH_2Br_2 because CH_2Cl_2 has a larger dipole moment.

　　　(b)　Just the opposite comparisons apply to CH_2F_2, which is less polarizable and has a higher dipole moment than CH_2Cl_2.

11.84　The GC base pair, with more hydrogen bonds, is more stable to heating. In order to break up a base pair by heating, sufficient thermal energy must be added to break the existing hydrogen bonds. With 50% more hydrogen bonds, the GC pair is definitely more stable (harder to break apart) than the AT pair.

11.85　The two O–H groups in ethylene glycol are involved in many hydrogen bonding interactions, leading to its high boiling point and viscosity, relative to pentane, which experiences only dispersion forces.

11.86　The more carbon atoms in the hydrocarbon, the longer the chain, the more polarizable the electron cloud, the higher the boiling point. A plot of the number of carbon atoms versus boiling point is shown below. For 8 C atoms, C_8H_{18}, the boiling point is approximately 130°C.

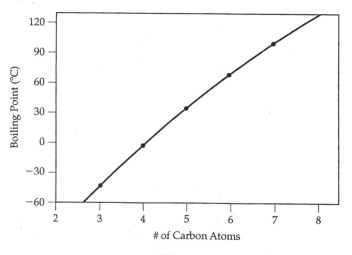

# of Carbon Atoms

11.87 For a substance above its critical temperature, as pressure increases beyond critical pressure, the solubility of a solute increases. The solubility of the solute is essentially zero below critical pressure. Supercritical CO_2 at very high pressure dissolves caffeine, and the solution leaves the extractor. The pressure reduction value reduces the pressure of CO_2 enough so that the caffeine becomes insoluble. The solid caffeine is deposited in the separator, and the low pressure CO_2 gas is recycled.

11.89 (a) If the Clausius-Clapeyron equation is obeyed, a graph of ln P vs 1/T(K) should be linear. Here are the data in a form form for graphing.

T(K)	1/T	P(torr)	ln P
280.0	3.571×10^{-3}	32.42	3.479
300.0	3.333×10^{-3}	92.47	4.527
320.0	3.125×10^{-3}	225.1	5.417
330.0	3.030×10^{-3}	334.4	5.812
340.0	2.941×10^{-3}	482.9	6.180

According to the graph, the Clausius-Clapeyron equation is obeyed, to a first approximation.

$$\Delta H_{vap} = -slope \times R; \ slope = \frac{3.479 - 6.180}{(3.571 - 2.941) \times 10^{-3}} = -\frac{2.701}{0.630 \times 10^{-3}} = -4.29 \times 10^3$$

$$\Delta H_{vap} = -(-4.29 \times 10^3) \times 8.314 \ J/K\text{-}mol = 35.7 \ kJ/mol$$

(b) The normal boiling point is the temperature at which the vapor pressure of the liquid equals atmospheric pressure, 760 torr. From the graph,

ln 760 = 6.63, 1/T for this vapor pressure = 2.828×10^{-3}; T = 353.6 K

11.90 (a) The Clausius-Clapeyron equation is $\ln P = \dfrac{-\Delta H_{vap}}{RT} + C$.

For two vapor pressures, P_1 and P_2, measured at corresponding temperatures T_1 and T_2, the relationship is

$$\ln P_1 - \ln P_2 = \left(\frac{-\Delta H_{vap}}{RT_1} + C \right) - \left(\frac{-\Delta H_{vap}}{RT_2} + C \right)$$

$$\ln P_1 - \ln P_2 = \frac{-\Delta H_{vap}}{R} \left(\frac{1}{T_1} - \frac{1}{T_2} \right) + C - C; \ \ln \frac{P_1}{P_2} = \frac{-\Delta H_{vap}}{R} \left(\frac{1}{T_1} - \frac{1}{T_2} \right)$$

(b) P_1 = 13.95 torr, T_1 = 298 K; P_2 = 144.78 torr, T_2 = 348 K

$$\ln \frac{13.95}{144.78} = \frac{-\Delta H_{vap}}{8.314 \ J/K\text{-}mol} \left(\frac{1}{298} - \frac{1}{348} \right)$$

$$-2.33974 \ (8.314 \ J/K\text{-}mol) = -\Delta H_{vap} \ (4.821 \times 10^{-4}/K)$$

$\Delta H_{vap} = 4.035 \times 10^4 = 4.0 \times 10^4$ J/mol = 40 kJ/mol

$[(1/T_1) - (1/T_2)]$ has 2 sig figs and so does the result.

(c)   The normal boiling point of a liquid is the temperature at which the vapor pressure of the liquid is 760 torr.

$P_1 = 144.78$ torr, $T_1 = 348$ K; $P_2 = 760$ torr, $T_2 = $ b.p. of octane

$$\ln\left(\frac{144.78}{760.0}\right) = \frac{-4.035 \times 10^4 \text{ J/mol}}{8.314 \text{ J/K-mol}}\left(\frac{1}{348 \text{ K}} - \frac{1}{T_2}\right)$$

$$\frac{-1.6581}{-4.8533 \times 10^3} = 2.874 \times 10^{-3} - \frac{1}{T_2}; \quad \frac{1}{T_2} = 2.874 \times 10^{-3} - 3.416 \times 10^{-4}$$

$$\frac{1}{T_2} = 2.532 \times 10^{-3} = 2.53 \times 10^{-3}; \quad T_2 = 394 \text{ K } (122°\text{C})$$

This is in good agreement with the literature boiling point of octane, 126°C.

(d) $P_1 = $ vp of octane at 30°C, $T_1 = 303$ K; $P_2 = 144.78$ torr, $T_2 = 348$ K

$$\ln\frac{P_1}{144.78 \text{ torr}} = \frac{-4.035 \times 10^4 \text{ J/mol}}{8.314 \text{ J/K-mol}}\left(\frac{1}{303} - \frac{1}{348}\right)$$

$$\ln\frac{P_1}{144.78 \text{ torr}} = \frac{-4.035 \times 10^4 \text{ J/mol}}{8.314 \text{ J/K-mol}} \times 4.268 \times 10^{-4} = -2.0710 = -2.1$$

$$\frac{P_1}{144.78 \text{ torr}} = e^{-2.0710}; \quad P_1 = 0.12606 \ (144.78) = 18.25 = 2 \times 10^1 \text{ torr}$$

[Strictly speaking, this result has 1 sig fig because (ln = –2.1) has 1 decimal place. In a ln or log, the places left of the decimal show order of magnitude, and places right of the decimal show sig figs in the real number.] The result, ~18 torr at 30°C, is reasonable, given that the vapor pressure of octane at 25°C is 13.95 torr. We expect vapor pressure to increase with temperature.

11.91   Physical data for the two compounds from the *Handbook of Chemistry and Physics*:

	MM	dipole moment	boiling point
CH_2Cl_2	85 g/mol	1.60 D	40.0°C
CH_3I	142 g/mol	1.62 D	42.4°C

(a)   The two substances have very similar molecular structures; each is an unsymmetrical tetrahedron with a single central carbon atom and no hydrogen bonding. Since the structures are very similar, the magnitudes of the dipole-dipole forces should be similar. This is verified by their very similar dipole moments. The heavier compound, CH_3I, will have slightly stronger London dispersion forces. Since the nature and magnitude of the intermolecular forces in the two compounds are nearly the same, it is very difficult to predict which will be more volatile [or which will have the higher boiling point as in part (b)].

(b)   Given the structural similarities discussed in part (a), one would expect the boiling points to be very similar, and they are. Based on its larger molar mass (and dipole-dipole forces being essentially equal) one might predict that CH_3I would have a slightly higher boiling point; this is verified by the known boiling points.

(c)   According to Equation 11.1, $\ln P = \dfrac{-\Delta H_{vap}}{RT} + C$

A plot of $\ln P$ vs. $1/T$ for each compound is linear. Since the order of volatility changes with temperature for the two compounds, the two lines must cross at some temperature; the slopes of the two lines, ΔH_{vap} for the two compounds, and the y-intercepts, C, must be different.

(d) CH_2Cl_2

ln P	T(K)	1/T
2.303	229.9	4.351×10^{-3}
3.689	250.9	3.986×10^{-3}
4.605	266.9	3.747×10^{-3}
5.991	297.3	3.364×10^{-3}

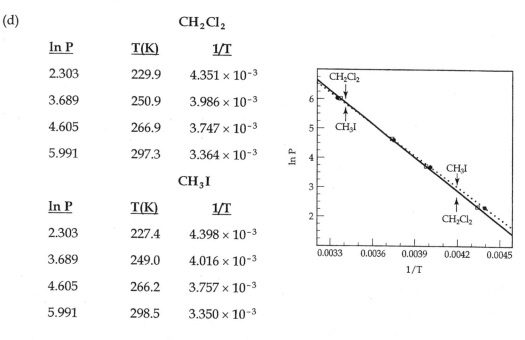

CH_3I

ln P	T(K)	1/T
2.303	227.4	4.398×10^{-3}
3.689	249.0	4.016×10^{-3}
4.605	266.2	3.757×10^{-3}
5.991	298.5	3.350×10^{-3}

For CH_2Cl_2, $-\Delta H_{vap}/R = \text{slope} = \dfrac{(5.991 - 2.303)}{(3.364 \times 10^{-3} - 4.350 \times 10^{-3})} = \dfrac{-3.688}{0.987 \times 10^{-3}}$

$$= -3.74 \times 10^3 = -\Delta H_{vap}/R$$

$\Delta H_{vap} = 8.314 \, (3.74 \times 10^3) = 3.107 \times 10^4 \text{ J/mol} = 31.1 \text{ kJ/mol}$

For CH_3I, $-\Delta H_{vap}/R = \text{slope} = \dfrac{(5.991 - 2.303)}{(3.350 \times 10^{-3} - 4.398 \times 10^{-3})} = \dfrac{-3.688}{1.048 \times 10^{-3}} = -3.519 \times 10^3$

$$= -\Delta H_{vap}/R$$

$\Delta H_{vap} = 8.314 \, (3.519 \times 10^3) = 2.926 \times 10^4 \text{ J/mol} = 29.3 \text{ kJ/mol}$

11.93   In the antifluorite structure, the Li^+ cations occupy the 8 interior positions in the unit cell shown in Figure 11.42(c), while the O^{2-} anions occupy the corners and face-centers.

(a) Li^+, coordination number = 4; O^{2-}, coordination number = 8. Each Li^+ is nearest three O^{2-} anions in face-centers, and one O^{2-} on a corner. Each O^{2-} anion is nearest four Li^+ ions above and four Li+ ions below it.

(b)   By analogy to the alkali metal chlorides (NaCl to CsCl), we expect the coordination number of the cation to increase as its ionic radius increases.

(c)   In the A_2O series, the anion remains constant while the cation changes. In order to maintain the 2:1 ratio of cations to anions, all 8 tetrahedral holes (interior of the unit cell) must be occupied by cations. As the cation size increases, the cation becomes too large to sit in the same lattice of O^{2-} anions. As the larger cations expand (push apart) the O^{2-} lattice, the antifluorite structure becomes unstable.

11.94   (a)   The face diagonal of a face-centered cubic unit cell has length $a\sqrt{2}$ and also $4r$, where a is the cubic cell dimension and r is the atomic radius.

$$4r = a\sqrt{2}; r = (a\sqrt{2})/4 = (4.078 \text{ Å})(\sqrt{2})/4 = 1.44179 = 1.442 \text{ Å}$$

(b)   Density is the mass of the unit cell contents divided by the unit cell volume (a^3). In a face-centered cubic unit cell, there are 4 Au atoms.

$$\text{density} = \frac{4 \text{ Au atoms}}{(4.078 \text{ Å})^3} \times \frac{196.97 \text{ g}}{6.0221 \times 10^{23} \text{ Au atoms}} \times \left(\frac{1 \text{ Å}}{1 \times 10^{-8} \text{ cm}}\right)^3 = 19.29 \text{ g/cm}^3$$

11.96   (a) $n\lambda = 2d \sin\theta$; $n = 1$, $\lambda = 1.54$ Å, $\theta = 14.22°$; calculate d.

$$d = \frac{n\lambda}{2 \sin\theta} = \frac{1 \times 1.54 \text{ Å}}{2 \sin(14.22)} = 3.1346 = 3.13 \text{ Å}$$

(b)   Assume the diffracted intensity at 14.22°C is the result of second-order diffraction, $n = 2$.

$$d = \frac{n\lambda}{2 \sin\theta} = \frac{2 \times 1.54 \text{ Å}}{2 \sin(14.22)} = 6.269 = 6.27 \text{ Å}$$

Note that this is exactly twice the d-spacing calculated assuming $n = 1$.

11.97   (a)   Both diamond ($d = 3.5$ g/cm^3) and graphite ($d = 2.3$ g/cm^3) are covalent-network solids with efficient packing arrangements in the solid state; there is relatively little empty space in their respective crystal lattices. Diamond, with bonded C–C distances of 1.54 Å in all directions, is more dense than graphite, with shorter C–C distances within carbon sheets but longer 3.41 Å separations between sheets (Figure 11.41). Buckminsterfullerene has much more empty space, both inside each C_{60} "ball" and between balls, than either diamond or graphite, so its density will be considerably less than 2.3 g/cm^3.

(b)   In a face-centered cubic unit cell, there are 4 complete C_{60} units.

$$\frac{4 \, C_{60} \text{ units}}{(14.2 \text{ Å})^3} \times \frac{720.66 \text{ g}}{6.022 \times 10^{23} \, C_{60} \text{ units}} \times \left(\frac{1 \text{ Å}}{1 \times 10^{-8} \text{ cm}}\right)^3 = 1.67 \text{ g/cm}^3$$

(1.67 g/cm^3 is the smallest density of the three allotropes, diamond, graphite, and buckminsterfullerene.)

Integrative Exercises

11.99   (a)   In Table 11.4, viscosity increases as the length of the carbon chain increases. Longer molecular chains become increasingly entangled, increasing resistance to flow.

(b)   Whereas viscosity depends on molecular chain length in a critical way, surface tension depends on the strengths of intermolecular interactions between molecules. These dispersion forces do not increase as rapidly with increasing chain length and molecular weight as viscosity does.

(c)   The –OH group in n-octyl alcohol gives rise to hydrogen bonding among molecules, which increases molecular entanglement and leads to greater viscosity and higher boiling point.

11.100   (a)   24 valence e^-, 12 e^- pairs

The geometry around the central C atom is trigonal planar, and around the two terminal C atoms, tetrahedral.

(b)   Polar. The C=O bond is quite polar and the dipoles in the trigonal plane around the central C atom do not cancel.

(c)   Dipole-dipole and London-dispersion forces

(d)   Since the molecular weights of acetone and 1-propanol are similar, the strength of the London-dispersion forces in the two compounds is also similar. The big difference is that 1-propanol has hydrogen bonding, while acetone does not. These relatively strong attractive forces lead to the higher boiling point for 1-propanol.

11.102   (a)   In order for butane to be stored as a liquid at temperatures above its boiling point (–5°C), the pressure in the tank must be greater than atmospheric pressure. In terms of the phase diagram of butane, the pressure must be high enough so that, at tank conditions, the butane is "above" the gas-liquid line and in the liquid region of the diagram.

The pressure of a gas is described by the ideal gas law as P = nRT/V; pressure is directly proportional to moles of gas. The more moles of gas present in the tank the greater the pressure, until sufficient pressure is achieved for the gas to liquify. At the point where liquid and gas are in equilibrium and temperature is constant, liquid will vaporize or condense to maintain the equilibrium vapor pressure. That is, as long as some liquid is present, the gas pressure in the tank will be constant.

(b)   If butane gas escapes the tank, butane liquid will vaporize (evaporate) to maintain the equilibrium vapor pressure. Vaporization is an endothermic process, so the butane will absorb heat from the surroundings. The temperature of the tank and the liquid butane will decrease.

(c) $250\,g\,C_4H_{10} \times \dfrac{1\,mol\,C_4H_{10}}{58.12\,g\,C_4H_{10}} \times \dfrac{21.3\,kJ}{mol} = 91.6\,kJ$

$$V = \frac{nRT}{P} = 250\,g \times \frac{1\,mol}{58.12\,g} \times \frac{0.08206\,L\text{-}atm}{mol\text{-}K} \times \frac{308\,K}{755\,torr} \times \frac{760\,torr}{1\,atm} = 109.44 = 109\,L$$

11.103 *Plan.*

(i)   Using thermochemical data from Appendix B, calculate the energy (enthalpy) required to melt and heat the H_2O.

(ii)   Using Hess's Law, calculate the enthalpy of combustion, ΔH_{comb}, for C_3H_8.

(iii)   Solve the stoichiometry problem.

Solve.

(i)   Heat $H_2O(s)$ from $-20°C$ to $0.0°C$; $5.50 \times 10^3\,g\,H_2O \times \frac{2.092\,J}{g-°C} \times 20°C = 2.301 \times 10^2$

$$= 2.3 \times 10^2\,kJ$$

Melt $H_2O(s)$; $5.50 \times 10^3\,g\,H_2O \times \frac{6.008\,kJ}{mol\,H_2O} \times \frac{1\,mol\,H_2O}{18.02\,g\,H_2O} = 1834 = 1.83 \times 10^3\,kJ$

Heat $H_2O(l)$ from $0°C$ to $75°C$; $5.50 \times 10^3\,g\,H_2O \times \frac{4.184\,J}{g-°C} \times 75°C = 1726 = 1.7 \times 10^3\,kJ$

Total energy = 230.1 kJ + 1834 kJ + 1726 kJ = 3790 = 3.8 $\times$ 10^3 kJ

(The result is significant to 100 kJ, limited by 1.7 $\times 10^3$ kJ)

(ii) $C_3H_8(g) + 5O_2(g) \rightarrow 3CO_2(g) + 4H_2O(l)$

Assume that one product is $H_2O(l)$, since this leads to a more negative ΔH_{comb} and fewer grams of $C_3H_8(g)$ required.

$$\Delta H_{comb} = 3\Delta H_f° \, CO_2(g) + 4\Delta H_f° \, H_2O(l) - \Delta H_f° \, C_3H_8(g) - 5\Delta H_f° \, O_2(g)$$

$$= 3(-393.5\,kJ) + 4\,(-285.83\,kJ) - (-103.85\,kJ) - 5(0) = -2219.97 = -2220\,kJ$$

(iii) $3.790 \times 10^3\,kJ$ required $\times \frac{1\,mol\,C_3H_8}{2219.97\,kJ} \times \frac{44.096\,g\,C_3H_8}{1\,mol\,C_3H_8} = 75\,g\,C_3H_8$

(3.8 $\times$ 10^3 kJ required has 2 sig figs and so does the result)

11.104 (a)   Low viscosity, low surface tension, and, especially, high thermal conductivity owing to metallic properties.

(b)   Metallic bonding between sodium atoms, which persists in the liquid state, is probably the main inhibitor of movement of atoms relative to one another. As temperature increases, thermal motion of the atoms increases and the liquid expands, weakening bonding interactions relative to thermal energies.

11.106 *Plan.* Relative humidity and v.p. of H_2O at given $T \rightarrow P_{H_2O} \rightarrow$ ideal-gas law $\rightarrow$ mol $H_2O(g) \rightarrow H_2O$ molecules. Change $°F \rightarrow °C$, volume of room from $ft^3 \rightarrow L$.

Solve. $°C = 5/9\,(°F - 32)$; $°C = 5/9\,(68\,°F - 32) = 20°C$;

r.h. $= (P_{H_2O}$ in air / v.p. of $H_2O) \times 100$

From Appendix B, v.p. of H_2O at 20°C = 17.54 torr

P_{H_2O} in air = r.h. × v.p. of H_2O/100 = 58 × 17.54 torr/100 = 10.173 = 10 torr

$$V = 12\,ft \times 10\,ft \times 8\,ft \times \frac{12^3\ in^3}{ft^3} \times \frac{2.54^3\ cm^3}{in^3} \times \frac{1\,L}{1000\ cm^3} = 2.718 \times 10^4 = 3 \times 10^4\ L$$

(The result has 1 sig fig, as does the measurement 8 ft.)

PV = nRT; n = PV/RT

$$n = 10.17\ torr \times \frac{1\,atm}{760\ torr} \times \frac{mol\text{-}K}{0.08206\ L\text{-}atm} \times \frac{2.718 \times 10^4\ L}{293\ K} = 15.13 = 2 \times 10^1\ mol\ H_2O$$

$$15.13\ mol\ H_2O \times \frac{6.022 \times 10^{23}\ molecules}{1\ mol} = 9.110 \times 10^{24} = 9 \times 10^{24}\ H_2O\ molecules$$

11.107 Data are taken from the 74th edition of the *Handbook of Chemistry and Physics*. T_m = melting point, T_b = boiling point

(a) W: T_m = 3410°C, T_b = 5660 °C; WF_6: T_m = 2.5°C, T_b = 17.5°C

W is a metal, with strong metallic bonding, and very high T_m and T_b. WF_6 is an octahedral, nonpolar molecule. Even though it has high molar mass, the spherical shape of the molecule prevents extensive molecular contacts. The resulting London dispersion forces are very weak, which leads to the low T_m and T_b.

(b) SO_2: T_m = –72.7°C, T_b = –10°C; SF_4: T_m = –124°C, T_b = –40°C (sublimes)

Both SO_2 and SF_4 are polar covalent molecules with a nonbonding electron pair on the central S atom. The electron-domain geometry in SO_2 is trigonal planar and the molecule shape is bent. The electron-domain geometry in SF_4 is trigonal bipyramidal and the molecular shape is see-saw. Both are gases at ambient temperature and pressure. SF_4 has higher molar mass but lower melting and boiling points than SO_2. This indicates that dipole-dipole forces are more influential on the properties of these molecules and that SF_4 has a smaller dipole moment than SO_2.

(c) SiO_2: T_m = 1723°C, T_b = 2230°C, MM = 60 g/mol

$SiCl_4$: T_m = –70°C, T_b = 57.57°C, MM = 170 g/mol

SiO_2 is a covalent-network substance, with high T_m and T_b. Covalent bonds hold SiO_2 units in a rigid lattice, and high energy is required to break these bonds and melt or boil the substance. $SiCl_4$ is a tetrahedral, nonpolar molecule. Weak dispersion forces between the approximately spherical molecules result in predictably low T_m and T_b.

12 Modern Materials

Visualizing Concepts

12.2 *Analyze/Plan.* Reverse the logic in Sample Exercise 12.1. That is, given the relative sizes of two band gaps, deduce the position of the component elements on the periodic table.

Solve. Material X has the larger band gap. The elements in X will be nearer the top of the chart, or have a greater horizontal separation. The possible materials are AlAs and GaAs; As is one component of each semiconductor. The second components, Al and Ga, are in the same family, and Al is above Ga. We expect AlAs to have the larger band gap. X is AlAs and Y is GaAs.

12.4 *Analyze/Plan.* Given diagrams with rod-like (columnar) molecules in different orientations, decide which diagram represents molecules in a liquid crystalline phase. Recall the molecular orientations that define solids, liquid crystals and isotropic liquids.

Solve. Liquid crystals exhibit molecular order in at least one dimension. Solids display three-dimensional order, and liquids are randomly ordered with close molecular contacts. Diagram (a) has columnar molecules with their long directions oriented parallel to the vertical direction of the box, a one-dimensional order characteristic of a liquid crystalline phase. Diagram (b) has the close but randomly oriented molecules characteristic of an isotropic liquid.

12.6 The plots show *resistivity*, resistance to the flow of electrical current. A superconductor loses all resistance to electric current flow at some relatively low temperature, the superconducting transition temperature, T_c. The top plot shows this behavior, with a T_c of approximately 19 K. The resistivity of a metal also decreases as temperature decreases, but lacks the sharp drop to zero of a superconductor. The bottom plot shows this behavior. By elimination, the middle plot must show the behavior of a semiconductor. According to the hint, thermal energy is needed to excite electrons and increase the conductivity of a semiconductor. That is, semiconductor conductivity increases and resistivity decreases as temperature increases. This is indeed the behavior shown in the middle plot.

Classes of Materials and Electronic Structure of Materials

12.8 (a) InAs—semiconductor, Gp 3A + Gp 5A

     (b) MgO—insulator; ionic compound

     (c) HgS—semiconductor, like CdS, average 4 e^- per atom

     (d) Sn—metal; p-block metal left of metalloid line

12.10

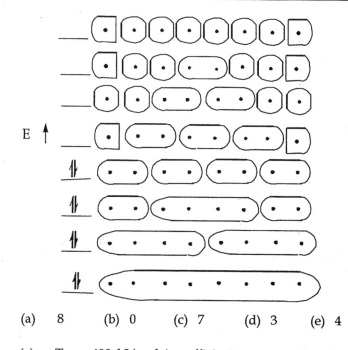

(a) 8   (b) 0   (c) 7   (d) 3   (e) 4

12.12   (a)   True. 400 kJ/mol is sufficient energy to break many chemical bonds. It is an energy barrier large enough to prohibit any charge mobility.

   (b)   True. The energy difference between valence and conduction can be small as in the case of metals, but the conduction band is always at the higher energy. Otherwise all materials would be conductors.

   (c)   False. Electrons can conduct only if they are in a partially filled band. If electrons in filled bands could conduct, there would be no insulators.

   (d)   False. "Holes" are empty electron sites. They are the absence of an electron where one would normally exist, leading to a region of at least partial positive charge.

Semiconductors

12.14   (a)   InP. P is in the same family and higher than As.

   (b)   AlP. Al and P are horizontally separated, resulting in greater bond polarity, and Al and P are in the row above Ge. Bond gap values in Table 12.2 confirm this order.

   (c)   AgI. The four elements are in the same row; Ag and I are farther apart than Cd and Te.

12.16   p-type semiconductors have a slight electron deficit. If the dopant replaces As, Group 5A, it should have fewer than five valence electrons. The dopant will be a 4A element, again probably Si or Ge. Si would be closer to As in bonding atomic radius.

12.18   Impurities can be thought of as dopants. Since dopants at concentrations of parts per million (ppm) can change the conductivity of Si, it needs to be more pure than the ppm level to ensure known and constant conductivity. Purity at ppm level would be 99.9999% pure. Purity of 99.999999999% is pure at nearly the parts per trillion level.

12.20   The bonding atomic radius of a Si atom is 1.11 Å (Figure 7.7), so the diameter is 2.22 Å.

$$65 \text{ nm} \times \frac{1 \times 10^{-9} \text{ m}}{1 \text{ nm}} \times \frac{1 \text{ Å}}{1 \times 10^{-10} \text{ m}} \times \frac{1 \text{ Si atom}}{2.22 \text{ Å}} = 2.9 \times 10^2 \text{ Si atoms.}$$

12.22   (a)   From Table 12.2, the band gap of CdTe is 145 kJ/mol, or 1.50 eV.

        (b) $\lambda = \dfrac{hc}{E} = 6.626 \times 10^{-34} \text{ J-S} \times \dfrac{2.998 \times 10^8 \text{ m}}{s} \times \dfrac{1}{1.50 \text{ eV}} \times \dfrac{1 \text{ eV}}{1.602 \times 10^{-19} \text{ J}}$

$$= 8.267 \times 10^{-7} = 8.27 \times \times 10^{-7} \text{ m}$$

        (c)   CdTe can absorb energies ≥ 1.50 eV or wavelengths $\leq 8.27 \times 10^{-7}$ m. The solar spectrum ranges from 30×10^{-7} m to 2×10^{-7} m, a span of 28.27×10^{-7} m. In this range CdTe can absorb from 8.27×10^{-7} m to 2×10^{-7} m, a span of $6.27 \times 10^{-7} = 6 \times 10^{-7}$ m.

$$\frac{6.27 \times 10^{-7}}{28 \times 10^{-7}} \times 100 = 22.38 = 22\% \text{ of the solar spectrum}$$

        (more than 22% of the total flux)

        (d)   CdTe can absorb 22% of the solar spectrum, while Si (see Solution 12.21) can absorb 32%. Whether comparing % of wavelengths or total flux, Si can absorb a larger portion of the solar spectrum than CdTe.

12.24   The band gap, ΔE of GaAs is 1.43 eV.

$$\lambda = \frac{hc}{E} = 6.626 \times 10^{-34} \text{ J-S} \times \frac{2.998 \times 10^8 \text{ m}}{s} \times \frac{1}{1.43 \text{ eV}} \times \frac{1 \text{ eV}}{1.602 \times 10^{-19} \text{ J}} = 8.67 \times 10^{-7} \text{ m (867 nm)}$$

The visible portion of the electromagnetic spectrum has wavelengths up to 750 nm (7.50×10^{-7} m). GaAs emits longer, 867 nm, light in the IR portion of the spectrum.

12.26   *Analyze/Plan.* Reverse the logic in Solution 12.25. Given λ, calculate E_g. Then solve for x assuming the value of E_g is a linear combination of the stoichiometric contributions of GaP and GaAs. That is, $2.26 x + 1.43(1 - x) = E_g$. *Solve.*

$$E_g = \frac{hc}{\lambda} = \frac{6.626 \times 10^{-34} \text{ J-S}}{6.60 \times 10^{-7} \text{ m}} \times \frac{2.998 \times 10^8 \text{ m}}{s} \times \frac{1 \text{ eV}}{1.602 \times 10^{-19} \text{ J}} = 1.8788 = 1.88 \text{ eV}$$

$2.26 x + 1.43(1 - x) = 1.8788;$ $2.26 x - 1.43 x + 1.43 = 1.8788$

$0.83 x = 0.4488;$ $x = 0.5407 = 0.54$

Check. From Solution 12.25, an E_g of 1.85 V corresponds to x = 0.5. $E_g = 1.88$ eV should have a very similar composition, and x = 0.54 is very close to x = 0.5. The P/As composition is very sensitive to wavelength, and provides a useful mechanism to precisely tune the wavelength of the diode.

Ceramics

12.28   Ceramics are not widely used to construct automobile parts because they are brittle and subject to catastrophic failure. They are not easily deformed under stress and tend to shatter when struck. Unlike metals, ceramics break rather than bend.

12.30   Since Zr and Ti are in the same family, assume that the stoichiometry of the compounds in a sol-gel process will be the same for the two metals.

   i.   Alkoxide formation: oxidation-reduction reaction

$$Zr(s) + 4CH_3CH_2OH(l) \rightarrow Zr(OCH_2CH_3)_4(s) + 2H_2(g)$$
$$\text{alkoxide}$$

   ii.   Sol formation: metathesis reaction

$$Zr(OCH_2CH_3)_4(soln) + 4H_2O(l) \rightarrow \underset{\substack{\text{"precipitate"}\\\text{sol}}}{Zr(OH)_4(s)} + \underset{\text{nonelectrolyte}}{4CH_3CH_2OH(l)}$$

   $Zr(OCH_2CH_3)_4(s)$ is dissolved in an alcohol solvent and then reacted with water. In general, reaction with water is called _hydrolysis_. The alkoxide anions $(CH_3CH_2O^-)$ combine with H^+ from H_2O to form the nonelectrolyte $CH_3CH_2OH(l)$, and Zr^{2+} cations combine with OH^- to form the $Zr(OH)_4$ solid. The product $Zr(OH)_4(s)$ is not a traditional coagulated precipitate, but a finely divided evenly dispersed collection of particles called a sol.

   iii. Gel formation: condensation reaction

$$(OH)_3Zr-O-H(s) + H-O-Zr(OH)_3(s) \rightarrow (HO)_3Zr-O-Zr(OH)_3(s) + H_2O(l)$$
$$\text{gel}$$

   Adjusting the acidity of the $Zr(OH)_4$ sol initiates condensation, the splitting-out of $H_2O(l)$ and formation of a zirconium-oxide network solid. The solid remains suspended in the solvent mixture and is called a gel.

   iv.   Processing: physical changes

   The gel is heated to drive off solvent and the resulting solid consists of dry, uniform and finely divided ZrO_2 particles.

12.32   (a)   Gallium arsenide, GaAs (Figure 12.3).

         (b)   Covalent. Silicon and carbon are both nonmetals, and their electronegativities are similar.

         (c)   In SiC, the C atoms form a face-centered cubic array with Si atoms occupying alternate tetrahedral holes in the lattice. This means that the coordination numbers of both Si and C are 4; each Si is bound to four C atoms in a tetrahedral arrangement, and each C is bound to four Si atoms in a tetrahedral arrangement, producing an extended three-dimensional network. SiC is high-melting because a great deal of chemical energy is stored in the covalent Si-C bonds. Melting requires breaking covalent Si-C bonds, which takes a huge amount of thermal energy. It is hard because the three-dimensional lattice resists any change that would weaken the Si-C bonding network.

Superconductors

12.34   A conductive metal such as Ag conducts electricity with a characteristic resistance to the flow of electrons given by Ohm's law, $E = IR$. A superconducting substance such as Nb_3Sn below its transition temperature conducts electricity with no resistance to the flow of electrons. Such a superconductor can transfer energy with no net loss, while a metallic conductor cannot be 100% efficient.

        Superconductors below their T_c also exhibit the Meisner effect; they exclude magnetic fields from their volume. As a result, a small permanent magnet is repelled by a superconductor below its T_c (Figure 12.17). A metal is attracted to a permanent magnet.

12.36   (a)   The superconducting transition temperature, T_c, is the temperature at which a material loses all resistance to the flow of electrical current, the temperature below which the material becomes superconducting.

         (b)   The temperature 77 K is significant because that is the temperature of liquid nitrogen, a readily available, inexpensive, and safe coolant. Materials with T_c temperatures above 77 K produce more financially viable devices than materials which must be cooled with liquid helium below 77 K to achieve superconductivity.

12.38   (a)   These metal nitrides are ceramics, rather than metals or metal alloys. By analogy to the conductivity of $YBa_2Cu_3O_7$, another ceramic superconductor, NbN will be a metallic conductor above its T_c, probably at temperatures as high as room temperature.

         (b)   Since ScN is an ionic network solid and not a superconductor at low temperatures, it will be an insulator at room temperature.

         (c)   Each of these ionic nitrides has a 1:1 stoichiometry so the charge on the metal cation is 3+. The group 4B cations have one d-electron, and the group 5B cations have two d-electrons. Again by analogy to $YBa_2Cu_3O_7$, covalent bonding involving unpaired d-electrons on the cation probably contributes to the conductivity and superconductivity of these nitrides. In ScN, bonding is primarily ionic, since Sc^{3+} has no d-electrons available for covalent bonding. The covalent bonding that appears to be one bonding characteristic of ceramic superconductors is not present in ScN, but is present in the group 4 and 5 nitrides.

Polymers

12.40   *n*-decane does not have a sufficiently high chain length or molecular mass to be considered a polymer.

12.42

12.44   (a)   By analogy to polyisoprene, Equation [12.7],

12.48   At the molecular level, the longer, unbranched chains of HDPE fit closer together and have more crystalline (ordered, aligned) regions than the shorter, branched chains of LDPE. Closer packing leads to higher density.

12.50   (a)   An *elastomer* is a polymer material that recovers its shape when released from a distorting force. A typical elastomeric polymer can be stretched to at least twice its original length and return to its original dimensions upon release.

(b)   A *thermoplastic material* can be shaped and reshaped by application of heat and/or pressure.

(c)   A *thermosetting* plastic can be shaped once, through chemical reaction in the shape-forming process, but cannot easily be reshaped, due to the presence of chemical bonds that cross-link the polymer chains.

(d)   A *plasticizer* is a substance of relatively low molar mass added to a polymer material to soften it.

Biomaterials

12.52   One structural characteristic of polymers that forms effective interfaces with biological systems is the presence of polar functional groups in the polymer backbone or as substituents. Polystyrene is a hydrocarbon; it has no polar functional groups and is a nonpolar substance. Polyurethane has polar carbon–oxygen, carbon–nitrogen, and nitrogen–hydrogen functional groups. The N–H groups mean that it can act as a hydrogen-bond donor as well as an acceptor. In fact, the polyurethane backbone is very similar to the protein backbone shown in this section. We expect polyurethane to be the superior biointerface.

12.54   Surface roughness in synthetic heart valves causes hemolysis, the breakdown of red blood cells. The surface of the valve implant was probably not smooth enough.

12.56   Polystyrene is an essentially nonpolar hydrocarbon, while polyethyleneterephthalate (PET) contains polar ester groups, as well as nonpolar hydrocarbon portions. PET is

more appropriate, because it provides polar ester groups $\left(\!-\overset{\overset{\displaystyle O}{\|}}{C}-O-C-\!\right)$ with

hydrogen bonding capabilities where the cells can attach. Also, the ester linkages are susceptible to hydrolysis (the reverse of condensation); this renders the synthetic matrix biodegradable when employed in the body.

Liquid Crystals

12.58   In an ordinary liquid, molecules are oriented randomly and their relative orientations are continuously changing. In liquid crystals, the molecules are aligned in at least one dimension. The relative orientations in the other two dimensions may change, but alignment in the oriented direction is maintained.

12.60   Reinitzer observed that cholesteryl benzoate has a phase that exhibits properties intermediate between those of the solid and liquid phases. This "liquid-crystalline" phase, formed by melting at 145°C, is opaque, changes color as the temperature increases, and becomes clear at 179°C.

12.62   Because order is maintained in at least one dimension, the molecules in a liquid-crystalline phase are not totally free to change orientation. This makes the liquid-crystalline phase more resistant to flow, more viscous, than the isotropic liquid.

12.64   The "LCD molecule" is long relative to its thickness. It has C=C and C≡N groups that promote rigidity and polarizability along the length of the molecule. The C≡N group also provides dipole-dipole interactions that encourage alignment. Unlike the molecules in Figure 12.34, which contain planar phenyl rings, the LCD molecule contains nonaromatic, nonplanar six-membered rings. These rings could contribute to specific physical properties such as the liquid crystal temperature range that make this molecule particularly functional in LCD displays.

12.66   As the temperature of a substance increases, the average kinetic energy of the molecules increases. More molecules have sufficient kinetic energy to overcome intermolecular attractive forces, so overall ordering of the molecules decreases as temperature increases. Melting provides kinetic energy sufficient to disrupt alignment in one dimension in the solid, producing a smectic phase with ordering in two dimensions. Additional heating of the smectic phase provides kinetic energy sufficient to disrupt alignment in another dimension, producing a nematic phase with one-dimensional order.

Nanomaterials

12.68   (a)   Calculate the wavelength of light that corresponds to 2.4 eV, then look at a visible spectrum such as Figure 6.4 to find the color that corresponds to this wavelength.

$$\lambda = \frac{hc}{E} = 6.626 \times 10^{-34} \text{ J-s} \times \frac{3.00 \times 10^8 \text{ m}}{s} \times \frac{1}{2.4 \text{ eV}} \times \frac{1 \text{ eV}}{1.602 \times 10^{-19} \text{ J}}$$

$$= 5.167 \times 10^{-7} = 5.2 \times 10^{-7} \text{ m (520 nm)}$$

The emitted 520 nm light is green.

(b)   Yes. The wavelength of blue light is shorter than the wavelength of green light. Emitting a shorter wavelength requires a larger band gap. As particle size decreases, band gap increases. Appropriately sized CdS quantum dots, far smaller than a large CdS crystal, would have a band gap greater than 2.4 eV and could emit shorter wavelength blue light.

(c)   No. The wavelength of red light is greater than the wavelength of green light. The large CdS crystal has plenty of AOs contributing to the MO scheme to ensure maximum delocalization. A bigger crystal will not reduce the size of the band gap. The 2.4 eV band gap of the large CdS crystal represents the minimum energy and maximum wavelength of light that can be emitted by this material.

12.70   True. Blue light has short wavelengths, corresponding to a relatively large band gap. As particle size decreases, band gap increases and wavelength decreases. We could begin with a semiconductor with a smaller band gap and make it a nanoparticle to increase E_g and decrease wavelength. (Nanoparticle size becomes one more way to tune the properties of semiconductors.)

12.72   There are four CdTe formula units in each cubic unit cell. Calculate the number of unit cells contained in a 5 nm cubic crystal. The volume of one unit cell is

$$(6.49)^3 \text{ Å}^3 \times \frac{1 \text{ nm}^3}{10^3 \text{ Å}^3} = 0.2734 = 0.273 \text{ nm}^3$$

The volume of the 5 nm crystal is 5^3 nm^3 = 125 = 1×10^2 nm^3

(The dimension 5 nm limits the volume to one sig fig.)

$$\frac{125 \text{ nm}^3/\text{crystal}}{0.2734 \text{ nm}^3/\text{unit cell}} \times \frac{4 \text{ CdTe units}}{\text{unit cell}} = 1829 = 2 \times 10^3 \text{ CdTe units}$$

That is, 2000 (or 1829) Cd atoms and 2000 (or 1829) Te atoms.

Additional Exercises

12.74   A dipole moment (permanent, partial charge separation) roughly parallel to the long dimension of the molecule would cause the molecules to reorient when an electric field is applied perpendicular to the usual direction of molecular orientation.

12.76   (a)   polymer   (b)   ceramic   (c)   ceramic   (d)   polymer

(e)   liquid crystal [an organic molecule with a characteristic long axis and the kinds of functional groups often found in compounds with liquid-crystalline phases (Figure 12.39); not enough repeating units to be a polymer]

12.77   Ceramics are usually three-dimensional network solids, whereas plastics most often consist of large, chain-like molecules (the chain may be branched) held loosely together by relatively weak van der Waals forces. Ceramics are rigid precisely because of the many strong bonding interactions intrinsic to the network. Once a crack forms, atoms near the defect are subject to great stress, and the crack is propagated. They are stable to high temperatures because tremendous kinetic energy (temperature) is required for an atom to break free from the bonding network. On the other hand, plastics are flexible because the molecules themselves are flexible (free rotation around the sigma bonds in the polymer chain), and it is easy for the molecules to move relative to one another (weak intermolecular forces). [However, recall that rigidity of the plastic can be increased in a number of ways, including cross-linking (Figure 12.28), or reinforcement with a second polymer (Figure 12.30).] Plastics are not thermally stable because their largely organic molecules are subject to oxidation and/or bond breaking at high temperatures.

12.78   In a liquid crystal display (Figure 12.42), the molecules must be free to rotate by 90°. The long directions of molecules remain aligned but any attractive forces between the ends of molecules are disrupted. At low Antarctic temperatures, the liquid crystalline phase is closer to its freezing point. The molecules have less kinetic energy due to temperature and the applied voltage may not be sufficient to overcome orienting forces among the ends of molecules. If some or all of the molecules do not rotate when the voltage is applied, the display will not function properly.

12.80   This phenomenon is similar to supercooling, Section 11.4. When the isotropic liquid is cooled below the liquid crystal-liquid transition temperature, the kinetic energy of the molecules has been decreased enough so that formation of the liquid crystalline phase is energetically favorable. However, the molecules may not be correctly organized so that long range ordering can take place.

12.81

Hydrogen bonding occurs between—C—N— amide groups of adjacent chains.

12.82

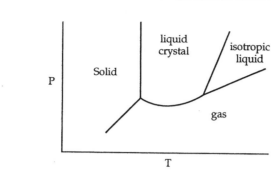

Integrative Exercises

12.85   (a)   sp³ hybrid orbitals at C, 109° bond angles around C

(b)

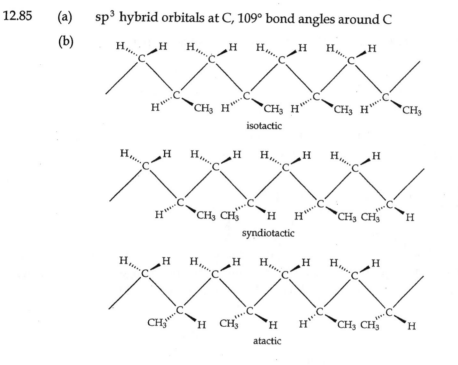

isotactic

syndiotactic

atactic

Isotactic polypropylene has the highest degree of crystallinity and highest melting point. The regular shape of the polymer backbone allows for close, orderly (almost zipper-like) contact between chains. This maximizes dispersion forces between chains and produces higher order (crystallinity) and melting point. Atactic polypropylene has the least order and the lowest melting point.

(c)   Cotton, with ——C—— groups and polyester, with—C—O—C

groups, both participate in hydrogen bonding interactions with H_2O molecules. These are strong intermolecular forces that hold the "moisture" at the surface of the fabric next to the skin. Polypropylene has no strong interactions with water, and capillary action "wicks" the moisture away from the skin.

12.86   If the expected oxidation states on Y and Ba are +3 and +2, respectively, the average oxidation state of Cu is +2 1/3. That is, two Cu ions are in the +2 state and one is in the +3 state. Y^{3+} and Ba^{2+} have the stable electron configurations of their nearest noble gases, while Cu has an incomplete d orbital set. Cu(II) is d^9, Cu(III) is d^8 and both have unpaired electrons. Although the mechanism by which copper 3d electrons interact through bridging oxygen atoms to form a superconducting state is still not clear, it is evident that the electronic structure of copper ions is essential to the observed superconductivity.

12.88   (a)   C—Cl   328 kJ/mol ⟵ lowest
C—C   348 kJ/mol
C—H   413 kJ/mol

(b)   C–Cl bonds are weakest, so they are most likely to break upon heating.

(c)   The repeating unit in polyvinyl chloride consists of two C atoms, each in a different environment. Consider the net changes in these two C atoms when the polymer is converted to diamond a high pressure.

Diamond is a covalent-network structure where each C atom is tetrahedrally bound to four other C atoms [Figure 11.41 (a)].

Assume that there is no net change to the C–C bonds in the structure, even though they may be broken and reformed. The net change to the 2-C vinyl chloride unit is then breaking three C–H bonds and one C–Cl bond, and making four C–C bonds.

ΔH = D(C–H) – D(C–I) – 4D(C–) = 3(413) + 328 – 4(348)

    = 523 kJ/vinyl choride unit

12.89   (a)

H—C—Ö—C ... C—N=N—C ... C—Ö—C—H

There are several other resonance structures involving alternate placement of the double bonds in the benzene rings.

(b)   Both N atoms are surrounded by 3 VSEPR electron domains, so the hybridization at both atoms is sp^2. The bond angles around the N attached to O will be approximately 120°.

(c)   When the $-OCH_3$ group is replaced by the $-CH_2CH_2CH_2CH_3$ group, a small rather compact group with some polarity is replaced by a larger, more flexible, nonpolar group. Thus, the molecules don't line up as well in the solid, and the melting point and liquid crystal temperature range are lower.

(d)   The density decreases going from solid to nematic liquid crystal to isotropic liquid. In the nematic liquid crystal, most of the long range order of the solid state is lost. The molecules are moving relative to one another with their long axes more or less aligned. The result is more empty space and a lower density than the solid. There is a further small decrease in density when the last degree of order is lost and the substance becomes an isotropic liquid.

12.90   (a)   Follow the logic outlined in Solution 12.71.

vol. of unit cell = $(5.43 \text{ Å})^3 = 160.1030 = 1.60 \times 10^2 \text{ Å}^3$

$$1 \text{ cm}^3 \times \frac{(1)^3 \text{ Å}^3}{(1 \times 10^{-8})^3 \text{ cm}^3} = 1 \times 10^{24} \text{ Å}^3 \text{ (volume of material)}$$

$$\frac{4 \text{ Si atoms}}{160.103 \text{ Å}^3} = \frac{x \text{ Si atoms}}{1 \times 10^{24} \text{ Å}^3}; x = 2.4984 \times 10^{22} = 2.50 \times 10^{22} \text{ Si atoms}$$

(To 1 sig fig, the result is 2×10^{22} Si atoms.)

(b)   1 ppm phosphorus = 1 P atom per 1×10^6 Si atoms

$$\frac{1 \text{ P atom}}{1 \times 10^6 \text{ Si atoms}} = \frac{x \text{ P atoms}}{2.4984 \times 10^{22} \text{ Si Atoms}}; x = 2.4984 \times 10^{22}$$

$$= 2.50 \times 10^{16} \text{ P atoms}$$

$$2.4984 \times 10^{16} \text{ P atom} \times \frac{1 \text{ mol}}{6.022 \times 10^{23} \text{ atoms}} \times \frac{30.97376 \text{ g P}}{\text{mol}} \times \frac{1 \text{ mg}}{1 \times 10^{-3} \text{ g}}$$

$$= 1.29 \times 10^{-3} \text{ mg P } (1.29 \mu \text{ g})$$

13 Properties of Solutions

Visualizing Concepts

13.2   Lattice energy is the main component of ΔH_1, the enthalpy required to separate solute particles. If ΔH_1 is too large, the dissolving process is prohibitively endothermic, and the substance is not very soluble.

13.3   (a) < (b) < (c). In Section 13.1, *entropy* is qualitatively defined as randomness or dispersal in space. Container (a) contains only one kind of particle and particles are close together, so (a) has the least entropy. In container (b), the particles occupy approximately the same volume as container (a). However, (b) contains two kinds of particles, homogeneously mixed, so the degree of dispersal and randomness is greater than in (a). Container (c) contains two kinds of homogeneously mixed particles and they occupy a larger volume than in (b), so (c) has the greatest entropy.

13.4   The pink solid is hydrated $CoCl_2$, $CoCl_2 \cdot xH_2O$, where x is a specific integer. The waters of hydration are either associated with Co^{2+}, Cl^-, or sit in specific sites in the crystal lattice. When heated in an oven, the water molecules incorporated into the crystal lattice gradually gain kinetic energy and vaporize. The blue solid is anhydrous $CoCl_2$, absent the waters of hydration and with a different solid-state structure than the pink hydrate.

13.6   Solubility increases in the order Ar, $1.50 \times 10^{-3}\ M$ < Kr, $2.79 \times 10^{-3}\ M$ < Xe, $5 \times 10^{-5}\ M$, the order of increasing polarizability. As the molar mass of the ideal gas increases, atomic size increases and the electron cloud is less tightly held by the nucleus, causing the cloud to be more polarizable. The greater the polarizability, the stronger the dispersion forces between the gas atoms and water, the more likely the gas atom is to stay dissolved rather than escape the solution, the greater the solubility of the gas.

13.7   Vitamin B_6 is likely to be largely water soluble. The three –OH groups and the —N̈— can enter into many hydrogen bonding interactions with water. The relatively small molecular size indicates that dispersion forces will not play a large role in intermolecular interactions and the hydrogen bonding will dominate. Vitamin E is likely to be largely fat soluble. The long, rod-like hydrocarbon chain will lead to strong dispersion forces among vitamin E and mostly nonpolar fats. Although vitamin E has one –OH and one —Ö— group, the long hydrocarbon chain prevents water from surrounding and separating the vitamin E molecules, reducing its water-solubility.

13.8   According to Figure 13.18, the solubility of CO at 25°C and 1 atm pressure is approximately 0.96 mM. By Henry's Law, $S_g = k\ P_g$. At the same temperature and pressure, k will be the same, so $S_1/P_1 = S_2/P_2$.

$$\frac{0.96\,mM}{1\,atm} = \frac{2.5\,mM}{x\,atm}; x = \frac{2.5\,mM \times 1\,atm}{0.96\,atm} = 2.6\,atm$$

13.10   (a)   The blue line represents the solution. According to Raoult's law, the presence of a nonvolatile solute lowers the vapor pressure of a volatile solvent. At any given temperature, the blue line has a lower vapor pressure and represents the solution.

(b)   The boiling point of a liquid is the temperature at which the vapor pressure of the liquid is equal to atmospheric pressure. Assuming atmospheric pressure of 1.0 atm, the boiling point of the solvent (red line) is approximately 64°C. The boiling point of the solution is approximately 70°C.

13.11   Ideally, 0.50 L. If the volume outside the balloon is very large compared to 0.25 L, solvent will flow across the semipermeable membrane until the molarities of the inner and outer solutions are equal, 0.10 M. This requires an "inner" solution volume twice as large as the initial volume, or 0.50 L. (In reality, osmosis across the balloon membrane is not perfect. The solution concentration inside the balloon will be slightly greater than 0.10 M and the volume of the balloon will be slightly less than 0.50 L.)

13.12   A detergent for solubilizing large hydrophobic proteins (or any other large nonpolar solute, such as greasy dirt) needs a hydrophobic part to interact with the solute, and a hydrophilic part to interact with water. In n-octyl glycoside, the eight-carbon n-octyl chain has strong dispersion interactions with the hydrophobic (nonpolar) protein. The –OH groups on the glycoside (sugar) ring form strong hydrogen bonds with water. This causes the glycoside to dissolve, dragging the hydrophobic protein along with it.

The Solution Process

13.14   (a)   For the same solute, NaCl, in different solvents, solute-solute interactions (ΔH_1) are the same. Because water experiences hydrogen bonding while benzene has only dispersion forces, solvent-solvent interactions (ΔH_2) are greater for water. On the other hand, solute-solvent interactions (ΔH_3) are much weaker between ionic NaCl and nonpolar benzene than between ionic NaCl and polar water. It is the large difference in ΔH_3 that causes NaCl to be soluble in water but not in benzene.

(b)   Ion-dipole forces between cations and water molecules and relatively small lattice energies (ion-ion forces between cations and anions) lead to strongly hydrated cations.

13.16   From weakest to strongest solvent-solute interactions:

(b), dispersion forces < (c), hydrogen bonding < (a), ion-dipole

13.18   Separation of solvent molecules, ΔH_2, will be smallest in this case, because hydrogen bonding is the weakest of the intermolecular forces involved. ΔH_1 involves breaking ionic bonds, and ΔH_3 involves formation of ion-dipole interactions, both stronger forces than hydrogen bonding.

13.20   KBr is quite soluble in water because of the sizeable increase in disorder of the system (ordered KBr lattice → freely moving hydrated ions) associated with the dissolving process. An increase in disorder or randomness in a process tends to make that process spontaneous.

Saturated Solutions; Factors Affecting Solubility

13.22   (a) $\dfrac{1.22\,\text{mol}\ MnSO_4 \bullet H_2O}{1\,\text{L soln}} \times \dfrac{169.0\,\text{g}\ MnSO_4 \bullet H_2O}{1\,\text{mol}} \times 0.100\,\text{L}$

$$= 20.6\,\text{g}\ MnSO_4 \bullet H_2O/100\,\text{mL}$$

The 1.22 M solution is unsaturated.

       (b)   Add a known mass, say 5.0 g, of $MnSO_4 \bullet H_2O$, to the unknown solution. If the solid dissolves, the solution is unsaturated. If there is undissolved $MnSO_4 \bullet H_2O$, filter the solution and weigh the solid. If there is less than 5.0 g of solid, some of the added $MnSO_4 \bullet H_2O$, dissolved and the unknown solution is unsaturated. If there is exactly 5.0 g, no additional solid dissolved and the unknown is saturated. If there is more than 5.0 g, excess solute has precipitated and the solution is supersaturated.

13.24   (a)   at 30°C, $\dfrac{10\,\text{g}\ KClO_3}{100\,\text{g}\ H_2O} \times 250\,\text{g}\ H_2O = 25\,\text{g}\ KClO_3$

       (b) $\dfrac{66\,\text{g}\ Pb(NO_3)_2}{100\,\text{g}\ H_2O} \times 250\,\text{g}\ H_2O = 165 = 1.7 \times 10^2\,\text{g}\ Pb(NO_3)_2$

       (c) $\dfrac{3\,\text{g}\ Ce_2(SO_4)_3}{100\,\text{g}\ H_2O} \times 250\,\text{g}\ H_2O = 7.5 = 8\,\text{g}\ Ce_2(SO_4)_3$

13.26   Immiscible means that oil and water do not mix homogeneously; they do not dissolve. Many substances are called "oil," but they are typically nonpolar carbon-based molecules with fairly high molecular weights. As such, there are fairly strong dispersion forces among oil molecules. The properties of water are dominated by its strong hydrogen bonding. The dispersion-dipole interactions between water and oil are likely to be weak. Thus, ΔH_1 and ΔH_2 are large and positive, while ΔH_3 is small and negative. The net ΔH_{soln} is large and positive, and mixing does not occur.

13.28   For small n values, the dominant interactions among acid molecules will be hydrogen-bonding. As n increases, dispersion forces between carbon chains become more important and eventually dominate. Thus, as n increases, water solubility decreases and hexane solubility increases.

13.30   *Analyze/Plan.* Water, H_2O, is a polar solvent that forms hydrogen bonds with other H_2O molecules. The more soluble solute in each case will have intermolecular interactions that are most similar to the hydrogen bonding in H_2O.   *Solve.*

       (a)   Glucose, $C_6H_{12}O_6$, is more soluble because it is capable of hydrogen bonding (Figure 13.12). Nonpolar C_6H_{12} is capable only of dispersion interactions and does not have strong intermolecular interactions with polar (hydrogen bonding) H_2O.

(b) Ionic sodium propionate, CH_3CH_2COONa, is more soluble. Sodium propionate is a crystalline solid, while propionic acid is a liquid. The increase in disorder or entropy when an ionic solid dissolves leads to significant water solubility, despite the strong ion-ion forces (large ΔH_1) present in the solute (see Solution 13.20).

(c) HCl is more soluble because it is a strong electrolyte and completely ionized in water. Ionization leads to ion-dipole solute-solvent interactions, and an increase in disorder. CH_3CH_2Cl is a molecular solute capable of relatively weak dipole-dipole solute-solvent interactions and is much less soluble in water.

13.32 Pressure has an effect on O_2 solubility in water because, at constant temperature and volume, pressure is directly related to the amount of O_2 available to dissolve. The greater the partial pressure of O_2 above water, the more O_2 molecules are available for dissolution, and the more molecules that strike the surface of the liquid.

Pressure does not affect the amount or physical properties of NaCl, or ionic solids in general, so it has little influence on the dissolving of NaCl in water.

13.34 $650 \text{ torr} \times \dfrac{1 \text{ atm}}{760 \text{ torr}} = 0.855 \text{ atm}; P_{O_2} = \chi_{O_2}(P_t) = 0.21(0.855 \text{ atm}) = 0.1796 = 0.18 \text{ atm}$

$$S_{O_2} = kP_{O_2} = \frac{1.38 \times 10^{-3} \text{ mol}}{\text{L - atm}} \times 0.1796 \text{ atm} = 2.5 \times 10^{-4} M$$

Concentrations of Solutions

13.36 (a) $\text{mass } \% = \dfrac{\text{mass solute}}{\text{total mass solution}} \times 100$

$\text{mass solute} = 0.035 \text{ mol } I_2 \times \dfrac{253.8 \text{ g } I_2}{1 \text{ mol } I_2} = 8.883 = 8.9 \text{ g } I_2$

$\text{mass } \% I_2 = \dfrac{8.883 \text{ g } I_2}{8.883 \text{ g } I_2 + 115 \text{ g } CCl_4} \times 100 = 7.170 = 7.2\% I_2$

(b) $\text{ppm} = \dfrac{\text{mass solute}}{\text{total mass solution}} \times 10^6 = \dfrac{0.0079 \text{ g } Sr^{2+}}{1 \times 10^3 \text{ g } H_2O} \times 10^6 = 7.9 \text{ ppm } Sr^{2+}$

13.38 (a) $\dfrac{25.5 \text{ g } C_6H_5OH}{94.11 \text{ g/mol}} = 0.2710 = 0.271 \text{ mol } C_6H_5OH$

$\dfrac{425 \text{ g } CH_3CH_2OH}{46.07 \text{ g/mol}} = 9.2251 = 9.23 \text{ mol } CH_3CH_2OH$

$\chi_{C_6H_5OH} = \dfrac{0.2710}{0.2710 + 9.2251} = 0.02853 = 0.0285$

(b) $\text{mass } \% = \dfrac{25.5 \text{ g } C_6H_5OH}{25.5 \text{ g } C_6H_5OH + 425 \text{ g } CH_3CH_2OH} \times 100 = 5.66\% C_6H_5OH$

(c) $m = \dfrac{0.2710 \text{ mol } C_6H_5OH}{0.425 \text{ kg } CH_3CH_2OH} = 0.6376 = 0.638 \text{ m } C_6H_5OH$

13.40   (a) $M = \dfrac{\text{mol solute}}{\text{L soln}}; \dfrac{15.0\,\text{g Al}_2(\text{SO}_4)_3}{0.350\,\text{L soln}} \times \dfrac{1\,\text{mol Al}_2(\text{SO}_4)_3}{342.2\,\text{g Al}_2(\text{SO}_4)_3} = 0.125\,M\,\text{Al}_2(\text{SO}_4)_3$

      (b) $\dfrac{5.25\,\text{g Mn(NO}_3)_2 \cdot 2\text{H}_2\text{O}}{0.175\,\text{L soln}} \times \dfrac{1\,\text{mol Mn(NO}_3)_2 \cdot 2\text{H}_2\text{O}}{215.0\,\text{g Mn(NO}_3)_2 \cdot 2\text{H}_2\text{O}} = 0.140\,M\,\text{Mn(NO}_3)_2$

      (c) $M_c \times L_c = M_d \times L_d$; $9.00\,M\,\text{H}_2\text{SO}_4 \times 0.0350\,\text{L} = ?M\,\text{H}_2\text{SO}_4 \times 0.500\,\text{L}$
           500 mL of 0.630 M H$_2$SO$_4$

13.42   (a) $16.0\,\text{mol H}_2\text{O} \times \dfrac{18.02\,\text{g H}_2\text{O}}{1\,\text{mol H}_2\text{O}} = 288.3\,\text{g H}_2\text{O} = 0.288\,\text{kg H}_2\text{O}$

          $m = \dfrac{1.25\,\text{mol KCl}}{0.2883\,\text{kg H}_2\text{O}} = 4.3358 = 4.34\,m\,\text{KCl}$

      (b) $m = \dfrac{\text{mol solute}}{\text{kg solute}}$; $\text{mol S}_8 = m \times \text{kg C}_{10}\text{H}_8 = 0.12\,m \times 0.1000\,\text{kg C}_{10}\text{H}_8 = 0.012\,\text{mol}$

          $0.012\,\text{mol S}_8 \times \dfrac{256.5\,\text{g S}_8}{1\,\text{mol S}_8} = 3.078 = 3.1\,\text{g S}_8$

13.44   (a) $\text{mass \%} = \dfrac{\text{mass C}_6\text{H}_8\text{O}_6}{\text{total mass solution}} \times 100$;

          $\dfrac{80.5\,\text{g C}_6\text{H}_8\text{O}_6}{80.5\,\text{g C}_6\text{H}_8\text{O}_6 + 210\,\text{g H}_2\text{O}} \times 100 = 27.71 = 27.7\%\,\text{C}_6\text{H}_8\text{O}_6$

      (b) $\text{mol C}_6\text{H}_8\text{O}_6 = \dfrac{80.5\,\text{g C}_6\text{H}_8\text{O}_6}{176.1\,\text{g/mol}} = 0.4571 = 0.457\,\text{mol C}_6\text{H}_8\text{O}_6$

          $\text{mol H}_2\text{O} = \dfrac{210\,\text{g H}_2\text{O}}{18.02\,\text{g/mol}} = 11.654 = 11.7\,\text{mol H}_2\text{O}$

          $\chi_{\text{C}_6\text{H}_8\text{O}_6} = \dfrac{0.4571\,\text{mol C}_6\text{H}_8\text{O}_6}{0.4571\,\text{mol C}_6\text{H}_8\text{O}_6 + 11.654\,\text{mol H}_2\text{O}} = 0.0377$

      (c) $m = \dfrac{0.4571\,\text{mol C}_6\text{H}_8\text{O}_6}{0.210\,\text{kg H}_2\text{O}} = 2.18\,m\,\text{C}_6\text{H}_8\text{O}_6$

      (d) $M = \dfrac{\text{mol C}_6\text{H}_8\text{O}_6}{\text{L solution}}$; $290.5\,\text{g soln} \times \dfrac{1\,\text{mL}}{1.22\,\text{g}} \times \dfrac{1\,\text{L}}{1000\,\text{mL}} = 0.2381 = 0.238\,\text{L}$

          $M = \dfrac{0.4571\,\text{mol C}_6\text{H}_8\text{O}_6}{0.2381\,\text{L soln}} = 1.92\,M\,\text{C}_6\text{H}_8\text{O}_6$

13.46   Given: 9.08 g C$_4$H$_4$S, 1.065 g/mL; 250.0 mL C$_7$H$_8$, 0.867 g/mL

      (a) $\text{mol C}_4\text{H}_4\text{S} = 9.08\,\text{g C}_4\text{H}_4\text{S} \times \dfrac{1\,\text{mol C}_4\text{H}_4\text{S}}{84.15\,\text{g C}_4\text{H}_4\text{S}} = 0.1079 = 0.108\,\text{mol C}_4\text{H}_4\text{S}$

          $\text{mol C}_7\text{H}_8 = \dfrac{0.867\,\text{g}}{1\,\text{mL}} \times 250.0\,\text{mL} \times \dfrac{1\,\text{mol C}_7\text{H}_8}{92.14\,\text{g C}_7\text{H}_8} = 2.352 = 2.35\,\text{mol}$

$$\chi_{C_4H_4S} = \frac{0.1079 \text{ mol } C_4H_4S}{0.1079 \text{ mol } C_4H_4S + 2.352 \text{ mol } C_7H_8} = 0.04386 = 0.0439$$

(b) $m_{C_4H_4S} = \dfrac{\text{mol } C_4H_4S}{\text{kg } C_7H_8}; 250.0 \text{ mL} \times \dfrac{0.867 \text{ g}}{1 \text{ mL}} \times \dfrac{1 \text{ kg}}{1000 \text{ g}} = 0.2168 = 0.217 \text{ kg } C_7H_8$

$$m_{C_4H_4S} = \frac{0.1079 \text{ mol } C_4H_4S}{0.2168 \text{ kg } C_7H_8} = 0.498 \; m \; C_4H_4S$$

(c) $9.08 \text{ g } C_4H_4S \times \dfrac{1 \text{ mL}}{1.065 \text{ g}} = 8.526 = 8.53 \text{ mL } C_4H_4S;$

$V_{soln} = 8.53 \text{ mL } C_4H_4S + 250.0 \text{ mL } C_7H_8 = 258.5 \text{ mL}$

$$M_{C_4H_4S} = \frac{0.1079 \text{ mol } C_4H_4S}{0.2585 \text{ L soln}} = 0.417 \; M \; C_4H_4S$$

13.48   (a) $\dfrac{1.50 \text{ mol } HNO_3}{1 \text{ L soln}} \times 0.185 \text{ L} = 0.2775 = 0.278 \text{ mol } HNO_3$

(b)   Assume that for dilute aqueous solutions, the mass of the solvent is the mass of solution.

$$\frac{1.25 \text{ mol } NaCl}{1 \text{ kg } H_2O} \times \frac{x \text{ mol}}{50.0 \times 10^{-6} \text{ kg}}; x = 6.25 \times 10^{-5} \text{ mol } NaCl$$

(c) $\dfrac{1.50 \text{ g } C_{12}H_{22}O_{11}}{100 \text{ g soln}} = \dfrac{x \text{ g } C_{12}H_{22}O_{11}}{75.0 \text{ g soln}}; x = 1.125 = 1.13 \text{ g } C_{12}H_{22}O_{11}$

$1.125 \text{ g } C_{12}H_{22}O_{11} \times \dfrac{1 \text{ mol } C_{12}H_{22}O_{11}}{342.3 \text{ g } C_{12}H_{22}O_{11}} = 3.287 \times 10^{-3} = 3.29 \times 10^{-3} \text{ mol } C_{12}H_{22}O_{11}$

13.50   (a) $\dfrac{0.110 \text{ mol } (NH_4)_2SO_4}{1 \text{ L soln}} \times 1.50 \text{ L} \times \dfrac{132.2 \text{ g } (NH_4)_2SO_4}{1 \text{ mol } (NH_4)_2SO_4} = 21.81 = 21.8 \text{ g } (NH_4)_2SO_4$

Weigh 21.8 g $(NH_4)_2SO_4$, dissolve in a small amount of water, continue adding water with thorough mixing up to a total solution volume of 1.50 L.

(b)   Determine the mass fraction of Na_2CO_3 in the solution:

$\dfrac{0.65 \text{ mol } Na_2CO_3}{1000 \text{ g } H_2O} \times \dfrac{106.0 \text{ g } Na_2CO_3}{1 \text{ mol } Na_2CO_3} = 68.9 \text{ g} = \dfrac{69 \text{ g } Na_2CO_3}{1000 \text{ g } H_2O}$

$\text{mass fraction} = \dfrac{68.9 \text{ g } Na_2CO_3}{1000 \text{ g } H_2O + 68.9 \text{ g } Na_2CO_3} = 0.06446 = 0.064$

In 120 g of solution, there are 0.06446(120) = 7.735 = 7.7 g Na_2CO_3.

Weigh out 7.7 g Na_2CO_3 and dissolve it in 120 – 7.7 = 112.3 g H_2O to make exactly 120 g of solution.

(112.3 g H_2O/0.997 g H_2O/mL @ 25° = 112.6 mL H_2O)

(c)　　$1.20\,L \times \dfrac{1000\,mL}{1\,L} \times \dfrac{1.16\,g}{1\,mL} = 1392\,g\ solution;\ 0.150(1392\,g\ soln) = 209\,g\ Pb(NO_3)_2$

Weigh 209 g $Pb(NO_3)$ and add $(1392 - 209) = 1183\,g\ H_2O$ to make exactly $(1392 = 1.39 \times 10^3)$ g or 1.20 L of solution.

$(1183\,g\ H_2O/0.997\,g/mL\ @\ 25°C = 1187\,mL\ H_2O)$

(d)　　Calculate the mol HCl necessary to neutralize 5.5 g $Ba(OH)_2$.

$Ba(OH)_2(s) + 2HCl(aq) \rightarrow BaCl_2(aq) + 2H_2O(l)$

$5.5\,g\ Ba(OH)_2 + \dfrac{1\,mol\ Ba(OH)_2}{171\,g\ Ba(OH)_2} \times \dfrac{2\,mol\ HCl}{1\,mol\ Ba(OH)_2} = 0.0643 = 0.064\,mol\ HCl$

$M = \dfrac{mol}{L};\ L = \dfrac{mol}{M} = \dfrac{0.0643\,mol\ HCl}{0.50\,M\ HCl} = 0.1287 = 0.13\,L = 130\,mL$

130 mL of 0.50 M HCl are needed.

$M_c \times L_c = M_d \times L_d;\ 6.0\,M \times L_c = 0.50\,M \times 0.1287\,L;\ L_c = 0.01072\,L = 11\,mL$

Using a pipette, measure exactly 11 mL of 6.0 M HCl and dilute with water to a total volume of 130 mL.

13.52　　*Analyze/Plan.* Assume 1.00 L of solution. Calculate mass of 1 L of solution using density. Calculate mass of NH_3 using mass %, then mol NH_3 in 1.00 L. *Solve.*

$1.00\,L\ soln \times \dfrac{1000\,mL}{1\,L} \times \dfrac{0.90\,g\ soln}{1\,mL\ soln} = 9.0 \times 10^2\ g\ soln/L$

$\dfrac{900\,g\ soln}{1.00\,L\ soln} \times \dfrac{28\,g\ NH_3}{100\,g\ soln} \times \dfrac{1\,mol\ NH_3}{17.03\,g\ NH_3} = 14.80 = 15\,mol\ NH_3/L\ soln = 15\,M\ NH_3$

13.54　　(a)　　$\dfrac{0.0750\,mol\ C_8H_{10}N_4O_2}{1\,kg\ CHCl_3} \times \dfrac{194.2\,g\ C_8H_{10}N_4O_2}{1\,mol\ C_8H_{10}N_4O_2} = 14.565$

$= 14.6\,g\ C_8H_{10}N_4O_2/kg\ CHCl_3$

$\dfrac{14.565\,g\ C_8H_{10}N_4O_2}{14.565\,g\ C_8H_{10}N_4O_2 + 1000.00\,g\ CHCl_3} \times 100 = 1.436 = 1.44\%\ C_8H_{10}N_4O_2\ by\ mass$

(b)　　$1000\,g\ CHCl_3 \times \dfrac{1\,mol\ CHCl_3}{119.4\,CHCl_3} = 8.375 = 8.38\,mol\ CHCl_3$

$\chi_{C_8H_{10}N_4O_2} = \dfrac{0.0750}{0.0750 + 8.375} = 0.00888$

13.56　　(a)　　For gases at the same temperature and pressure, volume % = mol %. The volume and mol % of CO_2 in this breathing air is 4.0%.

(b)　　$P_{CO_2} = \chi_{CO_2} \times P_t = 0.040\,(1\,atm) = 0.040\,atm$

$M_{CO_2} = \dfrac{P_{CO_2}}{RT} = \dfrac{0.040\,atm}{310\,K} \times \dfrac{K\text{-}mol}{0.08206\,L\text{-}atm} = 1.6 \times 10^{-3}\,M$

Colligative Properties

13.58   (a)   decrease   (b)   decrease

           (c)   increase   (d)   increase

13.60   (a)   An *ideal solution* is a solution that obeys Raoult's Law.

           (b)   *Analyze/Plan.* Calculate the vapor pressure predicted by Raoult's law and compare it to the experimental vapor pressure. Assume ethylene glycol (eg) is the solute. *Solve.*

$$\chi_{H_2O} = \chi_{eg} = 0.500; \; P_A = \chi_A P_A^\circ = 0.500(149) \text{ mm Hg} = 74.5 \text{ mm Hg}$$

                The experimental vapor pressure (P_A), 67 mm Hg, is less than the value predicted by Raoult's law for an ideal solution. The solution is not ideal.

                Check. An ethylene glycol-water solution has extensive hydrogen bonding, which causes deviation from ideal behavior. We expect the experimental vapor pressure to be less than the ideal value and it is.

13.62   (a) H_2O vapor pressure will be determined by the mole fraction of H_2O in the solution. The vapor pressure of pure H_2O at 343 K (70°C) = 233.7 torr.

$$\frac{32.5 \text{ g } C_3H_8O_3}{92.10 \text{ g/mol}} = 0.3529 = 0.353 \text{ mol}; \quad \frac{125 \text{ g } H_2O}{18.02 \text{ g/mol}} = 6.937 = 6.94 \text{ mol}$$

$$P_{H_2O} = \frac{6.937 \text{ mol } H_2O}{6.937 + 0.353} \times 233.7 \text{ torr} = 222.4 = 222 \text{ torr}$$

           (b)   Calculate χ_B by vapor pressure lowering; $\chi_B = \Delta P_A / P_A^\circ$ (see Solution 13.59(b)). Given moles solvent, calculate moles solute from the definition of mole fraction.

$$\chi_{C_2H_6O_2} = \frac{10.0 \text{ torr}}{100 \text{ torr}} = 0.100$$

$$\frac{1.00 \times 10^3 \text{ g } C_2H_5OH}{46.07 \text{ g/mol}} = 21.71 = 21.7 \text{ mol } C_2H_5OH; \text{ let } y = \text{mol } C_2H_6O_2$$

$$\chi_{C_2H_6O_2} = \frac{y \text{ mol } C_2H_6O_2}{y \text{ mol } C_2H_6O_2 + 21.71 \text{ mol } C_2H_5OH} = 0.100 = \frac{y}{y + 21.71}$$

$$0.100 \, y + 2.171 = y; \; 0.900 \, y = 2.171; \; y = 2.412 = 2.41 \text{ mol } C_2H_6O_2$$

$$2.412 \text{ mol } C_2H_6O_2 \times \frac{62.07 \text{ g}}{1 \text{ mol}} = 150 \text{ g } C_2H_6O_2$$

13.64   (a)   Since C_6H_6 and C_7H_8 form an ideal solution, we can use Raoult's Law. Since both components are volatile, both contribute to the total vapor pressure of 35 torr.

$$P_t = P_{C_6H_6} + P_{C_7H_8} ; P_{C_6H_6} = \chi_{C_6H_6} P_{C_6H_6}^\circ ; P_{C_7H_8} = \chi_{C_7H_8} P_{C_7H_8}^\circ$$

$$\chi_{C_7H_8} = 1 - \chi_{C_6H_6} ; P_T = \chi_{C_6H_6} P_{C_6H_6}^\circ + (1 - \chi_{C_6H_6}) P_{C_7H_8}^\circ$$

$$35 \text{ torr} = \chi_{C_6H_6}(75 \text{ torr}) + (1 - \chi_{C_6H_6})22 \text{ torr}$$

$$13 \text{ torr} = 53 \text{ torr} (\chi_{C_6H_6}); \; \chi_{C_6H_6} = \frac{13 \text{ torr}}{53 \text{ torr}} = 0.2453 = 0.25; \quad \chi_{C_7H_8} = 0.7547 = 0.75$$

(b) $P_{C_6H_6} = 0.2453(75 \text{ torr}) = 18.4 \text{ torr}; \quad P_{C_7H_8} = 0.7547(22 \text{ torr}) = 16.6 \text{ torr}$

In the vapor, $\quad \chi_{C_6H_6} = \dfrac{P_{C_6H_6}}{P_t} = \dfrac{18.4 \text{ torr}}{35 \text{ torr}} = 0.53; \quad \chi_{C_7H_8} = 0.47$

13.66   *Analyze/Plan.* ΔT_b depends on mol dissolved particles. Assume 100 g of each solution, calculate mol solute and mol dissolved particles. Glucose and sucrose are molecular solutes, but $NaNO_3$ dissociates into 2 mol particles per mol solute. *Solve.*

10% by mass means 10 g solute in 100 g solution. If we have 10 g of each solute, the one with the smallest molar mass will have the largest mol solute. The molar masses are: glucose, 180.2 g/mol; sucrose, 342.3 g/mol; $NaNO_3$, 85.0 g/mol. $NaNO_3$ has most mol solute, and twice as many dissolved particles, so it will have the highest boiling point. Sucrose has least mol solute and lowest boiling point. Glucose is intermediate.

In order of increasing boiling point: 10% sucrose < 10% glucose < 10% $NaNO_3$.

13.68   0.030 *m* phenol > 0.040 *m* glycerin = 0.020 *m* KBr. Phenol is very slightly ionized in water, but not enough to match the number of particles in a 0.040 *m* glycerin solution. The KBr solution is 0.040 *m* in particles, so it has the same freezing point as 0.040 *m* glycerin, which is a nonelectrolyte.

13.70 $\Delta T = K(m)$; first calculate the **molality** of the solute particles.

(a)   0.30 *m*

(b) $\dfrac{20.0 \text{ g } C_{10}H_{22}}{0.0455 \text{ kg CHCl}_3} \times \dfrac{1 \text{ mol } C_{10}H_{22}}{142.3 \text{ g } C_{10}H_{22}} = 3.089 = 3.09 \text{ m}$

(c) $m = \dfrac{0.45 \text{ mol eg} + 2(0.15) \text{ mol KBr}}{0.150 \text{ kg H}_2\text{O}} = \dfrac{0.75 \text{ mol particles}}{0.150 \text{ kg H}_2\text{O}} = 5.0 \text{ } m$

Then, f.p. $= T_f - K_f(m)$; b.p. $= T_b + K_b(m)$; T in °C

	m	T_f	$- K_f(m)$	f.p.	T_b	$+K_b(m)$	b.p.
(a)	0.30	−114.6	−1.99(0.30) = −0.60	−115.2	78.4	1.22(0.30) = 0.37	78.8
(b)	3.09	−63.5	−4.68(3.09) = −14.5	−78.0	61.2	3.63(3.09) = 11.2	72.4
(c)	5.0	0.0	−1.86(5.0) = −9.3	−9.3	100.0	0.51(5.0) = 2.6	102.6

13.72   Use ΔT_b = find *m* of aqueous solution, and then use *m* to calculate ΔT_f and freezing point. $K_b = 0.51$, $K_f = 1.86$.

b.p. $= 105.0°C$; $\Delta T_b = 105.0°C - 100.0°C = 5.0°C$

$\Delta T_b = K_b(m); \quad m = \dfrac{\Delta T_b}{K_b} = \dfrac{5.0°C}{0.51} = 9.804 = 9.8 \text{ } m$

$\Delta T_f = 1.86°C/m \times 9.804 \text{ } m = 18.24 = 18°C$;   freezing point $= 0.0°C - 18.24 °C = -18°C$

13.74 $\Pi = MRT$; $T = 20°C + 273 = 293 \text{ K}$

$M \text{ (of ions)} = \dfrac{\text{mol NaCl} \times 2}{\text{L soln}} = \dfrac{3.4 \text{ g NaCl}}{1 \text{ L soln}} \times \dfrac{1 \text{ mol NaCl}}{58.4 \text{ g NaCl}} \times \dfrac{2 \text{ mol ions}}{1 \text{ mol NaCl}} = 0.116 = 0.12 \text{ } M$

$$\Pi = \frac{0.116 \text{ mol}}{L} \times \frac{0.08206 \text{ L - atm}}{K \text{ - mol}} \times 293 \text{ K} = 2.8 \text{ atm}$$

13.76 $\Delta T_f = 5.5 - 4.1 = 1.4; \quad m = \frac{\Delta T_f}{K_f} = \frac{1.4}{5.12} = 0.273 = 0.27 \ m$

$$\text{MM lauryl alcohol} = \frac{\text{g lauryl alcohol}}{m \times \text{kg } C_6H_6} = \frac{5.00 \text{ g lauryl alcohol}}{0.273 \times 0.100 \text{ kg } C_6H_6}$$

$$= 1.8 \times 10^2 \text{ g/mol lauryl alcohol}$$

13.78 $M = n/RT = \frac{0/605 \text{ atm}}{298 \text{ K}} \times \frac{K \text{ - mol}}{0.08206 \text{ L - atm}} = 0.02474 = 0.0247 \ M$

$$MM = \frac{g}{M \times L} = \frac{2.35 \text{ g}}{0.02474 \ M \times 0.250 \ L} = 380 \text{ g/mol}$$

13.80   If these were ideal solutions, they would have equal ion concentrations and equal ΔT_f values. Data in Table 13.5 indicates that the van't Hoff factors (i) for both salts are less than the ideal values. For 0.030 m NaCl, i is between 1.87 and 1.94, about 1.92. For 0.020 m K_2SO_4, i is between 2.32 and 2.70, about 2.62. From Equation 13.14,

ΔT_f (measured) = $i \times \Delta T_f$ (calculated for nonelectrolyte)

NaCl: ΔT_f (measured) = $1.92 \times 0.030 \ m \times 1.86 \ °C/m = 0.11 \ °C$

K_2SO_4: ΔT_f (measured) = $2.62 \times 0.020 \ m \times 1.86°C/m = 0.097 \ °C$

0.030 m NaCl would have the larger ΔT_f.

The deviations from ideal behavior are due to ion-pairing in the two electrolyte solutions. K_2SO_4 has more extensive ion-pairing and a larger deviation from ideality because of the higher charge on SO_4^{2-} relative to Cl^-.

Colloids

13.82   (a)   Suspensions are classified as solutions or colloids according to the size of the dispersed particles. Solute particles have diameters less than 10 Å. Clearly a protein with a molecular mass of 30,000 amu will be longer than 10 Å. The aqueous suspensions are colloids because of the size of protein molecules.

     (b)   Emulsion. An emulsifying agent is one that aids in the formation of an emulsion. It usually has a polar part and a nonpolar part, to facilitate mixing of immiscible liquids with very different molecular polarities.

13.84   (a)   When the colloid *particle mass* becomes large enough so that gravitational and interparticle attractive forces are greater than the kinetic energies of the particles, settling and aggregation can occur.

     (b)   *Hydrophobic* colloids do not attract a sheath of water molecules around them and thus tend to aggregate from aqueous solution. They can be stabilized as colloids by adsorbing charges on their surfaces. The charged particles interact with solvent water, stabilizing the colloid.

(c) *Charges on colloid particles* can stabilize them against aggregation. Particles carrying like charges repel one another and are thus prevented from aggregating and settling out.

13.86 (a) The nonpolar hydrophobic tails of soap particles (the hydrocarbon chain of stearate ions) establish attractive intermolecular dispersion forces with the nonpolar oil molecules, while the charged hydrophilic head of the soap particles interacts with H_2O to keep the oil molecules suspended. (This is the mechanism by which laundry detergents remove greasy dirt from clothes.)

(b) Electrolytes from the acid neutralize surface charges of the suspended particles in milk, causing the colloid to coagulate.

Additional Exercises

13.88 In this equilibrium system, molecules move from the surface of the solid into solution, while molecules in solution are deposited on the surface of the solid. As molecules leave the surface of the small particles of powder, the reverse process preferentially deposits other molecules on the surface of a single crystal. Eventually, all molecules that were present in the 50 g of powder are deposited on the surface of a 50 g crystal; this can only happen if the dissolution and deposition processes are ongoing.

13.89 Assume that the density of the solution is 1.00 g/mL.

(a) $4 \text{ ppm } O_2 = \dfrac{4 \text{ mg } O_2}{1 \text{ kg soln}} = \dfrac{4 \times 10^{-3} \text{ g } O_2}{1 \text{ L soln}} \times \dfrac{1 \text{ mol } O_2}{32.0 \text{ g } O_2} = 1.25 \times 10^{-4} = 1 \times 10^{-4} \, M$

(b) $S_{O_2} = kP_{O_2}; \; P_{O_2} = S_{O_2}/k = \dfrac{1.25 \times 10^{-4} \text{ mol}}{L} \times \dfrac{\text{L-atm}}{1.71 \times 10^{-3} \text{ mol}} = 0.0731 = 0.07 \text{ atm}$

$0.0731 \text{ atm } \times \dfrac{760 \text{ mm Hg}}{1 \text{ atm}} = 55.6 = 60 \text{ mm Hg}$

13.91 0.10% by mass means 0.10 g glucose/100 g blood.

(a) $\text{ppm glucose} = \dfrac{\text{g glucose}}{\text{g solution}} \times 10^6 = \dfrac{0.10 \text{ g glucose}}{100 \text{ g blood}} \times 10^6 = 1000 \text{ ppm glucose}$

(b) m = mol glucose/kg solvent. Assume that the mixture of nonglucose components is the 'solvent'.

mass solvent = 100 g blood – 0.10 g glucose = 99.9 g solvent = 0.0999 kg solvent

$\text{mol glucose} = 0.10 \text{ g} \times \dfrac{1 \text{ mol}}{180.2 \text{ g } C_6H_{12}O_6} = 5.55 \times 10^{-4} = 5.6 \times 10^{-4} \text{ mol glucose}$

$m = \dfrac{5.55 \times 10^{-4} \text{ mol glucose}}{0.0999 \text{ kg solvent}} = 5.6 \times 10^{-3} \, m \text{ glucose}$

In order to calculate molarity, solution volume must be known. The density of blood is needed to relate mass and volume.

13.92 *Analyze.* The definition of ppb is (mass solute/mass solution) × 10^9. *Plan.* Use the definition to get g Pb and g solution. Change g Pb to mol Pb, g solution to L solution, calculate molarity. *Solve.*

(a) $9 \text{ ppb} = \dfrac{9 \text{ g Pb}}{1 \times 10^9 \text{ g soln}} \times 10^9$

For dilute aqueous solutions (drinking water) assume that the density of the solution is the density of H_2O.

$\dfrac{9 \text{ g Pb}}{1 \times 10^9 \text{ g soln}} \times \dfrac{1.0 \text{ g soln}}{\text{mL soln}} \times \dfrac{1000 \text{ mL}}{1 \text{ L}} \times \dfrac{1 \text{ mol Pb}}{207.2 \text{ g Pb}} = 4.34 \times 10^{-8} M = 4 \times 10^{-8} M$

(b) Change 60 m³ H_2O to cm³ (mL) H_2O to g H_2O (or g soln).

$60 \text{ m}^3 \times \dfrac{100^3 \text{ cm}^3}{\text{m}^3} \times \dfrac{1 \text{ g } H_2O}{\text{cm}^3 \ H_2O} = 6.0 \times 10^7 \text{ g } H_2O \text{ or soln}$

$\dfrac{9 \text{ g Pb}}{1 \times 10^9 \text{ g soln}} \times 6.0 \times 10^7 \text{ g soln} = 0.540 = 0.5 \text{ g Pb}$

13.94 Mole fraction ethyl alcohol, $\chi_{C_2H_5OH} = \dfrac{P_{C_2H_5OH}}{P^\circ_{C_2H_5OH}} = \dfrac{8 \text{ torr}}{100 \text{ torr}} = 0.08$

$\dfrac{620 \times 10^3 \text{ g } C_{24}H_{50}}{338.6 \text{ g/mol}} = 1.83 \times 10^3 \text{ mol } C_{24}H_{50}; \text{ let } y = \text{mol } C_2H_5OH$

$\chi_{C_2H_5OH} = 0.08 = \dfrac{y}{y + 1.83 \times 10^3}; \ 0.92 y = 146.4; \ y = 1.6 \times 10^2 \text{ mol } C_2H_5OH$

(Strictly speaking, y should have 1 sig fig because 0.08 has 1 sig fig, but this severely limits the calculation.)

$1.6 \times 10^2 \text{ mol } C_2H_5OH \times \dfrac{46 \text{ g } C_2H_5OH}{1 \text{ mol}} = 7.4 \times 10^3 \text{ g or } 7.4 \text{ kg } C_2H_5OH$

13.95 (a) The solvent vapor pressure over each solution is determined by the total particle concentrations present in the solutions. When the particle concentrations are equal, the vapor pressures will be equal and equilibrium established. The particle concentration of the nonelectrolyte is just 0.050 M, the ion concentration of the NaCl is 2 × 0.035 M = 0.070 M. Solvent will diffuse from the less concentrated nonelectrolyte solution. The level of the NaCl solution will rise, and the level of the nonelectrolyte solution will fall.

(b) Let x = volume of solvent transferred

$\dfrac{0.050 M \times 30.0 \text{ mL}}{(30.0 - x) \text{ mL}} = \dfrac{0.070 M \times 30.0 \text{ mL}}{(30.0 + x) \text{ mL}}; 1.5(30.0 + x) = 2.1(30.0 - x)$

$45 + 1.5 x = 63 - 2.1 x; \ 3.6 x = 18; \ x = 5.0 = 5 \text{ mL transferred}$

The volume in the nonelectrolyte beaker is (30.0 − 5.0) = 25.0 mL; in the NaCl beaker (30.0 + 5.0) = 35.0 mL.

13.97   Assume constant temperature of all liquids and vapors.

$P_{solution}$ = vapor pressure above the solution

$P^{o}_{solvent}$ = vapor pressure above pure solvent

$\chi_{solvent}$ = mol fraction solvent, χ_{solute} = $1 - \chi_{solvent}$

$\Delta P_{solvent}$ = vapor pressure reduction = $P^{o}_{solvent} - P_{solution}$

$P_{solution} = \chi_{solvent} P^{o}_{solvent}$ (Raoult's Law)

$P_{solution} = (1 - \chi_{solute}) P^{o}_{solvent}$

$P_{solution} = P^{o}_{solvent} - \chi_{solute} P^{o}_{solvent}$

$P_{solution} - P^{o}_{solvent} = -\chi_{solute} P^{o}_{solvent}$

$-(P_{solution} - P^{o}_{solvent}) = -(-\chi_{solute} P^{o}_{solvent})$

$P^{o}_{solvent} - P_{solution} = \chi_{solute} P^{o}_{solvent}$

$\Delta P_{solvent} = \chi_{solute} P^{o}_{solvent}$

13.98   In order to answer this question, you will need to find the physical properties (density, freezing point) of ethylene glycol in a source like the *CRC*. Calculate the freezing point of a solution that is 30% ethylene glycol, $C_2H_6O_2$, and 70% water. Since a car owner would typically use volume measurements to make this solution, assume this is a volume percent concentration. Volume percent is a volume ratio that is valid for any volume unit. For convenience, assume 30 mL $C_2H_6O_2$, and 70 mL H_2O. The density of $C_2H_6O_2$ is 1.1088 g/mL. The density of H_2O is 0.997 g/mL at 25°C. Find molality and then ΔT_f for water.

$$30 \text{ mL } C_2H_6O_2 \times \frac{1.1088 \text{ g } C_2H_6O_2}{\text{mL}} = 33.264 = 33 \text{ g } C_2H_6O_2$$

$$\frac{33.264 \text{ g } C_2H_6O_2}{70 \text{ mL } H_2O} \times \frac{1 \text{ mol } C_2H_6O_2}{62.1 \text{ g } C_2H_6O_2} \times \frac{1 \text{ mL } H_2O}{0.997 \text{ g } H_2O} \times \frac{1000 \text{ g}}{\text{kg}} = 7.6752 = 7.7 \text{ m}$$

$\Delta T_f = 1.86°C/m$ (7.6752 m) = 14.27°C

$T_f = 0.0°C - 14.27°C = -14.27 = -14°C$

In this 30% solution, ethylene glycol is the solute (present in lesser amount) and water is the solvent (present in greater amount). It is the freezing point of water, 0.0°C, that is depressed by the nonvolatile solute ethylene glycol. If pure ethylene glycol is used in the radiator, it freezes at its regular (not depressed) freezing point, -11.5°C. The freezing point of the solution, -14°C, is lower than the freezing point of pure ethylene glycol.

13.100 The compound with the larger i value is the stronger electrolyte.

$$i = \frac{\Delta T_f (\text{measured})}{\Delta T_f (\text{calculated})}$$ The idealized value is 3 for both salts.

$$Hg(NO_3)_2: \quad m = \frac{10.0\,g\,Hg(NO_3)_2}{1.00\,kg\,H_2O} \times \frac{1\,mol\,Hg(NO_3)_2}{324.6\,g\,Hg(NO_3)_2} = 0.0308\,m$$

$$\Delta T_f\,(\text{nonelectrolyte}) = -1.86(0.0308) = -0.0573°C$$

$$i = \frac{-0.162°C}{-0.0573°C} = 2.83$$

$$HgCl_2: \quad m = \frac{10.0\,g\,HgCl_2}{1.00\,kg\,H_2O} \times \frac{1\,mol\,HgCl_2}{271.5\,g\,HgCl_2} = 0.0368\,m$$

$$\Delta T_f\,(\text{nonelectrolyte}) = -1.86(0.0368) = -0.0685°C$$

$$i = \frac{-0.0685}{-0.0685} = 1.00$$

With an i value of 2.83, $Hg(NO_3)_2$ is almost completely dissociated into ions; with an i value of 1.00, the $HgCl_2$ behaves essentially like a nonelectrolyte. Clearly, $Hg(NO_3)_2$ is the stronger electrolyte.

13.101 (a) $\quad K_b = \dfrac{\Delta T_b}{m}; \quad \Delta T_b = 47.46°C - 46.30°C = 1.16°C$

$$m = \frac{\text{mol solute}}{\text{kg } CS_2} = \frac{0.250\,mol}{400.0\,mL\,CS_2} \times \frac{1\,mL\,CS_2}{1.261\,g\,CS_2} \times \frac{1000\,g}{1\,kg} = 0.4956 = 0.496\,m$$

$$K_b = \frac{1.16°C}{0.4956\,m} = 2.34°C/m$$

(b) $\quad m = \dfrac{\Delta T_b}{K_b} = \dfrac{(47.08 - 46.30)°C}{2.34°C/m} = 0.333 = 0.33\,m$

$$m = \frac{\text{mol unknown}}{\text{kg } CS_2}; \quad m \times kg\,CS_2 = \frac{g\,\text{unknown}}{MM\,\text{unknown}}; \quad MM = \frac{g\,\text{unknown}}{m \times kg\,CS_2}$$

$$50.0\,mL\,CS_2 \times \frac{1.261\,g\,CS_2}{1\,mL} \times \frac{1\,kg}{1000\,g} = 0.06305 = 0.0631\,kg\,CS_2$$

$$MM = \frac{5.39\,g\,\text{unknown}}{0.333\,m \times 0.06305\,kg\,CS_2} = 257 = 2.6 \times 10^2\,g/mol$$

13.103 $\quad M = \dfrac{\Pi}{RT} = \dfrac{57.1\,torr}{298\,K} \times \dfrac{1\,atm}{760\,torr} \times \dfrac{K \cdot mol}{0.08206\,L \cdot atm} = 3.072 \times 10^{-3} = 3.07 \times 10^{-3}\,M$

$$\frac{0.036\,g\,\text{solute}}{100\,g\,H_2O} \times \frac{1000\,g\,H_2O}{1\,kg\,H_2O} = 0.36\,g\,\text{solute/kg}\,H_2O$$

Assuming molarity and molality are the same in this dilute solution, we can then say 0.36 g solute $= 3.072 \times 10^{-3}$ mol; MM = 117 g/mol. Because the salt is completely ionized, the formula weight of the lithium salt is **twice** this calculated value, or **234 g/mol**. The organic portion, $C_nH_{2n+1}O_2^-$, has a formula weight of $234-7 = 227$ g. Subtracting 32 for the oxygens, and 1 to make the formula C_nH_{2n}, we have C_nH_{2n}, MM = 194 g/mol. Since each CH_2 unit has a mass of 14, $n \approx 194/14 \approx 14$. The formula for our salt is $LiC_{14}H_{29}O_2$.

Integrative Exercises

13.105 $\dfrac{0.015\,\text{g N}_2}{1\,\text{L blood}} \times \dfrac{1\,\text{mol N}_2}{28.01\,\text{g N}_2} = 5.355 \times 10^{-4} = 5.4 \times 10^{-4}\ \text{mol N}_2/\text{L blood}$

At 100 ft, the partial pressure of N_2 in air is 0.78 (4.0 atm) = 3.12 atm. This is just four times the partial pressure of N_2 at 1.0 atm air pressure. According to Henry's law, $S_g = kP_g$, a 4-fold increase in P_g results n a 4-fold increase in S_g, the solubility of the gas. Thus, the solubility of N_2 at 100 ft is $4(5.355 \times 10^{-4}\,M) = 2.142 \times 10^{-3} = 2.1 \times 10^{-3}\,M$. If the diver suddenly surfaces, the amount of N_2/L blood released is the difference in the solubilities at the two depths: $(2.142 \times 10^{-3}\,\text{mol/L} - 5.355 \times 10^{-4}\,\text{mol/L}) = 1.607 \times 10^{-3} = 1.6 \times 10^{-3}\ \text{mol N}_2/\text{L blood}$.

At surface conditions of 1.0 atm external pressure and 37°C = 310 K,

$$V = \dfrac{nRT}{P} = 1.607 \times 10^{-3}\ \text{mol} \times \dfrac{310\,\text{K}}{1.0\,\text{atm}} \times \dfrac{0.08206\,\text{L-atm}}{\text{mol-K}} = 0.041\,\text{L}$$

That is, 41 mL of tiny N_2 bubbles are released from each L of blood.

13.106   The stronger the intermolecular forces, the higher the heat (enthalpy) of vaporization.

(a)   None of the substances are capable of hydrogen bonding in the pure liquid, and they have similar molar masses. All intermolecular forces are van der Waals forces, dipole-dipole, and dispersion forces. In decreasing order of strength of forces:

acetone > acetaldehyde > ethylene oxide > cyclopropane

The first three compounds have dipole-dipole and dispersion forces, the last only dispersion forces.

(b)   The order of solubility in hexane should be the reverse of the order above. The least polar substance, propane, will be most soluble in hexane. Ethanol, CH_3CH_2OH, is capable of hydrogen bonding with the three polar compounds. Thus, acetaldehyde, acetone, and ethylene oxide should be more soluble than cyclopropane, but without further information we cannot distinguish among the polar molecules.

13.107   For ionic solids, the exothermic part of the solution process is step (3), surrounding the separated ions by solvent molecules. The released energy comes from the attractive interaction of the solvent with the separated ions. In hydrates, one or more water molecules are already associated with the ions, reducing the total energy released during solvation.

13.109   (a) $Zn(s) + H_2SO_4(aq) \rightarrow ZnSO_4(aq) + H_2(g)$

$$2.050\,\text{g Zn} \times \dfrac{1\,\text{mol Zn}}{65.39\,\text{g Zn}} = 0.03135\,\text{mol Zn}$$

$1.00\,M\ H_2SO_4 \times 0.0150\,\text{L} = 0.0150\ \text{mol H}_2SO_4$

Since Zn and H_2SO_4 react in a 1:1 mole ratio, H_2SO_4 is the limiting reactant; 0.0150 mol of $H_2(g)$ are produced.

(b) $P = \dfrac{nRT}{V} = \dfrac{0.0150\,mol}{0.122\,L} \times \dfrac{0.08206\,L\text{-}atm}{mol\text{-}K} \times 298\,K = 3.0066 = 3.01\,atm$

(c) $S_{H_2} = kP_{H_2} = \dfrac{7.8 \times 10^{-4}\,mol}{L\text{-}atm} \times 3.0066\,atm = 0.002345 = 2.3 \times 10^{-3}\,M$

$\dfrac{0.002345\,mol\,H_2}{L\,soln} \times 0.0150\,L = 3.518 \times 10^{-5} = 3.5 \times 10^{-5}\,mol\,dissolved\,H_2$

$\dfrac{3.5 \times 10^{-5}\,mol\,dissolved\,H_2}{0.0150\,mol\,H_2\,produced} \times 100 = 0.23\%\,dissolved\,H_2$

This is approximately 2.3 ppt; for every 10,000 H_2 molecules, 23 are dissolved. It was reasonable to ignore dissolved $H_2(g)$ in part (b).

13.110   (a) $\dfrac{1.3 \times 10^{-3}\,mol\,CH_4}{L\,soln} \times 4.0\,L = 5.2 \times 10^{-3}\,mol\,CH_4$

$V = \dfrac{nRT}{P} = \dfrac{5.2 \times 10^{-3}\,mol \times 273\,K}{1.0\,atm} \times \dfrac{0.08206\,L\text{-}atm}{K\text{-}mol} = 0.12\,L$

(b)   All three hydrocarbons are nonpolar; they have zero net dipole moment. In CH_4 and C_2H_6, the C atoms are tetrahedral and all bonds are σ bonds. C_2H_6 has a higher molar mass than CH_4, which leads to stronger dispersion forces and greater water solubility. In C_2H_4, the C atoms are trigonal planar and the π electron cloud is symmetric above and below the plane that contains all the atoms. The π cloud in C_2H_4 is an area of concentrated electron density that experiences attractive forces with the positive ends of H_2O molecules. These forces increase the solubility of C_2H_4 relative to the other hydrocarbons.

(c)   The molecules have similar molar masses. NO is most soluble because it is polar. The triple bond in N_2 is shorter than the double bond in O_2. It is more difficult for H_2O molecules to surround the smaller N_2 molecules, so they are less soluble than O_2 molecules.

(d) H_2S and SO_2 are polar molecules capable of hydrogen bonding with water. Hydrogen bonding is the strongest force between neutral molecules and causes the much greater solubility. H_2S is weakly acidic in water. SO_2 reacts with water to form H_2SO_3, a weak acid. The large solubility of SO_2 is a sure sign that a chemical process has occurred.

(e) N_2 and C_2H_4. N_2 is too small to be easily hydrated, so C_2H_4 is more soluble in H_2O.

NO (31) and C_2H_6 (30). The structures of these two molecules are very different, yet they have similar solubilities. NO is slightly polar, but too small to be easily hydrated. The larger C_2H_6 is nonpolar, but more polarizable (stronger dispersion forces).

NO (31) and O_2 (32). The slightly polar NO is more soluble than the slightly larger (longer O=O bond than $N \equiv O$ bond) but nonpolar O_2.

13.112   (a) $\Delta T_f = K_f m = K_f \times \dfrac{\text{mol } C_7H_6O_2}{\text{kg } C_6H_6} = K_f \times \dfrac{\text{g } C_7H_6O_2}{\text{kg } C_6H_6 \times M \, C_7H_6O_2}$

$MM = \dfrac{K_f \times \text{g } C_7H_6O_2}{\Delta T_f \times \text{kg } C_6H_6} = \dfrac{5.12 \times 0.55}{0.360 \times 0.032} = 2444.4 = 2.4 \times 10^2 \text{ g/mol}$

(b)   The formula weight of $C_7H_6O_2$ is 122 g/mol. The experimental molar mass is twice this value, indicating that benzoic acid is associated into dimers in benzene solution. This is reasonable, since the carboxyl group, –COOH, is capable of strong hydrogen bonding with itself. Many carboxylic acids exist as dimers in solution.

The structure of benzoic acid dimer in benzene solution is:

13.113 $\chi_{CHCl_3} = \chi_{C_3H_6O} = 0.500$

(a)   For an ideal solution, Raoult's Law is obeyed.

$P_t = P_{CHCl_3} + P_{C_3H_6O}; \quad P_{CHCl_3} = 0.5(300 \text{ torr}) = 150 \text{ torr}$

$P_{C_3H_6O} = 0.5(360 \text{ torr}) = 180 \text{ torr}; \quad P_t = 150 \text{ torr} + 180 \text{ torr} = 330 \text{ torr}$

(b)   The real solution has a lower vapor pressure, 250 torr, than an ideal solution of the same composition, 330 torr. Thus, fewer molecules escape to the vapor phase from the liquid. This means that fewer molecules have sufficient kinetic energy to overcome intermolecular attractions. Clearly, even weak hydrogen bonds such as this one are stronger attractive forces than dipole-dipole or dispersion forces. These hydrogen bonds prevent molecules from escaping to the vapor phase and result in a lower than ideal vapor pressure for the solution. There is essentially no hydrogen bonding in the individual liquids.

(c)   According to Coulomb's law, electrostatic attractive forces lead to an overall lowering of the energy of the system. Thus, when the two liquids mix and hydrogen bonds are formed, the energy of the system is decreased and $\Delta H_{soln} < 0$; the solution process is exothermic.

14 Chemical Kinetics

Visualizing Concepts

14.2   Chemical equation (d), B $\rightarrow$ 2A, is consistent with the data. The concentration of A increases with time, and concentration B decreases with time, so B must be a reactant and A must be a product. The ending concentration of A is approximately twice as large as the starting concentration of B, so mole ratio of A:B is 2:1. The reaction is B $\rightarrow$ 2A.

14.3   *Analyze/Plan.* Using the relationship rate = $k[A]^x$, determine the value of x that produces a rate law to match the described situation. *Solve.*

   (a)   x = 0. The rate of reaction does not depend on [A], so the reaction is zero-order in A.

   (b)   x = 2. When [A] increases by a factor of 3, rate increases by a factor of $(3)^2 = 9$.

   (c)   x = 3. When [A] increases by a factor of 2, rate increases by a factor of $(2)^3 = 8$.

14.5   *Plan.* For a first-order reaction, a plot of ln[A] vs. time is linear, as shown in the diagram. The slope is $-k$, and the intercept is $[A]_0$. According to the Arrhenius equation [14.19], k increases with increasing temperature. *Solve.*

   (a)   Lines 1 and 2 have the same slope, and thus the same rate constant, k. These experiments are done at the same temperature. The y-intercepts of the two lines are different; the experiments had different initial concentrations of A.

   (b)   Lines 2 and 3 have the same y-intercept and thus the same starting concentration of A. The slopes of the two lines are different, so their rate constants are different and they occur at different temperatures. Line 3, with the smaller slope and k value will occur at the lower temperature.

14.7   The reaction profile has a single high point (peak), so the reaction occurs in a single step. This step is necessarily the rate-determining step.

   (1)   The total potential energy of the reactants.

   (2)   The activation energy, E_a, of the reaction. This is the difference in energy between the potential energy of the activated complex (transition state) and the potential energy of the reactants.

   (3)   The net energy change (ΔE) for the reaction. This is the difference in energy between the products and reactants. (Under appropriate conditions, this could also be ΔH.) For this reaction, the energy of products is lower than the energy of reactants, and the reaction releases energy to the surroundings.

   (4)   The total potential energy of the products.

14.8   On a plot of ln k vs. 1/T, the slope is $-E_a$ and the y-intercept is ln A, where E_a is activation energy and A is the frequency factor.

(a)   If $E_a(2) > E_a(1)$ and $A_1 = A_2$, the lines will have the same y-intercept, negative slope direction, and the slope of line 2 will be steeper than the slope of line 1.

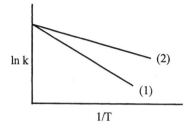

(b)   If $A_2 > A_1$ and $E_a(1) = E_a(2)$, the lines will be parallel with the same negative slopes and different y-intercepts.

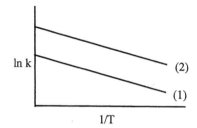

14.9   (a) $NO_2 + F_2 \rightarrow NO_2F + F$

$NO_2 + F \rightarrow NO_2F$

(b) $2NO_2 + F_2 \rightarrow 2NO_2F$

(c)   F is an intermediate, because it is produced and then consumed during the reaction.

(d)   rate = $k[NO_2][F_2]$

14.11   The most likely transition state shows the relative geometry of both reactants and products. It is reasonable to assume that multiple bonds, with greater total bond energy, remain intact at the expense of single bonds. In the black-and-white diagram below, open circles represent the red balls and closed circles represent the blue.

14.12   (a) $A_2 + AB + AC \rightarrow BA_2 + A + AC$
$BA_2 + A + AC \rightarrow A_2 + BA_2 + C$

net: $AB + AC \rightarrow BA_2 + C$

(b)   A is the intermediate; it is produced and consumed.

(c) A_2 is the catalyst; it is consumed and reproduced.

Reaction Rates

14.14   (a) M/s

       (b)   The hotter the oven, the faster the cake bakes. Milk sours faster in hot weather than cool weather.

       (c)   The *average rate* is the rate over a period of time, while the *instantaneous rate* is the rate at a particular time.

14.16

Time(s)	Mol A	(a) Mol B	Δ Mol A	(b) Rate $-(\Delta$ mol A/s)
0	0.100	0.000		
40	0.067	0.033	−0.033	8.3×10^{-4}
80	0.045	0.055	−0.022	5.5×10^{-4}
120	0.030	0.070	−0.015	3.8×10^{-4}
160	0.020	0.080	−0.010	2.5×10^{-4}

       (c)   The volume of the container must be known to report the rate in units of concentration (mol/L) per time.

14.18   (a)

Time (min)	Time Interval (min)	Concentration (M)	Δ M	Rate (M/s)
0.0		1.85		
54.0	54.0	1.58	−0.27	8.3×10^{-5}
107.0	53.0	1.36	−0.22	6.9×10^{-5}
215.0	108	1.02	−0.34	5.2×10^{-5}
430.0	215	0.580	−0.44	3.4×10^{-5}

       (b)   From the slopes of the lines in the figure at the right, the rates are: at 75.0 min, $4.2 \times 10^{-3} M/\text{min}$, or $7.0 \times 10^{-5} M/s$; at 250 min, $2.1 \times 10^{-3} M/\text{min}$ or $3.5 \times 10^{-5} M/s$

14.20   (a)   rate $= -\Delta[H_2O]/2\Delta t = \Delta[H_2]/2\Delta t = \Delta[O_2]/\Delta t$

       (b)   rate $= -\Delta[SO_2]/2\Delta t = -\Delta[O_2]/\Delta t = \Delta[SO_3]/2\Delta t$

       (c)   rate $= -\Delta[NO]/2\Delta t = -\Delta[H_2]/2\Delta t = \Delta[N_2]/\Delta t = \Delta[H_2O]/2\Delta t$

14.22 (a) $-\Delta[C_2H_4]/\Delta t = \Delta[CO_2]/2\Delta t = \Delta[H_2O]/2\Delta t$

$-2\Delta[C_2H_4]/\Delta t = \Delta[CO_2]/\Delta t = \Delta[H_2O]/\Delta t$

C_2H_4 is burning, $-\Delta[C_2H_4]/\Delta t = 0.025\ M/s$

CO_2 and H_2O are produced, at twice the rate that C_2H_4 is consumed.

$\Delta[CO_2]/\Delta t = \Delta[H_2O]/\Delta t = 2(0.025)\ M/s = 0.050\ M/s$

(b) In this reaction, pressure is a measure of concentration.

$-\Delta[N_2H_4]/\Delta t = -\Delta[H_2]/\Delta t = \Delta[NH_3]/2\Delta t$

N_2H_4 is consumed, $-\Delta[N_2H_4]/\Delta t = 63\ torr/hr$

H_2 is consumed, $-\Delta[H_2]/\Delta t = 63\ torr/hr$

NH_3 is produced at twice the rate that N_2H_4 and H_2 are consumed,

$\Delta[NH_3]/\Delta t = -2\Delta[N_2H_4]/\Delta t = 2(63)\ torr/hr = 126\ torr/hr$

$\Delta P_T/\Delta t = (+126\ torr/hr - 63\ torr/hr - 63\ torr/hr) = 0\ torr/hr$

Rate Laws

14.24 (a) rate = $k[A][C]^2$

(b) rate is proportional to [A], rate doubles

(c) rate is not affected by [B], no change

(d) rate changes as $[C]^2$, rate increases by a factor of 3^2 or 9

(e) rate increases by a factor of $(3)(3)^2 = 27$

14.26 (a) rate = $k[H_2][NO]^2$

(b) rate = $(6.0 \times 10^4\ M^{-2}\ s^{-1})(0.035\ M)^2(0.015\ M) = 1.1\ M/s$

(c) rate = $(6.0 \times 10^4\ M^{-2}\ s^{-1})\ (0.10\ M)^2\ (0.010\ M) = 6.0\ M/s$

14.28 (a, b) rate = $k[C_2H_5Br][OH^-]$; $k = \dfrac{rate}{[C_2H_5Br][OH^-]}$

at 298 K, $k = \dfrac{1.7 \times 10^{-7}\ M/s}{[0.0477\ M][0.100\ M]} = 3.6 \times 10^{-5}\ M^{-1}s^{-1}$

(c) Adding an equal volume of ethyl alcohol reduces both $[C_2H_5Br]$ and $[OH^-]$ by a factor of two. new rate = $(1/2)(1/2) = 1/4$ of old rate

14.30 (a) From the data given, when $[ClO_2]$ increases by a factor of 3 (experiment 2 to experiment 1), the rate increases by a factor of 9. When $[OH^-]$ increases by a factor of 3 (experiment 2 to experiment 3), the rate increases by a factor of 3. The reaction is second order in $[ClO_2]$ and first order in $[OH^-]$. rate = $k[ClO_2]^2[OH^-]$.

(b) Using data from Expt 2:

$k = \dfrac{rate}{[ClO_2]^2\ [OH^-]} = \dfrac{0.00276\ M/s}{(0.020\ M)^2\ (0.030\ M)} = 2.3 \times 10^2\ M^{-2}s^{-1}$

(c) rate = $2.3 \times 10^2\ M^{-2}\ s^{-1}\ (0.100\ M)^2(0.050\ M) = 0.115 = 0.12\ M/s$

14.32   *Analyze/Plan.* Follow the logic in Sample Exercise 14.6 to deduce the rate law. Rearrange the rate law to solve for k and deduce units. Calculate a k value for each set of concentrations and then average the three values.   *Solve.*

(a)   Doubling [NO] while holding $[O_2]$ constant increases the rate by a factor of 4 (experiments 1 and 2). Doubling $[O_2]$ while holding [NO] constant doubles the rate (experiments 2 and 3). The reaction is second order in [NO] and first order in $[O_2]$. rate = $k[NO]^2[O_2]$

(b, c)   From experiment 1: $k_1 = \dfrac{1.41 \times 10^{-2} \; M/s}{(0.0126 \; M)^2 (0.0125 \; M)} = 7105 = 7.11 \times 10^3 \; M^{-2}s^{-1}$

$k_2 = 0.113/(0.0252)^2(0.0250) = 7118 = 7.12 \times 10^3 \; M^{-2} \, s^{-1}$

$k_3 = 5.64 \times 10^{-2}/(0.0252)^2(0.125) = 7105 = 7.11 \times 10^3 \; M^{-2} \, s^{-1}$

$k_{avg} = (7105 + 7118 + 7105)/3 = 7109 = 7.11 \times 10^3 \; M^{-2} \, s^{-1}$

(d)   rate = $7.109 \times 10^3 \; M^{-2}s^{-1} \, (0.0750 \; M)^2(0.0100 \; M) = 0.3999 = 0.400 \; M/s$

(e)   The data are given in terms of the disappearance of NO. Use Equation [14.4] to relate the disappearance of NO to the disappearance of O_2.

$-\Delta[NO]/2\Delta t = -[O_2]/\Delta t$

For the concentrations given in part (d), $\Delta[NO]/\Delta t = 0.400 \; M/s$.

$\Delta[O_2]/\Delta t = \Delta[NO]/2\Delta t = 0.400 \; M/s/2 = 0.200 \; M/s$

14.34   (a)   Increasing $[S_2O_8^{2-}]$ by a factor of 1.5 while holding $[I^-]$ constant increases the rate by a factor of 1.5 (Experiments 1 and 2). Doubling $[S_2O_8^{2-}]$ and increasing $[I^-]$ by a factor of 1.5 triples the rate ($2 \times 1.5 = 3$, experiments 1 and 3). Thus the reaction is first order in both $[S_2O_8^{2-}]$ and $[I^-]$; rate = $k \, [S_2O_8^{2-}] \, [I^-]$.

(b) $k = rate/[S_2O_8^{2-}] \, [I^-]$

$k_1 = 2.6 \times 10^{-6} \; M/s/(0.018 \; M)(0.036 \; M) = 4.01 \times 10^{-3} = 4.0 \times 10^{-3} \; M^{-1}s^{-1}$

$k_2 = 3.9 \times 10^{-6} /(0.027)(0.036) = 4.01 \times 10^{-3} = 4.01 \times 10^{-3} = 4.0 \times 10^{-3} \; M^{-1}s^{-1}$

$k_3 = 7.8 \times 10^{-6} /(0.036)(0.054) = 4.01 \times 10^{-3} = 4.01 \times 10^{-3} = 4.0 \times 10^{-3} \; M^{-1}s^{-1}$

$k_4 = 1.4 \times 10^{-5} /(0.050)(0.072) = 3.89 \times 10^{-3} = 3.9 \times 10^{-3} \; M^{-1}s^{-1}$

$k_{avg} = 3.98 \times 10^{-3} = 4.0 \times 10^{-3} \; M^{-1}s^{-1}$

(c) $-\Delta[S_2O_8^{2-}]/\Delta t = -\Delta[I^-]/3\Delta t$; the rate of disappearance of $S_2O_8^{2-}$ is one-third the rate of disappearance of I^-.

(d)   Note that the data are given in terms of disappearance of $S_2O_8^{2-}$.

$\dfrac{-\Delta[I^-]}{\Delta t} = \dfrac{-3\Delta[S_2O_8^{2-}]}{\Delta t} = 3(3.98 \times 10^{-3} \; M^{-1}s^{-1})(0.025 \; M)(0.050 \; M) = 1.5 \times 10^{-5} \; M/s$

Change of Concentration with Time

14.36   (a)   A graph of 1/[A] vs time yields a straight line for a second-order reaction.

(b)   The half-life of a first-order reaction is independent of $[A]_0$, $t_{1/2} = 0.693/k$. Whereas, the half-life of a second-order reaction does depend on $[A]_0$, $t_{1/2} = 1/k[A]_0$.

14.38   (a)   For a first order reaction, $t_{1/2} = 0.693/k$.

$t_{1/2} = 0.693/0.271 \text{ s}^{-1} = 2.5572 = 2.56 \text{ s}$

(b)   For a first order reaction, $\ln[A]_t - \ln[A]_0 = -kt$. $\ln[A]_t = -kt + \ln[A]_0$

$[A]_0 = 0.050 \; M \; I_2, \; t = 5.12 \text{ s}, \; k = 0.271 \text{ s}^{-1}$

$\ln[I_2] = -0.271 \text{ s}^{-1} (5.12 \text{ s}) + \ln(0.050)$

$\ln[I_2] = -4.3833, \;\; [I_2] = 0.0125 \; M$

Check. 5.12 s is 2 half-lives. $[I_2]$ should be reduced by a factor of 4, and it is.

14.40   (a)   Using Equation [14.13] for a first order reaction: $\ln[A]_t = -kt + \ln[A]_0$

5.0 min = 300 s; $[N_2O_5]_0 = (0.0250 \text{ mol}/2.0 \text{ L}) = 0.0125 = 0.013 \; M$

$\ln[N_2O_5]_{300} = -(6.82 \times 10^{-3} \text{ s}^{-1})(300 \text{ s}) + \ln(0.0125)$

$\ln[N_2O_5]_{300} = -2.0460 + (-4.3820) = -6.4280 = -6.43$

$[N_2O_5]_{300} = 1.616 \times 10^{-3} = 1.6 \times 10^{-3} \; M; \text{ mol } N_2O_5 = 1.616 \times 10^{-3} \; M \times 2.0 \text{ L}$

$= 3.2 \times 10^{-3} \text{ mol}$

(b) $[N_2O_5]_t = 0.010 \text{ mol}/2.0 \text{ L} = 0.0050 \; M; \; [N_2O_5]_0 = 0.0125 \; M$

$\ln (0.0050) = -(6.82 \times 10^{-3} \text{ s}^{-1}) (t) + \ln(0.0125)$

$t = \dfrac{-[\ln(0.0050) - \ln(0.0125)]}{(6.82 \times 10^{-3} \text{s}^{-1})} = 134.35 = 1.3 \times 10^2 \text{ s} \times \dfrac{1 \text{ min}}{60 \text{ s}} = 2.24 = 2.2 \text{ min}$

(c) $t_{1/2} = 0.693/k = 0.693/6.82 \times 10^{-3} \text{ s}^{-1} = 101.6 = 102 \text{ s or } 1.69 \text{ min}$

14.42

t(s)	P_{CH_2NC}	$\ln P_{CH_3NC}$
0	502	6.219
2000	335	5.814
5000	180	5.193
8000	95.5	4.559
12000	41.7	3.731
15000	22.4	3.109

A graph of ln P vs t is linear with a slope of $-2.08 \times 10^{-4} \text{ s}^{-1}$. The rate constant k = – slope = $2.08 \times 10^{-4} \text{ s}^{-1}$. Half-life = $t_{1/2}$ = $0.693/k = 3.33 \times 10^3 \text{ s}$.

14.44   (a)   Make both first- and second-order plots to see which is linear. Moles is a satisfactory concentration unit, since volume is constant.

time(s)	mol A	ln (mol A)	1/mol A
0	0.1000	-2.303	10.00
40	0.067	-2.70	14.9
80	0.045	-3.10	22.2
120	0.030	-3.51	33.3
160	0.020	-3.91	50.0

The plot of ln (mol A) vs time is linear, so the reaction is first-order in A.

(b) $k = -\text{slope} = -[-3.91 - (-2.70)]/120 = 0.010083 = 0.0101 \text{ s}^{-1}$
(The best fit to this line yields the same value for the slope, $0.01006 = 0.0101 \text{ s}^{-1}$)

(c) $t_{1/2} = 0.693/k = 0.693/0.010083 \text{ s}^{-1} = 68.7 \text{ s}$

14.46   (a)   Make both first- and second-order plots to see which is linear.

time(min)	$[C_{12}H_{22}O_{11}](M)$	$\ln[C_{12}H_{22}O_{11}]$	$1/[C_{12}H_{22}O_{11}]$
0	0.316	-1.152	3.16
39	0.274	-1.295	3.65
80	0.238	-1.435	4.20
140	0.190	-1.661	5.26
210	0.146	-1.924	6.85

The plot of ln $[C_{12}H_{22}O_{11}]$ is linear, so the reaction is first order in $C_{12}H_{22}O_{11}$.

(b) $k = -\text{slope} = -[-1.924 - (-1.295)] / 171 \text{ min} = 3.68 \times 10^{-3} \text{ min}^{-1}$ (The slope of the best-fit line is $-3.67 \times 10^{-3} \text{ min}^{-1}$.)

Temperature and Rate

14.48   (a)   The orientation factor is less important in $H + Cl \rightarrow HCl$, because the reactants are monatomic and spherical (nondirectional); all collision orientations are equally effective.

(b)   The kinetic-molecular theory tells us that at some temperature T, there will be a distribution of molecular speeds and kinetic energies, and that the average kinetic energy of the sample is proportional to temperature. That is, as temperature of the sample increases, the average speed and kinetic energy of the molecules increases. At higher temperatures, there will be more molecular collisions (owing to greater speeds) and more energetic collisions (owing to greater kinetic energies). Overall there will be more collisions that have sufficient energy to form an activated complex, and the reaction rate will be greater.

14.50   (a) $f = e^{-E_a/RT}$ $E_a = 160 \text{ kJ/mol} = 1.60 \times 10^5 \text{ J/mol}$, $T = 500 \text{ K}$

$$-E_a/RT = -\frac{1.60 \times 10^5 \text{ J/mol}}{500 \text{ K}} \times \frac{\text{mol - K}}{8.314 \text{ J}} = -38.489 = -38.5$$

$$f = e^{-38.489} = 1.924 \times 10^{-17} = 2 \times 10^{-17}$$

(b) $$-E_a/RT = -\frac{1.60 \times 10^5 \text{ J/mol}}{510 \text{ K}} \times \frac{\text{mol - K}}{8.314 \text{ J}} = -37.735 = -37.7$$

$$f = e^{-37.735} = 4.093 \times 10^{-17} = 4.09 \times 10^{-17}$$

$$\frac{f \text{ at } 510 \text{ K}}{f \text{ at } 500 \text{ K}} = \frac{4.09 \times 10^{-17}}{1.92 \times 10^{-17}} = 2.13$$

An increase of 10 K means that 2.13 times more molecules have this energy.

14.52   *Analyze/Plan.* Use the definitions of activation energy ($E_{max} - E_{react}$) and ΔE ($E_{prod} - E_{react}$) to sketch the graph and calculate E_a for the reverse reaction. *Solve.*

(a)   (b) $E_a(\text{reverse}) = 18 \text{ kJ}$

14.54 E_a for the reverse reaction is:

(a) $45 - (-25) = 70 \text{ kJ}$ (b) $35 - (-10) = 45 \text{ kJ}$ (c) $55 - 10 = 45 \text{ kJ}$

Based on the magnitude of E_a, the reverse of reactions (b) and (c) occur at the same rate, which is faster than the reverse of reaction (a).

14.56 $T_1 = 737°C + 273 = 1010$ K, $k_1 = 0.0796$ $M^{-1}s^{-1}$;

$T_2 = 947°C + 273 = 1220$ K, $k_2 = 0.0815$ $M^{-1}s^{-1}$

$$\ln\left(\frac{k_1}{k_2}\right) = \frac{E_a}{R}\left(\frac{1}{T_2} - \frac{1}{T_1}\right)$$

$$\ln\left(\frac{0.0796}{0.0815}\right) = \frac{E_a}{8.314 \text{ J/mol}}\left(\frac{1}{1220} - \frac{1}{1010}\right)$$

$$-0.023589 = \frac{E_a \ (-1.704 \times 10^{-4})}{8.314 \text{ J/mol}}$$

$$E_a = \frac{8.314 \ (-0.023589) \text{ J/mol}}{(-1.704 \times 10^{-4})} = 1.151 \times 10^3 \text{ J/mol} = 1.15 \text{ kJ/mol}$$

14.58

k	ln k	T(K)	1/T($\times 10^3$)
0.028	-3.58	600	1.67
0.22	-1.51	650	1.54
1.3	0.26	700	1.43
6.0	1.79	750	1.33
23	3.14	800	1.25

Using the relationship $\ln k = \ln A - E_a/RT$, the slope, $-15.94 \times 10^3 = -16 \times 10^3$, is $-E_a/R$. $E_a = 15.94 \times 10^3 \times 8.314$ J/mol $= 1.3 \times 10^2$ kJ/mol. To calculate A, we will use the rate data at 700 K. From the equation given above, $0.262 = \ln A - 15.94 \times 10^3/700$; $\ln A = 0.262 + 22.771$. $A = 1.0 \times 10^{10}$.

14.60   (a) $T_1 = 77°F$; $°C = 5/9 \ (°F - 32) = 5/9 \ (77 - 32) = 25°C = 298$ K

$T_2 = 59°F$; $°C = 5/9 \ (59-32) = 15°C = 288$ K; $k_1/k_2 = 6$

$$\ln\left(\frac{k_1}{k_2}\right) = \frac{E_a}{R}\left[\frac{1}{T_2} - \frac{1}{T_1}\right]; \quad \ln(6) = \frac{E_a}{8.314 \text{ J/mol}}\left[\frac{1}{288} - \frac{1}{298}\right]$$

$$E_a = \frac{\ln(6)(8.314 \text{ J/mol})}{1.165 \times 10^{-4}} = 1.28 \times 10^5 \text{ J} = 1.3 \times 10^2 \text{ kJ/mol}$$

$T_1 = 77°F = 25°C = 298$ K; $T_2 = 41°F = 5°C = 278$ K, $k_1/k_2 = 40$

$$\ln(40) = \frac{E_a}{8.314 \text{ J/mol}}\left[\frac{1}{278} - \frac{1}{296}\right]; \quad E_a = \frac{\ln(40)(8.314 \text{ J/mol})}{2.414 \times 10^{-4}}$$

$$E_a = 1.27 \times 10^5 \text{ J} = 1.3 \times 10^2 \text{ kJ/mol}$$

The values are amazingly consistent, considering the precision of the data.

(b)   For a first order reaction, $t_{1/2} = 0.693/k$, $k = 0.693/t_{1/2}$

k_1 at 298 K = $0.693/2.7$ yr = $0.257 = 0.26$ yr^{-1}

$T_1 = 298$ K, $T_2 = 273 - 15°C = 258$ K

$$\ln\left(\frac{0.257}{k_2}\right) = \frac{1.27 \times 10^5 \text{ J}}{8.314 \text{ J/mol}}\left[\frac{1}{258} - \frac{1}{298}\right] = 7.9497$$

$0.257/k_2 = e^{7.9497} = 2.835 \times 10^3$; $k_2 = 0.257/2.835 \times 10^3 = 9.066 \times 10^{-5} = 9.1 \times 10^{-5}$ yr^{-1}

$t_{1/2} = 0.693/k = 0.693/9.066 \times 10^{-5} = 7.64 \times 10^3$ yr = 7.6×10^3 yr

Reaction Mechanisms

14.62   (a)   The *molecularity* of a process indicates the number of molecules that participate as reactants in the process. A unimolecular process has one reactant molecule, a bimolecular process has two reactant molecules and a termolecular process has three reactant molecules.

(b)   Termolecular processes are rare because it is highly unlikely that three molecules will simultaneously collide with the correct energy and orientation to form an activated complex.

(c)   An *intermediate* is a substance that is produced and then consumed during a chemical reaction. It does not appear in the balanced equation for the overall reaction.

14.64   (a)   bimolecular, rate = $k[NO]^2$

(b)   unimolecular, rate = $k[C_3H_6]$

(c)   unimolecular, rate = $k[SO_3]$

14.66   (a)   Two elementary reactions; two energy maxima

(b)   One intermediate; one energy minimum between reactants and products

(c)   The second step is rate-limiting; second energy maximum and E_a is larger.

(d)   Overall reaction is exothermic; energy of products is lower than energy of reactants.

14.68   *Analyze/Plan.* Follow the logic in Sample Exercise 14.14.   *Solve.*

(a)   First step: rate = $k_1[H_2O_2]$ $[I^-]$; second step: rate = $k_2[IO^-]$ $[H_2O_2]$

(b) $2H_2O_2(aq) \rightarrow 2H_2O(l) + O_2(g)$

(c) $IO^-(aq)$ is the intermediate.

(d)   rate = $k[H_2O_2]$ $[I^-]$

14.70   (a)   i. $HBr + O_2 \rightarrow HOOBr$

ii. $HOOBr + HBr \rightarrow 2HOBr$

iii. $2HOBr + 2HBr \rightarrow 2H_2O + 2Hr_2$
$$\overline{4HBr + O_2 \rightarrow 2H_2O + 2Br_2}$$

(b) The observed rate law is: rate = k[HBr][O₂], the rate law for the first elementary step. The first step must be rate-determining.

(c) HOOBr and HOBr are both intermediates; HOOBr is produced in i and consumed in ii and HOBr is produced in ii and consumed in iii.

(d) Since the first step is rate-determining, it is possible that neither of the intermediates accumulates enough to be detected. This does not disprove the mechanism, but indicates that steps ii and iii are very fast, relative to step i.

Catalysis

14.72 (a) The smaller the particle size of a solid catalyst, the greater the surface area. The greater the surface area, the more active sites and the greater the increase in reaction rate.

(b) Adsorption is the binding of reactants onto the surface of the heterogeneous catalyst. It is usually the first step in the catalyzed reaction.

14.74 (a) $2[NO(g) + N_2O(g) \rightarrow N_2(g) + NO_2(g)]$
$$\underline{2NO_2(g) \rightarrow 2NO(g) + O_2(g)}$$
$$2N_2O(g) \rightarrow 2N_2(g) + O_2(g)$$

(b) An intermediate is produced and then consumed during the course of the reaction. A catalyst is consumed and then reproduced. In other words, the catalyst is present when the reaction sequence begins and after the last step is completed. In this reaction, NO is the catalyst and NO₂ is an intermediate.

(c) No. The proposed mechanism cannot be ruled out, based on the behavior of NO₂. NO₂ functions as an intermediate; it is produced and then consumed during the reaction. That there is no measurable build-up of NO₂ indicates the first step is slow relative to the second; as soon as NO₂ is produced by the slow first step, it is consumed by the faster second step.

14.76 (a) Catalytic converters are heterogeneous catalysts that adsorb gaseous CO and hydrocarbons and speed up their oxidation to CO₂(g) and H₂O(g). They also adsorb nitrogen oxides, NOₓ, and speed up their reduction to N₂(g) and O₂(g). If a catalytic converter is working effectively, the exhaust gas should have very small amounts of the undesirable gases CO, (NO)ₓ and hydrocarbons.

(b) The high temperatures could increase the rate of the desired catalytic reactions given in part (a). It could also increase the rate of undesirable reactions such as corrosion, which decrease the lifetime of the catalytic converter.

(c) The rate of flow of exhaust gases over the converter will determine the rate of adsorption of CO, (NO)ₓ and hydrocarbons onto the catalyst and thus the rate of conversion to desired products. Too fast an exhaust flow leads to less than maximum adsorption. A very slow flow leads to back pressure and potential damage to the exhaust system. Clearly the flow rate must be adjusted to balance chemical and mechanical efficiency of the catalytic converter.

14.78   Just as the π electrons in C_2H_4 are attracted to the surface of a hydrogenation catalyst, the nonbonding electron density on S causes compounds of S to be attracted to these same surfaces. Strong interactions could cause the sulfur compounds to be permanently attached to the surface, blocking active sites and reducing adsorption of alkenes for hydrogenation.

14.80   The individual structure of each enzyme molecule leads to a unique coiling and folding pattern. The resulting shape and electronic properties of the active site in each enzyme leads to its substrate specificity.

14.82   Let k and E_a equal the rate constant and activation energy for the uncatalyzed reaction. Let k_c and E_{ac} equal the rate constant and activation energy of the catalyzed reaction. A is the same for the uncatalyzed and catalyzed reactions. $k_c/k = 1 \times 10^5$, $T = 37°C = 310$ K.

According to Equation [14.20], $\ln k = E_a/RT + \ln A$. Subtracting ln k from ln k_c

$$\ln k_c - \ln k = \left[\frac{-E_{ac}}{RT}\right] + \ln A - \left[\frac{-E_a}{RT}\right] - \ln A$$

$$\ln (k_c/k) = \frac{E_a - E_{ac}}{RT}; \quad E_a - E_{ac} = RT \ln (k_c/k)$$

$$E_a - E_{ac} = \frac{8.314\,J}{K\text{-}mol} \times 310\,K \times \ln(1 \times 10^5) = 2.966 \times 10^4\,J = 29.66\,kJ = 3 \times 10^1 kJ$$

The enzyme must lower the activation energy by 30 kJ in order to achieve a 1×10^5-fold increase in reaction rate.

Additional Exercises

14.84 $\text{rate} = \dfrac{-\Delta[H_2S]}{\Delta t} = \dfrac{\Delta[Cl^-]}{2\Delta t} = k[H_2S][Cl_2]$

$\dfrac{-\Delta[H_2S]}{\Delta t} = (3.5 \times 10^{-2}\,M^{-1}s^{-1})(2.0 \times 10^{-4}\,M)(0.025\,M) = 1.75 \times 10^{-7} = 1.8 \times 10^{-7}\,M/s$

$\dfrac{\Delta[Cl^-]}{\Delta t} = \dfrac{-2\Delta[H_2S]}{\Delta t} = 2(1.75 \times 10^{-7}\,M/s) = 3.5 \times 10^{-7}\,M/s$

14.85   (a) $\text{rate} = \dfrac{-\Delta[NO]}{2\Delta t} = \dfrac{-\Delta[O_2]}{\Delta t} = \dfrac{9.3 \times 10^{-5}\,M/s}{2} = 4.7 \times 10^{-5}\,M/s$

(b,c) $\text{rate} = k[NO]^2[O_2]; \quad k = \text{rate}/[NO]^2[O_2]$

$k = \dfrac{4.7 \times 10^{-5}\,M/s}{(0.040\,M)^2(0.035\,M)} = 0.8393 = 0.84\,M^{-2}\,s^{-1}$

(d)   Since the reaction is second order in NO, if the [NO] is increased by a factor of 1.8, the rate would increase by a factor of 1.8^2, or $(3.24) = 3.2$.

14.87   The units of rate are M/s. The reaction must be second order overall if the units of the rate constant are $M^{-1} s^{-1}$. If rate = $k[NO_2]^x$, then the cumulative units of $[NO_2]^x$ must be M^2, and x = 2.

If $[NO_2]_0 = 0.100\ M$ and $[NO_2]_t = 0.025\ M$, use the integrated form of the second order rate equation, $\dfrac{1}{[A]_t} = kt + \dfrac{1}{[A]_o}$, Equation [14.14], to solve for t.

$$\frac{1}{0.025\ M} = 0.63\ M^{-1}s^{-1}\ (t) + \frac{1}{0.100\ M};\ \frac{(40-10)\ M^{-1}}{0.63\ M^{-1}s^{-1}} = t = 47.62 = 48\ s.$$

14.88   For a first order reaction, $t_{1/2} = 0.633/k$. For a second order reaction, $t_{1/2} = 1/k[A]_0$. Both relationships show $t_{1/2}$ to be constant over the course of the reaction; neither is a function of time. The rate law for reaction (1) could be either first or second order. Reaction (2), where $t_{1/2}$ varies (gets larger) as time increases, has a rate law that is greater than second order.

14.90   *Analyze.* Given rate constants for the decay of two radioisotopes, determine half-lives, decay rates, and amount remaining after three half-lives. *Plan.* Determine reaction order. Based on reaction-order, select the appropriate relationships for (a) rate constant and half-life and (c) rate-constant, time and concentration. In this example, mass is a measure of concentration.

Solve. Decay of radioiosotopes is a first-order process, since only one species is involved and the decay is not initiated by collision.

(a)   For a first-order process, $t_{1/2} = 0.693/k$.

^{241}Am: $t_{1/2} = 0.693/1.6 \times 10^{-3}\ yr^{-1} = 433.1 = 4.3 \times 10^2\ yr$

^{125}I: $t_{1/2} = 0.693/0.011\ day^{-1} = 63.00 = 63\ days$

(b)   For a given sample size, half of the ^{241}Am sample decays in 433 years, whereas half of the ^{125}I sample decays in 63 days. ^{125}I decays at a much faster rate.

(c)   For a first order process, $\ln[A]_t - \ln[A]_0 = -kt$. $\ln[A]_t = -kt + \ln[A]_0$.

$[A]_0 = 1.0\ mg;\ t = 3\ t_{1/2}$.

^{241}Am: $t = 3\ t_{1/2} = 3(433.1\ yr) = 1.299 \times 10^3 = 1.3 \times 10^3\ yr$

$\ln[Am]_t = -1.6 \times 10^{-3}\ yr\ (1.299 \times 10^3 - \ln(1.0) = -2.079 - 0 = -2.08$

$[Am]_t = 0.125 = 0.13\ mg$

or, mass ^{241}Am remaining = $1.0\ mg/2^3 = 0.125 = 0.13\ mg$

^{125}I: For the same size starting sample and number of elapsed half-lives, the same mass, 0.13 mg ^{125}I, will remain. (The difference is that the elapsed time of 3 half-lives for ^{125}I is 3(63) = 189 days = 0.52 yr, vs. 433 yr for ^{241}Am.)

14.91   (a) $k = (8.56 \times 10^{-5}\ M/s)/(0.200\ M) = 4.28 \times 10^{-4}\ s^{-1}$

(b) $\ln[urea] = -(4.28 \times 10^{-4}s^{-1} \times 4.00 \times 10^3\ s) + \ln(0.500)$

$\ln[urea] = -1.712 - 0.693 = -2.405 = -2.41;\ [urea] = 0.0903 = 0.090\ M$

(c) $t_{1/2} = 0.693/k = 0.693/4.28 \times 10^{-4}\ s^{-1} = 1.62 \times 10^3\ s$

14.93

Time (s)	$[C_5H_6]$ (M)	$\ln[C_5H_6]$	$1/[C_5H_6]$
0	0.0400	–3.219	25.0
50	0.0300	–3.507	33.3
100	0.0240	–3.730	41.7
150	0.0200	–3.912	50.0
200	0.0174	–4.051	57.5

The plot of $1/[C_5H_6]$ vs time is linear and the reaction is second order.

The slope of this line is k. k = slope = $(50.0 - 25.0)\, M^{-1}/(150-0)s = 0.167\, M^{-1}\, s^{-1}$

(The best-fit slope and k value is 0.163 $M^{-1}\, s^{-1}$.)

14.94   (a)   No. The value of A, which is related to frequency and effectiveness of collisions, can be different for each reaction and k is proportional to A.

(b)   From Equation [14.20], reactions with different variations of k with respect to temperature have different activation energies, E_a. The fact that k for the two reactions is the same at a certain temperature is coincidental. The reaction with the higher rate at 35°C has the larger activation energy, because it was able to use the increase in energy more effectively.

14.96   (a)
$$NO(g) + NO(g) \rightarrow N_2O_2(g)$$

$$\frac{N_2O_2(g) + H_2(g) \rightarrow N_2O(g) + H_2O(g)}{2NO(g) + N_2O_2(g) + H_2(g) \rightarrow N_2O_2(g) + N_2O(g) + H_2O(g)}$$
$$2NO(g) + H_2(g) \rightarrow N_2O(g) + H_2O(g)$$

(b)   First reaction: $- \Delta [NO]/\Delta t = k[NO]\,[NO] = k[NO]^2$

Second reaction: $- \Delta [H_2]/\Delta t = k[H_2][N_2O_2]$

(c) N_2O_2 is the intermediate: it is produced in the first step and consumed in the second.

(d)   Since $[H_2]$ appears in the rate law, the second step must be slow relative to the first.

14.97   (a)

$$Cl(g) + O_3(g) \rightarrow ClO(g) + O_2(g)$$

$$ClO(g) + O(g) \rightarrow Cl(g) + O_2(g)$$

$$\overline{Cl(g) + O_3(g) + ClO(g) + O(g) \rightarrow ClO(g) + O_2(g) + Cl(g) + O_2(g)}$$

$$O_3(g) + O(g) \rightarrow 2O_2(g)$$

(b)   Cl(g) is the catalyst. It is consumed in the first step and reproduced in the second.

(c)   ClO(g) is the intermediate. It is produced in the first step and consumed in the second.

14.98   (a)

Reaction Pathway

14.30 = E_a for reaction [14.30]

14.31 = E_a for reaction [14.31]

(b)   The fact that Br$_2$ builds up during the reaction tells us that the appearance of Br$_2$ (reaction [14.30]) is faster than the disappearance of Br$_2$ (reaction [14.31]). This is the reason that E_a of [14.31] in the energy profile above is larger than E_a of [14.30].

14.100   (a) $(CH_3)_3AuPH_3 \rightarrow C_2H_6 + (CH_3)AuPH_3$

(b) $(CH_3)_3Au$, $(CH_3)Au$ and PH_3 are intermediates.

(c)   Reaction 1 is unimolecular, Reaction 2 is unimolecular, Reaction 3 is bimolecular.

(d)   Reaction 2, the slow one, is rate determining.

(e)   If Reaction 2 is rate determining, rate = $k_2[(CH_3)_3Au]$.

$(CH_3)_3Au$ is an intermediate formed in Reaction 1, an equilibrium. By definition, the rates of the forward and reverse processes in Reaction 1 are equal:

$k_1[(CH_3)_3AuPH_3] = k_{-1}[(CH_3)_3Au][PH_3]$; solving for $[(CH_3)_3Au]$,

$$[(CH_3)_3Au] = \frac{k_1[(CH_3)_3AuPH_3]}{k_{-1}[PH_3]}$$

Substituting into the rate law

$$\text{rate} = \left(\frac{k_2k_1}{k_{-1}}\right)\frac{[(CH_3)_3AuPH_3]}{[PH_3]} = \frac{k[(CH_3)_3AuPH_3]}{[PH_3]}$$

(f)   The rate is inversely proportional to [PH_3], so adding PH_3 to the $(CH_3)_3AuPH_3$ solution would decrease the rate of the reaction.

14.102   The rate law for the slow step is rate = k_2[Es], where ES is an intermediate. Use relationships from the fast equilibrium step to substitute for [ES].

rate of the forward reaction = k_1[E][S]

rate of the reverse reaction = k_{-1}[ES]

For an equilibrium, rate forward = rate reverse, k_1[E][S] = k_{-1}[ES]; [ES] = $\dfrac{k_1}{k_{-1}}$[E][S]

Rate = k_2[ES] = $\dfrac{k_2 k_1}{k_{-1}}$[E][S] = k[E][S]

14.103   (a)   The fact that the rate doubles with a doubling of the concentration of sugar tells us that the fraction of enzyme tied up in the form of an enzyme-substrate complex is small. A doubling of the substrate concentration leads to a doubling of the concentration of enzyme-substrate complex, because most of the enzyme molecules are available to bind substrates.

         (b)   The behavior of inositol suggests that it acts as a competitor with sucrose for binding at the active sites of the enzyme system. Such a competition results in a lower effective concentration of active sites for binding of sucrose, and thus results in a lower reaction rate.

Integrative Exercises

14.105   (a)   ln k = $-E_a/RT$ + ln A; E_a = 86.8 kJ/mol = 8.68×10^4 J/mol;

$$T = 35°C + 273 = 308 \text{ K}; A = 2.10 \times 10^{11} \, M^{-1} \, s^{-1}$$

$$\ln k = \frac{-8.68 \times 10^4 \text{ J/mol}}{308 \text{ K}} \times \frac{\text{mol - K}}{8.314 \text{ J}} + \ln (2.10 \times 10^{11} \, M^{-1} \, s^{-1})$$

ln k = –33.8968 + 26.0704 = –7.8264; k = $3.99 \times 10^{-4} \, M^{-1} \, s^{-1}$

         (b) $\dfrac{0.335 \text{ g KOH}}{0.250 \text{ L soln}} \times \dfrac{1 \text{ mol KOH}}{56.1 \text{ g KOH}} = 0.02389 = 0.0239 \, M \text{ KOH}$

$$\frac{1.453 \text{ g } C_2H_5I}{0.250 \text{ L soln}} \times \frac{1 \text{ mol } C_2H_5I}{156.0 \text{ g } C_2H_5I} = 0.03726 = 0.0373 \, M \, C_2H_5I$$

If equal volumes of the two solutions are mixed, the initial concentrations in the reaction mixture are 0.01194 M KOH and 0.01863 M C_2H_5I. Assuming the reaction is first order in each reactant:

rate = k[C_2H_5I][OH^-] = $3.99 \times 10^{-4} \, M^{-1}s^{-1}$ (0.01194 M)(0.01863 M) = $8.88 \times 10^{-8} \, M/s$

         (c)   Since C_2H_5I and OH^- react in a 1 : 1 mole ratio and equal volumes of the solutions are mixed, the reactant with the smaller concentration, KOH, is the limiting reactant.

14.107   **(a)** $\ln k = -E_a/RT + \ln A$, Equation [14.20]. $E_a = 6.3$ kJ/mol $= 6.3 \times 10^3$ J/mol

             T = 100°C + 273 = 373 K

$$\ln k = \frac{-6.3 \times 10^3 \text{ J/mol}}{8.314 \text{ J/K - mol} \times 373 \text{ K}} + \ln(6.0 \times 10^8 \ M^{-1}s^{-1})$$

             $\ln k = -2.032 + 20.212 = 18.181 = 18.2$; $k = 7.87 \times 10^7 = 8 \times 10^7 \ M^{-1}s^{-1}$

   (b)   NO, 11 valence e⁻, 5.5 e⁻ pair   ONF, 18 valence e⁻, 9 e⁻ pr
(Assume the less electronegative N atom will be electron deficient.)

$$\ddot{\text{O}}{=}\ddot{\text{N}}{-}\ddot{\text{F}}{:} \longleftrightarrow \left({:}\ddot{\text{O}}{-}\ddot{\text{N}}{=}\ddot{\text{F}} \right)$$

            $:\dot{\text{N}}{=}\ddot{\text{O}}:$ The resonance form on the right is a very minor contributor to the true bonding picture, due to high formal charges and the unlikely double bond involving F.

   (c)   ONF has trigonal planar electron domain geometry, which leads to a "bent" structure with a bond angle of approximately 120°.

   (d)

$$\left[\begin{array}{c} \text{O}{=}\text{N} \\ \text{F}{-}\text{F} \end{array} \right]$$

   (e)   The electron deficient NO molecule is attracted to electron-rich F_2, so the driving force for formation of the transition state is greater than simple random collisions.

14.108   **(a)** $\Delta H^{\circ}_{rxn} = 2\Delta H^{\circ}_f \ H_2O(g) + 2\Delta H^{\circ}_f \ Br_2(g) - 4\Delta H^{\circ}_f \ HBr(g) - \Delta H^{\circ}_f \ O_2(g)$

            $\Delta H^{\circ}_{rxn} = 2(-241.82) + 2(30.71) - 4(-36.23) - (0) = -277.30$ kJ

   (b)   Since the rate of the uncatalyzed reaction is very slow at room temperature, the magnitude of the activation energy for the rate-determining first step must be quite large. At room temperature, the reactant molecules have a distribution of kinetic energies (Chapter 10), but very few molecules even at the high end of the distribution have sufficient energy to form an activated complex. E_a for this step must be much greater than 3/2 RT, the average kinetic energy of the sample.

   (c)   20 e⁻, 10 e⁻ pr

          $\text{H}{-}\ddot{\text{O}}{-}\ddot{\text{O}}{-}\ddot{\text{B}}\text{r}:$

          The intermediate resembles hydrogen peroxide, H_2O_2.

14.110   **(a)**   If the reaction proceeds in a single elementary step, the coefficients in the balanced equation are the reaction orders for the respective reactants.

          rate = $k[Ce^{4+}]^2 [Tl^+]$

   (b)   If the uncatalyzed reaction occurs in a single step, it is **termolecular**. The activated complex requires collision of three particles with the correct energy and orientation for reaction. The probability of an effective three-particle collision is low and the rate is slow.

(c)   The first step is rate-determining.

(d)   The ability of Mn to adopt every oxidation state from +2 to +7 makes it especially suitable to catalyze this (and many other) reactions.

14.111   (a) $D(Cl–Cl) = 242$ kJ/mol Cl_2

$$\frac{242 \text{ kJ}}{\text{mol } Cl_2} \times \frac{1000 \text{ J}}{\text{kJ}} \times \frac{1 \text{ mol}}{6.022 \times 10^{23} \text{ molecules}} = 4.019 \times 10^{-19} = 4.02 \times 10^{-19} \text{ J}$$

$$\lambda = hc/E = \frac{6.626 \times 10^{-34} \text{ J-s} \times 2.998 \times 10^8 \text{ m/s}}{4.019 \times 10^{-19} \text{ J}} = 4.94 \times 10^{-7} \text{ m}$$

This wavelength, 494 nm, is in the visible portion of the spectrum.

(b)

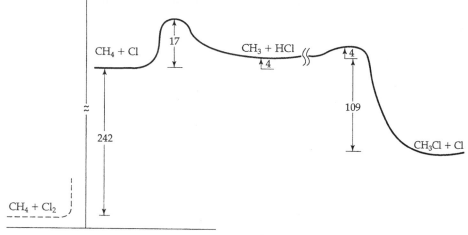

Reaction Pathway

(c)   Since $D(Cl–Cl)$ is 242 kJ/mol, $CH_4(g) + Cl_2(g)$ should be about 242 kJ below the starting point on the diagram. For the reaction $CH_4(g) + Cl_2(g) \rightarrow CH_3(g) + HCl(g) + Cl(g)$, E_a is $242 + 17 = 259$ kJ. (From bond dissociation enthalpies, ΔH for the overall reaction $CH_4(g) + Cl_2(g) \rightarrow CH_3Cl(g) + Cl(g)$ is –104 kJ, so the graph above is simply a sketch of the relative energies of some of the steps in the process.)

(d) CH_3, 7 valence e^-, odd electron species   H—Ċ—H
                                                      |
                                                      H

(e)   This sequence is called a chain reaction because Cl• radicals are regenerated in Reaction 4, perpetuating the reaction. Absence of Cl• terminates the reaction, so Cl• + Cl• $\rightarrow Cl_2$ is a termination step.

15 Chemical Equilibrium

Visualizing Concepts

15.2 Yes. The first box is pure reactant A. As the reaction proceeds, some A changes to B. In the fourth and fifth boxes, the relative amounts (concentrations) of A and B are constant. Although the reaction is ongoing the rates of A $\rightarrow$ B and B $\rightarrow$ A are equal, and the relative amounts of A and B are constant.

15.3 *Analyze.* Given box diagram and reaction type, determine whether K > 1 for the equilibrium mixture depicted in the box.

Plan. Assign species in the box to reactants and products. Write an equilibrium expression in terms of concentrations. Find the relationship between numbers of molecules and concentration. Calculate K.

Solve. Let red = A, blue = X, red and blue pairs = AX. (The colors of A and X are arbitrary.) There are 3A, 2B, and 8AX in the box.

M = mol/L. Since moles is a counting unit for particles, mol ratios and particle ratios are equivalent. We can use numbers of particles in place of moles in the molarity formula. V = 1 L, so in this case, [A] = number of A particles.

$$K = \frac{[AX]}{[A][X]}; \quad [AX] = 8/V = 8; [A] = 3/V = 3; [X] = 2/V = 2 . \quad K = \frac{8}{[3][2]} = \frac{8}{6} = 1.33$$

15.5 *Analyze/Plan.* The reaction with the largest equilibrium constant has the largest ratio of products to reactants. Count product and reactant molecules. Calculate ratios and compare. *Solve.*

$$K = \frac{[C_2H_4X_2]}{[C_2H_4][X_2]}. \quad \text{Use numbers of molecules as an adequate measure of concentration.}$$

(While the volume terms don't cancel, they are the same for all parts. For the purpose of comparison, we can ignore volume.) *Solve.*

(a)   8 $C_2H_4Cl_2$,   2 Cl_2,   2 C_2H_4. $K = \dfrac{8}{(2)(2)} = 2$

(b)   6 $C_2H_4Br_2$,   4 Br_2,   4 C_2H_4. $K = \dfrac{6}{(4)(4)} = 0.375 = 0.4$

(c)   3 $C_2H_4I_2$,   7 I_2,   7 C_2H_4. $K = \dfrac{3}{(7)(7)} = 0.0612 = 0.06$

From the smallest to the largest equilibrium constant, (c) < (b) < (a).

Check. By inspection, there are the fewest product molecules and the most reactant molecules in (c); most product and least reactant in (a).

15.7   For the reaction $A_2(g) + B(g) \rightleftharpoons A(g) + AB(g)$, $\Delta n = 0$ and $K_p = K_c$. We can evaluate the equilibrium expression in terms of concentration. Also since $\Delta n = 0$, the volume terms in the expression cancel and we can use number of particles as a measure of moles and molarity. The mixture contains 2A, 4AB and $2A_2$.

$$K_c = \frac{[A][AB]}{[A_2][B]} = \frac{(2)(4)}{(2)(B)} = 2; B = 2$$

2 B atoms should be added to the diagram.

15.8   *Analyze.* Given the diagram and reaction type, calculate the equilibrium constant K_c.

Plan. Analyze the contents of the cylinder. Express them as concentrations, using number of particles as a measure of moles, and V = 1 L. Write the equilibrium expression in terms of concentration and calculate K_c.   *Solve.*

(a)   The mixture contains $2A_2$, 2B, 4AB. $[A_2] = 2$, [B] = 2, [AB] = 4.

$$K_c = \frac{[AB]^2}{[A_2][B]^2} = \frac{(4)^2}{(2)(2)^2} = 2$$

(b)   A decrease in volume favors the reaction with fewer particles. This reaction has two particles in products and three in reactants, so a decrease in volume favors products. The number of AB (product) molecules will increase.

Note that a change in volume does not change the value of K_c. If V decreases, the number of AB molecules must increase in order to maintain the equilibrium value of K_c.

15.10   (a)   Exothermic. In both reaction mixtures (orange and blue), [AB] decreases as T increases.

(b)   In the reaction, there are fewer moles of gas in products than reactants, so greater pressure favors production of products. At any single temperature, [AB] is greater at P = y than at P = x. Since the concentration of the product, AB, is greater at P = y, P = y is the greater pressure.

Equilibrium; The Equilibrium Constant

15.12   (a) $K_c = \frac{[C][D]}{[A][B]}$; if K_c is large, the numerator of the K_c expression is much greater than the denominator and products will predominate at equilibrium.

(b) $K_c = k_f/k_r$; if K_c is large, k_f is larger than k_r and the forward reaction has the greater rate constant.

15.14   (a) $K_c = \dfrac{[O_2]^3}{[O_3]^2}$ (b) $K_c = \dfrac{1}{[Cl_2]^2}$

(c) $K_c = \dfrac{[C_2H_6]^2[O_2]}{[C_2H_4]^2[H_2O]^2}$ (d) $K_c = \dfrac{[CH_4]}{[H_2]^2}$

(e) $K_c = \dfrac{[Cl_2]^2}{[HCl]^4[O_2]}$

homogeneous: (a), (c); heterogeneous: (b), (d), (e)

15.16   (a)   equilibrium lies to right, favoring products ($K_c \gg 1$)

    (b)   equilibrium lies to left, favoring reactants ($K_c \ll 1$)

15.18 $SO_2(g) + Cl_2(g) \rightleftharpoons SO_2Cl_2(g)$, $K_p = 34.5$. $\Delta n = 1 - 2 = -1$

$K_p = K_c(RT)^{\Delta n}$; $34.5 = K_c(RT)^{-1} = K_c/RT$;

$K_c = 34.5\,RT = 34.5(0.08206)(303) = 857.81 = 858$

15.20 $2H_2(g) + S_2(g) \rightleftharpoons 2H_2S(g)$, $K_c = 1.08 \times 10^7$ at 700°C. $\Delta n = 2 - 3 = -1$.

    (a) $K_p = K_c(RT)^{\Delta n} = 700°C + 273 = 973$ K.

$K_p = 1.08 \times 10^7 (RT)^{-1} = \dfrac{1.08 \times 10^7}{(0.08206)(973)} = 1.35 \times 10^5$

    (b)   Both K_p and K_c are much greater than one, so the product, H_2S, is favored at equilibrium. The equilibrium mixture contains mostly H_2S.

15.22 $K_p = \dfrac{P_{HCl}^4 \times P_{O_2}}{P_{Cl_2}^2 \times P_{H_2O}^2} = 0.0752$

    (a) $K_p = \dfrac{P_{Cl_2}^2 \times P_{H_2O}^2}{P_{HCl}^4 \times P_{O_2}} = \dfrac{1}{0.0752} = 13.298 = 13.3$

    (b) $K_p = \dfrac{P_{HCl}^2 \times P_{O_2}^{1/2}}{P_{Cl_2} \times P_{H_2O}} = (0.0752)^{1/2} = 0.2742 = 0.274$

    (c) $K_p = K_c(RT)^{\Delta n}$; $\Delta n = 2.5 - 2 = 0.5$; $T = 480°C + 273 = 753$ K

$K_p = K_c(RT)^{1/2}$, $K_c = K_p/(RT)^{1/2} = 0.2742/[0.08206 \times 753]^{1/2} = 0.03488 = 0.0349$

15.24
$$2NO(g) + Br_2(g) \rightleftharpoons 2NOBr(g) \qquad\qquad K_1 = 2.0$$
$$N_2(g) + O_2(g) \rightleftharpoons 2NO(g) \qquad\qquad K_2 = \dfrac{1}{2.1 \times 10^{30}}$$
$$\overline{2NO(g) + Br_2(g) + N_2(g) + O_2(g) \rightleftharpoons 2NOBr(g) + 2NO(g)}$$
$$N_2(g) + O_2(g) + Br_2(g) \rightleftharpoons 2NOBr(g)$$

$K_c = K_1 \times K_2 = 2.0 \times \dfrac{1}{2.1 \times 10^{30}} = 9.524 \times 10^{-31} = 9.5 \times 10^{-31}$

15.26   (a) $K_p = 1/P_{SO_2}$

    (b) Na_2O is a pure solid. The molar concentration, the ratio of moles of a substance to volume occupied by the substance, is a constant for pure solids and liquids.

Calculating Equilibrium Constants

15.28 $[CH_3OH] = \dfrac{0.0406 \text{ mol}}{2.00 \text{ L}} = 0.0203 \text{ M}$

$[CO] = \dfrac{0.170 \text{ mol CO}}{2.00 \text{ L}} = 0.0850 \text{ M}; \quad [H_2] = \dfrac{0.302 \text{ mol H}_2}{2.00 \text{ L}} = 0.151 \text{ M}$

$K_c = \dfrac{[CH_3OH]}{[CO][H_2]^2} = \dfrac{0.0203}{(0.0850)(0.151)^2} = 10.4743 = 10.5$

15.30   (a) $K_p = \dfrac{P_{PCl_5}}{P_{PCl_3} \times P_{Cl_2}} = \dfrac{1.30 \text{ atm}}{0.124 \text{ atm} \times 0.157 \text{ atm}} = 66.8$

(b)   Since $K_p > 1$, products (the numerator of the K_p expression) are favored over reactants (the denominator of the K_p expression).

15.32   (a)   Calculate the concentrations of $H_2(g)$ and $Br_2(g)$ and the equilibrium concentration of $H_2(g)$. $M = \text{mol/L}$.

$[H_2]_{init} = 1.374 \text{ g H}_2 \times \dfrac{1 \text{ mol H}_2}{2.0159 \text{ g H}_2} \times \dfrac{1}{2.00 \text{ L}} = 0.34079 = 0.341 \, M$

$[Br_2] = 70.31 \text{ g Br}_2 \times \dfrac{1 \text{ mol Br}_2}{159.81 \text{ g Br}_2} \times \dfrac{1}{2.00 \text{ L}} = 0.21998 = 0.220 \, M$

$[H_2]_{equil} = 0.566 \text{ g H}_2 \times \dfrac{1 \text{ mol H}_2}{2.0159 \text{ g H}_2} \times \dfrac{1}{2.00 \text{ L}} = 0.14038 = 0.140 \, M$

	$H_2(g)$	+	$Br_2(g)$	$\rightleftharpoons$	$2HBr(g)$
initial	0.34079 M		0.21998 M		0
change	–0.20041 M		–0.20041 M		+2(0.20041) M
equil.	0.14038 M		0.01957 M		0.40082 M

The change in H_2 is (0.34079 – 0.14038 = 0.20041 = 0.200). The changes in $[Br_2]$ and [HBr] are set by stoichiometry, resulting in the equilibrium concentrations shown in the table.

(b) $K_c = \dfrac{[HBr]^2}{[H_2][Br_2]} = \dfrac{(0.40082)^2}{(0.14038)(0.01957)} = \dfrac{(0.401)^2}{(0.140)(0.020)} = 58.48 = 58$

The equilibrium concentration of Br_2 has 3 decimal places and 2 sig figs, so the value of K_c has 2 sig figs.

15.34   (a)

	$N_2O_4(g)$	$\rightleftharpoons$	$2NO_2(g)$
initial	1.500 atm		1.000 atm
change	+0.244 atm		–0.488 atm
equil	1.744 atm		0.512 atm

The change in P_{NO_2} is (1.000 – 0.512) = –0.488 atm, so the change in $P_{N_2O_4}$ is +(0.488/2) = +0.244 atm.

(b) $K_p = \dfrac{P_{NO_2}^2}{P_{N_2O_4}} = \dfrac{(0.512)^2}{(1.744)} = 0.1503 = 0.150$

Applications of Equilibrium Constants

15.36   (a)   If the value of Q_c equals the value of K_c, the system is at equilibrium.

          (b)   In the direction of less products (more reactants), to the left.

          (c) $Q_c = 0$ if the concentration of any product is zero.

15.38   Calculate the reaction quotient in each case, compare with

$$K_p = \dfrac{P_{NH_3}^2}{P_{N_2} \times P_{H_2}^3} = 4.51 \times 10^{-5}$$

          (a) $Q = \dfrac{(98)^2}{(45)(55)^3} = 1.3 \times 10^{-3}$

          Since $Q > K_p$, the reaction will shift toward reactants to achieve equilibrium.

          (b) $Q = \dfrac{(57)^2}{(143)(0)^3} = \infty$

          Since $Q > K_p$, reaction must shift shift toward reactants to achieve equilibrium. There must be some N_2 present to achieve equilibrium. In this example, the only source of N_2 is the decomposition of NH_3.

          (c) $Q = \dfrac{(13)^2}{(27)(82)^3} = 1.1 \times 10^{-5}$; Q is only slightly less than K_p, so the reaction will shift slightly toward products to achieve equilibrium.

15.40 $K_p = \dfrac{P_{SO_3}^2}{P_{SO_2}^2 \times P_{O_2}}$; $P_{SO_3} = \left(K_p \times P_{SO_2}^2 \times P_{O_2}\right)^{1/2} = [(0.345)(0.135)^2(0.455)]^{1/2} = 0.0535 \text{ atm}$

15.42   (a) $K_c = \dfrac{[I]^2}{[I_2]} = 3.1 \times 10^{-5}$

$$[I] = \dfrac{2.67 \times 10^{-2} \text{ g I}}{10.0 \text{ L}} \times \dfrac{1 \text{ mol I}}{126.9 \text{ g I}} = 2.1040 \times 10^{-5} = 2.10 \times 10^{-5} M$$

$$[I_2] = \dfrac{[I]^2}{K_c} = \dfrac{(2.104 \times 10^{-5})^2}{3.1 \times 10^{-5}} = 1.428 \times 10^{-5} = 1.43 \times 10^{-5} M$$

$$\dfrac{1.428 \times 10^{-5} \text{ mol } I_2}{L} \times 10.0 \text{ L} \times \dfrac{253.8 \text{ g } I_2}{\text{mol } I_2} = M = 0.0362 \text{ g } I_2$$

Check. $K_c = \dfrac{(2.104 \times 10^{-5})^2}{1.428 \times 10^{-5}} = 3.1 \times 10^{-5}$

(b) $PV = nRT; P = \dfrac{gRT}{MM\,V}$

$P_{SO_3} = \dfrac{1.17 \text{ g SO}_3}{80.06 \text{ g/mol}} \times \dfrac{0.08206 \text{ L-atm}}{\text{K-mol}} \times \dfrac{700 \text{ K}}{2.00 \text{ L}} = 0.4197 = 0.420 \text{ atm}$

$P_{O_2} = \dfrac{0.105 \text{ g O}_2}{32.00 \text{ g/mol}} \times \dfrac{0.08206 \text{ L-atm}}{\text{K-mol}} \times \dfrac{700 \text{ K}}{2.00 \text{ L}} = 0.09424 = 0.0942 \text{ atm}$

$K_p = 3.0 \times 10^4 = \dfrac{P_{SO_3}^2}{P_{SO_2}^2 \times P_{O_2}}; P_{SO_2} = \left[P_{SO_3}^2 / (K_p)(P_{O_2}) \right]^{1/2}$

$P_{SO_2} = [(0.4197)^2 / (3.0 \times 10^4)(0.09424)]^{1/2} = 7.894 \times 10^{-3} = 7.9 \times 10^{-3} \text{ atm}$

$g \text{ SO}_2 = \dfrac{MM\,PV}{RT} = \dfrac{64.06 \text{ g SO}_2}{\text{mol SO}_2} \times \dfrac{\text{K-mol}}{0.08206 \text{ L-atm}} \times \dfrac{7.894 \times 10^{-3} \text{ atm} \times 2.00 \text{ L}}{700 \text{ K}}$

$= 0.01761 = 0.018 \text{ g SO}_2$

Check. $K_p = [(0.4197)^2 / (7.894 \times 10^{-3})^2 (0.09424)] = 3.0 \times 10^4$

15.44 $[Br_2] = [Cl_2] = 0.30 \text{ mol} / 1.0 \text{ L} = 0.30 \, M$

	$Br_2(g)$	+	$Cl_2(g)$	$\rightleftharpoons$	$2BrCl(g)$	$K_c = \dfrac{[BrCl]^2}{[Br_2][Cl_2]} = 7.0$
initial	0.30 M		0.30 M		0	
change	–x		–x		+2x	
equil.	(0.30 – x)		(0.30 – x)		+2x	

$7.0 = \dfrac{(2x)^2}{(0.30 - x)^2}$ (We can solve this exactly by taking the square root of both sides.)

$(7.0)^{1/2} = \dfrac{2x}{0.30 - x}, 2.646(0.30 - x) = 2x, 0.7937 = 4.646x, x = 0.1709 = 0.17 \, M$

$[BrCl] = 2x = 0.3417 = 0.34 \, M; [Br_2] = [Cl_2] = 0.30 - x = 0.1291 = 0.13 \, M$

Check. $K_c = (0.3417)^2 / (0.1291)^2 = 7.0$

15.46 $K_c = [NH_3][H_2S] = 1.2 \times 10^{-4}$. Because of the stoichiometry, equilibrium concentrations of H_2S and NH_3 will be equal; call this quantity y. Then, $y^2 = 1.2 \times 10^{-4}$, $y = 0.010954 = 0.011 \, M$.

15.48   (a)   *Analyze/Plan.* If only $PH_3BCl_3(s)$ is present initially, the equation requires that the equilibrium concentrations of $PH_3(g)$ and $BCl_3(g)$ are equal. Write the K_c expression and solve for x = $[PH_3]$ = $[BCl_3]$.   *Solve.*

$K_c = [PH_3][BCl_3]; 1.87 \times 10^{-3} = x^2; x = 0.043243 = 0.0432 \, M \, PH_3$ and BCl_3

(b)   Since the mole ratios are 1:1:1, mol $PH_3BCl_3(s)$ required = mol PH_3 or BCl_3 produced.

$$\frac{0.043243 \text{ mol PH}_3}{L} \times 0.500 \text{ L} = 0.02162 = 0.0216 \text{ mol PH}_3 = 0.0216 \text{ mol PH}_3BCl_3$$

$$0.02162 \text{ mol PH}_3BCl_3 \times \frac{151.2 \text{ g PH}_3BCl_3}{1 \text{ mol PH}_3BCl_3} = 3.269 = 3.27 \text{ g PH}_3BCl_3$$

In fact, some $PH_3BCl_3(s)$ must remain for the system to be in equilibrium, so a bit more than 3.27 g PH_3BCl_3 is needed.

15.50 $CaCrO_4(s) \rightleftharpoons Ca^{2+}(aq) + CrO_4^{2-}(aq)$ $K_c = [Ca^{2+}][CrO_4^{2-}] = 7.1 \times 10^{-4}$

At equilibrium, $[Ca^{2+}] = [CrO_4^{2-}] = x$

$K_c = 7.1 \times 10^{-4} = x^2$, $x = 0.0266 = 0.027 \, M \, Ca^{2+}$ and CrO_4^{2-}

LeChâtelier's Principle

15.52 $4NH_3(g) + 5O_2(g) \rightleftharpoons 4NO(g) + 6 H_2O(g)$

      (a)   increase [NH₃], increase yield NO

      (b)   increase [H₂O], decrease yield NO

      (c)   decrease [O₂], decrease yield NO

      (d)   decrease container volume, decrease yield NO (fewer moles gas in reactants)

      (e)   add catalyst, no change

      (f)   increase temperature, decrease yield NO (reaction is exothermic)

15.54   (a)   The reaction must be endothermic ($+\Delta H$) if heating increases the fraction of products.

      (b)   There must be more moles of gas in the products if increasing the volume of the vessel increases the fraction of products.

15.56   (a) $\Delta H° = \Delta H_f^\circ \, CH_3OH(g) - \Delta H_f^\circ \, CO(g) - 2\Delta H_f^\circ \, H_2(g)$

                 $= -201.2 \text{ kJ} - (-110.5 \text{ kJ}) - 0 \text{ kJ}$

                 $= -90.7 \text{ kJ}$

      (b)   The reaction is exothermic; an increase in temperature would decrease the value of K and decrease the yield. A low temperature is needed to maximize yield.

      (c)   Increasing total pressure would increase the partial pressure of each gas, shifting the equilibrium toward products. The extent of conversion to CH_3OH increases as the total pressure increases.

Additional Exercises

15.58 $2A(g) \rightleftharpoons B(g)$, $K_c = 1$

      $\frac{[B]}{[A]^2} = 1$, $[B] = [A]^2$ and $[A] = [B]^{1/2}$

15.59 $CH_4(g) + H_2O(g) \rightarrow CO(g) + 3H_2(g)$

$$K_p = \frac{P_{CO} \times P_{H_2}^3}{P_{CH_4} \times P_{H_2O}} ; \quad P = \frac{g\, RT}{MM\, V} ; \quad T = 1000 \text{ K}$$

$$P_{CO} = \frac{8.62 \text{ g}}{28.01 \text{ g/mol}} \times \frac{0.08206 \text{ L-atm}}{\text{mol-K}} \times \frac{1000 \text{ K}}{5.00 \text{ L}} = 5.0507 = 5.05 \text{ atm}$$

$$P_{H_2} = \frac{2.60 \text{ g}}{2.016 \text{ g/mol}} \times \frac{0.08206 \text{ L-atm}}{\text{mol-K}} \times \frac{1000 \text{ K}}{5.00 \text{ L}} = 21.1663 = 21.2 \text{ atm}$$

$$P_{CH_4} = \frac{43.0 \text{ g}}{16.04 \text{ g/mol}} \times \frac{0.08206 \text{ L-atm}}{\text{mol-K}} \times \frac{1000 \text{ K}}{5.00 \text{ L}} = 43.9973 = 44.0 \text{ atm}$$

$$P_{H_2O} = \frac{48.4 \text{ g}}{18.02 \text{ g/mol}} \times \frac{0.08206 \text{ L-atm}}{\text{mol-K}} \times \frac{1000 \text{ K}}{5.00 \text{ L}} = 44.0811 = 44.1 \text{ atm}$$

$$K_p = \frac{(5.0507)(21.1663)^3}{(43.9973)(44.0811)} = 24.6949 = 24.7$$

$K_p = K_c(RT)^{\Delta n}$, $K_c = K_p/(RT)^{\Delta n}$; $\Delta n = 4 - 2 = 2$

$K_c = (24.6949)/[(0.08206)(1000)]^2 = 3.6673 \times 10^{-3} = 3.67 \times 10^{-3}$

15.61   (a) $H_2(g) + S(s) \rightleftharpoons H_2S(g)$ $K_c = [H_2S]/[H_2]$

(b)   Calculate the molarities of H_2S and H_2.

$$[H_2S] = \frac{0.46 \text{ g}}{34.1 \text{ g/mol}} \times \frac{1}{1.0 \text{ L}} = 0.01349 = 0.013 \ M$$

$$[H_2] = \frac{0.40 \text{ g}}{2.02 \text{ g/mol}} \times \frac{1}{1.0 \text{ L}} = 0.1980 = 0.20 \ M$$

$K_c = 0.01349/0.1980 = 0.06812 = 0.068$

(c)   Since S is a pure solid, its concentration doesn't change during the reaction, so [S] does not appear in the equilibrium expression.

15.62   (a) $K_p = \dfrac{P_{Br_2} \times P_{NO}^2}{P_{NOBr}^2} ; \quad P = \dfrac{gRT}{PV} ; \quad T = 100°C + 273 = 373$

$$P_{Br_2} = \frac{4.19 \text{ g}}{159.8 \text{ g/mol}} \times \frac{0.08206 \text{ L-atm}}{\text{K-mol}} \times \frac{373}{5.0 \text{ L}} = 0.16051 = 0.161 \text{ atm}$$

$$P_{NO} = \frac{3.08 \text{ g}}{30.01 \text{ g/mol}} \times \frac{0.08206 \text{ L-atm}}{\text{K-mol}} \times \frac{373}{5.0 \text{ L}} = 0.62828 = 0.628 \text{ atm}$$

$$P_{NOBr} = \frac{3.22 \text{ g NOBr}}{109.9 \text{ g/mol}} \times \frac{0.08206 \text{ L-atm}}{\text{K-mol}} \times \frac{373}{5.0 \text{ L}} = 0.17936 = 0.179 \text{ atm}$$

$$K_p = \frac{(0.16051)(0.62828)^2}{(0.17936)^2} = 1.9695 = 1.97 \qquad K_p = K_c(RT)^{\Delta n}, \ \Delta n = 3 - 2 = 1$$

$K_c = K_p/RT = 1.9695/(0.08206)(373) = 0.064345 = 0.0643$

(b) $P_t = P_{Br_2} + P_{NO} + P_{NOBr} = 0.16051 + 0.62828 + 0.17936 = 0.96815 = 0.968 \text{ atm}$

15.64   (a) $K_p = \dfrac{P_{NH_3}^2}{P_{N_2} \times P_{H_2}^3} = 4.34 \times 10^{-3}; \; T = 300°C + 273 = 573\,K$

$$P_{NH_3} = \dfrac{gRT}{MM \times V} = \dfrac{1.05\,g}{17.03\,g/mol} \times \dfrac{0.08206\,L\text{-}atm}{K\text{-}mol} \times \dfrac{573\,K}{1.00\,L} = 2.899 = 2.90\,atm$$

	$N_2(g) +$	$3H_2(g)$	$\rightleftharpoons$	$2NH_3(g)$
initial	0 atm	0 atm		?
change	x	3x		–2x
equil.	x atm	3x atm		2.899 atm

(Remember, only the change line reflects the stoichiometry of the reaction.)

$$K_p = \dfrac{(2.899)^2}{(x)(3x)^3} = 4.34 \times 10^{-3}; \; 27\,x^4 = \dfrac{(2.899)^2}{4.34 \times 10^{-3}}; \; x^4 = 71.725$$

$x = 2.910 = 2.91\,atm = P_{N_2}; \; P_{H_2} = 3x = 8.730 = 8.73\,atm$

$$g_{N_2} = \dfrac{MM \times PV}{RT} = \dfrac{28.02\,g\,N_2}{mol\,N_2} \times \dfrac{K\text{-}mol}{0.08206\,L\text{-}atm} \times \dfrac{2.910\,atm \times 1.00\,L}{573\,K} = 1.73\,g\,N_2$$

$$g_{H_2} = \dfrac{2.016\,g\,H_2}{mol\,H_2} \times \dfrac{K\text{-}mol}{0.08206\,L\text{-}atm} \times \dfrac{8.730\,atm \times 1.00\,L}{573\,K} = 0.374\,g\,H_2$$

(b)   The initial $P_{NH_3} = 2.899\,atm + 2(2.910\,atm) = 8.719 = 8.72\,atm$

$$g_{NH_3} = \dfrac{17.03\,g\,NH_3}{mol\,NH_3} \times \dfrac{K\text{-}mol}{0.08206\,L\text{-}atm} \times \dfrac{8.719\,atm \times 1.00\,L}{573\,K} = 3.16\,g\,NH_3$$

(c) $P_t = P_{N_2} + P_{H_2} + P_{NH_3} = 2.910\,atm + 8.730\,atm + 2.899\,atm = 14.54\,atm$

15.65

	$2IBr$	$\rightleftharpoons$	$I_2 +$	Br_2
initial	0.025 atm		0	0
change	–2x		x	x
equil.	(0.025 – 2x) atm		x	x

$$K_p = 8.5 \times 10^{-3} = \dfrac{P_{I_2} \times P_{Br_2}}{P_{IBr}^2} = \dfrac{x^2}{(0.025 - 2x)^2}; \quad \text{Taking the square root of both sides}$$

$$\dfrac{x}{0.025 - 2x} = (8.5 \times 10^{-3})^{1/2} = 0.0922; \; x = 0.0922(0.025 - 2x)$$

$x + 0.184x = 0.002305; \; 1.184x = 0.002305; \; x = 0.001947 = 1.9 \times 10^{-3}$

P_{IBr} at equilibrium $= 0.025 - 2(1.947 \times 10^{-3}) = 0.02111 = 0.021\,atm$

15.67 $K_p = P_{NH_3} \times P_{H_2S}; \; P_t = 0.614\,atm$

If the equilibrium amounts of NH_3 and H_2S are due solely to the decomposition of $NH_4HS(s)$, the equilibrium pressures of the two gases are equal, and each is 1/2 of the total pressure.

$$P_{NH_3} = P_{H_2S} = 0.614 \text{ atm}/2 = 0.307 \text{ atm}$$

$$K_p = (0.307)^2 = 0.0943$$

15.68   Initial $P_{SO_3} = \dfrac{gRT}{MM\, V} = \dfrac{0.831\, g}{80.07\, g/mol} \times \dfrac{0.08206\, L\text{-}atm}{K\text{-}mol} \times \dfrac{1100\, K}{1.00\, L} = 0.9368 = 0.937 \text{ atm}$

	2SO$_3$	⇌	2SO$_2$	+	O$_2$
initial	0.9368 atm		0		0
change	–2x		+2x		+x
equil.	0.9368–2x		2x		x
[equil.]	0.2104 atm		0.7264 atm		0.3632 atm

$P_t = (0.9368-2x) + 2x + x; \; 0.9368 + x = 1.300 \text{ atm}; \; x = 1.300 - 0.9368 = 0.3632 = 0.363 \text{ atm}$

$$K_p = \dfrac{P_{SO_2}^2 \times P_{O_2}}{P_{SO_3}^2} = \dfrac{(0.7264)^2\,(0.3632)}{(0.2104)^2} = 4.3292 = 4.33$$

$K_p = K_p = K_c(RT)^{\Delta n}; \; \Delta n = 3 - 2 = 1; \; K_p = K_c(RT)$

$K_c = K_p/RT = 4.3292/[(0.08206)(1100)] = 0.04796 = 0.0480$

15.70 $K_c = [CO_2] = 0.0108; [CO_2] = \dfrac{g\, CO_2}{44.01\, g/mol} \times \dfrac{1}{10.0\, L}$

In each case, calculate $[CO_2]$ and determine the position of the equilibrium.

(a) $[CO_2] = \dfrac{4.25\, g}{44.01\, g/mol} \times \dfrac{1}{10.0\, L} = 9.657 \times 10^{-3} = 9.66 \times 10^{-3}\, M$

$Q = 9.66 \times 10^{-3} > K_c$. The reaction proceeds to the right to achieve equilibrium and the amount of $CaCO_3(s)$ decreases.

(b) $[CO_2] = \dfrac{5.66\, g\, CO_2}{44.01\, g/mol} \times \dfrac{1}{10.0\, L} = 0.0129\, M$

$Q = 0.0129 > K_c$. The reaction proceeds to the left to achieve equilibrium and the amount of $CaCO_3(s)$ increases.

(c)   6.48 g CO_2 means $[CO_2] > 0.0129\, M$; $Q > 0.0129 > K_c$, the amount of $CaCO_3$ increases.

15.71

	CO$_2$(g)	+	H$_2$(g)	⇌	CO(g)	+	H$_2$O(g)
initial	1.50 mol		1.50 mol		0		0
change	–x		–x		+x		+x
equil.	(1.50 – x)mol		(1.50 – x)mol		x		x

Since $\Delta n = 0$, the volume terms cancel and we can use moles in place of molarity in the K expression.

$$K_c = 0.802 = \frac{[CO][H_2O]}{[CO_2][H_2O]} = \frac{x^2}{(1.50-x)^2}$$

Take the square root of both sides.

$(0.802)^{1/2} = x/(1.50-x);\ 0.8955(1.50-x) = x$

$1.3433 = 1.8955x,\ x = 0.7087 = 0.709$ mol

$[CO] = [H_2O] = 0.7087$ mol$/0.750$ L $= 0.945\ M$

$[CO_2] = [H_2] = (1.50 - 0.709)mol/0.750$ L $= 1.06\ M$

15.72   (a) $[CO_2] = \dfrac{25.0\ g\ CO_2}{44.01\ g/mol} \times \dfrac{1}{3.00\ L} = 0.18935 = 0.189\ M$

		C(s)	+	CO$_2$(g)	$\rightleftharpoons$	2CO(g)
initial		excess		0.189		0
change		$-x$		$-x$		$+2x$
equil.				0.189$- x$		$+2x$

$K_c = 1.9 = \dfrac{[CO]^2}{[CO_2]} = \dfrac{(2x)^2}{0.189-x};\quad 4x^2 = 1.9(0.189-x);\ 4x^2 + 1.9x - 0.36 = 0.$

Solve the quadratic for x.

$$x = \frac{-1.9 \pm \sqrt{(1.9)^2 - 4(4)(-0.36)}}{2(4)} = 0.14505 = 0.15\ M$$

$[CO] = 2x = 2(0.14505) = 0.2901 = 0.29\ M$

$$\frac{0.2901\ mol\ CO}{L} \times 3.00\ L \times \frac{28.01\ g\ CO}{mol} = 24.38 = 24\ g\ CO$$

(b)   The amount of C(s) consumed is related to x. Change M to mol to g C.

$$\frac{0.14505\ mol}{L} \times 3.00\ L \times 12.01\ g = 5.226 = 5.2\ g\ C\ consumed$$

(c)   A smaller vessel at the same temperature increases the total pressure of the mixture. The equilibrium shifts to form fewer total moles of gas, which favors reactants. The yield of CO product will be smaller in a smaller vessel.

(d)   If the reaction is endothermic, K is larger at higher temperature. This is confirmed by the two values for K$_c$, 0.133 at 298 k and 1.0 at 1000 K.

15.74   (a)

		CCl$_4$(g)	$\rightleftharpoons$	C(s) + 2Cl$_2$(g)
initial		2.00 atm		0 atm
change		$-x$ atm		$+2x$ atm
equil.		(2.00$-x$) atm		2x atm

$$K_p = 0.76 = \frac{P_{Cl_2}^2}{P_{CCl_4}} = \frac{(2x)^2}{(2.00-x)}$$

$$1.52 - 0.76x = 4x^2; \quad 4x^2 + 0.76x - 1.52 = 0$$

Using the quadratic formula, $a = 4$, $b = 0.76$, $c = -1.52$

$$x = \frac{-0.76 \pm \sqrt{(0.76)^2 - 4(4)(-1.52)}}{2(4)} = \frac{-0.76 + 4.99}{8} = 0.5287 = 0.53 \text{ atm}$$

$$\text{Fraction CCl}_4 \text{ reacted} = \frac{x \text{ atm}}{2.00 \text{ atm}} = \frac{0.53}{2.00} = 0.264 = 26\%$$

(b) $P_{Cl_2} = 2x = 2(0.5287) = 1.06 \text{ atm}$

$P_{CCl_4} = 2.00 - x = 2.00 - 0.5287 = 1.47 \text{ atm}$

15.75   (a) $Q = \dfrac{P_{PCl_5}}{P_{PCl_3} \times P_{Cl_2}} = \dfrac{(0.20)}{(0.50)(0.50)} = 0.80$

0.80 (Q) > 0.0870 (K), the reaction proceeds to the left.

(b)

	PCl₃(g)	+	Cl₂(g)	⇌	PCl₅(g)
initial	0.50 atm		0.50 atm		0.20 atm
change	+x atm		+x atm		–x atm
equil.	(0.50 + x) atm		(0.50 + x) atm		(0.20 – x) atm

(Since the reaction proceeds to the left, P_{PCl_5} must decrease and P_{PCl_3} and P_{Cl_2} must increase.)

$$K_p = 0.0870 = \frac{(0.20-x)}{(0.50+x)(0.50+x)}; \quad 0.0870 = \frac{(0.20-x)}{(0.250+1.00\,x+x^2)}$$

$$0.0870(0.250 + 1.00x + x^2) = 0.20 - x; \quad -0.17825 + 1.0870x + 0.0870\,x^2 = 0$$

$$x = \frac{-1.0870 \pm \sqrt{(1.0870)^2 - 4(0.0870)(-0.17825)}}{2(0.0870)} = \frac{-1.0870 + 1.1152}{0.174} = 0.162$$

$P_{PCl_3} = (0.50 + 0.162) \text{ atm} = 0.662$ $P_{Cl_2} = (0.50 + 0.162) \text{ atm} = 0.662 \text{ atm}$

$P_{PCl_5} = (0.20 - 0.162) \text{ atm} = 0.038 \text{ atm}$

To two decimal places, the pressures are 0.66, 0.66 and 0.04 atm, respectively. When substituting into the K_p expression, pressures to three decimal places yield a result much closer to 0.0870.

(c)   Increasing the volume of the container favors the process where more moles of gas are produced, so the reverse reaction is favored and the equilibrium shifts to the left; the mole fraction of Cl_2 increases.

(d)   For an exothermic reaction, increasing the temperature decreases the value of K; more reactants and fewer products are present at equilibrium and the mole fraction of Cl_2 increases.

15.77   (a)   Since the volume of the vessel = 1.00 L, mol = M. The reaction will proceed to the left to establish equilibrium.

$$A(g) + \quad 2B(g) \quad \rightleftharpoons \quad 2C(g)$$

initial	$0\ M$	$0\ M$	$1.00\ M$
change	$+x\ M$	$+2x\ M$	$-2x\ M$
equil.	$x\ M$	$2x\ M$	$(1.00 - 2x)\ M$

At equilibrium, $[C] = (1.00 - 2x)\ M$, $[B] = 2x\ M$.

    (b) x must be less than $0.50\ M$ (so that $[C]$, $1.00 - 2x$, is not less than zero).

    (c) $K_c = \dfrac{[C]^2}{[A][B]^2}$; $\dfrac{(1.00 - 2x)^2}{(x)(2x)^2} = 0.25$

$1.00 - 4x + 4x^2 = 0.25(4x)^3$; $x^3 - 4x^2 + 4x - 1 = 0$

    (d)

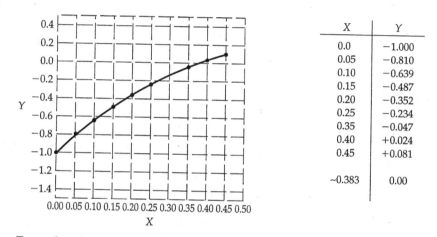

X	Y
0.0	-1.000
0.05	-0.810
0.10	-0.639
0.15	-0.487
0.20	-0.352
0.25	-0.234
0.35	-0.047
0.40	$+0.024$
0.45	$+0.081$
~0.383	0.00

    (e)   From the plot, $x \approx 0.383\ M$

$[A] = x = 0.383\ M$; $[B] = 2x = 0.766\ M$

$[C] = 1.00 - 2x = 0.234\ M$

Using the K_c expression as a check:

$K_c = 0.25$; $\dfrac{(0.234)^2}{(0.383)(0.766)^2} = 0.24$;   the estimated values are reasonable.

15.78 $K_p = \dfrac{P_{O_2} \times P_{CO}^2}{P_{CO_2}^2} \approx 1 \times 10^{-13}$; $P_{O_2} = (0.03)(1\ atm) = 0.03\ atm$

$P_{CO} = (0.002)(1\ atm) = 0.002\ atm$; $P_{CO_2} = (0.12)(1\ atm) = 0.12\ atm$

$Q = \dfrac{(0.03)(0.002)^2}{(0.12)^2} = 8.3 \times 10^{-6} = 8 \times 10^{-6}$

Since $Q > K_p$, the system will shift to the left to attain equilibrium. Thus a catalyst that promoted the attainment of equilibrium would result in a lower CO content in the exhaust.

Integrative Exercises

15.81   (a)   (i) $K_c = [Na^+]/[Ag^+]$

                 (ii) $K_c = [Hg^{2+}]^3 / [Al^{3+}]^2$

                 (iii) $K_c = [Zn^{2+}] [H_2] / [H^+]^2$

    (b)   According to Table 4.5, the activity series of the metals, a metal can be oxidized by any metal cation below it on the table.

                 (i) Ag^+ is far below Na, so the reaction will proceed to the right and K_c will be large.

                 (ii) Al^{3+} is above Hg, so the reaction will not proceed to the right and K_c will be small.

                 (iii) H^+ is below Zn, so the reaction will proceed to the right and K_c will be large.

    (c) $K_c < 1$ for this reaction, so Fe^{2+} (and thus Fe) is above Cd on the table. In other words, Cd is below Fe. The value of K_c, 0.06, is small but not extremely small, so Cd will be only a few rows below Fe.

15.82   (a) $AgCl(s) \rightleftharpoons Ag^+(aq) + Cl^-(aq)$

    (b) $K_c = [Ag^+][Cl^-]$

    (c)   Using thermodynamic data from Appendix C, calculate ΔH for the reaction in part (a).

$$\Delta H^\circ = \Delta H_f^\circ \; Ag^+(aq) + \Delta H_f^\circ \; Cl^-(aq) - \Delta H_f^\circ \; AgCl(s)$$

$$\Delta H^\circ = 105.90 \text{ kJ} - 167.2 \text{ kJ} - (-127.0 \text{ kJ}) = 65.7 \text{ kJ}$$

        The reaction is endothermic (heat is a reactant), so the solubility of AgCl(s) in $H_2O(l)$ will increase with increasing temperature.

15.84   Consider the energy profile for an exothermic reaction.

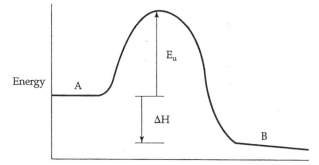

Reaction Pathway

The activation energy in the forward direction, E_{af}, equals E_u, and the activation energy in the reverse reaction, E_{ar}, equals $E_u - \Delta H$. (The same is true for an endothermic reaction because the sign of ΔH is the positive and $E_{ar} < E_{af}$). For the reaction in question,

$$K = \frac{k_f}{k_r} = \frac{A_f e^{-E_{af}/RT}}{A_r e^{-E_{ar}/RT}}$$

Since the ln form of the Arrhenius equation is easier to manipulate, we will consider ln K.

$$\ln K = \ln\left(\frac{k_f}{k_r}\right) = \ln k_f - \ln k_r = \frac{-E_{af}}{RT} + \ln A_f - \left[\frac{-E_{ar}}{RT} + \ln A_r\right]$$

Substituting E_u for E_{af} and $(E_u - \Delta H)$ for E_{ar}

$$\ln K = \frac{-E_u}{RT} + \ln A_f - \left[\frac{-(E_u - \Delta H)}{RT} + \ln A_r\right]; \ln K = \frac{-E_u + (E_u - \Delta H)}{RT} + \ln A_f - \ln A_r$$

$$\ln K = \frac{-\Delta H}{RT} + \ln \frac{A_f}{A_r}$$

For the catalyzed reaction, $E_{cat} < E_u$ and $E_{af} = E_{cat}$, $E_{ar} = E_{cat} - \Delta H$. The catalyst does not change the value of ΔH.

$$\ln K_{cat} = \frac{-E_{cat} + (E_{cat} - \Delta H)}{RT} + \ln A_f - \ln A_r$$

$$\ln K_{cat} = \frac{-\Delta H}{RT} + \frac{\ln A_f}{A_r}$$

Thus, assuming A_f and A_r are not changed by the catalyst, $\ln K = \ln K_{cat}$ and $K = K_{cat}$.

15.85   (a) $P = \dfrac{gRT}{MM\ V} = \dfrac{0.300\,g\,H_2S}{34.08\,g/mol\,H_2S} \times \dfrac{298\,K}{5.00\,L} \times \dfrac{0.08206\,L\text{-}atm}{mol\text{-}K} = 0.043053$

                                                        $= 0.0431\,atm$

   (b) $K_p = P_{NH_3} \times P_{H_2S}$. Before solid is added, $Q = P_{NH_3} \times P_{H_2S} = 0 \times 0.0431 = 0$.

        $Q < K$ and the reaction will proceed to the right. However, no $NH_4SH(s)$ is present to produce $NH_3(g)$, so the reaction cannot proceed.

   (c)

	NH$_4$SH(s) $\rightleftharpoons$	NH$_3$(g)	+	H$_2$S(g)
initial		0 atm		0.043053 atm
change		+x atm		+x atm
equil.		+x atm		(0.043053+x) atm

        Since $Q < K$ initially [part (a)], P_{H_2S} must increase along with P_{NH_3} until equilibrium is established.

        $K_p = P_{NH_3} \times P_{H_2S}; 0.120 = (x)(0.043053+x); 0 = x^2 + 0.043053\,x - 0.120$

        Solve for x using the quadratic formula.

$$x = \frac{-0.043053 \pm \sqrt{(0.043053)^2 - 4(1)(-0.120)}}{2(1)}; \quad x = 0.3256 = 0.326 \, atm$$

$P_{NH_3} = 0.326 \, atm; \quad P_{H_2S} = (0.043053 + 0.3256) \, atm = 0.3686 = 0.369 \, atm$

(d) $\quad \chi_{H_2S} = \dfrac{P_{H_2S}}{P_t} = \dfrac{0.3686 \, atm}{(0.3256 + 0.3686) \, atm} = 0.531$

(e) The minimum amount of $NH_4HS(s)$ required is slightly greater than the number of moles NH_3 present at equilibrium. We can calculate the mol NH_3 present at equilibrium using the ideal-gas equation.

$$n_{NH_3} = \frac{P_{NH_3}V}{RT} = 0.3256 \, atm \times \frac{K \cdot mol}{0.08206 \, L \cdot atm} \times \frac{5.00 \, L}{298 \, K} = 0.06657$$

$$= 0.0666 \, mol \, NH_3$$

$$0.06657 \, mol \, H_2S \times \frac{1 \, mol \, NH_4SH}{1 \, mol \, H_2S} \times \frac{51.12 \, g \, NH_4SH}{1 \, mol \, NH_4SH} = 3.40 \, g \, NH_4SH$$

The minimum amount is slightly greater than $3.40 \, g \, NH_4HS$.

15.87 (a) $\quad H_2O(l) \rightleftharpoons H_2O(g); \quad K_p = P_{H_2O}$

(b) At 30°C, the vapor pressure of $H_2O(l)$ is 31.82 torr. $K_P = P_{H_2O} = 31.82 \, torr$

$K_p = 31.82 \, torr \times 1 \, atm/760 \, torr = 0.041868 = 0.04187 \, atm$

(c) From part (b), the value of K_p is the vapor pressure of the liquid at that temperature. By definition, vapor pressure = atmospheric pressure = 1 atm at the normal boiling point. $K_p = 1 \, atm$

15.88 (a)

$$\text{C-C B.O.} = 1 \qquad\qquad \text{C=C B.O.} = 2$$

(b) $\Delta H = D$ (bond breaking) – D (bond making)

E1: $\quad \Delta H = D(C=Cl) - D(C-Cl) - D(C-C) - 2D(C-Cl)$

$\quad\quad \Delta H = 614 + 242 - 348 - 2(328) = -148 \, kJ$

E2: $\quad \Delta H = D(C-C) + D(C-H) + D(C-Cl) - D(C=C) - D(H-Cl)$

$\quad\quad = 348 + 413 + 328 - 614 - 431 = 44 \, kJ$

(c) E1 is exothermic with $\Delta n = -1$. The yield of $C_2H_4Cl_2(g)$ would decrease with increasing temperature and with increasing container volume.

(d) E2 is endothermic with $\Delta n = 1$. The yield of C_2H_3Cl would increase with increasing temperature and with increasing container volume.

(e) The boiling points of the reactants and products are: C_2H_4, –103.7°C; Cl_2, –34.6°C, $C_2H_4Cl_2$, +83.5°C; C_2H_3Cl, –13.4°C; HCl, –84.9°C.

Because the products of E1 and E2 are optimized by different conditions, carry out the two equilibria in separate reactors. Since E1 is exothermic and $\Delta n = -1$, the reactor on the left should be as small and cold as possible to maximize yield of $C_2H_4Cl_2$. At temperatures below 83.5°C, $C_2H_4Cl_2$ will condense to the liquid and it can be easily transferred to the second (right) reactor.

Since E_2 is endothermic, the reactor on the right should be as large and hot as possible to optimize production of C_2H_3Cl. The outlet stream will be a mixture of $C_2H_3Cl(g)$, $C_2H_4Cl_2(g)$, and $HCl(g)$. This mixture could be run through a heat exchanger to condense and subsequently recycle $C_2H_4Cl_2(l)$. Since HCl has a lower boiling point than C_2H_3Cl, it cannot be removed by condensation. The $C_2H_3Cl(g) / HCl(g)$ mixture could be bubbled through a basic aqueous solution such as $NaHCO_3(aq)$ or $NaOH(aq)$ to remove $HCl(g)$, leaving pure $C_2H_3Cl(g)$.

Other details of reactor design such as the use of catalysts to speed up these reactions, the exact costs and benefits of heat exchange, recycling unreacted components, and separation and recovery of products are issues best resolved by chemical engineers.

16 Acid-Base Equilibria

Visualizing Concepts

16.2   *Plan.* The stronger the acid, the greater the extent of ionization. The stronger the acid, the weaker its conjugate base. In an acid-base reaction, equilibrium will favor the side with the weaker acid and base.   *Solve.*

(a)   HY is stronger than HX. Starting with six HY molecules, four are dissociated; of six HX molecules, only two are dissociated. Because it is dissociated to a greater extent, HY is the stronger acid.

(b)   If HY is the stronger acid, Y^- is the weaker base and X^- is the stronger base.

(c)   HX and Y^-, the reactants, are the weaker acid and base. Equilibrium lies to the left, and $K_c < 1$.

16.4   *Analyze/Plan.* Consider how K_a and [HA] influence extent of ionization (or percent ionization).

Solve. The greater the extent of ionization, the less valid the approximation that x is small relative to [HA]. The greater the value of K_a, the stronger the acid and the greater the extent of ionization. On the other hand, the smaller the [HA], the greater the percent ionization. The approximation is then most valid for small K_a and large [HA]. All three cases have the same [HA], and case (a) has the smallest K_a. The approximation is most valid for case (a).

16.5   *Plan.* The definition of percent ionization is $\dfrac{[H^+]}{[HA]_{Initial}} \times 100$. *Solve.*

(a)   Curve C shows the effect of initial concentration on percent ionization of a weak acid.

(b)   The percent ionization is inversely related to initial acid concentration; only curve C shows a decrease in percent ionization as acid concentration increases.

16.7   *Plan.* Write the molecular formula so we can count the correct number of valence electrons. Use the atom connectivity shown to draw the Lewis structure.   *Solve.*

(a)   The molecular formula is $(CH_3)_2NH$, or C_2H_7N. The number of valence electrons is $2(4) + 7 + 5 = 20\ e^-$, $10\ e^-$ pr.

(b)   The compound is an amine. It is an ammonia molecule where two H atoms have been replaced by CH_3 groups.

16.8   *Plan.* Evaluate the interactions of Na$^+$ and X$^-$ with H$_2$O.

Solve. Na$^+$ does not affect the [H$^+$] or [OH$^-$] of an aqueous solution. It is a "negligible" acid in water (which can be thought of as the conjugate acid of the strong base NaOH).

X$^-$ is the conjugate base of HX. It is not a negligible base in water, because we see from the diagram that one X$^-$ has gained an H$^+$ to form HX. In this solution, H$_2$O acts as the Brønsted-Lowry acid, according to the hydrolysis equilibrium:

$$X^-(aq) + H_2O(l) \rightleftharpoons HX(aq) + OH^-(aq).$$

The missing ion is OH$^-$(aq). According to the equilibrium reaction, the number of HX molecules and OH$^-$ ions are equal. Since there is 1 HX molecule in the diagram, 1 OH$^-$ should be shown.

16.10   The carboxyl group is the H-atom group at the "top" of the molecule.

The group on the right of the molecule is <u>not</u> a carboxyl group because it contains no ionizable H.

16.11   *Analyze/Plan.* The acids listed are all carboxylic acids with two carbon atoms. They vary by having different substituents attached to the terminal (non-carbonyl) carbon atom. Consider how these substituents affect acidity, or acid strength.

Solve. The acids have different numbers and kinds of electronegative substituents on the terminal C atoms. Electronegative substituents affect acid strength in two ways. They withdraw electron density from the carbonyl group, making the O–H bond more polar and more likely to ionize. And, they stabilize the conjugate base, the carboxylate anion, by delocalizing negative charge. The more electronegative substituents, and the greater their electronegativity, the stronger the acid. Acetic acid, CH$_3$COOH, has 3 H atoms bond to the terminal C. As we replace H with Cl, acid strength increases going from CH$_2$ClCOOH to CHCl$_2$COOH to CCl$_3$COOH to CF$_3$COOH. Finally, F is more electronegative than Cl, so CF$_3$COOH is the most acidic. The order listed in the exercise is the order of increasing acidity.

$$CH_3COOH < CH_2ClCOOH < CHCl_2COOH < CCl_3COOH < CF_3COOH$$

16.12   (a)   *Plan.* Count valence electrons and draw the correct Lewis structures. Consider The definition of Lewis acids and bases. *Solve.*

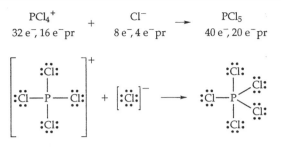

PCl_4^+ accepts an electron pair from Cl^-; PCl_4^+ is the Lewis acid and Cl^- is the Lewis base.

(b)   The hydrated cation is an oxyacid: the ionizable H is attached to O, which is bound to the central cation. As the charge on the cation increases, it attracts more electron density from the O–H bond, which becomes weaker and more polar. The degree of ionization increases and the equilibrium constant (K_a) increases.

Arrhenius and Brønsted-Lowry Acids and Bases

16.14   When NaOH dissolves in water, it completely dissociates to form $Na^+(aq)$ and $OH^-(aq)$. NH_3 is a Brønsted-Lowry base, an H^+ acceptor. When NH_3 dissolves in water, it participates in the equilibrium $NH_3(aq) + H_2O(l) \rightleftharpoons NH_4^+(aq) + OH^-(aq)$. Thus, the properties of both solutions are dominated by the presence of $OH^-(aq)$. Both solutions taste bitter, turn litmus paper blue (are basic), neutralize solutions of acids, and conduct electricity.

16.16   (a)   According to the Arrhenius definition, a *base* when dissolved in water increases $[OH^-]$. According to the Brønsted-Lowry theory, a *base* is an H^+ acceptor regardless of physical state. A Brønsted-Lowry base is not limited to aqueous solution and need not contain OH^- or produce it in aqueous solution.

(b) $NH_3(g) + H_2O(l) \rightleftharpoons NH_4^+(aq) + OH^-(aq)$ When NH_3 dissolves in water, it accepts H^+ from H_2O (B-L definition). In doing so, OH^- is produced (Arrhenius definition). Note that the OH^- produced was originally part of the H_2O molecule, not part of the NH_3 molecule.

16.18   A conjugate base has one less H^+ than its conjugate acid.   A conjugate acid has one more H^+ than its conjugate base.

(a)   (i) $C_6H_5COO^-$ (ii) PO_4^{3-}

(b)   (i) HCO_3^- (ii) $C_2H_5NH_3^+$ $(CH_3CH_2NH_3^+)$

16.20

	B-L acid	+	**B-L base**	$\rightleftharpoons$	**Conjugate acid**	+	**Conjugate base**
(a)	$HBrO(aq)$		$H_2O(l)$		$H_3O^+(aq)$		$BrO^-(aq)$
(b)	$HSO_4^-(aq)$		$HCO_3^-(aq)$		$H_2CO_3(aq)$		$SO_4^{2-}(aq)$
(c)	$H_3O^+(aq)$		$HSO_3^-(aq)$		$H_2SO_3(aq)$		$H_2O(l)$

16.22   (a) $H_2C_6H_7O_5^-(aq) + H_2O(l) \rightleftharpoons H_3C_6H_7O_5(aq) + OH^-(aq)$

(b) $H_2C_6H_7O_5^-(aq) + H_2O(l) \rightleftharpoons HC_6H_7O_5^{2-}(aq) + H_3O^+(aq)$

(c) $H_3C_6H_7O_5$ is the conjugate acid of $H_2C_6H_7O_5^-$

$HC_6H_7O_5^{2-}$ is the conjugate base of $H_2C_6H_7O_5^-$

16.24   Based on the chemical formula, decide whether the acid is strong, weak, or negligible. Is it one of the known seven strong acids (Section 16.5)? Also check Figure 16.4. Remove a single H and decrease the particle charge by one to write the formula of the conjugate base. *Solve.*

     (a) HNO_2, weak acid; NO_2^-, weak base

     (b) H_2SO_4, strong acid; HSO_4^-, negligible base

     (c) HPO_4^{2-}, weak acid; PO_4^{3-}, weak base

     (d) CH_4, negligible acid; CH_3^-, strong base

     (e) $CH_3NH_3^+$, weak acid; CH_3NH_2, weak base

16.26   (a) HNO_3. It is one of the seven strong acids (Section 16.5). Also, in a series of oxyacids with the same central atom (N), the acid with more O atoms is stronger (Section 16.10).

     (b) NH_3. When NH_3 and H_2O are combined, as in $NH_3(aq)$, NH_3 acts as the B-L base, accepting H^+ from H_2O. NH_3 has the greater tendency to accept H^+. For binary hydrides, base strength increases going to the left across a row of the periodic chart (Section 16.10).

16.28

Base	+	**Acid**	$\rightleftharpoons$	**Conjugate acid**	+	**Conjugate base**

     (a) $OH^-(aq)$ + $NH_4^+(aq)$ $\rightleftharpoons$ $H_2O(l)$ + $NH_3(aq)$

       OH^- is a stronger base than NH_3 (Figure 16.4), so the equilibrium lies to the right.

     (b) $CH_3COO^-(aq)$ + $H_3O^+(aq)$ $\rightleftharpoons$ $CH_3COOH(aq)$ + $H_2O(l)$

       H_3O^+ is a stronger acid than $HC_2H_3O_2$ (Figure 16.4), so the equilibrium lies to the right.

     (c) $F^-(aq)$ + $HCO_3^-(aq)$ $\rightleftharpoons$ $CO_3^{2-}(aq)$ + $HF(aq)$

       CO_3^{2-} is a stronger base than F^-, so the equilibrium lies to the left.

Autoionization of Water

16.30   (a) $H_2O(l) \rightleftharpoons H^+(aq) + OH^-(aq)$

     (b) $K_w = [H^+][OH^-]$. The $[H_2O(l)]$ is omitted because water is a pure liquid. The molarity (mol/L) of pure solids or liquids does not change as equilibrium is established, so it is usually omitted from equilibrium expressions.

     (c)   If a solution is basic, it contains more OH^- than H^+ ($[OH^-] > [H^+]$).

16.32   In pure water at 25°C, $[H^+] = [OH^-] = 1 \times 10^{-7} M$. If $[OH^-] > 1 \times 10^{-7} M$, the solution is basic; if $[OH^-] < 1 \times 10^{-7} M$, the solution is acidic.

     (a) $[OH^-] = \dfrac{K_W}{[H^+]} = \dfrac{1.0 \times 10^{-14}}{4.5 \times 10^{-3} M} = 2.2 \times 10^{-12} M < 1 \times 10^{-7} M$; acidic

     (b) $[OH^-] = \dfrac{K_w}{[H^+]} = \dfrac{1.0 \times 10^{-14}}{1.5 \times 10^{-9} M} = 6.7 \times 10^{-6} M > 1 \times 10^{-7} M$; basic

     (c) $[H^+] = 10[OH^-]$; $K_w = 10[OH^-][OH^-] = 10[OH^-]^2$

       $[OH^-] = (K_w/10)^{1/2} = 3.2 \times 10^{-8} M < 1 \times 10^{-7} M$; acidic

16.34 $K_w = [D^+][OD^-]$; for pure D_2O, $[D^+] = [OD^-]$; $8.9 \times 10^{-16} = [D^+]^2$;

        $[D^+] = [OD^-] = 3.0 \times 10^{-8}\ M$

The pH Scale

16.36 $[H^+]_A = 500\ [H^+]_B$ From Solution 16.35, $\Delta pH = -\log \dfrac{[H^+]_B}{[H^+]_A}$

       $\Delta pH = -\log \dfrac{[H^+]_B}{500\ [H^+]_B} = -\log\left(\dfrac{1}{500}\right) = 2.70$

       The pH of solution A is 2.70 pH units lower than the pH of solution B, because $[H^+]_A$ is 500 times greater than $[H^+]_B$. The greater $[H^+]$, the lower the pH of the solution.

16.38   (a) $K_w = [H^+][OH^-]$. If HNO_3 is added to water, it ionizes to form H^+ (aq) and NO_3^- (aq). This increases $[H^+]$ and necessarily decreases $[OH^-]$. When $[H^+]$ increases, pH decreases.

       (b)   On Figure 16.5, $1.4 \times 10^{-2}\ M\ OH^-$ is between pH = 12 ($1 \times 10^{-2}\ M\ OH^-$) and pH 13 ($1 \times 10^{-1}\ M\ OH^-$), slightly higher than pH = 12, so we estimate pH = 12.1. If pH > 7, the solution is basic.

       (c)   pH = 6.6 is midway between pH 6 and pH 7 on Figure 16.5.

           At pH = 7, $[H^+] = 1 \times 10^{-7}$; at pH = 6, $[H^+] = 1 \times 10^{-6} = 10 \times 10^{-7}$.

           A reasonable estimate is $5 \times 10^{-7}\ M\ H^+$. By calculation:

           pH = 6.6, $[H^+] = 10^{-pH} = 10^{-6.6} = 3 \times 10^{-7}$

           At pH = 6, $[OH^-] = 1 \times 10^{-8}$; at pH = 7, $[OH^-] = 1 \times 10^{-7} = 10 \times 10^{-8}$.

           A reasonable estimate is $5 \times 10^{-8}\ M\ OH^-$. By calculation:

           pOH = 14.0 – 6.6 = 7.4; $[OH^-] = 10^{-pOH} = 10^{-7.4} = 4 \times 10^{-8}\ M\ OH^-$.

16.40

pH	pOH	[H$^+$]	[OH$^-$]	acidic or basic
11.25	2.75	$5.6 \times 10^{-12}\ M$	$1.8 \times 10^{-3}\ M$	basic
7.98	6.02	$1.1 \times 10^{-8}\ M$	$9.6 \times 10^{-7}\ M$	basic
3.36	10.64	$4.4 \times 10^{-4}\ M$	$2.3 \times 10^{-11}\ M$	acidic
11.93	2.07	$1.2 \times 10^{-12}\ M$	$8.5 \times 10^{-3}\ M$	basic

16.42   The pH ranges from 5.2–5.6; pOH ranges from (14.0–5.2 =) 8.8 to (14.0–5.6 =) 8.4.

       $[H^+] = 10^{-pH}$, $[OH^-] = 10^{-pOH}$

       $[H^+] = 10^{-5.2} = 6.31 \times 10^{-6} = 6 \times 10^{-6}\ M$; $[H^+] = 10^{-5.6} = 2.51 \times 10^{-6} = 3 \times 10^{-6}\ M$

       The range of $[H^+]$ is $6 \times 10^{-6}\ M$ to $3 \times 10^{-6}\ M$.

       $[OH^-] = 10^{-8.8} = 1.58 \times 10^{-9} = 2 \times 10^{-9}\ M$; $[OH^-] = 10^{-8.4} = 3.98 \times 10^{-9} = 4 \times 10^{-9}\ M$.

       The range of $[OH^-]$ is $2 \times 10^{-9}\ M$ to $4 \times 10^{-9}\ M$.

       (The pH has one decimal place, so concentrations are reported to one sig fig.)

Strong Acids and Bases

16.44   (a)   A strong base is completely dissociated in aqueous solution; a strong base is a strong electrolyte.

        (b) $Sr(OH)_2$ is a soluble strong base.

$$Sr(OH)_2(aq) \rightarrow Sr^{2+}(aq) + 2OH^-(aq)$$

$$0.035 \; M \; Sr(OH)_2(aq) = 0.070 \; M \; OH^-$$

        (c)   False. Base strength should not be confused with solubility. Base strength describes the tendency of a dissolved molecule (formula unit for ionic compounds such as $Mg(OH)_2$) to dissociate into cations and hydroxide ions. $Mg(OH)_2$ is a strong base because each $Mg(OH)_2$ unit that dissolves also dissociates into $Mg^{2+}(aq)$ and $OH^-(aq)$. $Mg(OH)_2$ is not very soluble, so relatively few $Mg(OH)_2$ units dissolve when the solid compound is added to water.

16.46   For a strong acid, which is completely ionized, $[H^+]$ = the initial acid concentration.

        (a) $0.00135 \; M \; HNO_3 = 0.00135 \; M \; H^+$; $pH = -\log(0.00135) = 2.870$

        (b) $\dfrac{0.425 \; g \; HClO_4}{2.00 \; L \; soln} \times \dfrac{1 \; mol \; HClO_4}{100.5 \; g \; HClO_4} = 2.114 \times 10^{-3} = 2.11 \times 10^{-3} \; M \; HClO_4$

$[H^+] = 2.11 \times 10^{-3} \; M$; $pH = -\log(2.114 \times 10^{-3}) = 2.675$

        (c) $M_c \times V_c = M_d \times V_d$; $0.500 \; L = 500 \; mL$

$$1.00 \; M \; HCl \times 5.00 \; mL \; HCl = M_d \; HCl \times 500 \; mL \; HCl$$

$$M_d \; HCl = \dfrac{1.00 \; M \times 5.00 \; mL}{500 \; mL} = 1.00 \times 10^{-2} \; M \; HCl = 1.00 \times 10^{-2} \; M \, H^+$$

$$pH = -\log(1.00 \times 10^{-2}) = 2.000$$

        (d) $[H^+]_{total} = \dfrac{mol \; H^+ \; from \; HCl + mol \; H^+ \; from \; HI}{total \; L \; solution}$; $mol = M \times L$

$$[H^+]_{total} = \dfrac{(0.020 \; M \; HCl \times 0.0500 \; L) + (0.010 \; M \; HI \times 0.150 \; L)}{0.200 \; L}$$

$$[H^+]_{total} = \dfrac{1.0 \times 10^{-3} \; mol \; H^+ + 1.50 \times 10^{-3} \; mol \; H^+}{0.200 \; L} = 0.0125 = 0.013 \; M$$

$$pH = -\log(0.0125) = 1.90$$

16.48   For a strong base, which is completely dissociated, $[OH^-]$ = the initial base concentration. Then, $pOH = -\log[OH^-]$ and $pH = 14 - pOH$.

        (a) $0.082 \; M \; KOH = 0.082 \; M \; OH^-$; $pOH = -\log(0.082) = 1.09$; $pH = 14 - 1.09 = 12.91$

        (b) $\dfrac{1.065 \; g \; KOH}{0.5000 \; L} \times \dfrac{1 \; mol \; KOH}{56.106 \; g \; KOH} = 0.037964 = 0.03796 \; M = [OH^-]$

$$pOH = -\log(0.037964) = 1.4206; \; pH = 14 - pOH = 12.5794$$

(c) $M_c \times V_c = M_d \times V_d$

$0.0105\ M\ Ca(OH)_2 \times 10.0\ mL = M_d\ Ca(OH)_2 \times 500\ mL$

$$M_d\ Ca(OH)_2 = \frac{0.0105\ M\ Ca(OH)_2 \times 10.0\ mL}{500.0\ mL} = 2.10 \times 10^{-4}\ M\ Ca(OH)_2$$

$Ca(OH)_2(aq) \rightarrow Ca^{2+}(aq) + 2OH^-(aq)$

$[OH^-] = 2[Ca(OH)_2] = 2(2.10 \times 10^{-4}\ M) = 4.20 \times 10^{-4}\ M$

$pOH = -\log(4.20 \times 10^{-4}) = 3.377;\ pH = 14 - pOH = 10.623$

(d) $[OH^-]_{total} = \dfrac{mol\ OH^-\ from\ NaOH + mol\ OH^-\ from\ Ba(OH)_2}{total\ L\ solution}$

$$\frac{(7.5 \times 10^{-3}\ M \times 0.0400\ L) + 2(0.015\ M \times 0.0100\ L)}{0.0500\ L}$$

$$[OH^-]_{total} = \frac{3.00 \times 10^{-4}\ mol\ OH^- + 3.0 \times 10^{-4}\ mol\ OH^-}{0.0500\ L} = 0.01200 = 0.012\ M\ OH^-$$

$pOH = -\log(0.0120) = 1.92;\ pH = 14 - pOH = 12.08$

16.50 $pOH = 14 - pH = 14.00 - 12.05 = 1.95$

$pOH = 1.95 = -\log[OH^-];\ [OH^-] = 10^{-1.95} = 0.01122 = 1.1 \times 10^{-2}\ M$

$[OH^-] = 2[Ca(OH)_2];\ [Ca(OH)_2] = [OH^-]\ /\ 2 = 0.01122/2 = 5.6 \times 10^{-3}\ M$

Weak Acids

16.52   (a) $C_6H_5COOH(aq) \rightleftharpoons H^+(aq) + C_6H_5COO^-(aq);\ K_a = \dfrac{[H^+][C_6H_5COO^-]}{[HC_6H_5COOH]}$

$C_6H_5COOH(aq) + H_2O(l) \rightleftharpoons H_3O^+(aq) + C_6H_5COO^-(aq);$

$$K_a = \frac{[H_3O^+][C_6H_5COO^-]}{[HC_6H_5COOH]}$$

(b) $HCO_3^-(aq) \rightleftharpoons H^+(aq) + CO_3^{2-}(aq);\ K_a = \dfrac{[H^+][CO_3^{2-}]}{[HCO_3^-]}$

$HCO_3^-(aq) + H_2O(l) \rightleftharpoons H_3O^+(aq) + CO_3^{2-}(aq);\ K_a = \dfrac{[H_3O^+][CO_3^{2-}]}{[HCO_3^-]}$

16.54 $C_6H_5CH_2COOH(aq) \rightleftharpoons H^+(aq) + C_6H_5CH_2COO^-(aq);\ K_a = \dfrac{[H^+][C_6H_5CH_2COO^-]}{[C_6H_5CH_2COOH]}$

$[H^+] = [C_6H_5CH_2COO^-] = 10^{-2.68} = 2.09 \times 10^{-3} = 2.1 \times 10^{-3}\ M$

$[C_6H_5CH_2COOH] = 0.085 - 2.09 \times 10^{-3} = 0.0829 = 0.083\ M$

$K_a = \dfrac{(2.09 \times 10^{-3})^2}{0.0829} = 5.3 \times 10^{-5}$

16.56 $[H^+] = 0.132 \times [BrCH_2COOH]_{initial} = 0.0132\ M$

$$BrCH_2COOH(aq) \rightleftharpoons H^+(aq) + BrCH_2COO^-(aq)$$

initial	0.100 M	0	0
equil.	0.087	0.0132 M	0.0132 M

$$K_a = \frac{[H^+][BrCH_2COO^-]}{[BrCH_2COOH]} = \frac{(0.0132)^2}{0.087} = 2.0 \times 10^{-3}$$

16.58 $[H^+] = 10^{-pH} = 10^{-3.25} = 5.623 \times 10^{-4} = 5.62 \times 10^{-4}\ M$

$$K_a = 6.8 \times 10^{-4} = \frac{[H^+][F^-]}{[HF]} = \frac{(5.623 \times 10^{-4})^2}{x - 5.623 \times 10^{-4}}$$

$6.8 \times 10^{-4}(x - 5.623 \times 10^{-4}) = (5.623 \times 10^{-4})^2;$

$6.8 \times 10^{-4}\,x = 3.824 \times 10^{-7} + 3.162 \times 10^{-7} = 6.986 \times 10^{-7}$

$x = 1.027 \times 10^{-3} = 1.0 \times 10^{-3}\ M\ HF$

$mol = M \times L = 1.027 \times 10^{-3}\ M \times 0.200\ L = 2.055 \times 10^{-4} = 2.1 \times 10^{-4}\ mol\ HF$

16.60

$$HClO(aq) \rightleftharpoons H^+(aq) + ClO^-(aq)$$

initial	0.0090 M	0	0
equil.	(0.0090 – x) M	x M	x M

$$K_a = \frac{[H^+][ClO^-]}{[HClO]} = \frac{x^2}{(0.0090 - x)} \approx \frac{x^2}{0.0090} = 3.0 \times 10^{-8}$$

$x^2 = 0.0090\,(3.0 \times 10^{-8}); x = 1.6 \times 10^{-5}\ M = [H^+] = [H_3O^+] = [ClO^-]$

$[HClO] = 9.0 \times 10^{-3} - 1.6 \times 10^{-5} = 8.984 \times 10^{-3} = 9.0 \times 10^{-3}\ M$

Check. $\dfrac{1.6 \times 10^{-5}\ M\ H^+}{0.0090\ M\ HClO} \times 100 = 0.18\%$ ionization; the approximation is valid

16.62   (a)

$$HOCl(aq) \rightleftharpoons H^+(aq) + OCl^-(aq)$$

initial	0.095 M	0	0
equil	(0.095 – x) M	x M	x M

$$K_a = \frac{[H^+][OCl^-]}{[HOCl]} = \frac{x^2}{(0.095 - x)} \approx \frac{x^2}{0.095} = 3.0 \times 10^{-8}$$

$x^2 = 0.095\,(3.0 \times 10^{-8}); x = [H^+] = 5.3 \times 10^{-5}\ M, pH = 4.27$

Check. $\dfrac{5.3 \times 10^{-5}\ M\ H^+}{0.095\ M\ HOCl} \times 100 = 0.056\%$ ionization; the approximation is nearly valid.

To 2 sig figs, the quadratic formula gives the same $M\ H^+$.

(b) $K_a = \dfrac{[H^+][C_6H_5O^-]}{[C_6H_5OH]} = \dfrac{x^2}{(0.0085 - x)} \approx \dfrac{x^2}{0.0085} = 1.3 \times 10^{-10}$

$x^2 = 0.0085\,(1.3 \times 10^{-10}); \; x = [H^+] = 1.1 \times 10^{-6}\,M, \; pH = 5.98$

Check. Clearly $1.1 \times 10^{-6}\,M\,H^+$ is small compared to $8.5 \times 10^{-3}\,M\,C_6H_5OH$, and the approximation is valid.

(c)

	$HONH_2(aq) + H_2O(l)$	$\rightleftharpoons$	$HONH_3{}^+(aq)$	$+$	$OH^-(aq)$
initial	$0.095\,M$		0		0
equil	$(0.095 - x)\,M$		$x\,M$		$x\,M$

$K_b = \dfrac{[HONH_3{}^+][OH^-]}{[HONH_2]} = \dfrac{x^2}{(0.095 - x)} \approx \dfrac{x^2}{0.095} = 1.1 \times 10^{-8}$

$x^2 = 0.095\,(1.1 \times 10^{-8}); \; x = [OH^-] = 3.2 \times 10^{-5}\,M, \; pH = 9.51$

Check. $\dfrac{3.2 \times 10^{-5}\,M\,OH^-}{0.095\,M\,HONH_2} \times 100 = 0.034\%$ ionization; the approximation is valid

16.64   Calculate the initial concentration of $HC_9H_7O_4$.

$2\ \text{tablets} \times \dfrac{500\ \text{mg}}{\text{tablet}} \times \dfrac{1\ \text{g}}{1000\ \text{mg}} \times \dfrac{1\ \text{mol}\ HC_9H_7O_4}{180.2\ \text{g}\ HC_9H_7O_4} = 0.005549 = 0.00555\ \text{mol}\ HC_9H_7O_4$

$\dfrac{0.005549\ \text{mol}\ HC_9H_7O_4}{0.250\ \text{L}} = 0.02220 = 0.0222\ M\ HC_9H_7O_4$

	$HC_9H_7O_4\,(aq)$	$\rightleftharpoons$	$C_9H_7O_4{}^-\ +$	$H^+\,(aq)$
Initial	$0.0222\,M$		$0\,M$	$0\,M$
equil	$(0.0222 - x)$		$x\,M$	$x\,M$

$K_a = 3.3 \times 10^{-4} = \dfrac{[H^+][C_9H_7O_4{}^-]}{[HC_9H_7O_4]} = \dfrac{x^2}{(0.0222 - x)}$

Assuming x is small compared to 0.0222,

$x^2 = 0.0222\,(3.3 \times 10^{-4}); \; x = [H^+] = 2.7 \times 10^{-3}\,M$

$\dfrac{2.7 \times 10^{-3}\,M\,H^+}{0.0222\,M\,HC_9H_7O_4} \times 100 = 12\%$ ionization; the approximation is not valid

Using the quadratic formula, $x^2 + 3.3 \times 10^{-4}\,x - 7.325 \times 10^{-6} = 0$

$x = \dfrac{-3.3 \times 10^{-4} \pm \sqrt{(3.3 \times 10^{-4})^2 - 4(1)\,(-7.325 \times 10^{-6})}}{2(1)} = \dfrac{-3.3 \times 10^{-4} \pm \sqrt{2.941 \times 10^{-5}}}{2}$

$x = 2.547 \times 10^{-3} = 2.5 \times 10^{-3}\,M\,H^+; \; pH = -\log(2.547 \times 10^{-3}) = 2.594 = 2.59$

16.66   (a) $C_2H_5COOH(aq) \rightleftharpoons H^+(aq) + C_2H_5COO^-(aq)$

$K_a = 1.3 \times 10^{-5} = \dfrac{[H^+][C_2H_5COO^-]}{[C_2H_5COOH]} = \dfrac{x^2}{0.250 - x}$

$$x^2 \approx 0.250 \ (1.3 \times 10^{-5}); \ x = 1.803 \times 10^{-3} = 1.8 \times 10^{-3} \ M \ H^+$$

$$\% \ \text{ionization} = \frac{1.803 \times 10^{-3} \ M \ H^+}{0.250 \ M \ C_2H_5COOH} \times 100 = 0.721\%$$

(b) $\dfrac{x^2}{0.0800} \approx 1.3 \times 10^{-5}; \ x = 1.020 \times 10^{-3} = 1.0 \times 10^{-3} \ M \ H^+$

$$\% \ \text{ionization} = \frac{1.020 \times 10^{-3} \ M \ H^+}{0.0800 \ M \ C_2H_5COOH} \times 100 = 1.27\%$$

(c) $\dfrac{x^2}{0.0200} \approx 1.3 \times 10^{-5}; \ x = 5.099 \times 10^{-4} = 5.1 \times 10^{-4} \ M \ H^+$

$$\% \ \text{ionization} = \frac{5.099 \times 10^{-4} \ M \ H^+}{0.0200 \ M \ C_2H_5COOH} \times 100 = 2.55\%$$

16.68 $HX(aq) \rightleftharpoons H^+(aq) + X^-(aq); \ K_a = \dfrac{[H^+][X^-]}{[HX]}$

$[H^+] = [X^-]$; assume the % ionization is small; $K_a = \dfrac{[H^+]^2}{[HX]}; \ [H^+] = K_a^{1/2} [HX]^{1/2}$

$pH = -\log K_a^{1/2} [HX]^{1/2} = -\log K_a^{1/2} - \log [HX]^{1/2}; \ pH = -1/2 \log K_a - 1/2 \log [HX]$

This is the equation of a straight line, where the intercept is $-1/2 \log K_a$, the slope is $-1/2$, and the independent variable is $\log [HX]$.

16.70 $\begin{array}{llll} H_2C_4H_4O_6(aq) & \rightleftharpoons & H^+(aq) & + & HC_4H_4O_6^-(aq) & K_{a1} = 1.0 \times 10^{-3} \\ HC_4H_4O_6^-(aq & \rightleftharpoons & H^+(aq) & + & C_4H_4O_6^{2-}(aq) & K_{a2} = 4.6 \times 10^{-5} \end{array}$

Begin by calculating the $[H^+]$ from the first ionization. The equilibrium concentrations are $[H^+] = [HC_4H_4O_6^-] = x$, $[H_2C_4H_4O_6] = 0.25 - x$.

$$K_{a1} = \frac{[H^+][HC_4H_4O_6^-]}{[H_2C_4H_4O_6]} = \frac{x^2}{0.25 - x}; \ x^2 + 1.0 \times 10^{-3} \ x - 2.5 \times 10^{-4} = 0$$

Using the quadratic formula, $x = 1.532 \times 10^{-2} = 0.015 \ M \ H^+$ from the first ionization. Next calculate the H^+ contribution from the second ionization.

	$HC_4H_4O_6^-(aq)$	$\rightleftharpoons$	$H^+(aq)$	$+$	$C_4H_4O_6^{2-}(aq)$
initial	0.015		0.015		0
equil.	(0.015 – y)		(0.015 + y)		y

$$K_{a2} = \frac{(0.015 + y)(y)}{(0.015 - y)} = 4.6 \times 10^{-5}; \ \text{assuming y is small compared to } 0.015,$$

$y = 4.6 \times 10^{-5} \ M \ C_4H_4O_6^{2-}(aq)$

This approximation is reasonable, since 4.6×10^{-5} is only 0.3% of 0.015. $[H^+] = 0.015 \ M$ (first ionization) + 4.6×10^{-5} (second ionization). Since 4.6×10^{-5} is 0.3% of 0.015 M, it can be safely ignored when calculating total $[H^+]$. Thus, pH = $-\log(0.01532) = 1.18148 = 1.181$.

Assumptions:

1)   The ionization can be treated as a series of steps (valid by Hess' law).

2)   The extent of ionization in the second step (y) is small relative to that from the first step (valid for this acid and initial concentration). This assumption was used twice, to calculate the value of y from K_{a2} and to calculate total $[H^+]$ and pH.

Weak Bases

16.72   Organic amines (neutral molecules with nonbonded pairs on N atoms) and anions that are the conjugate bases of weak acids function as weak bases.

16.74   (a) $C_3H_7NH_2(aq) + H_2O(l) \rightleftharpoons C_3H_7NH_3^+(aq) + OH^-(aq)$; $K_b = \dfrac{[C_3H_7NH_3^+][OH^-]}{[C_3H_7NH_2]}$

         (b) $HPO_4^{2-}(aq) + H_2O(l) \rightleftharpoons H_2PO_4^-(aq) + OH^-(aq)$; $K_b = \dfrac{[H_2PO_4^-][OH^-]}{[HPO_4^{2-}]}$

         (c) $C_6H_5CO_2^-(aq) + H_2O(l) \rightleftharpoons C_6H_5CO_2H(aq) + OH^-(aq)$; $K_b = \dfrac{[C_6H_5CO_2H][OH^-]}{[C_6H_5CO_2^-]}$

16.76

	$BrO^-(aq)$	$+ H_2O(l)$	$\rightleftharpoons$	$HOBr(aq)$	$+$	$OH^-(aq)$
initial	$0.550\ M$			0		0
equil.	$(0.550 - x)\ M$			$x\ M$		$x\ M$

$$K_b = \frac{[HOBr][OH^-]}{[BrO^-]} = \frac{x^2}{0.550 - x} \approx \frac{x^2}{0.550} = 4.0 \times 10^{-6}$$

$x^2 = 0.550\ (4.0 \times 10^{-6})$; $x = [OH^-] = 1.48 \times 10^{-3} = 1.5 \times 10^{-3}\ M$; pH = 11.17

Check. $\dfrac{1.5 \times 10^{-3}\ M\ OH^-}{0.550\ M\ BrO^-} \times 100 = 0.27\%$ hydrolysis; the approximation is valid

16.78   (a)   pOH = 14.00 – 9.95 = 4.05; $[OH^-] = 10^{-4.05} = 8.91 \times 10^{-5} = 8.9 \times 10^{-5}\ M$

	$C_{18}H_{21}NO_3(aq)$	$+ H_2O(1)$	$\rightleftharpoons$	$C_{18}H_{21}NO_3H^+(aq)$	$+$	$OH^-(aq)$
initial	$0.0050\ M$			0		0
equil.	$(0.0050 - 8.9 \times 10^{-5})$			$8.9 \times 10^{-5}\ M$		$8.9 \times 10^{-5}\ M$

$$K_b = \frac{[C_{18}H_{21}NO_3H^+][OH^-]}{[C_{18}H_{21}NO_3]} = \frac{(8.91 \times 10^{-5})^2}{(0.0050 - 8.91 \times 10^{-5})} = 1.62 \times 10^{-6} = 1.6 \times 10^{-6}$$

         (b) $pK_b = -\log(K_b) = -\log(1.62 \times 10^{-6}) = 5.79$

The K_a – K_b Relationship; Acid-Base Properties of Salts

16.80   (a)   We need K_a for the conjugate acid of CO_3^{2-}, K_a for HCO_3^-. K_a for HCO_3^- is K_{a2}.

         (b) $K_b = K_w / K_a = 1.0 \times 10^{-14} / 5.6 \times 10^{-11} = 1.8 \times 10^{-4}$

         (c) K_b for CO_3^{2-} $(1.8 \times 10^{-4}) > K_b$ for NH_3 (1.8×10^{-5}).

               CO_3^{2-} is a stronger base than NH_3.

16.82   (a)   Ammonia is the stronger base because it has the larger K_b value.

       (b)   Hydroxylammonium is the stronger acid because the weaker base, hydroxylamine, has the stronger conjugate acid.

       (c) K_a for $NH_4^+ = K_w/K_b$ for $NH_3 = 1.0 \times 10^{-14}/1.8 \times 10^{-5} = 5.6 \times 10^{-10}$

            K_a for $HONH_3^+ = K_w/K_b$ for $HONH_2 = 1.0 \times 10^{-14}/1.1 \times 10^{-8} = 9.1 \times 10^{-7}$

            Note that K_a for $HONH_3^+$ is larger than K_a for NH_4^+.

16.84   (a)   Proceeding as in Solution 16.83(a):

$$F^-(aq) + H_2O(l) \rightleftharpoons HF(aq) + OH^-(aq)$$

          initial   0.105 M 0 M 0 M

          equil   (0.105 – x) M x M x M

          K_b for $F^- = \dfrac{[HF][OH^-]}{[F^-]} = \dfrac{K_w}{K_a \text{ for HF}} = \dfrac{1.0 \times 10^{-14}}{6.8 \times 10^{-4}} = 1.47 \times 10^{-11} = 1.5 \times 10^{-11}$

          $1.5 \times 10^{-11} = \dfrac{(x)(x)}{(0.105-x)}$; assume the amount of F^- that hydrolyzes is small

          $x^2 = 0.105(1.47 \times 10^{-11})$; $x = [OH^-] = 1.243 \times 10^{-6} = 1.2 \times 10^{-6} M$

          pOH = 5.91; pH = 14 – 5.91 = 8.09

       (b) $Na_2S(aq) \rightarrow S^{2-}(aq) + 2Na^+(aq)$

          $S^{2-}(aq) + H_2O(l) \rightleftharpoons HS^-(aq) + OH^-(aq)$

          As in part (a) above, $[OH^-] = [HS^-] = x$; $[S^{2-}] = 0.035 M$

          $K_b = \dfrac{[HS^-][OH^-]}{[S^{2-}]} = \dfrac{K_w}{K_a \text{ for HS}^-} = \dfrac{1.0 \times 10^{-14}}{1 \times 10^{-19}} = 1 \times 10^5$

          Since $K_b \gg 1$, the equilibrium above lies far to the right and $[OH^-] = [HS^-] = 0.035 M$. K_b for $HS^- = 1.05 \times 10^{-7}$; $[OH^-]$ produced by further hydrolysis of HS^- amounts to $6.1 \times 10^{-5} M$. The second hydrolysis step does not make a significant contribution to the total $[OH^-]$ and pH.

          $[OH^-] = 0.035 M$; pOH = 1.46, pH = 12.54

       (c)   As in Solution 16.83(c), calculate $[CH_3COO^-]$.

          $[CH_3COO^-]_t = [CH_3COO^-]$ from $CH_3COONa + [CH_3COO^-]$ from $(CH_3COO)_2Ba$
          $[CH_3COO^-]_t = 0.045 M + 2(0.055 M) = 0.155 M$

          The hydrolysis equilibrium is $CH_3COO^-(aq) + H_2O(l) \rightleftharpoons$
                                                $CH_3COOH(aq) + OH^-(aq)$

          $K_b = \dfrac{[CH_3COOH][OH^-]}{[CH_3COO^-]} = \dfrac{K_w}{K_a \text{ for CH}_3\text{COOH}} = \dfrac{1.0 \times 10^{-14}}{1.8 \times 10^{-5}} = 5.56 \times 10^{-10}$

                                                 $= 5.6 \times 10^{-10}$

$[OH^-] = [CH_3COOH] = x, [CH_3COO^-] = 0.155 - x$

$K_b = 5.56 \times 10^{-10} = \dfrac{x^2}{(0.155-x)}$; assume x is small compared to 0.155 M

$x^2 = 0.155 (5.56 \times 10^{-10}); x = [OH^-] = 9.280 \times 10^{-6} = 9.3 \times 10^{-6}$

$pH = 14 + log (9.280 \times 10^{-6}) = 8.97$

16.86   (a)   acidic; Cr^{3+} is a highly charged metal cation and a Lewis acid; Br^- is negligible.

(b)   neutral; both Li^+ and I^- are negligible.

(c)   basic; PO_4^{3-} is the conjugate base of HPO_4^{2-}; K^+ is negligible.

(d)   acidic; $CH_3NH_3^+$ is the conjugate acid of CH_3NH_2; Cl^- is negligible.

(e)   acidic; HSO_4^- is a negligible base, but a fairly strong acid ($K_a = 1.2 \times 10^{-2}$). K^+ is negligible.

16.88   *Plan*. Estimate pH of salt solution by evaluating the ions in the salts. Calculate to confirm if necessary. *Solve.*

KBr: salt of strong acid and strong base, neutral solution. The unknown is probably KBr. Check the others to be sure.

NH_4Cl: salt of a weak base and a strong acid, acidic solution

KCN: salt of a strong base and a weak acid, basic solution

K_2CO_3: salt of a strong base and a weak acid (HCO_3^-), basic solution

Only KBr fits the acid-base properties of the unknown.

16.90   The solution is basic because of the hydrolysis of PO_4^{3-}. The molarity of PO_4^{3-} is

$$\dfrac{35.0 \text{ g } Na_3PO_4}{1.00 \text{ L soln}} \times \dfrac{1 \text{ mol } Na_3PO_4}{163.9 \text{ g } Na_3PO_4} = 0.2135 = 0.214 \ M \ PO_4^{3-}$$

$PO_4^{3-}(aq) + H_2O(l) \rightleftharpoons HPO_4^{2-}(aq) + OH^-(aq)$

$$K_b = \dfrac{[HPO_4^{2-}][OH^-]}{[PO_4^{3-}]} = \dfrac{K_w}{K_a \text{ for } HPO_4^{2-}} = \dfrac{1.0 \times 10^{-14}}{4.2 \times 10^{-13}} = 0.0238 = 2.4 \times 10^{-2}$$

Ignoring the further hydrolysis of HPO_4^{2-},

$[OH^-] = [HPO_4^{2-}] = x, [PO_4^{3-}] = 0.214 - x$

$2.4 \times 10^{-2} = \dfrac{x^2}{(0.214-x)}; x^2 + 2.4 \times 10^{-2} x - 0.0051 = 0$

Since K_b is relatively large, we will not assume x is small compared to 0.275.

$$x = \dfrac{-0.024 \pm \sqrt{(0.024)^2 - 4(1)(-0.0051)}}{2(1)} = \dfrac{-0.024 \pm \sqrt{0.0210}}{2(1)}$$

$x = 0.060 \ M \ OH^-$; $pH = 14 + log (0.060) = 12.78$

Acid-Base Character and Chemical Structure

16.92   (a)   Acid strength increases as the polarity of the X–H bond increases and decreases as the strength of the X–H bond increases.

      (b)   Assuming the element, X, is more electronegative than H, as the electronegativity of X increases, the X–H bond becomes more polar and the strength of the acid increases. This trend holds true as electronegativity increases across a row of the periodic chart. However, as electronegativity decreases going down a family, acid strength increases because the strength of the H–X bond decreases, even though the H–X bond becomes less polar.

16.94   (a)   For binary hydrides, acid strength increases going across a row, so HCl is a stronger acid than H_2S.

      (b)   For oxyacids, the more electronegative the central atom, the stronger the acid, so H_3PO_4 is a stronger acid than H_3AsO_4.

      (c) $HBrO_3$ has one more nonprotonated oxygen and a higher oxidation number on Br, so it is a stronger acid than $HBrO_2$.

      (d)   The first dissociation of a polyprotic acid is always stronger because H^+ is more tightly held by an anion, so $H_2C_2O_4$ is a stronger acid than $HC_2O_4^-$.

      (e)   The conjugate base of benzoic acid, $C_6H_5COO^-$, is stabilized by resonance, while the conjugate base of phenol, $C_6H_5O^-$, is not. C_6H_5COOH has greater tendency to form its conjugate base and is the stronger acid.

16.96   (a) NO_2^- (HNO_3 is the stronger acid because it has more nonprotonated O atoms, so NO_2^- is the stronger base.)

      (b) PO_4^{3-} (K_a for $HAsO_4^{2-}$ is greater than K_a for HPO_4^{2-}, so K_b for PO_4^{3-} is greater and PO_4^{3-} is the stronger base. Note that P is more electronegative than As and H_3PO_4 is a stronger acid than H_3AsO_4, which could lead to the conclusion that AsO_4^{3-} is the stronger base. As in all cases, the measurement of base strength, K_b, supercedes the prediction. Chemistry is an experimental science.

      (c) CO_3^{2-} (The more negative the anion, the stronger the attraction for H^+.)

16.98   (a)   True.

      (b)   False. For oxyacids with the same structure but different central atom, the acid strength **increases** as the electronegativity of the central atom increases.

      (c)   False. HF is a weak acid, weaker than the other hydrogen halides, primarily because the H–F bond energy is exceptionally high.

Lewis Acids and Bases

16.100   No. If a substance is a Lewis acid, it is not necessarily a Brønsted or an Arrhenius acid. The Lewis definition of an acid, an electron pair acceptor, is most general. A Lewis acid does not necessarily fit the more narrow description of a Brønsted or Arrhenius acid. An electron pair acceptor isn't necessarily an H^+ donor, nor must it produce H^+ in aqueous solution. An example is Al^{3+}, which is a Lewis acid, but has no ionizable hydrogen.

16.102		**Lewis Acid**	**Lewis Base**
	(a)	HNO_2 (or H^+)	OH^-
	(b)	$FeBr_3$ (Fe^{3+})	Br^-
	(c)	Zn^{2+}	NH_3
	(d)	SO_2	H_2O

16.104 (a) $ZnBr_2$, smaller cation radius, same charge

      (b) $Cu(NO_3)_2$, higher cation charge

      (c) $NiBr_2$, smaller cation radius, same charge

Additional Exercises

16.106 (a) Correct.

      (b) Incorrect. A Brønsted acid must have ionizable hydrogen. Lewis acids are electron pair acceptors, but need not have ionizable hydrogen.

      (c) Correct.

      (d) Incorrect. K^+ is a negligible Lewis acid because it is the conjugate of strong base KOH. Its relatively large ionic radius and low positive charge render it a poor attractor of electron pairs.

      (e) Correct.

16.107 The equilibrium favors the weaker acid and base. That is, the stronger acid will be ionized; the weaker acid will not.

      (a) K > 1, the equilibrium lies to the right. The weaker acid, HPO_4^{2-}, and weaker base, NH_3, are present in greater concentrations than the stronger acid, NH_4^+, and stronger base, PO_4^{3-}.

      (b) K > 1, the equilibrium lies to the right. The weaker acid, HCN, and weaker base, CH_3COO^-, are present in greater concentrations than the stronger acid, CH_3COOH, and stronger base, CN^-.

16.109 (a) A higher O_2 concentration displaces protons from Hb, producing a more acidic solution, with lower pH in the lungs than in the tissues.

      (b) $[H^+]$ = antilog (–7.4) = 4.0×10^{-8} M. At body temperature, 37°C, $K_w = 2.4 \times 10^{-14}$ (see Solution 16.41). At this temperature, a "neutral" solution has $[H^+] = 1.5 \times 10^{-7}$ and pH 6.81. Even though the frame of reference is a bit different at this temperature, blood at pH 7.4 is slightly basic.

      (c) The equilibrium indicates that a high $[H^+]$ shifts the equilibrium toward the proton-bound form HbH^+, which means a lower concentration of HbO_2 in the blood. Thus the ability of hemoglobin to transport oxygen is impeded.

16.110   Upon dissolving, Li_2O dissociates to form Li^+ and O^{2-}. According to Equation 16.22, O^{2-} is completely protonated in aqueous solution.

$$Li_2O(s) + H_2O(l) \rightarrow 2Li^+(aq) + 2OH^-(aq)$$

Thus, initial $[Li_2O] = [O_2^-]$; $[OH^-] = 2[O^{2-}] = 2[Li_2O]$

$$[Li_2O] = \frac{mol\ Li_2O}{L\ solution} = 2.50\ g\ Li_2O \times \frac{1\ mol\ Li_2O}{29.88\ g\ Li_2O} \times \frac{1}{1.500\ L} = 0.0558 = 0.0558\ M$$

$[OH^-] = 0.11156 = 0.112\ M$; pOH $= 0.9525 = 0.953$; pH $= 14.00 - pOH = 13.0475 = 13.048$

16.112   Assume T = 25°C. For acid or base solute concentrations less than $1 \times 10^{-6}\ M$, we must consider the autoionization of water as a source of $[OH^-]$ and $[H^+]$.

	$H_2O(l)$	$\rightleftharpoons$	$[H^+]$	+	$[OH^-]$
initial	C		0		$2.5 \times 10^{-9}\ M$
equil	C		x		$(x + 2.5 \times 10^{-9})\ M$

$K_w = 1.0 \times 10^{-14} = [H^+][OH^-] = (x)(x + 2.5 \times 10^{-9})$; $x^2 + 2.5 \times 10^{-9}\ x - 1.0 \times 10^{-14} = 0$

From the quadratic formula, $x = \dfrac{-2.5 \times 10^{-9} \pm \sqrt{(2.5 \times 10^{-9})^2 - 4(1)(-1 \times 10^{-14})}}{2(1)}$

$$= 9.876 \times 10^{-8} = 9.9 \times 10^{-8}\ M\ H^+$$

$[H^+] = 9.9 \times 10^{-8}\ M$; $[OH^-] = (9.9 \times 10^{-8} + 2.5 \times 10^{-9}) = 1.013 \times 10^{-7} = 1.0 \times 10^{-7}\ M$

pH $= 7.0054 = 7.01$

Check: $[9.876 \times 10^{-8}][1.013 \times 10^{-7}] = 1.0 \times 10^{-14}$. Our answer makes sense. The very small concentration of OH^- from the solute raises the solution pH to slightly more than 7.

16.114   (a) $H_2X \rightarrow H^+ + HX^-$

          Assuming HX^- does not ionize, $[H^+] = 0.050\ M$, pH $= 1.30$

(b) $H_2X \rightarrow 2H^+ + X^-$; $0.050\ M\ H_2X = 0.10\ M\ H^+$; pH $= 1.00$

(c)   The observed pH of a $0.050\ M$ solution of H_2X is only slightly less than 1.30, the pH assuming no ionization of HX^-. HX^- is not completely ionized; H_2X, which is completely ionized, is a stronger acid than HX^-.

(d)   Since H_2X is a strong acid, HX^- has no tendency to act like a base. HX^- does act like a weak acid, so a solution of NaHX would be acidic.

16.115   (a) $K_b = K_w/K_a$; $pK_b = 14 - pK_a$; $pK_b = 14 - 4.84 = 9.16$

(b) K_a for butyric acid (buCOOH) is $10^{-4.84} = 1.4454 \times 10^{-5} = 1.4 \times 10^{-5}$

$$K_a = \frac{[H^+][buCOO^-]}{[buCOOH]};\ [H^+] = [buCOO^-] = x;\ [buCOOH] = 0.050 - x$$

$$1.4454 \times 10^{-5} = \frac{x^2}{0.050 - x}; \text{ assume x is small relative to } 0.050$$

$$x^2 = 7.227 \times 10^{-7}; \ x = [H^+] = 8.501 \times 10^{-4} = 8.5 \times 10^{-4} \, M \, H^+; \ pH = 3.07$$

(This represents 1.7% ionization, so the approximation is valid.)

(c) K_b for butyrate anion (buCOO$^-$) is $10^{-9.16} = 6.918 = 6.918 \times 10^{-10} = 6.9 \times 10^{-10}$

$$K_b = \frac{[OH^-][buCOOH]}{[buCOO^-]}; \ [OH^-] = [buCOOH] = x; \ [buCOO^-] = 0.050 - x$$

$$6.918 \times 10^{-10} = \frac{x^2}{0.050 - x}; \text{ assume x is small relative to } 0.050$$

$$x^2 = 3.459 \times 10^{-11}; \ x = [OH^-] = 5.881 \times 10^{-6} = 5.9 \times 10^{-6} \, M \, OH^-$$

$$pOH = 5.23, \ pH = 8.77$$

16.116   *Analyze/Plan.* Evaluate the acid-base properties of the cation and anion to determine whether a solution of the salt will be acidic, basic, or neutral.   *Solve.*

(i) NH_4NO_3: NH_4^+, weak conjugate acid of NH_3; NO_3^-, negligible conjugate base of HNO_3; acidic solution.

(ii) $NaNO_3$: Na^+, negligible conjugate acid of $NaOH$; NO_3^-, negligible conjugate base of HNO_3; neutral solution.

(iii) CH_3COONH_4: NH_4^+, weak conjugate acid of NH_3, $K_a = K_w/1.8 \times 10^{-5} = 5.6 \times 10^{-10}$; CH_3COO^-, weak conjugate base of CH_3COOH, $K_b = K_w/1.8 \times 10^{-5} = 5.6 \times 10^{-10}$; neutral solution ($K_a$ for the cation and K_b for the anion are accidentally equal, producing a neutral solution).

(iv) NaF: Na^+, negligible conjugate acid of $NaOH$; F^-, weak conjugate base of HF, $K_b = K_w/6.8 \times 10^{-4} = 1.5 \times 10^{-11}$; basic solution.

(v) CH_3COONa: Na^+, negligible; CH_3COO^-, weak base, $K_b = 5.6 \times 10^{-10}$, basic solution.

In order of increasing acidity and decreasing pH: 0.1 M CH_3COONa > 0.1 M NaF > 0.1 M CH_3COONH_4 = 0.1 M $NaNO_3$ > 0.1 M NH_4Cl; (v) > (iv) > (iii) ~ (ii) > (i)

(iv) and (v) are both bases, and (v) has the greater K_b value and higher pH. (ii) and (iii) are both neutral and (i) is acidic.

16.118   Call each compound in the neutral form Q.

Then, $Q(aq) + H_2O(l) \rightleftharpoons QH^+(aq) + OH^-$. $K_b = [QH^+][OH^-]/[Q]$

The ratio in question is $[QH^+]/[Q]$, which equals $K_b/[OH^-]$ for each compound. At pH = 2.5, pOH = 11.5, $[OH^-] = $ antilog $(-11.5) = 3.16 \times 10^{-12} = 3 \times 10^{-12} \, M$. Now calculate $K_b/[OH^-]$ for each compound:

Nicotine $\dfrac{[QH^+]}{[Q]} = 7 \times 10^{-7} / 3.16 \times 10^{-12} = 2 \times 10^5$

Caffeine $\dfrac{[QH^+]}{[Q]} = 4 \times 10^{-14} / 3.16 \times 10^{-12} = 1 \times 10^{-2}$

Strychnine $\dfrac{[QH^+]}{[Q]} = 1 \times 10^{-6} / 3.16 \times 10^{-12} = 3 \times 10^5$

Quinine $\dfrac{[QH^+]}{[Q]} = 1.1 \times 10^{-6} / 3.16 \times 10^{-12} = 3.5 \times 10^5$

For all the compounds except caffeine the protonated form has a much higher concentration than the neutral form. However, for caffeine, a very weak base, the neutral form dominates.

16.119   (a)   Consider the formation of the zwitterion as a series of steps (Hess' law).

$$NH_2-CH_2-COOH + H_2O \rightleftharpoons NH_2-CH_2-COO^- + H_3O^+ \qquad K_a$$

$$NH_2-CH_2-COOH + H_2O \rightleftharpoons {^+NH_3}-CH_2-COOH + OH^- \qquad K_b$$

$$H_3O^+ + OH^- \rightleftharpoons 2H_2O \qquad 1/K_w$$

$$NH_2-CH_2-COOH \rightleftharpoons {^+NH_3}-CH_2-COO^- \qquad \dfrac{K_a \times K_b}{K_w}$$

$$K = \dfrac{K_a \times K_b}{K_w} = \dfrac{(4.3 \times 10^{-3})(6.0 \times 10^{-5})}{1.0 \times 10^{-14}} = 2.6 \times 10^7$$

The large value of K indicates that formation of the zwitterion is favorable. The assumption is that the same NH_2-CH_2-COOH molecule is acting like an acid (–COOH → H^+ + –COO⁻) and a base (–NH_2 + H^+ → NH_3^+), simultaneously. Glycine is both a stronger acid and a stronger base than water, so the H^+ transfer should be intramolecular, as long as there are no other acids or bases in the solution.

(b)   Since glycine exists as the zwitterion in aqueous solution, the pH is determined by the equilibrium below.

$$^+NH_3-CH_2-COO^- + H_2O \rightleftharpoons NH_2-CH_2-COO^- + H_3O^+$$

$$K_a = \dfrac{[NH_2-CH_2-COO^-][H_3O^+]}{[^+NH_3-CH_2-COO^-]} = \dfrac{K_w}{K_b} = \dfrac{1.0 \times 10^{-14}}{6.0 \times 10^{-5}} = 1.67 \times 10^{-10} = 1.7 \times 10^{-10}$$

$$x = [H_3O^+] = [NH_2-CH_2-COO^-]; \quad K_a = 1.67 \times 10^{-10} = \dfrac{(x)(x)}{(0.050-x)} \approx \dfrac{x^2}{0.050}$$

$$x = [H_3O^+] = 2.89 \times 10^{-6} = 2.9 \times 10^{-6} \, M; \quad pH = 5.54$$

(c)   In a strongly acidic solution the –CO_2^- function would be protonated, so glycine would exist as $^+H_3NCH_2COOH$. In strongly basic solution the –NH_3^+ group would be deprotonated, so glycine would be in the form $H_2NCH_2CO_2^-$.

16.120 The general Lewis structures for these acids and their conjugate bases are shown below.

where X = H or Cl.

Replacement of H on the acid by the more electronegative chlorine atoms causes the central carbon to become more positively charged, thus withdrawing more electrons from the attached COOH group, in turn causing the O–H bond to be more polar, so that H^+ is more readily ionized.

For the conjugate base (two resonance structures), the electronegative X atoms delocalize negative charge and stabilize these forms relative to the unsubstituted anions. This favors products in the ionization equilibrium and increases the value of K_a.

To calculate pH proceed as usual, except that the full quadratic formula must be used for all but acetic acid.

Acid	pH
acetic	3.37
chloroacetic	2.51
dichloroacetic	2.09
trichloroacetic	2.0

Integrative Exercises

16.122 *Analyze.* Given mass % and density of concentrated HCl, calculate volume of concentrated solution required to produce 10.0 L of HCl with pH = 2.05. *Plan.* Calculate molarity of concentrated solution from density and mass %. Calculate molarity of dilute solution from pH. Use the dilution formula to calculate volume (mL) concentrated solution required.   *Solve.*

$$\frac{1.18 \text{ g conc. soln.}}{\text{mL conc. soln.}} \times \frac{36.0 \text{ g HCl}}{100 \text{ g conc. soln.}} \times \frac{1000 \text{ mL}}{1 \text{ L}} \times \frac{1 \text{ mol HCl}}{36.46 \text{ g HCl}} = 11.651 \text{ mol HCl/L}$$

$$= 11.7 \ M \text{ HCl/L}$$

For the dilute HCl solution, $[H^+] = 10^{-pH} = 10^{-2.05} = 8.913 \times 10^{-3} = 8.9 \times 10^{-3} \ M$ HCl

$M_c \times L_c = M_d \times M_d$; $11.651 \times L_c = 8.913 \times 10^{-3} \ M \times 10.0$ L;

$$L_c = 7.650 \times 10^{-3}; 7.650 \times 10^{-3} \text{ L} \times \frac{1000 \text{ mL}}{1 \text{ L}} = 7.65 = 7.7 \text{ mL conc. HCl}$$

16.123 $[H^+] = 10^{-pH} = 10^{-2} = 1 \times 10^{-2} \, M \, H^+; 1 \times 10^{-2} \, M \times 0.400 \, L = 4.0 \times 10^{-3} = 4 \times 10^{-3} \, mol \, H^+$

$$HCl(aq) + HCO_3^-(aq) \rightarrow Cl^-(aq) + H_2O(l) + CO_2(g)$$

$$4 \times 10^{-3} \, mol \, H^+ = 4 \times 10^{-3} \, mol \, HCO_3^- \times \frac{84.01 \, g \, NaHCO_3}{1 \, mol \, HCO_3^-} = 0.336 = 0.3 \, g \, NaHCO_3$$

16.125   (a)   24 valence e^-, 12 e^- pairs

The formal charges on all atoms are zero. Structures with multiple bonds lead to nonzero formal charges. There are three electron domains about Al. The electron-domain geometry and molecular structure are trigonal planar.

(b)   The Al atom in $AlCl_3$ has an incomplete octet and is electron deficient. It "needs" to accept another electron pair, to act like a Lewis acid.

(c)

Both the Al and N atoms in the product have tetrahedral geometry.

(d)   The Lewis theory is most appropriate. H^+ and $AlCl_3$ are both electron pair acceptors, Lewis acids.

16.126   *Plan.* Use acid ionization equilibrium to calculate the total moles of particles in solution. Use density to calculate kg solvent. From the molality (*m*) of the solution, calculate ΔT_b and T_b. *Solve.*

	$HSO_4^-(aq)$	$\rightleftharpoons$	$H^+(aq)$	$+$	$SO_4^{2-}(aq)$
Initial	0.10 *M*		0		0
equil.	0.10 – x *M*		x *M*		x *M*
	0.071 *M*		0.029 *M*		0.029 *M*

$K_a = 1.2 \times 10^{-2} = \dfrac{[H^+][SO_4^{2-}]}{[HSO_4^-]} = \dfrac{x^2}{0.10-x}$; K_a is relatively large, so use the quadratic.

$x^2 + 0.012 \, x - 0.0012 = 0$; $x = \dfrac{-0.012 \pm \sqrt{(0.012)^2 - 4(1)(-0.0012)}}{2}$; $x = 0.029 \, M \, H^+, SO_4^{2-}$

Total ion concentration = $0.10 \, M \, Na^+ + 0.071 \, M \, HSO_4^- + 0.029 \, M \, H^+ + 0.029 \, M \, SO_4^{2-}$
$$= 0.229 = 0.23 \, M.$$

Assume 100.0 mL of solution. 1.002 g/mL $\times$ 100.0 mL = 100.2 g solution.

$$0.10 \text{ } M \text{ NaHSO}_4 \times 0.1000 \text{ L} = 0.010 \text{ mol NaHSO}_4 \times \frac{120.1 \text{ g NaHSO}_4}{\text{mol NaHSO}_4}$$

$$= 1.201 = 1.2 \text{ g NaHSO}_4$$

$$100.2 \text{ g soln} - 1.201 \text{ g NaHSO}_4 = 99.0 \text{ g} = 0.099 \text{ kg H}_2\text{O}$$

$$m = \frac{\text{mol ions}}{\text{kg H}_2\text{O}} = \frac{0.229 \text{ } M \times 0.1000 \text{ L}}{0.0990 \text{ kg}} = 0.231 = 0.23 \text{ } m \text{ ions}$$

$$\Delta T_b = K_b(m) = 0.52°\text{C}/m \times (0.23 \text{ } m) = +0.12°\text{C}: T_b = 100.0 + 0.12 = 100.1°\text{C}$$

16.128   (a)   rate = $k[\text{IO}_3^-][\text{SO}_3^{2-}][\text{H}^+]$

       (b) $\Delta\text{pH} = \text{pH}_2 - \text{pH}_1 = 3.50 - 5.00 = -1.50$

              $\Delta\text{pH} = -\log [\text{H}^+]_2 - (-\log [\text{H}^+]_1); -\Delta\text{pH} = \log [\text{H}^+]_2 - \log [\text{H}^+]_1$

              $-\Delta\text{pH} = \log [\text{H}^+]_2 / [\text{H}^+]_1; [\text{H}^+]_2 / [\text{H}^+]_1 = 10^{-\Delta\text{pH}}$

              $[\text{H}^+]_2/[\text{H}^+]_1 = 10^{1.50} = 31.6 = 32$. The rate will increase by a factor of 32 if $[\text{H}^+]$ increases by a factor of 32. The reaction goes faster at lower pH.

       (c)   Since H^+ does not appear in the overall reaction, it is either a catalyst or an intermediate. An intermediate is produced and then consumed during a reaction, so its contribution to the rate law can usually be written in terms of concentrations of other reactants (Sample Exercise 14.15). A catalyst is present at the beginning and end of a reaction and can appear in the rate law if it participates in the rate-determining step (Solution 14.72). This reaction is pH dependent because H^+ is a homogeneous catalyst that participates in the rate-determining step.

16.130   (a)   The structures of the two acids are similar, but lactic acid has an –OH group on the C atom adjacent to the –COOH group. This electronegative substituent withdraws electron density from the –COOH group, and stabilizes its conjugate base, increasing the strength of lactic acid relative to propionic acid. The stronger the acid, the larger the K_a value and the smaller the pK_a.

       (b) $pK_a = 3.85, K_a = 10^{-3.85} = 1.4 \times 10^{-4}; [\text{H}^+] = [\text{CH}_3\text{CH(OH)COO}^-] = x;$
              $[\text{CH}_3\text{CH(OH)COOH}] = 0.050 - x$

$$K_a = 1.4 \times 10^{-4} = \frac{[\text{H}^+][\text{CH}_3\text{CH(OH)COO}^-]}{[\text{CH}_3\text{CH(OH)COOH}]} = \frac{x^2}{0.050 - x} \approx \frac{x^2}{0.050}$$

              $x^2 = 0.050(1.4 \times 10^{-4}) = 2.646 \times 10^{-3} = 2.6 \times 10^{-3} \text{ } M \text{ CH}_3\text{CH(OH)COO}^-$

              (This represents 5.3% dissociation; solution by the quadratic yields essentially the same result.)

       (c)   Strategy: Assume a 100 g sample. Calculate mol Cu in sample. Use mole ratios from formula to calculate mass of O and H not due to H_2O. Subtract masses of Cu, C, H, and O from 100 g to get mass of H_2O. Calculate mol H_2O and X.

Assume a 100 g sample, 22.9 g Cu.

$$22.9 \text{ g Cu} \times \frac{1 \text{ mol Cu}}{63.55 \text{ g Cu}} = 0.3603 = 0.360 \text{ mol Cu}$$

mole ratios of Cu, O, and H (not due to H_2O): 1 Cu:6 O:10 H

$$\text{g O} = 0.3603 \text{ mol Cu} \times \frac{6 \text{ mol O}}{1 \text{ mol Cu}} \times \frac{16.00 \text{ g O}}{1 \text{ mol O}} = 34.59 = 34.6 \text{ g O}$$

$$\text{g H} = 0.3603 \text{ mol Cu} \times \frac{10 \text{ mol H}}{1 \text{ mol Cu}} \times \frac{1.008 \text{ g H}}{1 \text{ mol H}} = 3.632 = 3.63 \text{ g H}$$

$$\text{g } H_2O = 100 \text{ g sample} - [22.9 \text{ g Cu} + 26.0 \text{ g C} + 34.6 \text{ g O} + 3.63 \text{ g H}]$$

$$= 12.87 = 12.9 \text{ g } H_2O$$

$$12.87 \text{ g } H_2O \times \frac{1 \text{ mol } H_2O}{18.02 \text{ g } H_2O} = 0.7142 \text{ mol } H_2O/0.3603 = 1.98 \approx 2$$

X = 2 mol H_2O in the hydrate.

(d)   Compare K_a for Cu^{2+}(aq) and K_b for $CH_3CH(OH)COO^-$(aq).

The ion with the larger K value will undergo hydrolysis to the greater extent and will determine the pH of the solution.

pK_b for $CH_3CH(OH)COO^- = 14 - pK_a$ for $CH_3CH(OH)COOH = 14.00 - 3.85 = 10.15$

$K_b = 10^{-10.15} = 7.1 \times 10^{-11}$.

Since K_a for Cu^{2+} (1.0×10^{-8}) is greater than K_b for $CH_3CH(OH)COO^-$ (7.1×10^{-11}) the solution will be slightly acidic.

17 Additional Aspects of Aqueous Equilibria

Visualizing Concepts

17.2 (a) According to Figure 16.7, methyl orange is yellow above pH 4.5 and red (really pink) below pH 3.5. The beaker on the left has a pH greater than 4.5, and the one on the right has pH less than 3.5. (By calculation, pH of left beaker = 4.7, pH of right beaker = 2.9.) The right beaker, with lower pH and greater [H^+], is pure acetic acid. The left beaker contains equal amounts of the weak acid and its conjugate base, acetic acid and acetate ion. Adding the "common-ion" acetate (in the form of sodium acetate) shifts the acid ionization equilibrium to the left, decreases [H^+], and raises pH.

(b) When small amounts of NaOH are added, the left beaker is better able to maintain its pH. For solutions of the same weak acid, pH depends on the **ratio** of conjugate base to conjugate acid. Small additions of base (or acid) have the least effect when this ratio is close to one. The left beaker is a buffer because it contains a weak conjugate acid-conjugate base pair and resists rapid pH change upon addition of small amounts of strong base or acid.

17.3 [HX] > [X^-]. Buffers prepared from weak acids (HX) and their conjugate bases (X^-, usually in the form of a salt) have pH values in a range of approximately 2 pH units, centered around pK_a for the weak acid. If concentration of the weak acid is greater than concentration of the conjugate base, pH < pK_a. If concentration of the conjugate base is greater than concentration of the weak acid, pH > pK_a. This is generally true for buffers containing a weak conjugate acid (CA) and conjugate base (CB) pair.

[CA] > [CB], pH of buffer < pK_a of CA

[CA] < [CB], pH of buffer > pK_a of CA

17.5 *Analyze/Plan.* Consider the reaction $HA + OH^- \rightarrow A^- + H_2O$. What are the major species present in solution at the listed stages of the titration? Which diagram represents these species? *Solve.*

(a) *Before addition of NaOH,* the solution is mostly HA. The only A^- is produced by the ionization equilibrium of HA and is too small to appear in the diagram. This situation is shown in diagram (iii), which contains only HA.

(b) *After addition of NaOH but before the equivalence point,* some, but not all, HA has been converted to A^-. The solution contains a mixture of HA and A^-; this is shown in diagram (i).

276

(c) *At the equivalence point,* all HA has been converted to A^-, with no excess HA or OH^- present. This is shown in diagram (iv).

(d) *After the equivalence point,* the same amount of A^- is present as at the equivalence point, plus some excess OH^-. This is diagram (ii).

17.6 *Analyze/Plan.* In each case, the first substance is in the buret, and the second is in the flask. If acid is in the flask, the initial pH is low; with base in the flask, the pH starts high. Strong acids have lower pH than weak acids; strong bases have higher pH than weak bases. Polyprotic acids and bases have more than one "jump" in pH.

(a) Strong base in flask, pH starts high, ends low as acid is added. Only diagram (ii) fits this description.

(b) Weak acid in flask, pH starts low, but not extremely low. Diagrams (i), (iii), and (iv) all start at low pH and get higher. Diagram (i) has very low initial pH, and likely has strong acid in the flask. Diagram (iv) has two pH jumps, so it has a polyprotic acid in the flask. Diagram (iii) best fits the profile of adding a strong base to a weak acid.

(c) Strong acid in the flask, pH starts very low, diagram (i).

(d) Polyprotic acid, more than one pH jump, diagram (iv).

17.8 *Analyze/Plan.* The product of the ion concentrations in a saturated solution equals K_{sp}. Use numbers of anions and cations as a measure of concentration to calculate relative "K_{sp}" values. Counting cations is not adequate, because excess anions in some of the boxes drive down the cation concentrations. Ion-products must be considered. *Solve.*

AgX: $(4 \, Ag^+)(4 X^-) = 16$

AgY: $(1 \, Ag^+)(9 \, Y^-) = 9$

AgZ: $(3 \, Ag^+)(6 \, Y^-) = 18$

AgY has the smallest K_{sp}.

17.9 *Analyze/Plan.* Consider the solubility equilibrium $MX(s) \rightleftharpoons M^{2+}(aq) + X^{2-}(aq)$, as well as the effects of common ions and competing equilibria.

(a) In pure water, the dissociation of MX is the only source of M^{2+} and X^{2-}, so the numbers of each ion are equal (ignoring any hydrolysis of X^{2-}). This describes figure (b).

(b) If MX(s) is dissolved in $Na_2X(aq)$, $X^{2-}(aq)$ is a common ion. There are two sources of X^{2-} dissociation of MX(s) and $Na_2X(aq)$. There will be more X^{2-} than M^{2+} in the resulting solution. This describes figure (c).

(c) If X^{2-} is a basic anion, it will combine with $2H^+(aq)$ to form H_2X molecules, which are not shown in the figures. This will reduce the number of free X^{2-} ions, and the solution will contain more M^{2+} than X^{2-}. This describes figure (a).

(d) K_{sp} will have the same value in each of the three scenarios. While the presence of a common ion or hydrolysis of an anion changes the solubility of MX, the value of K_{sp} is constant at a specified temperature.

17.11   A metal hydroxide that is soluble at very low and very high pH's, that is, in strong acid or strong base, is called amphoteric.

17.12   According to Figure 17.22, the two precipitating agents are 6 M HCl (first) and H_2S in 0.2 M HCl (second).

Cation A = Ag^+ (precipitates as AgCl)

Cation B = Cu^+ (precipitates as Cu_2S, acid insoluble)

Cation C = Ni^{2+} (remains in acidic solution)

Common-Ion Effect

17.14   (a)   For a generic weak base B, $K_b = \dfrac{[HB^+][OH^-]}{[B]}$. If an external source of HB^+ such

as HB^+Cl^- is added to a solution of B(aq), $[HB^+]$ increases, decreasing $[OH^-]$ and increasing [B], effectively suppressing the ionization (hydrolysis) of B.

   (b) NH_4Cl

17.16   *Analyze/Plan.* Follow the logic in Sample Exercise 17.1.   *Solve.*

   (a)   HCOOH is a weak acid, and HCOONa contains the common ion $HCOO^-$, the conjugate base of HCOOH. Solve the common-ion equilibrium problem.

	HCOOH(aq)	$\rightleftharpoons$	H^+(aq)	+	$HCOO^-$(aq)
i	0.200 M				0.150 M
c	$-x$		$+x$		$+x$
e	$(0.200 - x)$ M		$+x$ M		$(0.150 + x)$ M

$$K_a = 1.8 \times 10^{-4} = \frac{[H^+][HCOO^-]}{[HCOOH]} = \frac{(x)(0.150 + x)}{(0.200 - x)} \approx \frac{0.150\, x}{0.200}$$

$x = 2.40 \times 10^{-4} = 2.4 \times 10^{-4}\, M = [H^+]$, pH = 3.62

Check. Since the extent of ionization of a weak acid or base is suppressed by the presence of a conjugate salt, the 5% rule usually holds true in buffer solutions.

   (b) C_5H_5N is a weak base, and C_5H_5NHCl contains the common ion $C_5H_5NH^+$, which is the conjugate acid of C_5H_5N. Solve the common ion equilibrium problem.

	C_5H_5N(aq) + H_2O(l)	$\rightleftharpoons$	$C_5H_5NH^+$(aq)	+	OH^-(aq)
i	0.210 M		0.350 M		
c	$-x$		$+x$		$+x$
e	$(0.210 - x)$ M		$(0.350 + x)$ M		$+x$ M

$$K_b = 1.7 \times 10^{-9} = \frac{[C_5H_5NH^+][OH^-]}{[C_5H_5N]} = \frac{(0.350 + x)(x)}{(0.210 - x)} \approx \frac{0.350\, x}{0.210}$$

$x = 1.02 \times 10^{-9} = 1.0 \times 10^{-9}\, M = [OH^-]$, pOH = 8.991, pH = 14.00 − 8.991 = 5.01

Check. In a buffer, if [conj. acid] > [conj. base], pH < pK_a of the conj. acid. If [conj. acid] < [conj. base], pH > pK_a of the conj. acid. In this buffer, pK_a of $C_5H_5NH^+$ is 5.23. $[C_5H_5NH^+] > [C_5H_5N]$ and pH = 5.01, less than 5.23.

(c)   mol = $M \times L$; mol HF = 0.050 $M \times$ 0.125 L = 6.25×10^{-3} = 6.3×10^{-3} mol;

mol F^- = 0.10 $M \times$ 0.0500 L = 0.0050 mol

	HF(aq)	$\rightleftharpoons$	H^+(aq)	+	F^-(aq)
i	6.25×10^{-3} mol		0		0.0050 mol
c	$-x$		$+x$		$+x$
e	$(6.25 \times 10^{-3} - x)$ mol		$+x$		$(0.0050 + x)$ mol

$[HF] = (6.25 \times 10^{-3} + x)/0.175$ L; $[F^-] = (0.0050 + x)/0.175$ L

Note that the volumes will cancel when substituted into the K_a expression.

$$K_a = 6.8 \times 10^{-4} = \frac{[H^+][F^-]}{[HF]} = \frac{x(0.0050 + x)/0.175}{(6.25 \times 10^{-3} - x)/0.175} \approx \frac{x(0.0050)}{0.00625}$$

$x = 8.50 \times 10^{-4} = 8.5 \times 10^{-4}$ M H^+; pH = 3.07

Check. pK_a for HF = 3.17. [HF] > [F^-], pH of buffer = 3.07, less than 3.17.

17.18 $CH_3CH(OH)COOH \rightleftharpoons H^+(aq) + CH_3CH(OH)COO^-$

equil (a) $0.085 - x$ M x M x M

equil (b) $0.095 - x$ M x M $0.0075 + x$ M

$$K_a = \frac{[H^+][CH_3CH(OH)COO^-]}{[CH_3CH(OH)COOH]} = 1.4 \times 10^{-4}$$

(a) $K_a = 1.4 \times 10^{-4} = \dfrac{x^2}{0.085 - x} \approx \dfrac{x^2}{0.085}$; $x = [H^+] = 3.45 \times 10^{-3}$ M = 3.5×10^{-3} M H^+

% ionization = $\dfrac{3.5 \times 10^{-3} \, M \, H^+}{0.085 \, M \, CH_3CH(OH)COOH} \times 100 = 4.1\%$ ionization

(b) $K_a = 1.4 \times 10^{-4} = \dfrac{(x)(0.0075 + x)}{0.095 - x} \approx \dfrac{0.0075\,x}{0.095}$; $x = 1.8 \times 10^{-3}$ M H^+

% ionization = $\dfrac{1.8 \times 10^{-3} \, M \, H^+}{0.095 \, M \, CH_3CH(OH)COOH} \times 100 = 1.9\%$ ionization

Buffers

17.20   NaOH is a strong base and will react with CH_3COOH to form CH_3COONa. As long as CH_3COOH is present in excess, the resulting solution will contain both the conjugate acid CH_3COOH(aq) and the conjugate base CH_3COO^-(aq), the requirements for a buffer.

mmol = $M \times$ mL; mmol $CH_3COOH = 1.00\ M \times 100$ mL = 10.0 mmol

mmol NaOH = 0.100 $M \times 50$ mL = 5.0 mmol

	CH_3COOH (aq)	+	NaOH(aq)	$\rightarrow$	CH_3COONa (aq) + $H_2O(l)$
initial	10.0 mmol		5.0 mmol		
after rx	5.0 mmol		0		5.0 mmol

Mixing these two solutions has created a buffer by partial neutralization of the weak acid CH_3COOH.

17.22   Assume that % ionization is small in these buffers (Solutions 17.17 and 17.18).

(a)   The conjugate acid in this buffer is HCO_3^-, so use K_{a2} for H_2CO_3, 5.6×10^{-11}

$$K_a = \frac{[H^+][CO_3^{2-}]}{[HCO_3^-]} ; [H^+] = \frac{K_a[HCO_3^-]}{[CO_3^{2-}]} = \frac{5.6 \times 10^{-11}\,(0.105)}{(0.125)}$$

$[H^+] = 4.70 \times 10^{-11} = 4.7 \times 10^{-11}\ M;\ pH = 10.33$

(b)   mol = $M \times$ L; total volume = 140 mL = 0.140 L

$$[H^+] = \frac{K_a\,(0.20\ M \times 0.065\ L)/0.140\ L}{(0.15\ M \times 0.075\ L)/0.140\ L} = \frac{5.6 \times 10^{-11}\,(0.20 \times 0.065)}{(0.15 \times 0.075)}$$

$[H^+] = 6.47 \times 10^{-11} = 6.5 \times 10^{-11}\ M;\ pH = 10.19$

17.24 NH_4^+/NH_3 is a basic buffer. Either the hydrolysis of NH_3 or the dissociation of NH_4^+ can be used to determine the pH of the buffer. Using the dissociation of NH_4^+ leads directly to $[H^+]$ and facilitates use of the Henderson-Hasselbach relationship.

(a) $NH_4^+(aq) \rightleftharpoons H^+(aq) + NH_3(aq)$

$$K_a = \frac{K_w}{K_b} = \frac{1.0 \times 10^{-14}}{1.8 \times 10^{-5}} = 5.56 \times 10^{-10} = 5.6 \times 10^{-10}$$

$$[NH_3] = \frac{7.0\ g\ NH_3}{2.50\ L\ soln} \times \frac{1\ mol\ NH_3}{17.0\ g\ NH_3} = 0.1647 = 0.16\ M\ NH_3$$

$$[NH_4^+] = \frac{20.0\ g\ NH_4Cl}{2.50\ L} \times \frac{1\ mol\ NH_4Cl}{53.50\ g\ NH_4Cl} = 0.1495 = 0.15\ M\ NH_4^+$$

$$K_a = \frac{[H^+][NH_3]}{[NH_4^+]} ; [H^+] = \frac{K_a[NH_4^+]}{[NH_3]} = \frac{5.56 \times 10^{-10}\,(0.1495 - x)}{(0.1647 + x)} \approx \frac{5.56 \times 10^{-10}\,(0.1495)}{(0.1647)}$$

$[H^+] = 5.047 \times 10^{-10} = 5.0 \times 10^{-10}\ M,\ pH = 9.30$

(b) $NH_3(aq) + H^+(aq) + NO_3^-(aq) \rightarrow NH_4^+(aq) + NO_3^-(aq)$

(c) $NH_4^+(aq) + Cl^-(aq) + K^+(aq) + OH^-(aq) \rightarrow NH_3(aq) + H_2O(l) + Cl^-(aq) + K^+(aq)$

17.26 $CH_3CH(OH)COOH(aq) \rightleftharpoons H^+(aq) + CH_3CH(OH)COO^-(aq)$

$$[H^+] = \frac{K_a[CH_3CH(OH)COOH]}{[CH_3CH(OH)COO^-]}; [H^+] = 10^{-4.00} = 1.0 \times 10^{-4}$$

$[CH_3CH(OH)COOH] = 0.150\ M$; calculate $[CH_3CH(OH)COO^-]$

$$[CH_3CH(OH)COO^-] = \frac{K_a[CH_3CH(OH)COOH]}{[H^+]} = \frac{1.4 \times 10^{-4}\ (0.150)}{1.0 \times 10^{-4}} = 0.2100 = 0.21\ M$$

$$\frac{0.210\ \text{mol } CH_3CH(OH)COONa}{1.00\ \text{L}} \times \frac{112.1\ \text{g } CH_3CH(OH)COONa}{1\ \text{mol } CH_3CH(OH)COONa}$$

$$= 23.54 = 24\ \text{g } CH_3CH(OH)COONa$$

17.28   (a) $$K_a = \frac{[H^+][C_2H_5COO^-]}{[C_2H_5COOH]}; [H^+] = \frac{K_a[C_2H_5COOH]}{[C_2H_5COO^-]}$$

Since this expression contains a ratio of concentrations, we can ignore total volume and work directly with moles.

$$[H^+] = \frac{1.3 \times 10^{-5}\ (0.10 - x)}{(0.13 + x)} \approx \frac{1.3 \times 10^{-5}\ (0.10)}{0.13} = 1.00 \times 10^{-5} = 1.0 \times 10^{-5}\ M,\ pH = 5.00$$

(b) $C_2H_5COOH(aq)$ + $OH^-(aq)$ $\rightarrow$ $C_2H_5COO^-(aq) + H_2O(l)$

0.10 mol	0.01 mol	0.13 mol
−0.01 mol	−0.01 mol	+0.01 mol
0.09 mol	0 mol	0.14 mol

$$[H^+] \approx \frac{1.3 \times 10^{-5}\ (0.09)}{(0.14)} = 8.357 \times 10^{-6} = 8 \times 10^{-6}\ M;\ pH = 5.1$$

(c) $C_2H_5COO^-(aq)$ + $HI(aq)$ $\rightarrow$ $C_2H_5COOH(aq) + I^-(aq)$

0.13 mol	0.01 mol	0.10 mol
−0.01 mol	−0.01 mol	+0.01 mol
0.12 mol	0 mol	0.11 mol

$$[H^+] \approx \frac{1.3 \times 10^{-5}\ (0.11)}{(0.12)} = 1.192 \times 10^{-5} = 1.19 \times 10^{-5}\ M;\ pH = 4.92$$

17.30 $$\frac{6.5\ \text{g } NaH_2PO_4}{0.355\ \text{L soln}} \times \frac{1\ \text{mol } NaH_2PO_4}{120\ \text{g } NaH_2PO_4} = 0.153 = 0.15\ M$$

$$\frac{8.0\ \text{g } Na_2HPO_4}{0.355\ \text{L soln}} \times \frac{1\ \text{mol } Na_2HPO_4}{142\ \text{g } Na_2HPO_4} = 0.159 = 0.16\ M$$

Use Equation [17.9] to find the pH of the buffer. K_a for $H_2PO_4^-$ is K_{a2} for H_3PO_4, 6.2×10^{-8}

$$pH = -\log(6.2 \times 10^{-8}) + \log \frac{0.159}{0.153} = 7.2076 + 0.0167 = 7.22$$

17.32 The solutes listed contain three possible conjugate acid/conjugate base (CA/CB) pairs. These are:

HCOOH/HCOONa, $pK_a = 3.74$

C_2H_5COOH/C_2H_5COONa, $pK_a = 4.89$

H_3PO_4/NaH_2PO_4, $pK_a = 2.12$

For maximum buffer capacity, pK_a should be within 1 pH unit of the buffer. The propionic acid/propionate pair are most appropriate for a buffer with pH 4.80.

$$pH = pK_a + \log \frac{[CB]}{[CA]}; \ 4.80 = 4.886 + \log \frac{[C_2H_5COONa]}{[C_2H_5COOH]}$$

$$\log \frac{[C_2H_5COONa]}{[C_2H_5COOH]} = -0.0861; \frac{[C_2H_5COONa]}{[C_2H_5COOH]} = 0.8202 = 0.82$$

Since we are making a total of 1 L of buffer, let y = vol C_2H_5COONa and

(1 − y) = vol C_2H_5COOH.

$$0.8202 = \frac{[C_2H_5COONa]}{[C_2H_5COOH]} = \frac{(0.10 \ M \times y)/1.0 \ L}{[0.10 \ M \times (1-y)]/1.0 \ L} = \frac{0.10 \ y}{0.10 - 0.10 \ y}$$

0.8202(0.10 − 0.10 y) = 0.10 y; 0.08202 = 0.18202 y; y = 0.4506 = 0.45 L

450 mL of 0.10 M C_2H_5COONa, 550 mL C_2H_5COOH

Check. pH (buffer) < pK_a (CA) and the calculated amount of CA in the buffer is greater than the amount of CB.

Acid-Base Titrations

17.34 (a) The quantity of base required to reach the equivalence point is the same in the two titrations, assuming both sample solutions contain equal moles of acid.

(b) The pH is higher initially in the titration of a weak acid.

(c) The pH is higher at the equivalence point in the titration of a weak acid.

(d) The pH in excess base is essentially the same for the two cases.

(e) In titrating a weak acid, one needs an indicator that changes at a higher pH than for the strong acid titration. The choice is more critical because the change in pH close to the equivalence point is smaller for the weak acid titration.

17.36 (a) HCOOH(aq) + NaOH(aq) → HCOONa(aq) + H_2O(l)

At the equivalence point, the major species are Na^+ and $HCOO^-$. Na^+ is negligible and $HCOO^-$ is the CB of HCOOH. The solution is basic, above pH 7.

(b) $Ca(OH)_2$(aq) + 2$HClO_4$(aq) → $Ca(ClO_4)_2$(aq) + 2H_2O(l)

At the equivalence point, the major species are Ca^{2+} and ClO_4^-; both are negligible. The solution is at pH 7.

(c) $C_5H_5N(aq) + HNO_3(aq) \rightarrow C_5H_5NH^+NO_3^-(aq)$

At the equivalence point, the major species are $C_5H_5NH^+$ and NO_3^-. NO_3^- is negligible and $C_5H_5NH^+$ is the CA of C_5H_5N. The solution is acidic, below pH 7.

17.38 (a) At the equivalence point, moles HX added = moles B initially present =
$0.10\ M \times 0.0300\ L = 0.0030$ moles HX added.

(b) $BH^+(aq)$

(c) Both K_a for BH^+ and concentration BH^+ determine pH at the equivalence point.

(d) Because the pH at the equivalence point will be less than 7, methyl red would be more appropriate.

17.40 (a) $45.0\ \text{mL NaOH} \times \dfrac{0.0950\ \text{mol NaOH}}{1000\ \text{mL soln}} \times \dfrac{1\ \text{mol HCl}}{1\ \text{mol NaOH}} \times \dfrac{1000\ \text{mL soln}}{0.105\ \text{mol HCl}}$
$$= 40.7\ \text{mL HCl soln}$$

(b) $22.5\ \text{mL NH}_3 \times \dfrac{0.118\ \text{mol NH}_3}{1000\ \text{mL soln}} \times \dfrac{1\ \text{mol HCl}}{1\ \text{mol NH}_3} \times \dfrac{1000\ \text{mL soln}}{0.105\ \text{mol HCl}}$
$$= 25.3\ \text{mL HCl soln}$$

(c) $125.0\ \text{mL} \times \dfrac{1.35\ \text{g NaOH}}{1000\ \text{mL}} \times \dfrac{1\ \text{mol NaOH}}{40.00\ \text{g NaOH}} \times \dfrac{1\ \text{mol HCl}}{1\ \text{mol NaOH}} \times \dfrac{1000\ \text{mL soln}}{0.105\ \text{mol HCl}}$
$$= 40.2\ \text{mL HCl soln}$$

17.42 moles $OH^- = M_{KOH} \times L_{KOH} = 0.150\ M \times 0.0300\ L = 4.50 \times 10^{-3}\ \text{mol}$

moles $H^+ = M_{HClO_4} \times L_{HClO_4} = 0.125\ M \times L_{HClO_4}$

	mL_{KOH}	mL_{HClO_4}	Total Volume	Moles OH^-	Moles H^+	Molarity Excess Ion	pH
(a)	30.0	30.0	60.0	4.50×10^{-3}	3.75×10^{-3}	$0.0125(OH^-)$	12.10
(b)	30.0	35.0	65.0	4.50×10^{-3}	4.38×10^{-3}	$1.9 \times 10^{-3}(OH^-)$	11.28
(c)	30.0	36.0	66.0	4.50×10^{-3}	4.50×10^{-3}	$1.0 \times 10^{-7}(OH^-)$	7.00
(d)	30.0	37.0	67.0	4.50×10^{-3}	4.63×10^{-3}	$1.9 \times 10^{-3}(H^+)$	2.73
(e)	30.0	40.0	70.0	4.50×10^{-3}	5.00×10^{-3}	$7.1 \times 10^{-3}(H^+)$	2.15

$$\text{molarity of excess ion} = \dfrac{\text{moles ion}}{\text{total vol in L}}$$

(a) $\dfrac{4.50 \times 10^{-3}\ \text{mol OH}^- - 3.75 \times 10^{-3}\ \text{mol H}^+}{0.0600\ \text{L}} = 0.0125 = 0.013\ M\ OH^-$

(b) $\dfrac{4.50 \times 10^{-3} \text{ mol OH}^- - 4.38 \times 10^{-3} \text{ mol H}^+}{0.0650 \text{ L}} = 1.92 \times 10^{-3} = 1.9 \times 10^{-3} \text{ M OH}^-$

(c) equivalence point, mol H^+ = mol OH^-

   $KClO_4$ does not hydrolyze, so $[H^+] = [OH^-] = 1 \times 10^{-7}$

(d) $\dfrac{4.63 \times 10^{-3} \text{ mol H}^+ - 4.50 \times 10^{-3} \text{ mol OH}^-}{0.0670 \text{ L}} = 1.87 \times 10^{-3} = 1.9 \times 10^{-3} \text{ M H}^+$

(e) $\dfrac{5.00 \times 10^{-3} \text{ mol H}^+ - 4.50 \times 10^{-3} \text{ mol OH}^-}{0.0700 \text{ L}} = 7.14 \times 10^{-3} = 7.1 \times 10^{-3} \text{ M H}^+$

17.44 (a) Weak base problem: $K_b = 1.8 \times 10^{-5} = \dfrac{[NH_4^+][OH^-]}{[NH_3]}$

   At equilibrium, $[OH^-] = x$, $[NH_3] = (0.030 - x)$; $[NH_4^+] = x$

   $1.8 \times 10^{-5} = \dfrac{x^2}{(0.030 - x)} \approx \dfrac{x^2}{0.030}$; $x = [OH^-] = 7.348 \times 10^{-4} = 7.3 \times 10^{-4} \ M$

   $pH = 14.00 - 3.13 = 10.87$

(b–f) Calculate mol NH_3 and mol NH_4^+ after the acid-base reaction takes place. 0.030 $M\ NH_3 \times 0.0300 \text{ L} = 9.0 \times 10^{-4} \text{ mol } NH_3$ present initially.

		NH_3(aq)	+	HCl(aq)	→	NH_4^+(aq) + Cl^-(aq)
				$(0.025\ M \times 0.0100 \text{ L}) =$		
(b)	before rx	9.0×10^{-4} mol		2.5×10^{-4} mol		0 mol
	after rx	6.5×10^{-4} mol		0 mol		2.5×10^{-4} mol
				$(0.025\ M \times 0.0200 \text{ L}) =$		
(c)	before rx	9.0×10^{-4} mol		5.0×10^{-4} mol		0 mol
	after rx	4.0×10^{-4} mol		0 mol		5.0×10^{-4} mol
				$(0.025\ M \times 0.0350 \text{ L}) =$		
(d)	before rx	9.0×10^{-4} mol		8.75×10^{-4} mol		0 mol
	after rx	0.25×10^{-4} mol		0 mol		8.75×10^{-4} mol
				$(0.025\ M \times 0.0360 \text{ L}) =$		
(e)	before rx	9.0×10^{-4} mol		9.0×10^{-4} mol		0 mol
	after rx	0 mol		0 mol		9.0×10^{-4} mol
				$(0.025\ M \times 0.0370 \text{ L}) =$		
(f)	before rx	9.0×10^{-4} mol		9.25×10^{-4} mol		0 mol
	after rx	0 mol		0.25×10^{-4} mol		9.0×10^{-4} mol

(b) Using the acid dissociation equilibrium for NH_4^+ (so that we calculate $[H^+]$ directly), $NH_4^+(aq) \rightleftharpoons H^+(aq) + NH_3(aq)$

$$K_a = \frac{[H^+][NH_3]}{[NH_4^+]} = \frac{K_w}{K_b \text{ for } NH_3} = \frac{1.0 \times 10^{-14}}{1.8 \times 10^{-5}} = 5.56 \times 10^{-10} = 5.6 \times 10^{-10}$$

$$[NH_3] = \frac{6.5 \times 10^{-4} \text{ mol}}{0.0400 \text{ L}} = 0.01625 \, M; [NH_4^+] = \frac{2.50 \times 10^{-4} \text{ mol}}{0.0400 \text{ L}} = 6.25 \times 10^{-3} \, M$$

$$[H^+] = \frac{5.56 \times 10^{-10} [NH_4^+]}{[NH_3]} \approx \frac{5.56 \times 10^{-10} (6.25 \times 10^{-3})}{(0.01625)} = 2.14 \times 10^{-10}; pH = 9.67$$

(We will assume $[H^+]$ is small compared to $[NH_3]$ and $[NH_4^+]$.)

(c) $[NH_3] = \dfrac{4.0 \times 10^{-4} \text{ mol}}{0.0500 \text{ L}} = 0.0080 \, M; [NH_4^+] = \dfrac{5.0 \times 10^{-4} \text{ mol}}{0.0500 \text{ L}} = 0.010 \, M$

$$[H^+] = \frac{5.56 \times 10^{-10} (0.010)}{(0.0080)} = 6.94 \times 10^{-10} = 6.9 \times 10^{-10} \, M; pH = 9.16$$

(d) $[NH_3] = \dfrac{0.25 \times 10^{-4} \text{ mol}}{0.0650 \text{ L}} = 3.846 \times 10^{-4} = 4 \times 10^{-4} \, M; [NH_4^+] = \dfrac{8.75 \times 10^{-4} \text{ mol}}{0.0650 \text{ L}}$

$$= 0.01346 = 0.013 \, M$$

$$[H^+] = \frac{5.56 \times 10^{-10} (0.01346)}{3.846 \times 10^{-4}} = 1.946 \times 10^{-8} = 2 \times 10^{-8} \, M; pH = 7.7$$

(e) At the equivalence point, $[H^+] = [NH_3] = x$

$$[NH_4^+] = \frac{9.0 \times 10^{-4} \, M}{0.0660 \text{ L}} = 0.01364 = 0.014 \, M$$

$$5.56 \times 10^{-10} = \frac{x^2}{0.01364}; x = [H^+] = 2.754 \times 10^{-6} = 2.8 \times 10^{-6}; pH = 5.56$$

(f) Past the equivalence point, $[H^+]$ from the excess HCl determines the pH.

$$[H^+] = \frac{0.25 \times 10^{-4} \text{ mol}}{0.0670 \text{ L}} = 3.731 \times 10^{-4} = 4 \times 10^{-4} \, M; pH = 3.4$$

17.46 The volume of NaOH solution required in all cases is

$$V_{base} = \frac{V_{acid} \times M_{acid}}{M_{base}} = \frac{(0.100) \, V_{acid}}{(0.080)} = 1.25 \, V_{acid}$$

The total volume at the equivalence point is $V_{base} + V_{acid} = 2.25 \, V_{acid}$.

The concentration of the salt at the equivalence point is $\dfrac{M_{acid} \, V_{acid}}{2.25 \, V_{acid}} = \dfrac{0.100}{2.25} = 0.0444 \, M$

(a) $0.0444 \, M$ NaBr, pH = 7.00

(b) $0.0444\,M\,Na^+\,C_3H_5O_3^-$; $C_3H_5O_3^-(aq) + H_2O(l) \;\rightleftharpoons\; HC_3H_5O_3(aq) + OH^-(aq)$

$$K_b = \frac{[CH_3CH(OH)COOH][OH^-]}{[CH_3CH(OH)COO^-]} = \frac{K_w}{K_a} = \frac{1.0 \times 10^{-14}}{1.4 \times 10^{-4}} = 7.14 \times 10^{-11} = 7.1 \times 10^{-11}$$

$[CH_3CH(OH)COOH] = [OH^-]$; $[CH_3CH(OH)COO^-] \approx 0.0444$

$[OH^-]^2 \approx 0.0444(7.14 \times 10^{-11})$; $[OH^-] = 1.78 \times 10^{-6} = 1.8 \times 10^{-6}\,M$, pOH = 5.75;

$$pH = 8.25$$

(c) $CrO_4^{2-}(aq) + H_2O(l) \;\rightleftharpoons\; HCrO_4^-(aq) + OH^-(aq)$

$$K_b = \frac{[HCrO_4^-][OH^-]}{[CrO_4^{2-}]} = \frac{K_w}{K_a} = \frac{1.0 \times 10^{-14}}{3.0 \times 10^{-7}} = 3.33 \times 10^{-8} = 3.3 \times 10^{-8}$$

$[OH^-]^2 \approx 0.0444(3.33 \times 10^{-8})$; $[OH^-] = 3.849 \times 10^{-5} = 3.8 \times 10^{-5}$, pH = 9.59

Solubility Equilibria and Factors Affecting Solubility

17.48 (a) Solubility is the amount (grams, moles) of solute that will dissolve in a certain volume of solution. Solubility-product constant is an equilibrium constant, the product of the molar concentrations of all the dissolved ions in solution.

(b) $K_{sp} = [Mn^{2+}][CO_3^{2-}]$; $K_{sp} = [Hg^{2+}][OH^-]^2$; $K_{sp} = [Cu^{2+}]^3[PO_4^{3-}]^2$

17.50 (a) $PbBr_2(s) \;\rightleftharpoons\; Pb^{2+}(aq) + 2Br^-(aq)$

$K_{sp} = [Pb^{2+}][Br^-]^2$; $[Pb^{2+}] = 1.0 \times 10^{-2}\,M$, $[Br^-] = 2.0 \times 10^{-2}\,M$

$K_{sp} = (1.0 \times 10^{-2}\,M)(2.0 \times 10^{-2}\,M)^2 = 4.0 \times 10^{-6}$

(b) $AgIO_3(s) \;\rightleftharpoons\; Ag^+(aq) + IO_3^-(aq)$; $K_{sp} = [Ag^+][IO_3^-]$

$$[Ag^+] = [IO_3^-] = \frac{0.0490\,g\,AgIO_3}{1.00\,L\,soln} \times \frac{1\,mol\,AgIO_3}{282.8\,g\,AgIO_3} = 1.733 \times 10^{-4} = 1.73 \times 10^{-4}\,M$$

$K_{sp} = (1.733 \times 10^{-4}\,M)\,(1.733 \times 10^{-4}\,M) = 3.00 \times 10^{-8}$

(c) $Cu(OH)_2(s) \;\rightleftharpoons\; Cu^{2+}(aq) + 2OH^-(aq)$; $K_{sp} = [Cu^{2+}][OH^-]^2$

$[Cu^{2+}] = x$, $[OH^-] = 2x$; $K_{sp} = 4.8 \times 10^{-20} = (x)(2x)^2$

$4.8 \times 10^{-20} = 4x^3$; $x = [Cu^{2+}] = 2.290 \times 10^{-7} = 2.3 \times 10^{-7}\,M$

$$\frac{2.290 \times 10^{-7}\,mol\,Cu(OH)_2}{1\,L} \times \frac{97.56\,g\,Cu(OH)_2}{1\,mol\,Cu(OH)_2} = 2.2 \times 10^{-5}\,g\,Cu(OH)_2$$

However, $[OH^-]$ from $Cu(OH)_2 = 4.58 \times 10^{-7}\,M$; this is similar to $[OH^-]$ from the autoionization of water.

$K_w = [H^+][OH^-]$; $[H^+] = y$, $[OH^-] = (4.58 \times 10^{-7} + y)$

$1.0 \times 10^{-14} = y(4.58 \times 10^{-7} + y)$; $y^2 + 4.58 \times 10^{-7}\,y - 1.0 \times 10^{-14}$

$$y = \frac{-4.58 \times 10^{-7} \pm \sqrt{(4.58 \times 10^{-7})^2 - 4(1)(-1.0 \times 10^{-14})}}{2}; \; y = 2.09 \times 10^{-8}$$

$[OH^-]_{total} = 4.58 \times 10^{-7} \, M + 0.209 \times 10^{-7} \, M = 4.79 \times 10^{-7} \, M$

Recalculating $[Cu^{2+}]$ and thus molar solubility of $Cu(OH)_2(s)$:

$4.8 \times 10^{-20} = x(4.79 \times 10^{-7})^2; \; x = 2.09 \times 10^{-7} \, M \, Cu^{2+}$

$$\frac{2.09 \times 10^{-7} \text{ mol } Cu(OH)_2(s)}{1 \, L} \times \frac{97.56 \text{ g } Cu(OH)_2}{1 \text{ mol } Cu(OH)_2} = 2.0 \times 10^{-5} \text{ g } Cu(OH)_2$$

Note that the presence of OH^- as a common ion decreases the water solubility of $Cu(OH)_2$.

17.52 $PbI_2(s) \rightleftharpoons Pb^{2+}(aq) + 2I^-(aq); \; K_{sp} = [Pb^{2+}][I^-]^2$

$$[Pb^{2+}] = \frac{0.54 \text{ g } PbI_2}{1.00 \text{ L soln}} \times \frac{1 \text{ mol } PbI_2}{461.0 \text{ g } PbI_2} = 1.17 \times 10^{-3} = 1.2 \times 10^{-3} \, M$$

$[I^-] = 2[Pb^{2+}]; \; K_{sp} = [Pb^{2+}](2[Pb^{2+}])^2 = 4[Pb^{2+}]^3 = 4(1.17 \times 10^{-3})^3 = 6.4 \times 10^{-9}$

17.54 $LaF_3(s) \rightleftharpoons La^{3+}(aq) + 3F^-(aq); \; K_{sp} = [La^{3+}][F^-]^3$

(a)   molar solubility = $x = [La^{3+}]; \; [F^-] = 3x$

$K_{sp} = 2 \times 10^{-19} = (x)(3x)^3; \; 2 \times 10^{-19} = 27 \, x^4; \; x = (7.41 \times 10^{-21})^{1/4}, \; x = 9.28 \times 10^{-6}$

$= 9 \times 10^{-6} \, M \, La^{3+}$

$$\frac{9.28 \times 10^{-6} \text{ mol } LaF_3}{1 \, L} \times \frac{195.9 \text{ g } LaF_3}{1 \text{ mol}} = 1.82 \times 10^{-3} = 2 \times 10^{-3} \text{ g } LaF_3/L$$

(b)   molar solubility = $x = [La^{3+}]$

There are two sources of F^-: $KF(0.010 \, M)$ and $LaF_3 \, (3x \, M)$

$K_{sp} = (x)(0.010 + 3x)^3;$ assume x is small compared to 0.010 M.

$2 \times 10^{-19} = (0.010)^3 \, x; \; x = 2 \times 10^{-19}/1.0 \times 10^{-6} = 2 \times 10^{-13} \, M \, La^{3+}$

$$\frac{2 \times 10^{-13} \text{ mol } LaF_3}{1 \, L} \times \frac{195.9 \text{ g } LaF_3}{1 \text{ mol}} = 3.92 \times 10^{-11} = 4 \times 10^{-11} \text{ g } LaF_3/L$$

(c)   molar solubility = $x, \; [F^-] = 3x, \; [La^{3+}] = 0.050 \, M + x$

$K_{sp} = (0.050 + x)(3x)^3;$ assume x is small compared to 0.050 M.

$2 \times 10^{-19} = (0.050)(27 \, x^3) = 1.35 \, x^3; \; x = (1.48 \times 10^{-19})^{1/3} = 5.29 \times 10^{-7} = 5 \times 10^{-7} \, M$

$$\frac{5.29 \times 10^{-7} \text{ mol } LaF_3}{1 \, L} \times \frac{195.9 \text{ g } LaF_3}{1 \text{ mol}} = 1.04 \times 10^{-4} = 1 \times 10^{-4} \text{ g } LaF_3/L$$

17.56 $Fe(OH)_2(s) \rightleftharpoons Fe^{2+}(aq) + 2OH^-(aq); \; K_{sp} = 8.0 \times 10^{-16}$

Since the $[OH^-]$ is set by the pH of the solution, the solubility of $Fe(OH)_2$ is just $[Fe^{2+}]$.

(a) $pH = 8.0$, $pOH = 14 - pH = 6.0$, $[OH^-] = 10^{-pOH} = 1 \times 10^{-6} \, M$

$$K_{sp} = 7.9 \times 10^{-16} = [Fe^{2+}](1.0 \times 10^{-6})^2; [Fe^{2+}] = \frac{7.9 \times 10^{-16}}{1.0 \times 10^{-12}} = 7.9 \times 10^{-4} = 8 \times 10^{-4} \, M$$

(b) $pH = 10.0$, $pOH = 4.0$, $[OH^-] = 1.0 \times 10^{-4} = 1 \times 10^{-4} \, M$

$$K_{sp} = 7.9 \times 10^{-16} = [Fe^{2+}][1.0 \times 10^{-4}]^2; [Fe^{2+}] = \frac{7.9 \times 10^{-16}}{1.0 \times 10^{-8}} = 7.9 \times 10^{-8} = 8 \times 10^{-8} \, M$$

(c) $pH = 12.0$, $pOH = 2.0$, $[OH^-] = 1.0 \times 10^{-2} = 1 \times 10^{-2} \, M$

$$K_{sp} = 7.9 \times 10^{-16} = [Fe^{2+}][1.0 \times 10^{-2}]^2; [Fe^{2+}] = \frac{7.9 \times 10^{-16}}{1.0 \times 10^{-4}} = 7.9 \times 10^{-12} = 8 \times 10^{-12} \, M$$

17.58 If the anion in the slightly soluble salt is the conjugate base of a strong acid, there will be no reaction.

(a) $MnS(s) + 2H^+(aq) \rightarrow H_2S(aq) + Mn^{2+}(aq)$

(b) $PbF_2(s) + 2H^+(aq) \rightarrow 2HF(aq) + Pb^{2+}(aq)$

(c) $AuCl_3(s) + H^+(aq) \rightarrow$ no reaction

(d) $Hg_2C_2O_4(s) + 2H^+(aq) \rightarrow H_2C_2O_4(aq) + Hg_2^{2+}(aq)$

(e) $CuBr(s) + H^+(aq) \rightarrow$ no reaction

17.60 $NiC_2O_4(s) \rightleftharpoons Ni^{2+}(aq) + C_2O_4^{2-}(aq); \quad K_{sp} = [Ni^{2+}][C_2O_4^{2-}] = 4 \times 10^{-10}$

When the salt has just dissolved, $[C_2O_4^{2-}]$ will be 0.020 M. Thus $[Ni^{2+}]$ must be less than $4 \times 10^{-10} / 0.020 = 2 \times 10^{-8} \, M$. To achieve this low $[Ni^{2+}]$ we must complex the Ni^{2+} ion with NH_3: $Ni^{2+}(aq) + 6NH_3(aq) \rightleftharpoons Ni(NH_3)_6^{2+}(aq)$. Essentially all Ni(II) is in the form of the complex, so $[Ni(NH_3)_6^{2+}] = 0.020$. Find K_f for $Ni(NH_3)_6^{2+}$ in Table 17.1.

$$K_f = \frac{[Ni(NH_3)_6^{2+}]}{[Ni^{2+}][NH_3]^6} = \frac{(0.020)}{(2 \times 10^{-8})[NH_3]^6} = 1.2 \times 10^9; [NH_3]^6 = 8.33 \times 10^{-4};$$

$$[NH_3] = 0.307 = 0.3 \, M$$

17.62
$$Ag_2S(s) \rightleftharpoons 2Ag^+(aq) + S^{2-}(aq) \qquad\qquad K_{sp}$$

$$S^{2-}(aq) + 2H^+(aq) \rightleftharpoons H_2S(aq) \qquad\qquad 1/(K_{a1} \times K_{a2})$$

$$2[Ag^+(aq) + 2Cl^-(aq) \rightleftharpoons AgCl_2^-(aq)] \qquad\qquad K_f^2$$

$$Ag_2S(s) + 2H^+(aq) + 4Cl^-(aq) \rightleftharpoons 2AgCl_2^-(aq) + H_2S(aq)$$

$$K = \frac{K_{sp} \times K_f^2}{K_{a1} \times K_{a2}} = \frac{(6 \times 10^{-51})(1.1 \times 10^5)^2}{(9.5 \times 10^{-8})(1 \times 10^{-19})} = 7.64 \times 10^{-15} = 8 \times 10^{-15}$$

Precipitation; Qualitative Analysis

17.64   (a) $Co(OH)_2(s) \rightleftharpoons Co^{2+}(aq) + 2OH^-(aq)$; $K_{sp} = [Co^{2+}][OH^-]^2 = 1.3 \times 10^{-15}$

$pH = 8.5$; $pOH = 14 - 8.5 = 5.5$; $[OH^-] = 10^{-5.5} = 3.16 \times 10^{-6} = 3 \times 10^{-6}\,M$

$Q = (0.020)(3.16 \times 10^{-6})^2 = 2 \times 10^{-13}$; $Q > K_{sp}$, $Co(OH)_2$ will precipitate

     (b) $AgIO_3(s) \rightleftharpoons Ag+(aq) + IO_3^-(aq)$; $K_{sp} = [Ag+][IO_3^-] = 3.1 \times 10^{-8}$

$$[Ag^+] = \frac{0.010\,M\,Ag^+ \times 0.020\,L}{0.030\,L} = 6.667 \times 10^{-3} = 6.7 \times 10^{-3}\,M$$

$$[IO_3^-] = \frac{0.015\,M\,IO_3^- \times 0.010\,L}{0.030\,L} = 5.000 \times 10^{-3} = 5.0 \times 10^{-3}\,M$$

$Q = (6.667 \times 10^{-3})(5.00 \times 10^{-3}) = 3.3 \times 10^{-5}$; $Q > K_{sp}$, $AgIO_3$ will precipitate

17.66 $AgCl(s) \rightleftharpoons Ag^+(aq) + Cl^-(aq)$; $K_{sp} = [Ag^+][Cl^-] = 1.8 \times 10^{-10}$

$$[Ag^+] = \frac{0.10\,M \times 0.2\,mL}{10.2\,mL} = 2 \times 10^{-3}\,M; \quad [Cl^-] = \frac{1.8 \times 10^{-10}}{2 \times 10^{-3}\,M} = 9 \times 10^{-8}\,M$$

$$\frac{9 \times 10^{-8}\,mol\,Cl^-}{1\,L} \times \frac{35.45\,g\,Cl^-}{1\,mol\,Cl^-} \times 0.0102\,L = 3.25 \times 10^{-8}\,g\,Cl^- = 3 \times 10^{-8}\,g\,Cl^-$$

17.68   (a)   Precipitation will begin when $Q = K_{sp}$.

$BaSO_4$: $K_{sp} = [Ba^{2+}][SO_4^{2-}] = 1.1 \times 10^{-10}$

$1.1 \times 10^{-10} = (0.010)[SO_4^{2-}]$; $[SO_4^{2-}] = 1.1 \times 10^{-8}\,M$

$SrSO_4$: $K_{sp} = [Sr^{2+}][SO_4^{2-}] = 3.2 \times 10^{-7}$

$3.2 \times 10^{-7} = (0.010)[SO_4^{2-}]$; $[SO_4^{2-}] = 3.2 \times 10^{-5}\,M$

The $[SO_4^{2-}]$ necessary to begin precipitation is the smaller of the two values, $1.1 \times 10^{-8}\,M\,SO_4^{2-}$.

     (b) Ba^{2+} precipitates first, because it requires the smaller $[SO_4^{2-}]$.

     (c) Sr^{2+} will begin to precipitate when $[SO_4^{2-}]$ in solution (not bound in $BaSO_4$) reaches $3.2 \times 10^{-5}\,M$.

17.70   Initial solubility in water rules out CdS and HgO. Formation of a precipitate on addition of HCl indicates the presence of $Pb(NO_3)_2$ (formation of $PbCl_2$). Formation of a precipitate on addition of H_2S at pH 1 probably indicates $Cd(NO_3)_2$ (formation of CdS). (This test can be misleading because enough Pb^{2+} can remain in solution after filtering $PbCl_2$ to lead to visible precipitation of PbS.) Absence of a precipitate on addition of H_2S at pH 8 indicates that $ZnSO_4$ is not present. The yellow flame test indicates presence of Na^+. In summary, $Pb(NO_3)_2$ and Na_2SO_4 are definitely present, $Cd(NO_3)_2$ is probably present, and CdS, HgO and $ZnSO_4$ are definitely absent.

17.72   (a)   Make the solution slightly acidic and saturate with H_2S; CdS will precipitate, Na^+ remains in solution.

(b) Make the solution acidic, saturate with H_2S; CuS will precipitate, Mg^{2+} remains in solution.

(c) Add HCl, $PbCl_2$ precipitates. (It is best to carry out the reaction in an ice-water bath to reduce the solubility of $PbCl_2$.)

(d) Add dilute HCl; AgCl precipitates, Hg^{2+} remains in solution.

17.74 The addition of $(NH_4)_2HPO_4$ could result in precipitation of salts from metal ions of the other groups. The $(NH_4)_2HPO_4$ will render the solution basic, so metal hydroxides could form as well as insoluble phosphates. It is essential to separate the metal ions of a group from other metal ions before carrying out the specific tests for that group.

Additional Exercises

17.76 Benzene sulfonic acid = BSOH, sodium benzene sulfonate = BSONa

Use Equation 17.9, $pH = pK_a + \log [CB]/[CA]$.

$pH = 2.25 + \log(0.125\ M\ BSONa/0.150\ M\ BSOH)$

$pH = 2.25 + \log(0.8333) = 2.25 + (-0.0792) = 2.17$

17.78 $K_a = \dfrac{[H^+][In^-]}{[HIn]}$; at $pH = 4.68$, $[HIn] = [In^-]$; $[H^+] = K_a$; $pH = pK_a = 4.68$

17.79 (a) $HA(aq) + B(aq) \rightleftharpoons HB^+(aq) + A^-(aq)$ $K_{eq} = \dfrac{[HB^+][A^-]}{[HA][B]}$

(b) Note that the solution is slightly basic because B is a stronger base than HA is an acid. (Or, equivalently, that A^- is a stronger base than HB^+ is an acid.) Thus, a little of the A^- is used up in reaction: $A^-(aq) + H_2O(l) \rightleftharpoons HA(aq) + OH^-(aq)$. Since pH is not very far from neutral, it is reasonable to assume that the reaction in part (a) has gone far to the right, and that $[A^-] \approx [HB^+]$ and $[HA] \approx [B]$. Then

$K_a = \dfrac{[A^-][H^+]}{[HA]} = 8.0 \times 10^{-5}$; when $pH = 9.2$, $[H^+] = 6.31 \times 10^{-10} = 6 \times 10^{-10}\ M$

$\dfrac{[A^-]}{[HA]} = 8.0 \times 10^{-5}/6.31 \times 10^{-10} = 1.268 \times 10^5 = 1 \times 10^5$

From the assumptions above, $\dfrac{[A^-]}{[HA]} = \dfrac{[HB^+]}{[B]}$, so $K_{eq} \approx \dfrac{[A^-]^2}{[HA]^2} = 1.608 \times 10^{10} = 2 \times 10^{10}$

(c) K_b for the reaction $B(aq) + H_2O(l) \rightleftharpoons BH^+(aq) + OH^-(aq)$ can be calculated by noting that the equilibrium constant for the reaction in part (a) can be written as $K = K_a\ (HA) \times K_b\ (B) / K_w$. (You should prove this to yourself.) Then,

$K_b\ (B) = \dfrac{K \times K_w}{K_a\ (HA)} = \dfrac{(1.608 \times 10^{10})(1.0 \times 10^{-14})}{8.0 \times 10^{-5}} = 2.010 = 2$

$K_b\ (B)$ is larger than $K_a\ (HA)$, as it must be if the solution is basic.

17.81 $\dfrac{0.15 \text{ mol CH}_3\text{COOH}}{1 \text{ L soln}} \times 0.750 \text{ L} = 0.1125 = 0.11 \text{ mol CH}_3\text{COOH}$

$0.1125 \text{ mol CH}_3\text{COOH} \times \dfrac{60.05 \text{ g CH}_3\text{COOH}}{1 \text{ mol CH}_3\text{COOH}} \times \dfrac{1 \text{ g gl acetic acid}}{0.99 \text{ g CH}_3\text{COOH}} \times \dfrac{1.00 \text{ mL gl acetic acid}}{1.05 \text{ g gl acetic acid}}$

$= 6.5 \text{ mL glacial acetic acid}$

At pH 4.50, $[\text{H}^+] = 10^{-4.50} = 3.16 \times 10^{-5} = 3.2 \times 10^{-5} \, M$; this is small compared to $0.15 \, M \text{ CH}_3\text{COOH}$.

$K_a = \dfrac{(3.16 \times 10^{-5})[\text{CH}_3\text{COO}^-]}{0.15} = 1.8 \times 10^{-5}; [\text{CH}_3\text{COO}^-] = 0.0854 = 0.085 \, M$

$\dfrac{0.0854 \text{ mol CH}_3\text{COONa}}{1 \text{ L soln}} \times 0.750 \text{ L} \times \dfrac{82.03 \text{ g CH}_3\text{COONa}}{1 \text{ mol CH}_3\text{COONa}} = 5.253 = 5.25 \text{ g CH}_3\text{COONa}$

17.83 At the equivalence point of a titration, moles strong base added equals moles weak acid initially present. $M_B \times V_B = \text{mol base added} = \text{mol acid initial}$

At the half-way point, the volume of base is one-half of the volume required to reach the equivalence point, and the moles base delivered equals one-half of the mol acid initially present. This means that one-half of the weak acid HA is converted to the conjugate base A^-. If exactly half of the acid reacts, mol HA = mol A^- and [HA] = [A^-] at the half-way point.

From Equation [17.9], $\text{pH} = \text{p}K_a + \log\dfrac{[\text{conj. base}]}{[\text{conj. acid}]} = \text{p}K_a + \log\dfrac{[\text{A}^-]}{[\text{HA}^-]}$.

If $[\text{A}^-]/[\text{HA}] = 1$, $\log(1) = 0$ and $\text{pH} = \text{p}K_a$ of the weak acid being titrated.

17.84 (a) $\dfrac{0.4885 \text{ g KHP}}{0.100 \text{ L}} \times \dfrac{1 \text{ mol KHP}}{204.2 \text{ g KHP}} = 0.02392 = 0.0239 \, M \text{ P}^{2-}$ at the equivalence point

The pH at the equivalence point is determined by the hydrolysis of P^{2-}.

$\text{P}^{2-}(aq) + \text{H}_2\text{O}(l) \rightleftharpoons \text{HP}^-(aq) + \text{OH}^-(aq)$

$K_b = \dfrac{[\text{HP}^-][\text{OH}^-]}{[\text{P}^{2-}]} = \dfrac{K_w}{K_a \text{ for HP}^-} = \dfrac{1.0 \times 10^{-14}}{3.1 \times 10^{-6}} = 3.23 \times 10^{-9} = 3.2 \times 10^{-9}$

$3.23 \times 10^{-9} = \dfrac{x^2}{(0.02392 - x)} \approx \dfrac{x^2}{0.02392}; X = [\text{OH}^-] = 8.8 \times 10^{-6} \, M$

pH = 14 – 5.06 = 8.94. From Figure 16.7, either phenolphthalein (pH 8.2 – 10.0) or thymol blue (pH 8.0 – 9.6) could be used to detect the equivalence point. Phenolphthalein is usually the indicator of choice because the colorless to pink change is easier to see.

(b) $0.4885 \text{ g KHP} \times \dfrac{1 \text{ mol KHP}}{204.2 \text{ g KHP}} \times \dfrac{1 \text{ mol NaOH}}{1 \text{ mol KHP}} \times \dfrac{1}{0.03855 \text{ L NaOH}} = 0.06206 \, M \text{ NaOH}$

17.86 The reaction involved is $HA(aq) + OH^-(aq) \rightleftharpoons A^-(aq) + H_2O(l)$. We thus have 0.080 mol A^- and 0.12 mol HA in a total volume of 1.0 L, so the "initial" molarities of A^- and HA are 0.080 M and 0.12 M, respectively. The weak acid equilibrium of interest is

$$HA(aq) \rightleftharpoons H^+(aq) + A^-(aq)$$

(a) $K_a = \dfrac{[H^+][A^-]}{[HA]}$; $[H^+] = 10^{-4.80} = 1.58 \times 10^{-5} = 1.6 \times 10^{-5} M$

Assuming $[H^+]$ is small compared to [HA] and $[A^-]$,

$$K_a \approx \frac{(1.58 \times 10^{-5})(0.080)}{(0.12)} = 1.06 \times 10^{-5} = 1.1 \times 10^{-5}, pK_a = 4.98$$

(b)   At pH = 5.00, $[H^+] = 1.0 \times 10^{-5} M$. Let b = extra moles NaOH.

[HA] = 0.12 – b, $[A^-]$ = 0.080 + b

$$1.06 \times 10^{-5} \approx \frac{(1.0 \times 10^{-5})(0.080 + b)}{(0.12 - b)}; 2.06 \times 10^{-5} b = 4.72 \times 10^{-7};$$

b = 0.023 mol NaOH

17.87 Assume that H_3PO_4 will react with NaOH in a stepwise fashion: (This is not unreasonable, since the three K_a values for H_3PO_4 are significantly different.)

	$H_3PO_4(aq)$	+	$NaOH(aq)$	$\rightarrow$	$H_2PO_4^-(aq) + Na^+(aq) + H_2O(l)$
before	0.20 mol		0.30 mol		0 mol
after	0 mol		0.10 mol		0.20 mol

	$H_2PO_4^-(aq)$	+	$NaOH(aq)$	$\rightarrow$	$HPO_4^-(aq) + Na^+(aq) + H_2O(l)$
before	0.20 mol		0.10 mol		0.25 mol
after	0.10 mol		0		0.35 mol

Thus, after all NaOH has reacted, the resulting 1.00 L solution is a buffer containing 0.10 mol $H_2PO_4^-$ and 0.35 mol HPO_4^{2-}. $H_2PO_4^-(aq) \rightleftharpoons H^+(aq) + HPO_4^{2-}(aq)$

$$K_a = 6.2 \times 10^{-8} = \frac{[HPO_4^{2-}][H^+]}{[H_2PO_4^-]}; [H^+] = \frac{6.2 \times 10^{-8}(0.10\ M)}{0.35\ M} = 1.77 \times 10^{-8} = 1.8 \times 10^{-8}\ M;$$

$$pH = 7.75$$

17.89 $CH_3CH(OH)COO^-$ will be formed by reaction $CH_3CH(OH)COOH$ with NaOH. $0.1000\ M \times 0.02500\ L = 2.500 \times 10^{-3}$ mol $CH_3CH(OH)COOH$; b = mol NaOH needed

	$CH_3CH(OH)COOH$	+	$NaOH$	$\rightarrow$	$CH_3CH(OH)COO^- + H_2O + Na^+$
initial	2.500×10^{-3} mol		b mol		
rx	–b mol		–b mol		+b mol
after rx	$(2.500 \times 10^3 - b)$ mol		0		b mol

$$K_a = \frac{[H^+][CH_3CH(OH)COO^-]}{[CH_3CH(OH)COOH]}; K_a = 1.4 \times 10^{-4}; [H^+] = 10^{-pH} = 10^{-3.75} = 1.778 \times 10^{-4} = 1.8 \times 10^{-4} M$$

Since solution volume is the same for reaction $CH_3CH(OH)COOH$ and $CH_3CH(OH)COO^-$, we can use moles in the equation for $[H^+]$.

$$K_a = 1.4 \times 10^{-4} = \frac{1.778 \times 10^{-4}(b)}{(2.500 \times 10^{-3} - b)}; 0.7874(2.500 \times 10^{-3} - b) = b, 1.969 \times 10^{-3} = 1.7874\,b,$$

$$b = 1.10 \times 10^{-3} = 1.1 \times 10^{-3} \text{ mol OH}^-$$

(The precision of K_a dictates that the result has 2 sig figs.)

Substituting this result into the K_a expression gives $[H^+] = 1.8 \times 10^{-4}$. This checks and confirms our result. Calculate volume NaOH required from $M = \text{mol/L}$.

$$1.10 \times 10^{-3} \text{ mol OH}^- \times \frac{1\,L}{1.000\,\text{mol}} \times \frac{1\,\mu L}{1 \times 10^{-6}\,L} = 1.1 \times 10^3\,\mu L \ (1.1\,mL)$$

17.90   (a) $H^+(aq) + HCO_3^-(aq) \rightleftharpoons H_2CO_3(aq) \rightleftharpoons H_2O(l) + CO_2(g)$

A person breathing normally exhales $CO_2(g)$. Rapid breathing causes excess $CO_2(g)$ to be removed from the blood. By LeChatelier's principle, this causes both equilibria above to shift right, reducing $[H^+]$ in the blood and raising blood pH.

(b)   Breathing in a paper bag traps the exhaled CO_2; the gas in the bag contains more CO_2 than ambient air. When a person inhales gas from the bag, a greater amount (partial pressure) of $CO_2(g)$ in the lungs shifts the equilibria left, increasing $[H^+]$ and lowering blood pH.

17.92   Consider the effects of common ions and competing equilibria on the solubility equilibrium.

$$CaCO_3 (s) \rightleftharpoons Ca^{2+}(aq) + CO_3^{2-}(aq)$$

(a)   0.10 M NaCl, same. Neither Na^+ nor Cl^- are ions common to $CaCO_3$. To a first approximation, the solubility of $CaCO_3$ is the same in 0.010 M NaCl as in water. (Ionic strength, an expression of the total number of ions in a solution, does influence solubility somewhat.)

(b)   0.10 M $Ca(NO_3)_2$, less soluble. Ca^{2+} is common to $CaCO_3$. The $Ca^{2+}(aq)$ shifts the solubility equilibrium to the left and decreases the solubility of $CaCO_3$ relative to its solubility in water.

(c)   0.10 M Na_2CO_3, less soluble. CO_3^{2-} is common to $CaCO_3$. The $CO_3^{2-}(aq)$ shifts the solubility equilibrium to the left and decreases the solubility of $CaCO_3$ relative to pure water.

(d)   0.10 M HCl, more soluble. H^+ reacts with CO_3^{2-}, removing it from solution and shifting the solubility equilibrium to the right. $CaCO_3$ is more soluble in 0.10 M HCl (or any acid) than in water.

17.93   (a)   Hydroxyapatite: $K_{sp} = [Ca^{2+}]^5[PO_4^{3-}]^3[OH^-]$

Fluoroapatite: $K_{sp} = [Ca^{2+}]^5[PO_4^{3-}]^3[F^-]$

(b)   For each mole of apatite dissolved, one mole of OH^- or F^- is formed. Express molar solubility, s, in terms of $[OH^-]$ and $[F^-]$.

Hydroxyapatite: $[OH^-] = s$, $[Ca^{2+}] = 5s$, $[PO_4^{3-}] = 3s$

$K_{sp} = 6.8 \times 10^{-27} = (5s)^5 (3s)^3 (s) = 84{,}375\, s^9$

$s^9 = 8.059 \times 10^{-32} = 8.1 \times 10^{-32}$.

Use logs to find s. $s = 3.509 \times 10^{-4} = 3.5 \times 10^{-4}\, M\, Ca_5(PO_4)_3OH$.

Fluoroapatite: $[F^-] = s$, $[Ca^{2+}] = 5s$, $[PO_4^{3-}] = 3s$

$K_{sp} = 1.0 \times 10^{-60} = (5s)^5 (3s)^3 (s) = 84{,}375\, s^9$

$s^9 = 1.185 \times 10^{-65} = 1.2 \times 10^{-65}$; $s = 6.109 \times 10^{-8} = 6.1 \times 10^{-8}\, M\, Ca_5(PO_4)_3F$

17.95   Calcium oxide dissolves in and reacts with water to form calcium hydroxide.

$$CaO(s) + H_2O(l) \rightarrow Ca(OH)_2(aq)$$

$Ca(OH)_2$ provides $OH^-(aq)$ to precipitate $Mg(OH)_2(s)$, removing $Mg^{2+}(aq)$ from seawater. Consider this reaction in two steps.

$Ca(OH)_2(s) + \rightleftharpoons Ca^{2+}(aq) + 2OH^-(aq)$	$K_1 = K_{sp} = 6.5 \times 10^{-6}$
$Mg^{2+}(aq)\ 2OH^-(aq) \rightleftharpoons Mg(OH)_2(s)$	$K_2 = 1/K_{sp} = 1/1.8 \times 10^{-11}$
	$= 5.6 \times 10^{10}$

$Ca(OH)_2(s) + Mg^{2+}(aq) \rightleftharpoons Ca^{2+}(aq) + Mg(OH)_2(s)$ $K = K_1 \times K_2 = 3.6 \times 10^5$

$$K = \frac{[Ca^{2+}]}{[Mg^{2+}]} = 3.6 \times 10^5$$

The reaction goes to completion. A stoichiometric amount of CaO quantitatively removes $Mg^{2+}(aq)$ from seawater.

17.97 $[Ca^{2+}][CO_3^{2-}] = 4.5 \times 10^{-9}$; $[Fe^{2+}][CO_3^{2-}] = 2.1 \times 10^{-11}$

Since $[CO_3^{2-}]$ is the same for both equilibria:

$$[CO_3^{2-}] = \frac{4.5 \times 10^{-9}}{[Ca^{2+}]} = \frac{2.1 \times 10^{-11}}{[Fe^{2+}]}; \text{ rearranging } \frac{[Ca^{2+}]}{[Fe^{2+}]} = \frac{4.5 \times 10^{-9}}{2.1 \times 10^{-11}} = 214 = 2.1 \times 10^2$$

17.98 $PbSO_4(s) \rightleftharpoons Pb^{2+}(aq) + SO_4^{2-}(aq)$; $K_{sp} = 6.3 \times 10^{-7} = [Pb^{2+}][SO_4^{2-}]$

$SrSO_4(s) \rightleftharpoons Sr^{2+}(aq) + SO_4^{2-}(aq)$; $K_{sp} = 3.2 \times 10^{-7} = [Sr^{2+}][SO_4^{2-}]$

Let $x = [Pb^{2+}]$, $y = [Sr^{2+}]$, $x + y = [SO_4^{2-}]$

$$\frac{x(x+y)}{y(x+y)} = \frac{6.3 \times 10^{-7}}{3.2 \times 10^{-7}}; \frac{x}{y} = 1.9688 = 2.0; x = 1.969\, y = 2.0\, y$$

$y(1.969\, y + y) = 3.2 \times 10^{-7}$; $2.969\, y^2 = 3.2 \times 10^{-7}$; $y = 3.283 \times 10^{-4} = 3.3 \times 10^{-4}$

$x = 1.969\,y; \ x = 1.969(3.283 \times 10^{-4}) = 6.464 \times 10^{-4} = 6.5 \times 10^{-4}$

$[Pb^{2+}] = 6.5 \times 10^{-4}\,M, \ [Sr^{2+}] = 3.3 \times 10^{-4}\,M, \ [SO_4^{2-}] = (3.283 + 6.464) \times 10^{-4} = 9.7 \times 10^{-4}\,M$

17.100 The student failed to account for the hydrolysis of the AsO_4^{3-} ion. If there were no hydrolysis, $[Mg^{2+}]$ would indeed be 1.5 times that of $[AsO_4^{3-}]$. However, as the reaction $AsO_4^{3-}(aq) + H_2O(l) \rightleftharpoons HAsO_4^{2-}(aq) + OH^-(aq)$ proceeds, the ion product $[Mg^{2+}]^3[AsO_4^{3-}]^2$ falls below the value for K_{sp}. More $Mg_3(AsO_4)_2$ dissolves, more hydrolysis occurs, and so on, until an equilibrium is reached. At this point $[Mg^{2+}]$ in solution is much greater than 1.5 times free $[AsO_4^{3-}]$. However, it is exactly 1.5 times the **total** concentration of all arsenic-containing species. That is,

$[Mg^{2+}] = 1.5\,([AsO_4^{3-}] + [HAsO_4^{2-}] + [H_2AsO_4^-] + [H_3AsO_4])$

17.101
$$Zn(OH)_2(s) \rightleftharpoons Zn^{2+}(aq) + 2OH^-(aq) \qquad K_{sp} = 3.0 \times 10^{-16}$$
$$Zn^{2+}(aq) + 4OH^-(aq) \rightleftharpoons Zn(OH)_4^{2-}(aq) \qquad K_f = 4.6 \times 10^{17}$$

$$Zn(OH)_2(s) + 2OH^-(aq) \rightleftharpoons Zn(OH)_4^{2-}(aq) \qquad K = K_{sp} \times K_f = 138 = 1.4 \times 10^2$$

$$K = 138 = 1.4 \times 10^2 = \frac{[Zn(OH)_4^{2-}]}{[OH^-]^2}$$

If 0.015 mol $Zn(OH)_2$ dissolves, 0.015 mol $Zn(OH)_4^{2-}$ should be present at equilibrium.

$$[OH^-]^2 = \frac{(0.015)}{138}; [OH^-] = 1.043 \times 10^{-2}\,M; \ [OH^-] \geq 1.0 \times 10^{-2}\,M \ \text{or pH} \geq 12.02$$

Integrative Exercises

17.103 (a) For a monoprotic acid (one H^+ per mole of acid), at the equivalence point

moles OH^- added = moles H^+ originally present

$M_B \times V_B$ = g acid/molar mass

$$MM = \frac{\text{g acid}}{M_B \times V_B} = \frac{0.1044\,\text{g}}{0.0500\,M \times 0.02210\,\text{L}} = 94.48 = 94.5\,\text{g/mol}$$

(b) 11.05 mL is exactly half-way to the equivalence point (22.10 mL). When half of the unknown acid is neutralized, $[HA] = [A^-]$, $[H^+] = K_a$ and pH = pK_a.

$K_a = 10^{-4.89} = 1.3 \times 10^{-5}$

(c) From Appendix D, Table D.1, acids with K_a values close to 1.3×10^{-5} are

name	K_a	formula	molar mass
propionic	1.3×10^{-5}	C_2H_5COOH	74.1
butanoic	1.5×10^{-5}	C_3H_7COOH	88.1
acetic	1.8×10^{-5}	CH_3COOH	60.1
hydroazoic	1.9×10^{-5}	HN_3	43.0

Of these, butanoic has the closest match for K_a and molar mass, but the agreement is not good.

17.104 $n = \dfrac{PV}{RT} = 735 \text{ torr} \times \dfrac{1 \text{ atm}}{760 \text{ torr}} \times \dfrac{7.5 \text{ L}}{295 \text{ K}} \times \dfrac{\text{K-mol}}{0.08206 \text{ L-atm}} = 0.300 = 0.30 \text{ mol NH}_3$

$0.40 \, M \times 0.50 \text{ L} = 0.20 \text{ mol HCl}$

	HCl(aq) +	NH$_3$(g) $\rightarrow$	NH$_4^+$(aq) +	Cl$^-$(aq)
Before	0.20 mol	0.30 mol		
after	0	0.10 mol	0.20 mol	0.20 mol

The solution will be a buffer because of the substantial concentrations of NH$_3$ and NH$_4^+$ present. Use K$_a$ for NH$_4^+$ to describe the equilibrium.

$$\text{NH}_4^+(aq) \rightleftharpoons \text{NH}_3(aq) + \text{H}^+(aq)$$

equil. $0.20 - x$ $0.10 + x$ x

$K_a = \dfrac{1.0 \times 10^{-14}}{1.8 \times 10^{-5}} = 5.56 \times 10^{-10} = 5.6 \times 10^{-10}; \ K_a = \dfrac{[\text{NH}_3][\text{H}^+]}{[\text{NH}_4^+]}; [\text{H}^+] = \dfrac{K_a[\text{NH}_4^+]}{[\text{NH}_3]}$

Since this expression contains a ratio of concentrations, volume will cancel and we can substitute moles directly. Assume x is small compared to 0.10 and 0.20.

$[\text{H}^+] = \dfrac{5.56 \times 10^{-10} \, (0.20)}{(0.10)} = 1.111 \times 10^{-9} = 1.1 \times 10^{-9} \, M, \text{pH} = 8.95$

17.106 According to Equation [13.4], $S_g = kP_g$

$S_{CO_2} = 3.1 \times 10^{-2} \dfrac{\text{mol}}{\text{L-atm}} \times 1.10 \text{ atm} = 0.0341 = \dfrac{0.034 \text{ mol}}{\text{L}} = 0.034 \, M \, CO_2$

$CO_2(g) + H_2O(l) \rightarrow H_2CO_3(aq); \ 0.0341 \, M \, CO_2 = 0.0341 \, M \, H_2CO_3$

Consider the stepwise dissociation of H$_2$CO$_3$(aq).

	H$_2$CO$_3$(aq) $\rightleftharpoons$	H$^+$(aq) +	HCO$_3^-$(aq)
initial	0.0341 M	0	0
equil.	(0.0341-x) M	x	x

$K_{a1} = \dfrac{[\text{H}^+][\text{HCO}_3^-]}{[\text{H}_2\text{CO}_3]} = \dfrac{x^2}{(0.0341 - x)} \approx \dfrac{x^2}{0.0341} \approx 4.3 \times 10^{-7}$

$x^2 = 1.47 \times 10^{-8}; \ x = 1.2 \times 10^{-4} \, M \, \text{H}^+; \text{pH} = 3.92$

$K_{a2} = 5.6 \times 10^{-11}$; assume the second ionization does not contribute significantly to [H$^+$].

17.107 $Ca(OH)_2(aq) + 2HCl(aq) \rightarrow CaCl_2(aq) + 2H_2O$

mmol HCl = $M \times$ mL = 0.0983 $M \times$ 11.23 mL = 1.1039 = 1.10 mmol HCl

mmol Ca(OH)$_2$ = mmol HCl/2 = 1.1039/2 = 0.55195 = 0.552 mmol Ca(OH)$_2$

$[\text{Ca}^{2+}] = \dfrac{0.55195 \text{ mmol}}{50.00 \text{ mL}} = 0.01104 = 0.0110 \, M$

$[OH^-] = 2[Ca^{2+}] = 0.02208 = 0.0221\ M$

$K_{sp} = [Ca^{2+}][OH^-]^2 = (0.01104)(0.02208)2 = 5.38 \times 10^{-6}$

The value in Appendix D is 6.5×10^{-6}, a difference of 17%. Since a change in temperature does change the value of an equilibrium constant, the solution may not have been kept at 25 °C. It is also possible that experimental errors led to the difference in K_{sp} values.

17.109   For very dilute aqueous solutions, assume the solution density is 1 g/mL.

$$\text{ppb} = \frac{\text{g solute}}{10^9\ \text{g solution}} = \frac{1 \times 10^{-6}\ \text{g solute}}{1 \times 10^3\ \text{g solution}} = \frac{\mu\text{g solute}}{\text{L solution}}$$

(a) $K_{sp} = [Ag^+][Cl^-] = 1.8 \times 10^{-10}$; $[Ag^+] = (1.8 \times 10^{-10})^{1/2} = 1.34 \times 10^{-5} = 1.3 \times 10^{-5}\ M$

$$\frac{1.34 \times 10^{-5}\ \text{mol Ag}^+}{L} \times \frac{107.9\ \text{g Ag}^+}{1\ \text{mol Ag}^+} \times \frac{1\ \mu\text{g}}{1 \times 10^{-6}\ \text{g}} = \frac{1.4 \times 10^3\ \mu\text{g Ag}^+}{L}$$

$$= 1.4 \times 10^3\ \text{ppb} = 1.4\ \text{ppm}$$

(b) $K_{sp} = [Ag^+][Br^-] = 5.0 \times 10^{-13}$; $[Ag^+] = (5.0 \times 10^{-13})^{1/2} = 7.07 \times 10^{-7} = 7.1 \times 10^{-7}\ M$

$$\frac{7.07 \times 10^{-7}\ \text{mol Ag}^+}{L} \times \frac{107.9\ \text{g Ag}^+}{1\ \text{mol Ag}^+} \times \frac{1\ \mu\text{g}}{1 \times 10^{-6}\ \text{g}} = 76\ \text{ppb}$$

(c) $K_{sp} = [Ag^+][I^-] = 8.3 \times 10^{-17}$; $[Ag^+] = (8.3 \times 10^{-17})^{1/2} = 9.11 \times 10^{-9} = 9.1 \times 10^{-9}\ M$

$$\frac{9.11 \times 10^{-9}\ \text{mol Ag}^+}{L} \times \frac{107.9\ \text{g Ag}^+}{1\ \text{mol Ag}^+} \times \frac{1\ \mu\text{g}}{1 \times 10^{-6}\ \text{g}} = 0.98\ \text{ppb}$$

AgBr(s) would maintain $[Ag^+]$ in the correct range.

17.110   To determine precipitation conditions, we must know K_{sp} for $CaF_2(s)$ and calculate Q under the specified conditions. $K_{sp} = 3.9 \times 10^{-11} = [Ca^{2+}][F^-]^2$

$[Ca^{2+}]$ and $[F^-]$: The term 1 ppb means 1 part per billion or 1 g solute per billion g solution. Assuming that the density of this very dilute solution is the density of water:

$$1\ \text{ppb} = \frac{1\ \text{g solute}}{1 \times 10^9\ \text{g solution}} \times \frac{1\ \text{g solution}}{1\ \text{mL solution}} \times \frac{1 \times 10^3\ \text{mL}}{1\ L} = \frac{1 \times 10^{-6}\ \text{g solute}}{1\ \text{L solution}}$$

$$\frac{1 \times 10^{-6}\ \text{g solute}}{1\ \text{L solution}} \times \frac{1\ \mu\text{g}}{1 \times 10^{-6}\ \text{g}} = 1\ \mu\text{g}/1\ L$$

$$8\ \text{ppb Ca}^{2+} \times \frac{1\ \mu\text{g}}{1\ L} = \frac{8\ \mu\text{g Ca}^{2+}}{1\ L} = \frac{8 \times 10^{-6}\ \text{g Ca}^{2+}}{1\ L} \times \frac{1\ \text{mol Ca}^{2+}}{40\ \text{g}} = 2 \times 10^{-7}\ M\ \text{Ca}^{2+}$$

$$1\ \text{ppb F}^- \times \frac{1\ \mu\text{g}}{1\ L} = \frac{1\ \mu\text{g F}^-}{1\ L} = \frac{1 \times 10^{-6}\ \text{g F}^-}{1\ L} \times \frac{1\ \text{mol F}^-}{19.0\ \text{g}} = 5 \times 10^{-8}\ M\ \text{F}^-$$

$Q = [Ca^{2+}][F^-]^2 = (2 \times 10^{-7})(5 \times 10^{-8})^2 = 5 \times 10^{-22}$

$5 \times 10^{-22} < 3.9 \times 10^{-11}$, $Q < K_{sp}$, no CaF_2 will precipitate

18 Chemistry of the Environment

Visualizing Concepts

18.2 Molecules in the upper atmosphere tend to have multiple bonds because they have sufficiently high bond dissociation enthalpies (Table 8.4) to survive the incoming high energy radiation from the sun. According to Table 8.4, for the same two bonded atoms, multiple bonds have higher bond dissociation enthalpies than single bonds. Molecules with single bonds are likely to undergo photodissociation in the presence of the high energy, short wavelength solar radiation present in the upper atmosphere.

18.4 *Analyze.* Given granite, marble, bronze, and other solid materials, what observations and measurements indicate whether the material is appropriate for an outdoor sculpture? If the material changes (erodes) over time, what chemical processes are responsible?

Plan. An appropriate material resists chemical and physical changes when exposed to environmental conditions. An inappropriate material undergoes chemical reactions with substances in the troposphere, degrading the structural strength of the material and the sculpture. *Solve.*

(a) The appearance and mass of the material upon environmental exposure are both indicators of chemical and physical changes. If the appearance and mass of the material are unchanged after a period of time, the material is well-suited for the sculpture because it is inert to chemical and physical changes. Changes in the color or texture of the material's surface indicate that a chemical reaction has occurred, because a different substance with different properties has formed. A decrease in mass indicates that some of the material has been lost, either by chemical reaction or physical change. An increase in mass indicates corrosion. If the mass of the material is unchanged, it is probably inert to chemical and physical environmental changes and suitable for sculpture.

(b) The two main chemical processes that lead to erosion are reaction with acid rain and corrosion or air oxidation, which is encouraged by acid conditions (see Section 20.8).

Acid rain is primarily H_2SO_3 and/or H_2SO_4, which reacts directly with carbonate minerals such as marble and limestone. Acidic conditions created by acid rain encourage corrosion of metals such as iron, steel, and bronze. Corrosion produces metal oxides which may or may not cling to the surface of the material. If the oxides are washed away, the material will lose mass after corrosion. Physical erosion due to the effects of wind and rain on soft materials such as sandstone also causes mass to decrease.

18.6   *Analyze/Plan.* Explain how an ion-exchange column "softens" water. See the Closer Look box on "Water Softening" in Section 18.6.

Solve. The plastic beads in an ion-exchange column contain covalently bound anionic groups such as $R\text{–}COO^-$ and $R\text{–}SO_3^-$. These groups have Na^+ cations associated with them for charge balance. When "hard" water containing Ca^{2+} and other divalent cations passes over the beads, the 2+ cations are attracted to the anionic groups and Na^+ is displaced. The higher charge on the divalent cations leads to greater electrostatic attractions, which promote the cation exchange. The "soft" water that comes out of the column contains two Na^+ ions in place of each divalent cation, mostly Ca^{2+} and Mg^{2+}, that remains in the column associated with the plastic beads.

18.8   Some of the missing CO_2 is absorbed by "land plants" (vegetation other than trees) and incorporated into the soil. Soil is the largest land-based carbon reservoir. The amount of carbon-storing capacity of soil is affected by erosion, soil fertility, and other complex factors. For more details, search the internet for "carbon budget."

Earth's Atmosphere

18.10   (a)   Boundaries between regions of the atmosphere are at maxima and minima (peaks and valleys) in the atmospheric temperature profile. For example, in the troposphere, temperature decreases with altitude, while in the stratosphere, it increases with altitude. The temperature minimum is the tropopause boundary.

(b)   Atmospheric pressure in the troposphere ranges from 1.0 atm to 0.4 atm, while pressure in the stratosphere ranges from 0.4 atm to 0.001 atm. Gas density (g/L) is directly proportional to pressure. The much lower density of the stratosphere means it has the smaller mass, despite having a larger volume than the troposphere.

18.12 $P_{Ar} = \chi_{Ar} \times P_{atm}$; $P_{Ar} = 0.00934\,(96.5\text{ kPa}) = 0.901\text{ kPa}$; $0.901\text{ kPa} \times \dfrac{760\,\text{torr}}{101.325\text{ kPa}} = 6.76\text{ torr}$

$P_{CO_2} = \chi_{CO_2} \times P_{atm}$; $P_{CO_2} = 0.000382\,(96.5\text{ kPa}) = 0.0369\text{ kPa}$; $0.0369\text{ kPa} \times \dfrac{760\text{ torr}}{101.325\text{ kPa}}$
$$= 0.276\text{ torr}$$

18.14   (a)   ppm Ne = mol Ne/1×10^6 mol air; $\chi_{Ne} = 1.818 \times 10^{-5}$ mol Ne/mol air

$$\frac{1.818 \times 10^{-5}\text{ mol Ne}}{1\text{ mol air}} = \frac{x\text{ mol Ne}}{1 \times 10^6\text{ mol air}}; x = 18.18\text{ ppm Ne}$$

(b) $P_{Ne} = \chi_{Ne} \times P_{atm} = 1.818 \times 10^{-5} \times 733\text{ torr} \times \dfrac{1\text{ atm}}{760\text{ torr}} = 1.7534 \times 10^{-5}$

$$= 1.75 \times 10^{-5}\text{ atm}$$

T = 292 K

$$\frac{n_{Ne}}{V} = \frac{P_{Ne}}{RT} = \frac{1.7534 \times 10^{-5}\text{ atm}}{292\text{ K}} \times \frac{K\text{-mol}}{0.08206\text{ L-atm}} = 7.3176 \times 10^{-7} = 7.32 \times 10^{-7}\text{ mol/L}$$

$$\frac{7.3176 \times 10^{-7} \text{ mol Ne}}{L} \times \frac{6.022 \times 10^{23} \text{ atoms}}{\text{mol}} = 4.4067 \times 10^{17}$$

$$= 4.41 \times 10^{17} \text{ Ne atoms/L}$$

The Upper Atmosphere; Ozone

18.16 $\quad \dfrac{339 \times 10^3 \text{ J}}{1 \text{ mol}} \times \dfrac{1 \text{ mol}}{6.022 \times 10^{23} \text{ molecules}} = 5.6294 \times 10^{-19} = 5.63 \times 10^{-19} \text{ J/molecule}$

$$\lambda = \frac{hc}{E} = \frac{(6.626 \times 10^{-34} \text{ J-sec})(3.00 \times 10^8 \text{ m/sec})}{5.6294 \times 10^{-19} \text{ J}} = 3.53 \times 10^{-7} \text{ m} = 353 \text{ nm}$$

$$\frac{293 \times 10^3 \text{ J}}{1 \text{ mol}} \times \frac{1 \text{ mol}}{6.022 \times 10^{23} \text{ molecules}} = 4.8655 \times 10^{-19} = 4.87 \times 10^{-19} \text{ J/molecule}$$

$$\lambda = \frac{(6.626 \times 10^{-34} \text{ J-sec})(3.00 \times 10^8 \text{ m/sec})}{4.8655 \times 10^{-19} \text{ J}} = 4.09 \times 10^{-7} \text{ m} = 409 \text{ nm}$$

Photons of wavelengths longer than 409 nm cannot cause rupture of the C–Cl bond in either CF_3Cl or CCl_4. Photons with wavelengths between 409 and 353 nm can cause C–Cl bond rupture in CCl_4, but not in CF_3Cl.

18.18   Photodissociation of N_2 is relatively unimportant compared to photodissociation of O_2 for two reasons. The bond dissociation energy of N_2, 941 kJ/mol, is much higher than that of O_2, 495 kJ/mol. Photons with a wavelength short enough to photodissociate N_2 are not as abundant as the ultraviolet photons that lead to photodissociation of O_2. Also, N_2 does not absorb these photons as readily as O_2 so even if a short-wavelength photon is available, it may not be absorbed by an N_2 molecule.

18.20   32 e^-, 16 e^- pr

$$\begin{array}{c} :\!\ddot{F}\!: \\ | \\ :\!\ddot{C}l\!-\!C\!-\!\ddot{C}l\!: \\ | \\ :\!\ddot{C}l\!: \end{array}$$

CFC–11, $CFCl_3$, contains C–Cl bonds that can be cleaved by UV light in the stratosphere to produce Cl atoms. It is chlorine in atomic form that catalyzes the destruction of stratospheric ozone. CFC–11 is chemically inert and resists decomposition in the troposphere, so that it eventually reaches the stratosphere in molecular form.

18.22   Yes. Assuming $CFBr_3$ reaches the stratosphere intact, it contains C–Br bonds that are even more susceptible to cleavage by UV light than C–Cl bonds. According to Table 8.4, the average C–Br bond dissociation energy is 276 kJ/mol, compared to 328 kJ/mol for C–Cl bonds. Once in atomic form, Br atoms catalyze the destruction of ozone by a mechanism similar to that of Cl atoms.

Chemistry of the Troposphere

18.24   Rainwater is naturally acidic due to the presence of $CO_2(g)$ in the atmosphere. All oxides of nonmetals produce acidic solutions when dissolved in water. Even in the absence of polluting gases such as SO_2, SO_3, NO, and NO_2, CO_2 causes rainwater to be acidic. The important equilibria are:

$$CO_2(g) + H_2O(l) \rightleftharpoons H_2CO_3(aq) \rightleftharpoons H^+(aq) + HCO_3^-(aq).$$

18.26   (a) $Fe(s) + O_2(g) + 4H_3O^+(aq) \rightarrow Fe^{2+}(aq) + 6H_2O(l)$

(b)   No. Silver is a "noble" metal. It is relatively resistant to oxidation, and much more resistant than iron. In Table 4.5, The Activity Series of Metals in Aqueous Solution, Ag is much, much lower than Fe and it is below hydrogen, while Fe is above hydrogen. This means that Fe is susceptible to oxidation by acid, while Ag is not.

18.28   (a)   Visible (Figure 6.4)

(b) $E_{photon} = hc/\lambda = \dfrac{6.626 \times 10^{-34} \text{ J-s} \times 3.00 \times 10^8 \text{ m/s}}{420 \times 10^{-9} \text{ m}} = 4.733 \times 10^{-19}$

$$= 4.73 \times 10^{-19} \text{ J/photon}$$

$$\dfrac{4.733 \times 10^{-19} \text{ J}}{1 \text{ photon}} \times \dfrac{6.022 \times 10^{23} \text{ photons}}{1 \text{ mol}} \times \dfrac{1 \text{ kJ}}{1000 \text{ J}} = 285 \text{ kJ/mol}$$

(c) $\ddot{\underset{..}{O}} = \dot{N} - \ddot{\underset{..}{O}}: + h\nu \longrightarrow \ddot{\underset{..}{O}} = \ddot{N}\cdot + :\ddot{\underset{..}{O}}\cdot$

18.30   (a)   A *greenhouse gas* absorbs energy in the 10,000–30,000 nm or infrared region. It absorbs wavelengths of radiation emitted by earth and returns it as heat. A non-greenhouse gas is transparent to radiation in this wavelength range.

(b) $Ar(g)$ is monatomic, while $CH_4(g)$ contains 4 C–H bonds. Infrared radiation has insufficient energy to cause electron transitions or bond cleavage; but it has an appropriate amount of energy to cause molecular deformations, bond stretching, and angle bending. Monatomic gases such as Ar cannot "use" infrared radiation and are transparent to it.

The World Ocean

18.32   If the phosphorous is present as phosphate, there is a 1:1 ratio between the molarity of phosphorus and molarity of phosphate. Thus, we can calculate the molarity based on the given mass of P.

$$\dfrac{0.07 \text{ g P}}{1 \times 10^6 \text{ g H}_2\text{O}} \times \dfrac{1 \text{ mol P}}{31 \text{ g P}} \times \dfrac{1 \text{ mol PO}_4^{3-}}{1 \text{ mol P}} \times \dfrac{1 \times 10^3 \text{ g H}_2\text{O}}{1 \text{ L H}_2\text{O}} = 2.26 \times 10^{-6} = 2 \times 10^{-6} \text{ } M \text{ PO}_4^{3-}$$

18.34   0.05 ppb Au = 0.05 g Au/1×10^9 g seawater

$$\$1,000,000 \times \dfrac{1 \text{ oz Au}}{\$800} \times \dfrac{1 \text{ lb}}{16 \text{ oz}} \times \dfrac{453.6 \text{ g}}{1 \text{ lb}} = 3.5438 \times 10^4 \text{ g} = 3.54 \times 10^4 \text{ g Au needed}$$

$$3.5438 \times 10^4 \text{ g Au} \times \frac{1 \times 10^9 \text{ g seawater}}{0.05 \text{ g Au}} \times \frac{1 \text{ mL seawater}}{1.03 \text{ g seawater}} \times \frac{1 \text{ L}}{1000 \text{ mL}} = 6.8811 \times 10^{11}$$

$$= 7 \times 10^{11} \text{ L seawater}$$

7×10^{11} L seawater are needed if the process is 100% efficient; since it is only 50% efficient, twice as much seawater is needed.

$6.8811 \times 10^{11} \times 2 = 1.3762 \times 10^{12} = 1 \times 10^{12}$ L seawater

Note that the 1 sig fig in 0.05 ppb Au limits the precision of the calculation.

18.36   Calculate the total ion concentration of sea water by summing the molarities given in Table 18.6. Then use $\Pi = \Delta MRT$ to calculate pressure.

$$M_{total} = 0.55 + 0.47 + 0.028 + 0.054 + 0.010 + 0.010 + 2.3 \times 10^{-3} + 8.3 \times 10^{-4}$$

$$+ 4.3 \times 10^{-4} + 9.1 \times 10^{-5} + 7.0 \times 10^{-5} = 1.1257 = 1.13 \text{ } M$$

$$\Pi = \frac{(1.1257 - 0.02) \text{ mol}}{L} \times \frac{0.08206 \text{ L} \times \text{atm}}{\text{mol} \cdot \text{K}} \times 297 \text{ K} = 26.948 = 26.9 \text{ atm}$$

Check. The largest numbers in the molarity sum have 2 decimal places, so M_{total} has 2 decimal places and 3 sig figs. ΔM also has 2 decimal places and 3 sig figs so the calculated pressure has 3 sig figs. Units are correct.

Fresh Water

18.38   (a)   Decomposition of organic matter by aerobic bacteria depletes dissolved O_2. A low dissolved oxygen concentration indicates the presence of organic pollutants.

         (b)   According to Section 13.3, the solubility of $O_2(g)$ (or any gas) in water decreases with increasing temperature.

18.40 $120,000 \text{ persons} \times \dfrac{59 \text{ g } O_2}{1 \text{ person}} \times \dfrac{1 \times 10^6 \text{ g } H_2O}{9 \text{ g } O_2} \times \dfrac{1 \text{ L } H_2O}{1 \times 10^3 \text{ g } H_2O} = 7.9 \times 10^8 = 8 \times 10^8 \text{ L } H_2O$

18.42   (a) $Ca^{2+}, Mg^{2+}, Fe^{2+}$

         (b)   Divalent cations (ions with 2+ charges) contribute to water hardness. These ions react with soap to form scum on surfaces or leave undesirable deposits on surfaces, particularly inside pipes, upon heating.

18.44 $Ca(OH)_2$ is added to remove Ca^{2+} as $CaCO_3(s)$, and Na_2CO_3 removes the remaining Ca^{2+}.

         $Ca^{+2}(aq) + 2HCO_3^-(aq) + [Ca^{2+}(aq) + 2OH^-(aq)] \rightarrow 2CaCO_3(s) + 2H_2O(l)$.

One mole $Ca(OH)_2$ is needed for each 2 moles of $HCO_3^-(aq)$ present.

$$5.0 \times 10^7 \text{ L } H_2O \times \frac{1.7 \times 10^{-3} \text{ mol } HCO_3^-}{1 \text{ L } H_2O} \times \frac{1 \text{ mol } Ca(OH)_2}{2 \text{ mol } HCO_3^-} \times \frac{74 \text{ g } Ca(OH)_2}{1 \text{ mol } Ca(OH)_2}$$

$$= 3.1 \times 10^6 \text{ g } Ca(OH)_2$$

Half of the native HCO_3^- precipitates the added Ca^{2+} so this operation reduces the Ca^{2+} concentration from 5.7×10^{-3} M to $(5.7 \times 10^{-3} - 8.5 \times 10^{-4})$ $M = 4.85 \times 10^{-3} = 4.9 \times 10^{-3}$ M. Next we must add sufficient Na_2CO_3 to further reduce $[Ca^{2+}]$ to

$1.1 \times 10^{-3} M$ (20% of the original $[Ca^{2+}]$). We thus need to reduce $[Ca^{2+}]$ by $(4.85 \times 10^{-3} - 1.1 \times 10^{-3}) M = 3.75 \times 10^{-3} = 3.8 \times 10^{-3} M$

$Ca^{2+}(aq) + CO_3^{-2}(aq) \rightarrow CaCO_3(s)$.

$$5.0 \times 10^7 \text{ L H}_2\text{O} \times \frac{3.75 \times 10^{-3} \text{ mol Ca}^{2+}}{1 \text{ L H}_2\text{O}} \times \frac{1 \text{ mol Na}_2\text{CO}_3}{1 \text{ mol Ca}^{2+}} \times \frac{106 \text{ g Na}_2\text{CO}_3}{1 \text{ mol Na}_2\text{CO}_3}$$
$$= 2.0 \times 10^7 \text{ g Na}_2\text{CO}_3$$

18.46　$Al_2(SO_4)_3$ is a typical coagulant in municipal water purification. It reacts with OH^- in a slightly basic solution to form a gelatinous precipitate that occludes very small particles and bacteria. The precipitate settles slowly and is removed by sand filtration.

Properties of $Al_2(SO_4)_3$ and other useful coagulants are:

- They react with low concentrations of $OH^-(aq)$. That is, K_{sp} of the hydroxide precipitate is very small. The capacity to form a hydroxide precipitates means that no extra salts must be added to form the precipitate. Also, the $[OH^-]$ can be easily adjusted by $Ca(OH)_2$ and other reagents that are part of the purification process.

- The hydroxide precipitate is composed of very small, evenly dispersed particles that do not settle quickly. This is required to remove very small bacteria and viruses from all parts of the liquid, not just the sites of solid formation.

Green Chemistry

18.48　Catalysts increase the rate of a reaction by lowering activation energy, E_a. For an uncatalyzed reaction that requires extreme temperatures and pressures to generate product at a viable rate, finding a suitable catalyst reduces the required temperature and/or pressure, which reduces the amount of energy used to run the process. A catalyst can also increase rate of production, which would reduce the net time and thus energy required to generate a certain amount of product.

18.50　- In either solvent, the reaction is catalyzed, which usually leads to decreased processing temperatures and times, and greater energy efficiency.

- $scCO_2$ is the preferred solvent. It achieves maximum conversion much faster than CH_2Cl_2 solvent. $scCO_2$ reduces processing time, temperature, and energy requirements. It also results in fewer unwanted by-products to be separated and processed. While use of $scCO_2$ increases the amount of a greenhouse gas released to the environment, it eliminates use of CH_2Cl_2, which is implicated in stratospheric ozone depletion. Use of $scCO_2$ rather than CH_2Cl_2 is a good green trade-off.

Additional Exercises

18.52　MM_{avg} at the surface $= 83.8(0.17) + 16.0(0.38) + 32.0(0.45) = 34.73 = 35$ g/mol.

Next, calculate the percentage composition at 200 km. The fractions can be "normalized" by saying that the 0.45 fraction of O_2 is converted into **two** 0.45 fractions of O atoms, then dividing by the total fractions, $0.17 + 0.38 + 0.45 + 0.45 = 1.45$:

$$MM_{avg} = \frac{83.8(0.17) + 16.0(0.38) + 16.0(0.90)}{1.45} = 23.95 = 24 \text{ g/mol}$$

18.54

$$2[Cl(g) + O_3(g) \rightarrow ClO(g) + O_2(g)] \qquad [18.7]$$
$$2Cl(g) + 2O_3(g) \rightarrow 2ClO(g) + 2O_2(g)$$
$$\underline{2ClO(g) \qquad \rightarrow O_2(g) + 2Cl(g) \qquad\qquad [18.9]}$$
$$2Cl(g) + 2O_3(g) + 2ClO(g) \rightarrow 2ClO(g) + 3O_2(g) + 2Cl(g)$$
$$2O_3(g) \xrightarrow{\text{Cl}} 3O_2(g) \qquad\qquad [18.10]$$

Note that $Cl(g)$ fits the definition of a catalyst in this reaction.

18.55 Chlorofluorocarbons (CFCs), primarily $CFCl_3$ and CF_2Cl_2, are chemically inert and water insoluble. These properties make them valuable as propellants, refrigerants and foaming agents because they are virtually unreactive in the *troposphere* (lower atmosphere) and do not initiate or propagate undesirable reactions. Further, they are water-insoluble and not removed from the atmosphere by rain; they do not end up in the fresh water supply.

These properties render CFCs a long-term problem in the *stratosphere*. Because CFCs are inert and water-insoluble, they are not removed from the troposphere by reaction or dissolution and have very long lifetimes. Virtually the entire mass of released CFCs eventually diffuses into the stratosphere where conditions are right for photo-dissociation and the production of Cl atoms. Cl atoms catalyze the destruction of ozone, O_3.

18.57 In an HFC, C–Cl bonds are replaced by C–F bonds. The bond dissociation enthalpy of a C–F bond is 485 kJ/mol, much more than for a C–Cl bond, 328 kJ/mol (Table 8.4). Although HFCs have long lifetimes in the stratosphere, it is infrequent that light with energy sufficient to dissociate a C–F bond will reach an HFC molecule. F atoms, the bad actors in ozone destruction, are much less likely than Cl atoms to be produced by photodissociation in the stratosphere.

18.58 (a) $\cdot \ddot{O}$—H

(b) HNO_3 is a major component in acid rain.

(c) While it removes CO, the reaction produces NO_2. The photodissociation of NO_2 to form O atoms is the first step in the formation of tropospheric ozone and photochemical smog.

(d) Again, NO_2 is the initiator of photochemical smog. Also, methoxyl radical, OCH_3, is a reactive species capable of initiating other undesirable reactions.

18.60 Oxygen is present in the atmosphere to the extent of 209,000 parts per million. If CO binds 210 times more effectively than O_2, then the **effective** concentration of CO is 210×125 ppm = 26,250 = 26,300 ppm. The fraction of carboxyhemoglobin in the blood leaving the lungs is thus $\frac{26,250}{26,250 + 209,000} = 0.112$. Thus, 11.2 percent of the blood is in the form of carboxyhemoglobin, 88.8 percent as the O_2-bound oxyhemoglobin.

18.62　(a)　According to Section 13.3, the solubility of gases in water decreases with increasing temperature. Thus, the solubility of $CO_2(g)$ in the ocean would decrease if the temperature of the ocean increased.

(b)　If the solubility of $CO_2(g)$ in the ocean decreased because of global warming, more $CO_2(g)$ would be released into the atmosphere, perpetuating a cycle of increasing temperature and concomitant release of $CO_2(g)$ from the ocean.

18.63　Most of the 390 watts/m^2 radiated from Earth's surface is in the infrared region of the spectrum. Tropospheric gases, particularly $H_2O(g)$ and $CO_2(g)$, absorb much of this radiation and prevent it from escaping into space (Figure 18.12). The energy absorbed by these so-called "greenhouse gases" warms the atmosphere close to Earth's surface and makes the planet livable.

18.65　(a)　$NO(g) + hv \rightarrow N(g) + O(g)$

(b)　$NO(g) + hv \rightarrow NO^+(g) + e^-$

(c)　$NO(g) + O_3(g) \rightarrow NO_2(g) + O_2(g)$

(d)　$3NO_2(g) + H_2O(l) \rightarrow 2HNO_3(aq) + NO(g)$

18.66　(a)　CO_3^{2-} is a relatively strong Brønsted-Lowry base and produces OH^- in aqueous solution according to the hydrolysis reaction:

$$CO_3^{2-}(aq) + H_2O(l) \rightleftharpoons HCO_3^-(aq) + OH^-(aq), \quad K_b = 1.8 \times 10^{-4}$$

If $[OH^-(aq)]$ is sufficient for the reaction quotient, Q, to exceed K_{sp} for $Mg(OH)_2$, the solid will precipitate.

(b)　$$\frac{125 \text{ mg Mg}^{2+}}{1 \text{ kg soln}} \times \frac{1 \text{ g Mg}^{2+}}{1000 \text{ mg Mg}^{2+}} \times \frac{1.00 \text{ kg soln}}{1.00 \text{ L soln}} \times \frac{1 \text{ mol Mg}^{2+}}{24.305 \text{ g Mg}^{2+}} = 5.143 \times 10^{-3}$$

$$= 5.14 \times 10^{-3} M \text{ Mg}^{2+}$$

$$\frac{4.0 \text{ g Na}_2CO_3}{1.0 \text{ L soln}} \times \frac{1 \text{ mol CO}_3^{2-}}{106.0 \text{ g Na}_2CO_3} = 0.03774 = 0.038 M \text{ CO}_3^{2-}$$

$$K_b = 1.8 \times 10^{-4} = \frac{[HCO_3^-][OH^-]}{[CO_3^{2-}]} \approx \frac{x^2}{0.03774}; x = [OH^-] = 2.606 \times 10^{-3}$$

$$= 2.6 \times 10^{-3} M$$

(This represents 6.9% hydrolysis, but the result will not be significantly different using the quadratic formula.)

$Q = [Mg^{2+}][OH^-]^2 = (5.143 \times 10^{-3})(2.606 \times 10^{-3})^2 = 3.5 \times 10^{-8}$

K_{sp} for $Mg(OH)_2 = 1.6 \times 10^{-12}$; $Q > K_{sp}$, so $Mg(OH)_2$ will precipitate.

18.68　*Plan.* Calculate the volume of air above Los Angeles and the volume of pure O_3 that would be present at the 84 ppb level. For gases at the same temperature and pressure, volume fractions equal mole fractions. *Solve.*

$$V_{air} = 4000 \text{ mi}^2 \times \frac{(1.6093)^2 \text{ km}^2}{\text{mi}^2} \times \frac{(1000)^2 \text{ m}^2}{1 \text{ km}^2} \times 10 \text{ m} \times \frac{1 L}{1 \times 10^{-3} \text{ m}^3} = 1.036 \times 10^{14}$$

$$= 1.0 \times 10^{14} \text{ L air}$$

$$84 \text{ ppb O}_3 = \frac{84 \text{ mol O}_3}{1 \times 10^9 \text{ mol air}} = 8.4 \times 10^{-8} = \chi_{O_3}$$

V (pure O_3) = 8.4×10^{-8} (1.036×10^{14} L air) = 8.702×10^6 = 8.7×10^6 L O_3

Values for P and T are required to calculate mol O_3 from volume O_3, using the ideal-gas law. Since these are not specified in the exercise, we will make a reasonable assumption for a sunny April day in Los Angeles. The city is near sea level and temperatures are moderate throughout the year, so P = 1 atm and T = 25°C (78°F) are reasonable values.

PV = nRT, n = PV/RT

$$n = 1.000 \text{ atm} \times \frac{8.702 \times 10^6 \text{ L}}{298 \text{ K}} \times \frac{\text{K-mol}}{0.08206 \text{ L-atm}} = 3.558 \times 10^5 = 3.6 \times 10^5 \text{ mol } O_3$$

Check. Using known conditions to make reasonable estimates and assumptions is a valuable skill for problem solving. Knowing when assumptions are required is an important step in the learning process.

Integrative Exercises

18.70 (a) $8{,}376{,}726 \text{ tons coal} \times \dfrac{83 \text{ ton C}}{100 \text{ ton coal}} \times \dfrac{44.01 \text{ ton } CO_2}{12.01 \text{ ton C}} = 2.5 \times 10^7 \text{ ton } CO_2$

$8{,}376{,}726 \text{ tons coal} \times \dfrac{2.5 \text{ ton S}}{100 \text{ ton coal}} \times \dfrac{64.07 \text{ ton } SO_2}{32.07 \text{ ton S}} = 4.2 \times 10^5 \text{ ton } SO_2$

(b) $CaO(s) + SO_2(g) \rightarrow CaSO_3(s)$

$4.18 \times 10^5 \text{ ton } SO_2 \times \dfrac{55 \text{ ton } SO_2 \text{ removed}}{100 \text{ ton } SO_2 \text{ produced}} \times \dfrac{120.15 \text{ ton } CaSO_3}{64.07 \text{ ton } SO_2}$
$$= 4.3 \times 10^5 \text{ ton } CaSO_3$$

18.71 *Coarse sand* is removed by coarse sand filtration. *Finely divided particles* and some *bacteria* are removed by precipitation with aluminum hydroxide. Remaining *harmful bacteria* are removed by ozonation. *Trihalomethanes* are removed by either aeration or activated carbon filtration; use of activated carbon might be preferred because it does not involve release of TCMs into the atmosphere. *Dissolved organic substances* are oxidized (and rendered less harmful, but not removed) by both aeration and ozonation. Dissolved *nitrates* and *phosphates* are not removed by any of these processes, but are rendered less harmful by adequate aeration.

18.73 According to Equation [14.12], $\ln([A]_t / [A]_o) = -kt$. $[A]_t = 0.10 [A]_o$

$\ln(0.10 [A]_o / [A]_o) = \ln(0.10) = -(2 \times 10^{-6} \text{ s}^{-1}) t$

$t = -\ln(0.10) / 2 \times 10^{-6} \text{ s}^{-1} = 1.151 \times 10^6 \text{ s}$

$1.151 \times 10^6 \text{ s} \times \dfrac{1 \text{ min}}{60 \text{ s}} \times \dfrac{1 \text{ hr}}{60 \text{ min}} \times \dfrac{1 \text{ day}}{24 \text{ hr}} = 13.3 \text{ days } (1 \times 10 \text{ days})$

The value of the rate constant limits the result to 1 sig fig. This implies that there is minimum uncertainty of ± 1 in the tens place of our answer. Realistically, the remediation could take anywhere from 1 to 20 days.

18.74   (i) $ClO(g) + O_3(g) \rightarrow ClO_2(g) + O_2(g)$

$\Delta H_i = \Delta H_f^\circ \, ClO_2(g) + \Delta H_f^\circ \, O_2(g) - \Delta H_f^\circ \, ClO(g) - \Delta H_f^\circ \, O_3(g)$

$\Delta H_i = 102 + 0 - 101 - (142.3) = -141$ kJ

      (ii) $ClO_2(g) + O(g) \rightarrow ClO(g) + O_2(g)$

$\Delta H_{ii} = \Delta H_f^\circ \, ClO(g) + \Delta H_f^\circ \, O_2(g) - \Delta H_f^\circ \, ClO_2(g) - \Delta H_f^\circ \, O(g)$

$\Delta H_{ii} = 101 + 0 - 102 - (247.5) = -249$ kJ

(overall) $ClO(g) + O_3(g) + ClO_2(g) + O(g) \rightarrow ClO_2(g) + O_2(g) + ClO(g) + O_2(g)$

$$O_3(g) + O(g) \rightarrow 2O_2(g)$$

$\Delta H = \Delta H_i + \Delta H_{ii} = -141$ kJ $+ (-249)$ kJ $= -390$ kJ

Because the enthalpies of both (i) and (ii) are distinctly exothermic, it is possible that the $ClO - ClO_2$ pair could be a catalyst for the destruction of ozone.

18.76   (a)   A rate constant of $M^{-1}s^{-1}$ is indicative of a reaction that is second order overall. For the reaction given, the rate law is probably rate = $k[O][O_3]$. (Although rate = $k[O]^2$ or $k[O_3]^2$ are possibilities, it is difficult to envision a mechanism consistent with either one that would result in two molecules of O_2 being produced.)

      (b)   Yes. Most atmospheric processes are initiated by collision. One could imagine an activated complex of four O atoms collapsing to form two O_2 molecules. Also, the rate constant is large, which is less likely for a multistep process. The reaction is analogous to the destruction of O_3 by Cl atoms (Equation [18.7]), which is also second order with a large rate constant.

      (c)   According to the Arrhenius equation, $k = Ae^{-Ea/RT}$. Thus, the larger the value of k, the smaller the activation energy, E_a. The value of the rate constant for this reaction is large, so the activation energy is small.

      (d) $\Delta H^\circ = 2\Delta H_f^\circ \, O_2(g) - \Delta H_f^\circ \, O(g) - \Delta H_f^\circ \, O_3(g)$

$\Delta H^\circ = 0 - 247.5$ kJ $- 142.3$ kJ $= -389.8$ kJ

The reaction is exothermic, so energy is released; the reaction would raise the temperature of the stratosphere.

18.77   (a)   17 e^-, 8.5 e^- pairs

$$\ddot{O}\!=\!\dot{N}\!-\!\ddot{\underset{..}{O}}\colon \longleftrightarrow \colon\!\ddot{\underset{..}{O}}\!-\!\dot{N}\!=\!\ddot{O}$$

Owing to its lower electronegativity, N is more likely to be electron deficient and to accommodate the odd electron.

      (b)   The fact that NO_2 is an electron deficient molecule indicates that it will be highly reactive. Dimerization results in formation of a N—N single bond which completes the octet of both N atoms. NO_2 and N_2O_4 exist in equilibrium in a closed system. The reaction is exothermic, Equation [22.48]. In an urban environment, NO_2 is produced from hot automobile combustion. At these temperatures, equilibrium favors the monomer because the reaction is exothermic.

(c) $2NO_2(g) + 4CO(g) \rightarrow N_2(g) + 4CO_2(g)$

$NO_2(g) + CO(g) \rightarrow NO(g) + CO_2(g)$

NO_2 is an oxidizing agent and CO is a reducing agent, so we expect products to contain N in a more reduced form, NO or N_2, and C in a more oxidized form, CO_2.

(d) No. Because it is an odd-electron molecule, NO_2 is very reactive. We expect it to undergo chemical reactions or photodissociate before it can migrate to the stratosphere. The expected half-life of an NO_2 molecule is short.

18.79   rate = k[CF_3CH_2F][OH]. k = 1.6×10^8 M^{-1} s^{-1} at 4°C.

[CF_3CH_2F] = 6.3×10^8 molecules/cm³, [OH] = 8.1×10^5 molecules/cm³

Change molecules/cm³ to mol/L (M) and substitute into the rate law.

$$\frac{6.3 \times 10^8 \text{ molecules}}{cm^3} \times \frac{1 \text{ mol}}{6.022 \times 10^{23} \text{ molecules}} \times \frac{1000 \text{ cm}^3}{1 \text{ L}} =$$

$$1.0462 \times 10^{-12} = 1.0 \times 10^{-12} \, M \, CF_3CH_2F$$

$$\frac{8.1 \times 10^5 \text{ molecules}}{cm^3} \times \frac{1 \text{ mol}}{6.022 \times 10^{23} \text{ molecules}} \times \frac{1000 \text{ cm}^3}{1 \text{ L}} =$$

$$1.3451 \times 10^{-15} = 1.3 \times 10^{-15} \, M \, OH$$

$$\text{rate} = \frac{1.6 \times 10^8}{M\text{-s}} \times 1.0462 \times 10^{-12} \, M \times 1.3451 \times 10^{-15} \, M = 2.2515 \times 10^{-19} = 2.3 \times 10^{-19} \, M/s$$

18.80   (a) According to Table 18.1, the mole fraction of CO_2 in air is 0.000375.

$$P_{CO_2} = \chi_{CO_2} \times P_{atm} = 0.000375\,(1.00 \text{ atm}) = 3.75 \times 10^{-4} \text{ atm}$$

$$C_{CO_2} = kP_{CO_2} = 3.1 \times 10^{-2} \, M/atm \times 3.75 \times 10^{-4} \text{ atm} = 1.16 \times 10^{-5} = 1.2 \times 10^{-5} \, M$$

(b) H_2CO_3 is a weak acid, so the [H⁺] is regulated by the equilibria:

$H_2CO_3(aq) \rightleftharpoons H^+(aq) + HCO_3^-(aq)$ $K_{a1} = 4.3 \times 10^{-7}$

$HCO_3^-(aq) \rightleftharpoons H^+(aq) + CO_3^{2-}(aq)$ $K_{a2} = 5.6 \times 10^{-11}$

Since the value of K_{a2} is small compared to K_{a1}, we will assume that most of the $H^+(aq)$ is produced by the first dissociation.

$$K_{a1} = 4.3 \times 10^{-7} = \frac{[H^+][HCO_3^-]}{[H_2CO_3]}; [H^+] = [HCO_3^-] = x, [H_2CO_3] = 1.2 \times 10^{-5} - x$$

Since K_{a1} and [H_2CO_3] have similar values, we cannot assume x is small compared to 1.2×10^{-5}.

$$4.3 \times 10^{-7} = \frac{x^2}{(1.2 \times 10^{-5} - x)}; 5.00 \times 10^{-12} - 4.3 \times 10^{-7} \, x = x^2$$

$$0 = x^2 + 4.3 \times 10^{-7}x - 5.00 \times 10^{-12}$$

$$x = \frac{-4.3 \times 10^{-7} \pm \sqrt{(4.3 \times 10^{-7})^2 - 4(1)(-5.00 \times 10^{-12})}}{2(1)}$$

$$x = \frac{-4.3 \times 10^{-7} \pm \sqrt{1.85 \times 10^{-13} + 2.00 \times 10^{-11}}}{2} = \frac{-4.3 \times 10^{-7} \pm 4.49 \times 10^{-6}}{2}$$

The negative result is meaningless; $x = 2.03 \times 10^{-6} = 2.0 \times 10^{-6} \, M \, H^+$; pH = 5.69

Since this $[H^+]$ is quite small, the $[H^+]$ from the autoionization of water might be significant. Calculation shows that for $[H^+] = 2.0 \times 10^{-6} \, M$ from H_2CO_3, $[H^+]$ from $H_2O = 5.2 \times 10^{-9} \, M$, which we can ignore.

18.82   (a) $Al(OH)_3(s) \rightleftharpoons Al^{3+}(aq) + 3OH^-(aq)$ $K_{sp} = 1.3 \times 10^{-33} = [Al^{3+}][OH^-]^3$

This is a precipitation conditions problem. At what $[OH^-]$ (we can get pH from $[OH^-]$) will $Q = 1.3 \times 10^{-33}$, the requirement for the onset of precipitation?

$Q = 1.3 \times 10^{-33} = [Al^{3+}][OH^-]^3$. Find the molar concentration of $Al_2(SO_4)_3$ and thus $[Al^{3+}]$.

$$\frac{5.0 \text{ lb } Al_2(SO_4)_3}{2000 \text{ gal } H_2O} \times \frac{453.6 \text{ g}}{1 \text{ lb}} \times \frac{1 \text{ mol } Al_2(SO_4)_3}{342.2 \text{ g } Al_2(SO_4)_3} \times \frac{1 \text{ gal}}{4 \text{ qt}} \times \frac{1 \text{ qt}}{0.946 \text{ L}}$$

$$= 8.758 \times 10^{-4} \, M \, Al_2(SO_4)_3 = 1.752 \times 10^{-3} = 1.8 \times 10^{-3} \, M \, Al^{3+}$$

$Q = 1.3 \times 10^{-33} = (1.752 \times 10^{-3})[OH^-]^3$; $[OH^-]^3 = 7.42 \times 10^{-31}$

$[OH^-] = 9.054 \times 10^{-11} = 9.1 \times 10^{-11} \, M$; pOH = 10.04; pH = 14 – 10.04 = 3.96

(b) $CaO(s) + H_2O(l) \rightarrow Ca^{2+}(aq) + 2OH^-(aq)$; $[OH^-] = 9.054 \times 10^{-11}$ mol/L

$$\text{mol } OH^- = \frac{9.054 \times 10^{-11} \text{ mol}}{1 \text{ L}} \times 2000 \text{ gal} \times \frac{4 \text{ qt}}{1 \text{ gal}} \times \frac{0.946 \text{ L}}{1 \text{ qt}} = 6.852 \times 10^{-7}$$

$$= 6.9 \times 10^{-7} \text{ mol } OH^-$$

$$6.852 \times 10^{-7} \text{ mol } OH^- \times \frac{1 \text{ mol } CaO}{2 \text{ mol } OH^-} \times \frac{56.1 \text{ g } CaO}{1 \text{ mol } CaO} \times \frac{1 \text{ lb}}{453.6 \text{ g}} = 4.2 \times 10^{-8} \text{ lb } CaO$$

This is a **very** small amount of CaO, about 20 μg.

19 Chemical Thermodynamics

Visualizing Concepts

19.2 (a) The process depicted is a change of state from a solid to a gas. ΔS is positive because of the greater motional freedom of the particles. ΔH is positive because both melting and boiling are endothermic processes.

(b) Since $\Delta G = \Delta H - T\Delta S$, and both ΔH and ΔS are positive, the sign of ΔG depends on temperature. This is true for all phase changes. If the temperature of the system is greater than the boiling point of the substance, the process is spontaneous and ΔG is negative. If the temperature is lower than the boiling point, the process is not spontaneous and ΔG is positive.

(c) If the process is spontaneous, the second law states that $\Delta S_{univ} \geq 0$. Since ΔS_{sys} is positive, ΔS_{surr} must be negative. If the change occurs via a reversible pathway, $\Delta S_{univ} = 0$ and $\Delta S_{surr} = -\Delta S_{sys}$. If the pathway is irreversible, the magnitude of ΔS_{sys} is greater than the magnitude of ΔS_{surr}, but the sign of ΔS_{surr} is still negative.

19.3 In the depicted reaction, both reactants and products are in the gas phase (they are far apart and randomly placed). There are twice as many molecules (or moles) of gas in the products, so ΔS is positive for this reaction.

19.5 (a) At 300 K, $\Delta H = T\Delta S$. Since $\Delta G = \Delta H - T\Delta S$, $\Delta G = 0$ at this point. When $\Delta G = 0$, the system is at equilibrium.

(b) The reaction is spontaneous when ΔG is negative. This condition is met when $T\Delta S > \Delta H$. From the diagram, $T\Delta S > \Delta H$ when $T > 300$ K. The reaction is spontaneous at temperatures above 300 K.

19.7 (a) *Analyze.* The boxes depict three different mixtures of reactants and products for the reaction $A_2 + B_2 \rightleftharpoons 2AB$.

Plan. $K_c = 1 = \dfrac{[AB]^2}{[A][B]}$. Calculate Q for each box, using number of molecules as a measure of concentration. If $Q = 1$, the system is at equilibrium. *Solve.*

Box 1: $K = \dfrac{(3)^2}{(3)(3)} = 1$.

Box 2: $Q = \dfrac{(1)^2}{(4)(4)} = \dfrac{1}{16} = 0.0625 = 0.06$

Box 3: $Q = \dfrac{(7)^2}{(1)(1)} = \dfrac{49}{1} = 49$

Box 1 is at equilibrium.

(b) All nonequilibrium systems move spontaneously to achieve equilibrium. As the reaction is written, the sign of ΔG indicates the direction in which the reaction is spontaneous. If ΔG is negative, the reaction is spontaneous in the forward direction, toward products. This is the case for Box 2. If ΔG is positive, the reaction is spontaneous in the reverse direction, toward reactants. This is the case for Box 3.

(c) Qualitatively, Box 3 is farthest from equilibrium, so it has the largest magnitude of ΔG (driving force to reach equilibrium), then Box 2, then Box 1, where $\Delta G = 0$. Box 1 < Box 2 < Box 3.

Quantitatively, $\Delta G = \Delta G° - RT\ln Q$. For Box 1, $\Delta G = 0$ and $K = 1$, so $\Delta G° = 0$.

Box 2: $\Delta G = 0 - RT\ln(0.0625) = 2.77\ RT$

Box 3: $\Delta G = 0 - RT\ln(49) = -3.89\ RT$

Quantitative treatment confirms the order for magnitude of ΔG as Box 1 < Box 2 < Box 3.

19.8 (a) The minimum in the plot is the equilibrium position of the reaction, where $\Delta G = 0$.

(b) X is the difference in free energy between reactant and products in their standard states, $\Delta G°$.

Spontaneous Processes

19.10 (a) Spontaneous; at 1 atm, ice melts spontaneously at a temperature above its normal melting point.

(b) Nonspontaneous; a mixture cannot be separated without outside intervention.

(c) Spontaneous.

(d) Spontaneous; see Figure 8.2.

(e) Spontaneous; the very polar HCl molecules readily dissolve in water to form concentrated HCl(aq).

19.12 Yes. Bertholet postulated that all spontaneous processes are exothermic. The spontaneous endothermic reaction described in this exercise is one of many exception's to Bertholet's generalization.

19.14 (a) Exothermic. If melting requires heat and is endothermic, freezing must be exothermic.

(b) At 1 atm (indicated by the term "normal" freezing point), the freezing of 1-propanol is spontaneous at temperatures below −127°C.

(c) At 1 atm, the freezing of 1-propanol is nonspontaneous at temperatures above −127°C.

(d) At 1 atm and −127°C, the normal freezing point of 1-propanol, the solid and liquid phases are in equilibrium. That is, at the freezing point, 1-propanol molecules escape to the liquid phase at the same rate as liquid 1-propanol solidifies, assuming no heat is exchanged between 1-propanol and the surroundings.

19.16 (a) A process is *irreversible* if the system cannot be returned to its original state by the same path that the forward process took place.

(b) Since the system returned to its initial state via a different path (different q_r and w_r than q_f and w_f), there is a net change in the surroundings.

(c) The condensation of a liquid will be irreversible if it occurs at any temperature other than the boiling point of the liquid, at a specified pressure.

19.18 (a) $\Delta E (1 \rightarrow 2) = -\Delta E (2 \rightarrow 1)$

(b) We can say nothing about the values of q and w because we have no information about the paths.

(c) If the changes of state are reversible, the two paths are the same and $w (1 \rightarrow 2) = -w (2 \rightarrow 1)$. This is the maximum realizable work from this system.

19.20 (a) The detonation of an explosive is definitely spontaneous, once it is initiated.

(b) The quantity q is related to ΔH. Since the detonation is highly exothermic, q is large and negative.

If only PV-work is done and P is constant, $\Delta H = q$. Although these conditions probably do not apply to a detonation, we can still predict the sign of q, based on ΔH, if not its exact magnitude.

(c) The sign (and magnitude) of w depend on the path of the process, the exact details of how the detonation is carried out. It seems clear, however, that work will be done by the system on the surroundings in almost all circumstances (buildings collapse, earth and air are moved), so the sign of w is probably negative.

(d) $\Delta E = q + w$. If q and w are both negative, then the sign of ΔE is negative, regardless of the magnitudes of q and w.

Entropy and the Second Law of Thermodynamics

19.22 Both vaporizations are *isothermal*; they occur at constant temperature. For an isothermal process, : $\Delta S = q_{rev}/T$.

(a) Assuming that q_{rev} is closely related to enthalpy of vaporization and is about the same at the two temperatures, ΔS is larger at 25°C than at 100°C.

(b) No. Because ΔS is a state function, it is independent of path. We can calculate ΔS for a reversible pathway, even if the change does not occur that way.

19.24 (a) $Cs(l) \rightarrow Cs(s)$, ΔS is negative. [ΔH_{fus} is positive, but ΔH for freezing is negative.]

(b) $\Delta H = 15.0\,g\,Cs \times \dfrac{1\,mol\,Cs}{132.9\,g\,Cs} \times \dfrac{-2.09\,kJ}{mol\,Cs} = -0.2359 = -0.236\,kJ$

$\Delta S = \dfrac{\Delta H}{T} = -0.2359\,kJ \times \dfrac{1000\,J}{1\,kJ} \times \dfrac{1}{(273.15+28.4)K} = -0.782\,J/K$

19.26   (a)   For a spontaneous process, $\Delta S_{universe} > 0$. For a reversible process, $\Delta S_{universe} = 0$.

        (b) $\Delta S_{surroundings}$ is positive and greater than the magnitude of the decrease in ΔS_{system}.

        (c) $\Delta S_{system} = 78\,J/K$.

19.28   According to Boyle's law, $P_1 V_1 = P_2 V_2$ at constant n and T.

1.25 atm $\times V_1$ = 2.75 atm $\times V_2$; V_2/V_1 = 1.25 atm/2.75 atm = 0.45454 = 0.455

ΔS_{sys} = nR ln (V_2/V_1) = 0.850 mol (8.314 J/mol-K)(ln 0.45454) = -5.5719 = -5.57 J/K

Check. An increase in pressure results in a decrease in volume at constant T, so we expect ΔS to be negative, and it is.

The Molecular Interpretation of Entropy

19.30   (a) ΔH_{vap} for H_2O at 25°C = 44.02 kJ/mol; at 100°C = 40.67 kJ/mol

$\Delta S = \dfrac{q_{rev}}{T} = \dfrac{44.02\,kJ}{mol} \times \dfrac{1000\,J}{kJ} \times \dfrac{1}{298\,K} = 148\,J/mol\text{-}K$

$\Delta S = \dfrac{q_{rev}}{T} = \dfrac{40.67\,kJ}{mol} \times \dfrac{1000\,J}{kJ} \times \dfrac{1}{373\,K} = 109\,J/mol\text{-}K$

        (b)   At both temperatures, the liquid $\rightarrow$ gas phase transition is accompanied by an increase in entropy, as expected. That the magnitude of the increase is greater at the lower temperature requires some explanation.

        In the liquid state, there are significant hydrogen bonding interactions between H_2O molecules. This reduces the number of possible molecular positions and the number of microstates. Liquid water at 100° has sufficient kinetic energy to have broken many hydrogen bonds, so the number of microstates for $H_2O(l)$ at 100° is greater than the number of microstates for $H_2O(l)$ at 25°C. The difference in the number of microstates upon vaporization at 100°C is smaller, and the magnitude of ΔS is smaller.

19.32   (a)   Solids are much more ordered than gases, so ΔS is negative.

        (b)   The entropy of the system increases in Exercise 19.10 (a) and (e). There is more motional freedom for the system in both cases. In (b), (c) and (d), there is less motional freedom after the change and the entropy of the system decreases.

19.34   (a)   When temperature increases, the range of accessible molecular speeds and kinetic energies increases. This produces more microstates and an increase in entropy.

        (b)   When the volume of a gas increases (even at constant T), there are more possible positions for the particles, more microstates, and greater entropy.

(c) When equal volumes of two miscible liquids are mixed, the volume of the sample and therefore the number of possible arrangements increases. This produces more microstates and an increase in entropy.

19.36 (a) Since CO_2 has more than one atom, the thermal energy can be distributed as translational, vibrational, or rotational motion.

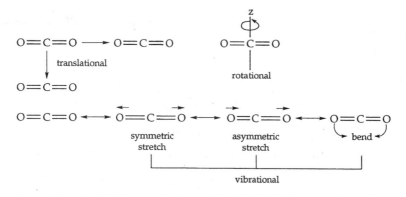

(b) According to Boltzmann's Law, $S = k \ln W$. The number of microstates, W, is directly proportional to entropy, S. Thus, if the number of microstates for a system increases, the entropy of the system increases.

19.38 Melting = –126.5°C; boiling = 97.4°C.

(graph with vertical axis labeled $S°$ and horizontal axis labeled T (°C) with markings at 200, 100, 0, −100, −200)

19.40 (a) 1 mol of $As_4(g)$ at 300°C, 0.01 atm (As_4 has more massive atoms in a comparable system at the same temperature.)

(b) 1 mol $H_2O(g)$ at 100°C, 1 atm (larger volume occupied by $H_2O(g)$)

(c) 0.5 mol $CH_4(g)$ at 298 K, 20-L volume (more complex molecule, more rotational and vibrational degrees of freedom)

(d) 100 g of $Na_2SO_4(aq)$ at 30°C (more motional freedom in aqueous solution)

19.42 (a) $Ar(g) \rightarrow Ar(l)$; negative ΔS, moles of gas decrease

(b) $N_2O_4(g) \rightarrow 2NO_2(g)$; positive ΔS, moles of gas increase

(c) $K(s) + O_2(g) \rightarrow KO_2(s)$; negative ΔS, moles of gas decrease

(d) $Pb(NO_3)_2(aq) + 2KBr(aq) \rightarrow PbBr_2(s) + 2KNO_3(aq)$; ΔS is negative, less motional freedom, fewer moles of ions in aqueous solution.

Entropy Changes in Chemical Reactions

19.44   Propylene will have a higher $S°$ at 25°C. At this temperature, both are gases, so there are no lattice effects. Since they have the same molecular formula, only the details of their structures are different. In propylene, there is free rotation around the C—C single bond, while in cyclopropane the 3-membered ring severely limits rotation. The greater motional freedom of the propylene molecule leads to a higher absolute entropy.

19.46   (a) $CuO(s)$, 42.59 J/mol-K; $Cu_2O(s)$, 92.36 J/mol-K. Molecules in the solid state have only vibrational motion available to them. The more complex Cu_2O molecule has more vibrational degrees of freedom and a larger standard entropy.

        (b)   1 mol $N_2O_4(g)$, 304.3 J/K; 2 mol $NO_2(g)$, 2(240.45) = 480.90 J/K. More particles have a greater number of arrangements.

        (c) $SiO_2(s)$, 41.84 J/mol-K; $CO_2(g)$, 213.6 J/mol-K. Molecules in the gas phase have a larger volume and more motional freedom than molecules in the solid state. SiO_2 is a covalent network solid so its molecular motion is even more restrained than a typical molecular solid.

        (d) $CO(g)$, 197.9 J/mol-K; $CO_2(g)$, 213.6 J/mol-K. The more complex CO_2 mole-cule has more vibrational degrees of freedom and a slightly higher entropy.

19.48   (a)   C(diamond), $S°$ = 2.43 J/mol-K; C(graphite), $S°$ = 5.69 J/mol-K. Diamond is a network covalent solid with each C atom tetrahedrally bound to four other C atoms. Graphite consists of sheets of fused planar 6-membered rings with each C atom bound in a trigonal planar arrangement to three other C atoms. The internal entropy in graphite is greater because there is translational freedom among the planar sheets of C atoms, while there is very little translational or vibrational freedom within the covalent-network diamond lattice.

        (b) $S°$ for buckminsterfullerene will be $\geq$ 10 J/mol-K. $S°$ for graphite is twice $S°$ for diamond, and $S°$ for the fullerene should be higher than that of graphite. The 60-atom "bucky" balls have more flexibility than graphite sheets. Also, the balls have translational freedom in three dimensions, while graphite sheets have it in only two directions. Because of the ball structure, there is more empty space in the fullerene lattice than in graphite or diamond; essentially, 60 C-atoms in fullerene occupy a larger volume than 60 C-atoms in graphite or diamond. Thus, the fullerene has additional "molecular" complexity, more degrees of translational freedom, and occupies a larger volume, all features that point to a higher absolute entropy.

19.50   (a) $\Delta S° = 2S° \, NH_3(g) - S° \, N_2H_4(g) - S° \, H_2(g)$

                = 2(192.5) - 238.5 - 130.58 = +15.9 J/K

        $\Delta S°$ is small because there are the same number of moles of gas in the products as in reactants. The slight increase is due to the relatively small $S°$ value of $H_2(g)$, which has fewer degrees of freedom than molecules with more than two atoms.

(b) $\Delta S° = S° \, KO_2(s) - S° \, K(s) - S° \, O_2(g)$

         $= 122.5 - 64.67 - 205.0 = -147.2 \, J/K$

$\Delta S°$ is negative because there are fewer moles of gas in the products.

(c) $\Delta S° = S° \, MgCl_2(s) + 2S° H_2O(l) - S° \, Mg(OH)_2(s) - 2S° \, HCl(g)$

         $= 89.6 + 2(69.91) - 63.24 - 2(186.69) = -207.2 \, J/K$

$\Delta S°$ is negative because the products contain fewer (no) moles of gas.

(d) $\Delta S° = S° \, CH_3OH(g) - S° \, CO(g) - 2S° \, H_2(g)$

         $= 237.6 - 197.9 - 2(130.58) = -221.5 \, J/K$

$\Delta S°$ is negative because there are fewer moles of gas in the products.

Gibbs Free Energy

19.52   (a)   The *standard* free energy change, $\Delta G°$, represents the free energy change for the process when all reactants and products are in their standard states. When any or all reactants or products are not in their standard states, the free energy is represented simply as ΔG. The value for ΔG thus depends on the specific states of all reactants and products.

        (b)   When $\Delta G = 0$, the system is at equilibrium.

        (c)   The sign and magnitude of ΔG give no information about rate; we cannot predict whether the reaction will occur rapidly.

19.54   (a) $\Delta H°$ is negative; the reaction is exothermic.

        (b) $\Delta S°$ is positive; the reaction leads to an increase in disorder.

        (c) $\Delta G° = \Delta H° - T\Delta S° = -19.5 \, kJ - 298 \, K \, (0.0427 \, kJ/K) = -32.225 = -32.2 \, kJ$

        (d)   At 298 K, $\Delta G°$ is negative. If all reactants and products are present in their standard states, the reaction is spontaneous (in the forward direction) at this temperature.

19.56   (a)   Data for $CrBr_3$ does not appear in Appendix C.

          Consider the reaction $4Cr(s) + 3O_2(g) \rightarrow 2Cr_2O_3(s)$.

          $\Delta H° = 2(-1139.7) - 4(0) + 3(0) = -2279.4 \, kJ$

          $\Delta S° = 2(81.2) - 4(23.6) - 3(205.0) = -547.0 \, J/K$

          $\Delta G° = 2(-1058.1) - 4(0) - 3(0) = -2116.2 \, kJ$

          $\Delta G° = -2279.4 \, kJ - 298 \, K(-0.5470 \, kJ/K) = -2116.4 \, kJ$

        (b) $\Delta H° = -553.5 - 393.5 - (-1216.3) = 269.3 \, kJ$

          $\Delta S° = 70.42 + 213.6 - 112.1 = 171.92 = 171.9 \, J/K$

          $\Delta G° = -525.1 - 394.4 - (-1137.6) = 218.1 \, kJ$

          $\Delta G° = 269.3 \, kJ - 298 \, K \, (0.1719 \, kJ/K) = 218.1 \, kJ$

(c)   Assume the reactant is $P(g)$, not $P(s)$.

$\Delta H° = 2(-1594.4) + 5(0) - 2(316.4) - 10(-268.61) = -1135.5$ kJ

$\Delta S° = 2(300.8) + 5(130.58) - 2(163.2) - 10(173.51) = -807.0$ J/K

$\Delta G° = 2(-1520.7) + 5(0) - 2(280.0) - 10(-270.70) = -894.4$ kJ

$\Delta G° = -1135.5$ kJ $- 298$ K$(-0.8070$ kJ/K$) = -895.014 = -895.0$ kJ

(The small discrepancy in $\Delta G°$ values is due to experimental uncertainties in tabulated thermodynamic data.)

(d) $\Delta H° = -284.5 - (0) - (0) = -284.5$ kJ

$\Delta S° = 122.5 - 64.67 - 205.0 = -147.2$ J/K

$\Delta G° = -240.6 - (0) - (0) = -240.6$ kJ

$\Delta G° = -284.5$ kJ $- 298$ K $(-0.1472$ kJ/K$) = -240.634 = -240.6$ kJ

19.58   (a) $\Delta G° = 2\Delta G°\ HCl(g) - [\Delta G°\ H_2(g) + \Delta G°\ Cl_2(g)]$

             $= 2(-95.27$ kJ$) - 0 - 0 = -190.5$ kJ, spontaneous

(b) $\Delta G° = \Delta G°\ MgO(s) + 2\Delta G°\ HCl(g) - [\Delta G°\ MgCl_2(s) + \Delta G°\ H_2O(l)]$

             $= -569.6 + 2(-95.27) - [-592.1 + (-237.13)] = +69.1$ kJ, nonspontaneous

(c) $\Delta G° = \Delta G°\ N_2H_4(g) + \Delta G°\ H_2(g) - 2\Delta G°\ NH_3(g)$

             $= 159.4 + 0 - 2(-16.66) = +192.7$ kJ, nonspontaneous

(d) $\Delta G° = 2\Delta G°\ NO(g) + \Delta G°\ Cl_2(g) - 2\Delta G°\ NOCl(g)$

             $= 2(86.71) + 0 - 2(66.3) = +40.8$ kJ, nonspontaneous

19.60   (a) $\Delta G°$ should be less negative than $\Delta H°$. Products contain fewer moles of gas, so $\Delta S°$ is negative. $\Delta G° = \Delta H° - T\Delta S°$; $-T\Delta S°$ is positive so $\Delta G°$ is less negative than $\Delta H°$.

(b)   We can estimate $\Delta S°$ using a similar reaction and then use $\Delta G° = \Delta H° - T\Delta S°$ (estimate) to get a ballpark figure. There are no sulfite salts listed in Appendix C, so use a reaction such as $CO_2(g) + CaO(s) \rightarrow CaCO_3(s)$ or $CO_2(g) + BaO(s) \rightarrow BaCO_3(s)$. Or calculate both $\Delta S°$ values and use the average as your estimate.

19.62 $\Delta G° = \Delta H° - T\Delta S°$

(a) $\Delta G° = -844$ kJ $- 298$ K$(-0.165$ kJ/K$) = -795$ kJ, spontaneous

(b) $\Delta G° = +572$ kJ $- 298$ K$(0.179$ kJ/K$) = +519$ kJ, nonspontaneous

To be spontaneous, ΔG must be negative ($\Delta G < 0$).

Thus, $\Delta H° - T\Delta S° < 0$; $\Delta H° < T\Delta S°$; $T > \Delta H°/\Delta S°$; $T > \dfrac{572\ \text{kJ}}{0.179\ \text{kJ/K}} = 3.20 \times 10^3$ K

19.64   At $-25°C$ or 248 K, $\Delta G > 0$. $\Delta G = \Delta H - T\Delta S > 0$

$\Delta H - 248$ K $(95$ J/K$) > 0$; $\Delta H > +2.4 \times 10^4$ J; $\Delta H > +24$ kJ

19.66 ΔG is negative when $T\Delta S > \Delta H$ or $T > \Delta H/\Delta S$.

$\Delta H^\circ = \Delta H^\circ\ CH_3OH + \Delta H^\circ\ CO(g) - \Delta H^\circ\ CH_3COOH(l)$

$= -201.2 - 110.5 - (-487.0) = 175.3\ kJ$

$\Delta S^\circ = S^\circ\ CH_3OH + S^\circ\ CO(g) - S^\circ\ CH_3COOH(l) = 237.6 + 197.9 - 159.8 = 275.7\ J/K$

$T > \dfrac{175.3\ kJ}{0.2757\ kJ/K} = 635.8\ K$

The reaction is spontaneous above 635.8 K (363°C)

19.68   (a) $\Delta H^\circ = \Delta H_f^\circ\ CH_3OH(g) - \Delta H_f^\circ\ CH_4(g) - 1/2\ \Delta H_f^\circ\ O_2(g)$

$= -201.2 - (-74.8) - (1/2)(0) = -126.4\ kJ$

$\Delta S^\circ = S^\circ\ CH_3OH(g) - S^\circ\ CH_4(g) - 1/2\ S^\circ\ O_2(g)$

$= 237.6 - 186.3 - 1/2(205.0) = -51.2\ J/K = -0.0512\ kJ/K$

(b) $\Delta G^\circ = \Delta H^\circ - T\Delta S^\circ$. $-T\Delta S^\circ$ is positive, so ΔG° becomes more positive as temperature increases.

(c) $\Delta G^\circ = \Delta H^\circ - T\Delta S^\circ = -126.4\ kJ - 298\ K(-0.0512\ kJ/K) = -111.1\ kJ$

The reaction is spontaneous at 298 K because ΔG° is negative at this temperature. In this case, ΔG° could have been calculated from ΔG_f° values in Appendix C, since these values are tabulated at 298 K.

(d)   The reaction is at equilibrium when $\Delta G^\circ = 0$.

$\Delta G^\circ = \Delta H^\circ - T\Delta S^\circ = 0$. $\Delta H^\circ = T\Delta S^\circ$, $T = \Delta H^\circ/\Delta S^\circ$

$T = -126.4\ kJ/-0.0512\ kJ/K = 2469 = 2470\ K$.

This temperature is so high that the reactants and products are likely to decompose. At standard conditions, equilibrium is functionally unattainable for this reaction.

19.70   (a)   As in Sample Exercise 19.10, $T_{sub} = \Delta H_{sub}^\circ/\Delta S_{sub}^\circ$

Use Data from Appendix C to calculate ΔH_{sub}° and ΔH_{sub}° for $I_2(s)$.

$I_2(s) \rightarrow I_2(l)$ melting

$\underline{I_2(l) \rightarrow I_2(g)\ boiling}$

$I_2(s) \rightarrow I_2(g)$ sublimation

$\Delta H_{sub}^\circ = \Delta H_f^\circ I_2(g) - \Delta H_f^\circ\ I_2(s) = 62.25 - 0 = 62.25\ kJ$

$\Delta S_{sub}^\circ = S^\circ\ I_2(g) - S^\circ\ I_2(s) = 260.57 - 116.73 = 143.84\ J/K = 0.14384\ kJ/K$

$T_{sub} = \dfrac{\Delta H_{sub}^\circ}{\Delta S_{sub}^\circ} = \dfrac{62.25\ kJ}{0.14384\ kJ/K} = 432.8\ K = 159.6°C$

(b) T_m for $I_2(s) = 386.85\ K = 113.7°C$; $T_b = 457.4\ K = 184.3°C$
(from WebElements™, 2005)

(c) The boiling point of I_2 is closer to the sublimation temperature. Both boiling and sublimation begin with molecules in a condensed phase (little space between molecules) and end in the gas phase (large intermolecular distances). Separation of the molecules is the main phenomenon that determines both ΔH and ΔS, so it is not surprising that the ratio of $\Delta H / \Delta S$ is similar for sublimation and boiling.

19.72 (a) $C_2H_6(g) + 7/2\, O_2(g) \rightarrow 2\, CO_2(g) + 3\, H_2O(l)$

$\Delta H^\circ = 2\Delta H^\circ_f\, CO_2(g) + 3\Delta H^\circ_f\, H_2O(l) - \Delta H^\circ_f\, C_2H_6(g) - 7/2\Delta H^\circ_f\, O_2(g)$

$= 2(-393.5) + 3(-285.83) - (-84.68) - 7/2(0) = -1559.8 \text{ kJ/mol } C_2H_4 \text{ burned}$

(b) $w_{max} = \Delta G^\circ = 2\Delta G^\circ_f\, CO_2(g) + 3\Delta G^\circ_f\, H_2O(l) - \Delta G^\circ_f\, C_2H_6(g) - 7/2\Delta G^\circ_f\, O_2(g)$

$= 2(-394.4) + 3(-237.13) - (-32.89) - 7/2(0) = -1467.3 \text{ kJ}$

The system can accomplish at most 1467.3 kJ of work per mole of C_2H_6 on the surroundings.

Free Energy and Equilibrium

19.74 Consider the relationship $\Delta G = \Delta G^\circ + RT \ln Q$, where Q is the reaction quotient.

(a) $H_2(g)$ appears in the denominator of Q for this reaction. An increase in pressure of H_2 decreases Q and ΔG becomes smaller or more negative. Increasing the concentration of a reactant increases the tendency for a reaction to occur.

(b) $H_2(g)$ appears in the numerator of Q for this reaction. Increasing the pressure of H_2 increases Q and ΔG becomes more positive. Increasing the concentration of a product decreases the tendency for the reaction to occur.

(c) $H_2(g)$ appears in the denominator of Q for this reaction. An increase in pressure of H_2 decreases Q and ΔG becomes smaller or more negative.

19.76 (a) $\Delta G^\circ = 4\Delta G^\circ\, PH_3(g) - 6\Delta G^\circ\, H_2(g) - \Delta G^\circ\, P_4(g) = 4(13.4) - 6(0) - 24.4 = 29.2 \text{ kJ}$

(b) $\Delta G = \Delta G^\circ + RT \ln[P^4_{PH_3} / P^6_{H_2} \times P_{P_4}]$

$= 29.2 + \dfrac{8.314 \times 10^{-3} \text{ kJ}}{K\text{-mol}} \times 298 \text{ K} \times \ln[(0.22)^4/(8.0)^6 \times (0.050)]$

$= 29.2 - 38.495 = -9.295 = -9.3 \text{ kJ}$

19.78 $\Delta G^\circ = -RT \ln K$; $\ln K = -\Delta G^\circ / RT$; at 298 K, RT = 2.4776 = 2.478 kJ

(a) $\Delta G^\circ = \Delta G^\circ\, NaOH(s) + \Delta G^\circ\, CO_2(g) - \Delta G^\circ\, NaHCO_3(s)$

$= -379.5 + (-394.4) - (-851.8) = +77.9 \text{ kJ}$

$\ln K = \dfrac{-\Delta G^\circ}{RT} = \dfrac{-77.9 \text{ kJ}}{2.478 \text{ kJ}} = -31.442 = -31.4; \quad K = 2 \times 10^{-14}$

$K = P_{CO_2} = 2 \times 10^{-14}$

(b) $\Delta G^\circ = 2\Delta G^\circ\, HCl(g) + \Delta G^\circ\, Br_2(g) - 2\Delta G^\circ\, HBr(g) - \Delta G^\circ\, Cl_2(g)$

$= 2(-95.27) + 3.14 - 2(-53.22) - 0 = -80.96 \text{ kJ}$

$$\ln K = \frac{-(-80.96)}{2.4776} = +32.68; \quad K = 1.6 \times 10^{14}$$

$$K = \frac{P_{HCl}^2 \times P_{Br_2}}{P_{HBr}^2 \times P_{Cl_2}} = 1.6 \times 10^{14}$$

(c) From Solution 19.57(a), $\Delta G°$ at 298 K = –140.0 kJ.

$$\ln K = \frac{-\Delta G°}{RT} = \frac{-(-140.0)}{2.4776} = 56.51; \quad K = 3.5 \times 10^{24}$$

$$K = \frac{P_{SO_3}^2}{P_{SO_2}^2 \times P_{O_2}} = 3.5 \times 10^{24}$$

19.80 $K = P_{CO_2}$. Calculate $\Delta G°$ at the two temperatures using $\Delta G° = \Delta H° - T\Delta S°$ and then calculate K and P_{CO_2}.

$\Delta H° = \Delta H° \; PbO(s) + \Delta H° \; CO_2(g) - \Delta H° \; PbCO_3(s)$

    $= -217.3 - 393.5 + 699.1 = 88.3$ kJ

$\Delta S° = S° \; PbO(s) + S° \; CO_2(g) - S° \; PbCO_3(s)$

    $= 68.70 + 213.6 - 131.0 = 151.3$ J/K or 0.1513 kJ/K

(a) $\Delta G° = \Delta H° - T\Delta S°$. At 673 K, $\Delta G° = 88.3$ kJ $- 673$ K(0.1513 kJ/K) = -13.525

                                                         $= -13.5$ kJ

$$\ln K = \frac{-\Delta G°}{RT} = \frac{-(-13.525 \times 10^3)J}{8.314 \; J/K \times 673 \; K} = 2.4172 = 2.42$$

$$K = P_{CO_2} = 11.214 = 11 \text{ atm}$$

(b) $\Delta G° = \Delta H° - T\Delta S°$. At 453 K, $\Delta G° = 88.3$ kJ $- 453$ K (0.1513 kJ) = 19.7611

                                                         $= 19.8$ kJ

$$\ln K = \frac{-(19.7611 \times 10^3 \; J)}{8.314 \; J/K \times 453 \; K} = -5.2469 = -5.25; \quad K = P_{CO_2} = 5.3 \times 10^{-3} \text{ atm}$$

19.82   (a) $CH_3NH_2(aq) + H_2O(l) \; \rightleftharpoons \; CH_3NH_3^+(aq) + OH^-(aq)$

      (b) $\Delta G° = -RT \ln K_b = -(8.314 \times 10^{-3})(298) \ln (4.4 \times 10^{-4}) = 19.148 = 19.1$ kJ

      (c) $\Delta G = 0$ at equilibrium

      (d) $\Delta G = \Delta G° + RT \ln Q; \; [OH^-] = 1 \times 10^{-14}/1.5 \times 10^{-8} = 6.7 \times 10^{-7}$

$$= 19.148 + (8.314 \times 10^{-3})(298) \ln \frac{(5.5 \times 10^{-4})(6.67 \times 10^{-7})}{0.120} = -29.43 = -29 \text{ kJ}$$

Additional Exercises

19.83   (a)   False. The essential question is whether the reaction proceeds far to the right before arriving at equilibrium. The position of equilibrium, which is the essential aspect, is not only dependent on ΔH but on the entropy change as well.

      (b)   True.

(c)   True.

(d)   False. **Nonspontaneous** processes in general require that work be done to force them to proceed. Spontaneous processes occur without application of work.

(e)   False. Such a process **might** be spontaneous, but would not necessarily be so. Spontaneous processes are those that are exothermic and/or that lead to increased disorder in the system.

19.85   There is no inconsistency. The second law states that in any spontaneous process there is an increase in the entropy of the universe. While there may be a decrease in entropy of the system, as in the present case, this decrease is more than offset by an increase in entropy of the surroundings.

19.86   If $NH_4NO_3(s)$ dissolves spontaneously in water, $\Delta G = \Delta H - T\Delta S$. If ΔG is negative and ΔH is positive, the sign of ΔS must be positive. Furthermore, $T\Delta S > \Delta H$ at room temperature.

19.87   At the normal boiling point of a liquid, $\Delta G = 0$ and $\Delta H_{vap} = T\Delta S_{vap}$; $T = \Delta H_{vap}/\Delta S_{vap}$. By Trouton's rule, $\Delta S_{vap} = 88$ J/mol-K. The process of vaporization is:

(a) $Br_2(l) \rightleftharpoons Br_2(g)$

$\Delta H_{vap} = \Delta H_f^o \, Br_2(g) - \Delta H_f^o \, Br_2(l) = 30.71 \text{ kJ} - 0 = 30.71 \text{ kJ}$

$T_b = \dfrac{\Delta H_{vap}}{\Delta S_{vap}} = \dfrac{30.71 \text{ kJ}}{88 \text{ J/mol-K}} \times \dfrac{1000 \text{ J}}{\text{kJ}} = 349 = 3.5 \times 10^2 \text{ K}$

(b)   According to WebElements™ 2005, the normal boiling point of $Br_2(l)$ is 332 K. Trouton's rule provides a good "ballpark" estimate.

19.89   (a)   (i) $Ti(s) + 2Cl_2(g) \rightarrow TiCl_4(g)$

$\Delta H^\circ = \Delta H^\circ \, TiCl_4(g) - \Delta H^\circ \, Ti(s) - 2\Delta H^\circ \, Cl_2(g)$

$= -763.2 - 0 - 2(0) = -763.2 \text{ kJ}$

$\Delta S^\circ = 354.9 - 30.76 - 2(222.96) = -121.78 = -121.8 \text{ J/K}$

$\Delta G^\circ = -726.8 - 0 - 2(0) = -726.8 \text{ kJ}$

(ii) $C_2H_6(g) + 7Cl_2(g) \rightarrow 2CCl_4(g) + 6HCl(g)$

$\Delta H^\circ = 2\Delta H^\circ \, CCl_4(g) + 6\Delta H^\circ \, HCl(g) - \Delta H^\circ \, C_2H_6(g) - 7\Delta H^\circ \, Cl_2(g)$

$= 2(-106.7) + 6(-92.30) - (-84.68) - 7(0) = -682.52 = -682.5 \text{ kJ}$

$\Delta S^\circ = 2(309.4) + 6(186.69) - 229.5 - 7(222.96) = -51.28 = -51.4 \text{ J/K}$

$\Delta G^\circ = 2(-64.0) + 6(-95.27) - (-32.89) - 7(0) = -666.73 \text{ kJ}$

(iii) $BaO(s) + CO_2(g) \rightarrow BaCO_3(s)$

$\Delta H^\circ = \Delta H^\circ \, BaCO_3(s) - \Delta H^\circ \, BaO(s) - \Delta H^\circ \, CO_2(g)$

$= -1216.3 - (-553.5) - (-393.5) = -269.3 \text{ kJ}$

$\Delta S^\circ = 112.1 - 70.42 - 213.6 = -171.9 \text{ J/K}$

$\Delta G^\circ = -1137.6 - (-525.1) - (-394.4) = -218.1 \text{ kJ}$

(b)   (i), (ii) and (iii) all have negative $\Delta G°$ values and are spontaneous at standard conditions and 25°C.

(c) $\Delta G° = \Delta H° - T\Delta S°$. All three reactions have $-\Delta H°$ and $-\Delta S°$. They all have $-\Delta G°$ at 25°C, and $\Delta G°$ becomes more positive as T increases.

19.90 $\Delta G = \Delta G° + RT \ln Q$

(a) $Q = \dfrac{P_{NH_3}^2}{P_{N_2} \times P_{H_2}^3} = \dfrac{(1.2)^2}{(2.6)(5.9)^3} = 2.697 \times 10^{-3} = 2.7 \times 10^{-3}$

$\Delta G° = 2\Delta G° \, NH_3(g) - \Delta G° \, N_2(g) - 3\Delta G° \, H_2(g)$

$= 2(-16.66) - 0 - 3(0) = -33.32 \text{ kJ}$

$\Delta G = -33.32 \text{ kJ} + \dfrac{8.314 \times 10^{-3} \text{ kJ}}{\text{K-mol}} \times 298 \text{ K} \times \ln(2.69 \times 10^{-3})$

$\Delta G = -33.32 - 14.66 = -47.98 = -48.0 \text{ kJ}$

(b) $Q = \dfrac{P_{N_2}^3 \times P_{H_2O}^4}{P_{N_2H_4}^2 \times P_{NO_2}^2} = \dfrac{(0.5)^3(0.3)^4}{(5.0 \times 10^{-2})^2(5.0 \times 10^{-2})^2} = 162 = 2 \times 10^2$

$\Delta G° = 3\Delta G° \, N_2(g) + 4\Delta G° \, H_2O(g) - 2\Delta G° \, N_2H_4(g) - 2\Delta G° \, NO_2(g)$

$= 3(0) + 4(-228.57) - 2(159.4) - 2(51.84) = -1336.8 \text{ kJ}$

$\Delta G = -1336.8 \text{ kJ} + 2.478 \ln 162 = -1324.2 = -1.32 \times 10^3 \text{ kJ}$

(c) $Q = \dfrac{P_{N_2} \times P_{H_2}^2}{P_{N_2H_4}} = \dfrac{(1.5)(2.5)^2}{0.5} = 18.75 = 2 \times 10^1$

$\Delta G° = \Delta G° \, N_2(g) + 2\Delta G° \, H_2(g) - \Delta G° \, N_2H_4(g)$

$= 0 + 2(0) - 159.4 = -159.4 \text{ kJ}$

$\Delta G = -159.4 \text{ kJ} + 2.478 \ln 18.75 = -152.1 = -152 \text{ kJ}$

19.91

Reaction	(a) Sign of $\Delta H°$	(a) Sign of $\Delta S°$	(b) K > 1?	(c) Variation in K as Temp. Increases
(i)	–	–	yes	decrease
(ii)	+	+	no	increase
(iii)	+	+	no	increase
(iv)	+	+	no	increase

(a)   Note that at a particular temperature, positive $\Delta H°$ leads to a smaller value of K, while positive $\Delta S°$ increases the value of K.

19.93   (a)   First calculate $\Delta G°$ for each reaction:

For $C_6H_{12}O_6(s) + 6O_2(g) \;\rightleftharpoons\; 6CO_2(g) + 6H_2O(l)$ (A)

$\Delta G° = 6(-237.13) + 6(-394.4) - (-910.4) + 6(0) = -2878.8 \text{ kJ}$

For $C_6H_{12}O_6(s) \rightleftharpoons 2C_2H_5OH(l) + 2CO_2(g)$ (B)

$\Delta G° = 2(-394.4) + 2(-174.8) - (-910.4) = -228.0$ kJ

For (A), $\ln K = 2879 \times 10^3 / (8.314)(298) = 1162; K = 5 \times 10^{504}$

For (B), $\ln K = 228 \times 10^3 / (8.314)(298) = 92.026 = 92.0; K = 9 \times 10^{39}$

(b)   Both these values for K are unimaginably large. However, K for reaction (A) is larger, because $\Delta G°$ is more negative. The magnitude of the work that can be accomplished by coupling a reaction to its surroundings is measured by ΔG. According to the calculations above, considerably more work can in principle be obtained from reaction (A), because $\Delta G°$ is more negative.

19.94   (a) $\Delta G° = -RT \ln K$ (Equation [19.17]); $\ln K = -\Delta G° / RT$

Use $\Delta G° = \Delta H° - T\Delta S°$ to get $\Delta G°$ at the two temperatures. Calculate $\Delta H°$ and $\Delta S°$ using data in Appendix C.

$2CH_4(g) \rightarrow C_2H_6(g) + H_2(g)$

$\Delta H° = \Delta H° \, C_2H_6(g) + \Delta H° \, H_2(g) - 2\Delta H° \, CH_4(g) = -84.68 + 0 - 2(-74.8) = 64.92$

$= 64.9$ kJ

$\Delta S° = S° \, C_2H_6(g) + S° \, H_2(g) - 2S° \, CH_4(g) = 229.5 + 130.58 - 2(186.3) = -12.52$

$= -12.5$ J/K

at 298 K, $\Delta G = 64.92$ kJ $- 298$ K$(-12.52 \times 10^{-3}$ kJ/K$) = 68.65 = 68.7$ kJ

$\ln K = \dfrac{-68.65 \text{ kJ}}{(8.314 \times 10^{-3} \text{ kJ/K})(298 \text{ K})} = -27.709 = -27.7, K = 9.25 \times 10^{-13} = 9 \times 10^{-13}$

at 773 K, $\Delta G = 64.9$ kJ $- 773$ K$(-12.52 \times 10^{-3}$ J/K$) = 74.598 = 74.6$ kJ

$\ln K = \dfrac{-74.598 \text{ kJ}}{(8.314 \times 10^{-3} \text{ kJ/K})(773 \text{ K})} = -11.607 = -11.6, \; K = 9.1 \times 10^{-6}$

Because the reaction is endothermic, the value of K increases with an increase in temperature.

$2CH_4(g) + 1/2 \, O_2(g) \rightarrow C_2H_6(g) + H_2O(g)$

$\Delta H° = \Delta H° \, C_2H_6(g) + \Delta H° \, H_2O(g) - 2\Delta H° \, CH_4(g) - 1/2 \, \Delta H° \, O_2(g)$

$= -84.68 + (-241.82) - 2(-74.8) - 1/2 \, (0) = -176.9$ kJ

$\Delta S° = S° \, C_2H_6(g) + S° \, H_2O(g) - 2S° \, CH_4(g) - 1/2 \, S° \, O_2(g)$

$= 229.5 + 188.83 - 2(186.3) - 1/2 \, (205.0) = -56.77 = -56.8$ J/K

at 298 K, $\Delta G = -176.9$ kJ $- 298$ K$(-56.77 \times 10^{-3}$ kJ/K$) = -159.98 = -160.0$ kJ

$\ln K = \dfrac{-(-159.98 \text{ kJ})}{(8.314 \times 10^{-3} \text{ kJ/K})(298 \text{ K})} = 64.571 = 64.57; \; K = 1.1 \times 10^{28}$

at 773 K, $\Delta G = -176.9$ kJ $- 773$ K $(-56.77 \times 10^{-3}$ kJ/K$) = -133.02 = -133.0$ kJ

$$\ln K = \frac{-(-133.02\ kJ)}{(8.314\times10^{-3}\ kJ/K)(773\ K)} = 20.698 = 20.70;\ K = 9.750\times10^8 = 9.8\times10^8$$

Because this reaction is exothermic, the value of K decreases with increasing temperature.

(b) The difference in $\Delta G°$ for the two reactions is primarily enthalpic; the first reaction is endothermic and the second exothermic. Both reactions have $-\Delta S°$, which inhibits spontaneity.

(c) This is an example of coupling a useful but nonspontaneous reaction with a spontaneous one to spontaneously produce a desired product.

$$2CH_4(g)\ \rightarrow\ C_2H_6(g)+H_2(g) \qquad \Delta G_{298}^{o} = +68.7\ kJ,\ nonspontaneous$$
$$H_2(g)+1/2\,O_2(g)\ \rightarrow\ H_2O(g) \qquad \Delta G_{298}^{o} = -228.57\ kJ,\ spontaneous$$
$$2CH_4(g)+1/2\,O_2(g)\ \rightarrow\ C_2H_6(g)+H_2O(g) \qquad \Delta G_{298}^{o} = -159.9\ kJ,\ spontaneous$$

(d) $CH_4(g) + 2O_2(g) \rightarrow CO_2(g) + 2H_2O(g)$

19.95 $\Delta G°$ for the metabolism of glucose is:

$6\Delta G°\ CO_2(g) + 6\Delta G°\ H_2O(l) - \Delta G°\ C_6H_{12}O_6(s) - 6\Delta G°\ O_2(g)$

$\Delta G° = 6(-394.4) + 6(-237.13) - (-910.4) + 6(0) = -2878.8\ kJ$

moles ATP = $-2878.8\ kJ \times 1$ mol ATP / $(-30.5\ kJ) = 94.4$ mol ATP / mol glucose

Note that this calculation is done at standard conditions, not metabolic conditions. A more accurate answer would be obtained using ΔG values that reflect actual concentration, partial pressure, and pH in a cell.

19.97 (a) To obtain $\Delta H°$ from the equilibrium constant data, graph lnK at various temperatures vs 1/T, being sure to employ absolute temperature. The slope of the linear relationship that should result is $-\Delta H°/R$; thus, $\Delta H°$ is easily calculated.

(b) Use $\Delta G° = \Delta H° - T\Delta S°$ and $\Delta G° = -RT \ln K$. Substituting the second expression into the first, we obtain

$$-RT \ln K = \Delta H° - T\Delta S°; \quad \ln K = \frac{-\Delta H°}{RT} - \frac{-\Delta S°}{R} = \frac{-\Delta H°}{RT} + \frac{\Delta S°}{R}$$

Thus, the constant in the equation given in the exercise is $\Delta S°/R$.

19.98 $S = k \ln W$ (Equation [19.5]), $k = R/N$, $W \propto V^m$

$\Delta S = S_2 - S_1; S_1 = k \ln W_1, S_2 = k \ln W_2$

$\Delta S = k \ln W_2 - k \ln W_1; W_2 = cV_2^{\,m}; W_1 = cV_1^{\,m}$

(The number of particles, m, is the same in both states.)

$\Delta S = k \ln cV_2^{\,m} - k \ln cV_1^{\,m}; \ln a^b = b \ln a$

$\Delta S = k\,m \ln cV_2 - k\,m \ln cV_1; \ln a - \ln b = \ln (a/b)$

$$\Delta S = k \, m \ln\!\left(\frac{cV_2}{cV_1}\right) = k \, m \ln\!\left(\frac{V_2}{V_1}\right) = \frac{R}{N} m \ln\!\left(\frac{V_2}{V_1}\right)$$

$$\frac{m}{N} = \frac{particles}{6.022 \times 10^{23}} = n(mol); \quad \Delta S = nR \ln\!\left(\frac{V_2}{V_1}\right)$$

19.99   (a) $(T_{high} - T_{low})/T_{high} = (700 - 288)/700 = 0.58857 = 58.9\%$ efficiency

       (b)   The cooler the exit temperature of the engine or generator, the more efficient the engine. If a body of water can be used to naturally reduce the exit temperature, the efficiency of energy production increases.

       (c)   The closer the exit temperature to 0 K, the more efficient the heat engine.

       (d)   (Refer to Figure 5.9)

Integrative Exercises

19.101   (a)   Polymerization is the process of joining many small molecules (monomers) into a few very large molecules (polymers). Polyethylene in particular can have extremely high molecular weights. In general, reducing the number of particles in a system reduces entropy, so ΔS_{poly} is expected to be negative.

       (b) $\Delta G_{poly} = \Delta H_{poly} - T\Delta S_{poly}$. If the polymerization of ethylene is spontaneous, ΔG_{poly} is negative. If ΔS_{poly} is negative, $-T\Delta S_{poly}$ is positive, so ΔH_{poly} must be negative for ΔG_{poly} to be negative. The enthalpy of polymerization must be exothermic.

       (c)   According to Equation [12.4], polymerization of ethylene requires breaking one C=C and forming 2C—C per monomer (1C—C between the C-atoms of the monomer and $2 \times 1/2$ C—C to two other monomers).

$$\Delta H = D(C\!=\!C) - 2D(C\!-\!C) = 614 - 2(348) = -82 \text{ kJ/mol C}_2H_4$$

$$\frac{-82 \text{ kJ}}{mol \text{ C}_2H_4} \times \frac{1 \, mol}{6.022 \times 10^{23} \, molecules} \times \frac{1000 \text{ J}}{1 \text{ kJ}} = 1.36 \times 10^{-19} \text{ J/C}_2H_4 \text{ monomer}$$

(d)   The products of a condensation polymerization are the polymer and a small molecule, typically H_2O; there is usually one small molecule formed per monomer unit. Unlike addition polymerization, the total number of particles is not reduced. A condensation polymer does impose more order on the monomer or monomers than an addition polymer. If there is a single monomer, it has different functional groups at the two ends and only one end can react to join the polymer, so orientation is required. If there are two different monomers, as in nylon, the monomers alternate in the polymer, so only the correct monomer can react to join the polymer. In terms of structure, the condensation polymer imposes more order on the monomer(s) than an addition polymer. But, condensation polymerization does not lead to a reduction in the number of particles in the system, so ΔS_{poly} will be less negative than for addition polymerization.

19.102   The activated complex in Figure 14.13 is a single "particle" or entity that contains four atoms. It is formed from an atom A and a triatomic molecule, ABC, that must collide with exactly the correct energy and orientation to form the single entity. There are many fewer degrees of freedom for the activated complex than the separate reactant particles, so the *entropy of activation* is negative. [Figure 14.6 shows a unimolecular activated complex.]

19.104   (a)   16 e^-, 8 e^- pairs. The C-S bond order is approximately 2.

$$\ddot{\underset{..}{S}}{=}C{=}\ddot{\underset{..}{S}}$$

(b)   2 e^- domains around C, linear e^- domain geometry, linear molecular structure

(c) $CS_2(l) + 3O_2(g) \rightarrow CO_2(g) + 2SO_2(g)$

(d) $\Delta H° = \Delta H° \, CO_2(g) + 2\Delta H° \, SO_2(g) - \Delta H° \, CS_2(l) - 3\Delta H° \, O_2(g)$

$= -393.5 + 2(-296.9) - (89.7) - 3(0) = -1077.0 \text{ kJ}$

$\Delta G° = \Delta G° \, CO_2(g) + 2\Delta G° \, SO_2(g) - \Delta G° \, CS_2(l) - 3\,\Delta G° \, O_2(g)$

$= -394.4 + 2(-300.4) - (65.3) - 3(0) = -1060.5 \text{ kJ}$

The reaction is exothermic ($-\Delta H°$) and spontaneous ($-\Delta G°$) at 298 K.

(e)   vaporization: $CS_2(l) \rightarrow CS_2(g)$

$\Delta G°_{vap} = \Delta H°_{vap} - T\Delta S°_{vap}; \quad \Delta S°_{vap} = (\Delta H°_{vap} - \Delta G°_{vap})/T$

$\Delta G°_{vap} = \Delta G° \, CS_2(g) - \Delta G° \, CS_2(l) = 67.2 - 65.3 = 1.9 \text{ kJ}$

$\Delta H°_{vap} = \Delta H° \, CS_2(g) - \Delta H° \, CS_2(l) = 117.4 - 89.7 = 27.7 \text{ kJ}$

$\Delta S°_{vap} = (27.7 - 1.9) \text{ kJ}/298 \text{ K} = 0.086577 = 0.0866 \text{ kJ/K} = 86.6 \text{ J/K}$

ΔS_{vap} is always positive, because the gas phase occupies a greater volume, has more motional freedom and a larger absolute entropy than the liquid.

(f)   At the boiling point, $\Delta G = 0$ and $\Delta H_{vap} = T_b\Delta S_{vap}$.

$T_b = \Delta H_{vap}/\Delta S_{vap} = 27.7 \text{ kJ}/0.086577 \text{ kJ/K} = 319.9 = 320 \text{ K}$

$T_b = 320 \text{ K} = 47°C. \, CS_2$ is a liquid at 298 K, 1 atm

19.105 (a) $Ag(s) + 1/2\,N_2(g) + 3/2\,O_2(g) \rightarrow AgNO_3(s)$; S decreases because there are fewer moles of gas in the product.

(b) $\Delta G_f^\circ = \Delta H_f^\circ - T\Delta S^\circ$; $\Delta S^\circ = (\Delta G_f^\circ - \Delta H_f^\circ)/(-T) = (\Delta H_f^\circ - \Delta G_f^\circ)/T$

$\Delta S^\circ = -124.4\,kJ - (-33.4\,kJ)\,/\,298\,K = -0.305\,kJ/K = -305\,J/K$

ΔS° is relatively large and negative, as anticipated from part (a).

(c)   Dissolving of $AgNO_3$ can be expressed as

$AgNO_3(s) \rightarrow AgNO_3\,(aq,\,1\,M)$

$\Delta H^\circ = \Delta H^\circ\,AgNO_3(aq) - \Delta H^\circ\,AgNO_3(s) = -101.7 - (-124.4) = +22.7\,kJ$

$\Delta H^\circ = \Delta H^\circ\,MgSO_4(aq) - \Delta H^\circ\,MgSO_4(s) = -1374.8 - (-1283.7) = -91.1\,kJ$

Dissolving $AgNO_3(s)$ is endothermic $(+\Delta H^\circ)$, but dissolving $MgSO_4(s)$ is exothermic $(-\Delta H^\circ)$.

(d) $AgNO_3$: $\Delta G^\circ = \Delta G_f^\circ\,AgNO_3(aq) - \Delta G_f^\circ\,AgNO_3(s) = -34.2 - (-33.4) = -0.8\,kJ$

$\Delta S^\circ = (\Delta H^\circ - \Delta G^\circ)\,/\,T = [22.7\,kJ - (-0.8\,kJ)]\,/\,298\,K = 0.0789\,kJ/K = 78.9\,J/K$

$MgSO_4$: $\Delta G^\circ = \Delta G_f^\circ\,MgSO_4(aq) - \Delta G_f^\circ\,MgSO_4(s) = -1198.4 - (-1169.6) = -28.8\,kJ$

$\Delta S^\circ = (\Delta H^\circ - \Delta G^\circ)\,/\,T = [-91.1\,kJ - (-28.8\,kJ)]\,/\,298\,K = -0.209\,kJ/K = -209\,J/K$

(e)   In general, we expect dissolving a crystalline solid to be accompanied by an increase in positional disorder and an increase in entropy; this is the case for $AgNO_3$ ($\Delta S^\circ = +78.9\,J/K$). However, for dissolving $MgSO_4(s)$, there is a substantial decrease in entropy ($\Delta S = -209\,J/K$). According to Section 13.5, ion-pairing is a significant phenomenon in electrolyte solutions, particularly in concentrated solutions where the charges of the ions are greater than 1. According to Table 13.5, a 0.1 m $MgSO_4$ solution has a van't Hoff factor of 1.21. That is, for each mole of $MgSO_4$ that dissolves, there are only 1.21 moles of "particles" in solution instead of 2 moles of particles. For a 1 m solution, the factor is even smaller. Also, the exothermic enthalpy of mixing indicates substantial interactions between solute and solvent. Substantial ion-pairing coupled with ion-dipole interactions with H_2O molecules lead to a decrease in entropy for $MgSO_4(aq)$ relative to $MgSO_4(s)$.

19.107 (a) $\Delta G^\circ = 3\Delta G_f^\circ\,S(s) + 2\Delta G_f^\circ\,H_2O(g) - \Delta G_f^\circ\,SO_2(g) - 2\Delta G_f^\circ\,H_2S(g)$

$= 3(0) + 2(-228.57) - (-300.4) - 2(-33.01) = -90.72 = -90.7\,kJ$

$\ln K = \dfrac{-\Delta G^\circ}{RT} = \dfrac{-(-90.72\,kJ)}{(8.314 \times 10^{-3}\,kJ/K)(298\,K)} = 36.6165 = 36.6;\ K = 7.99 \times 10^{15}$

$= 8 \times 10^{15}$

(b)   The reaction is highly spontaneous at 298 K and feasible in principle. However, use of $H_2S(g)$ produces a severe safety hazard for workers and the surrounding community.

(c) $P_{H_2O} = \dfrac{25 \text{ torr}}{760 \text{ torr/atm}} = 0.033 \text{ atm}$

$K = \dfrac{P_{H_2O}^2}{P_{SO_2} \times P_{H_2S}^2}; \quad P_{SO_2} = P_{H_2S} = x \text{ atm}$

$K = 7.99 \times 10^{15} = \dfrac{(0.033)^2}{x(x)^2}; \quad x^3 = \dfrac{(0.033)^2}{7.99 \times 10^{15}}$

$x = 5 \times 10^{-7} \text{ atm}$

(d) $\Delta H^\circ = 3\Delta H_f^\circ \, S(s) + 2\Delta H_f^\circ \, H_2O(g) - \Delta H_f^\circ \, SO_2(g) - 2\Delta H_f^\circ \, H_2S(g)$

$= 3(0) + 2(-241.82) - (-296.9) - 2(-20.17) = -146.4 \text{ kJ}$

$\Delta S^\circ = 3S^\circ \, S(s) + 2S^\circ \, H_2O(g) - S^\circ \, SO_2(g) - 2S^\circ \, H_2S(g)$

$= 3(31.88) + 2(188.83) - 248.5 - 2(205.6) = -186.4 \text{ J/K}$

The reaction is exothermic ($-\Delta H$), so the value of K_{eq} will decrease with increasing temperature. The negative ΔS° value means that the reaction will become nonspontaneous at some higher temperature. The process will be less effective at elevated temperatures.

19.108   (a)   When the rubber band is stretched, the molecules become more ordered, so the entropy of the system decreases, ΔS_{sys} is negative.

(b) $\Delta S_{sys} = q_{rev}/T$. Since ΔS_{sys} is negative, q_{rev} is negative and heat is evolved by the system.

20 Electrochemistry

Visualizing Concepts

20.2 (a) If a Zn(s) strip was placed in a $CdSO_4$(aq) solution, Cd(s) would form on the strip. Although E°_{red} for Zn^{2+}(aq), –0.763 V, and Cd^{2+}(aq), –0.403 V, are both negative, the value for Cd^{2+} is larger (less negative), so it will be the reduced species in the redox reaction.

(b) If a Cu(s) strip was placed in a $AgNO_3$(aq) solution, Ag(s) would form on the strip. Although E°_{red} for Cu^{2+}(aq), 0.337 V, and Ag^+(aq), 0.799 V, are both positive, the species with the larger E°_{red} value, Ag^+, will be reduced in the reaction.

20.3 *Analyze/Plan.* Apply the definitions of oxidation, reduction, anode and cathode to the diagram. Recall relationship between atomic and ionic size from Chapter 7. *Solve.*

(a) Oxidation. The gray spheres are uniformly sized and closely aligned; they represent an elemental solid. The diagram shows atoms from the surface of the solid going into solution. In a voltaic cell, this happens when metal atoms on an electrode surface are oxidized. They lose electrons, form cations and move into solution.

(b) Anode. Oxidation occurs at the anode.

(c) When a neutral atom loses a valence electron, Z_{eff} for the remaining electrons increases, and the radius of the resulting cation is smaller than the radius of the neutral atom. The neutral atoms in the electrode are represented by larger spheres than the cations moving into solution.

20.5 The species with the largest E°_{red} is easiest to reduce, while the species with the smallest, most negative E°_{red} is easiest to oxidize.

(a) The species easiest to oxidize is at the bottom of Figure 20.14.

(b) The species easiest to reduce is at the top of Figure 20.14.

20.6 $A(aq) + B(aq) \rightarrow A^-(aq) + B^+(aq)$

(a) A gains electrons; it is being reduced. B loses electrons; it is being oxidized.

(b) Reduction occurs at the cathode; oxidation occurs at the anode.

$A(aq) + 1e^- \rightarrow A^-(aq)$ occurs at the cathode.

$B(aq) \rightarrow B^+(aq) + 1e^-$ occurs at the anode.

(c)   In a voltaic cell, the anode is at higher potential energy than the cathode. The anode reaction, $B(aq) \rightarrow B^+(aq) + 1e^-$, is higher in potential energy.

(d) $\Delta G° = -nFE°$; the signs of $\Delta G°$ and $E°$ (or ΔG and E) are opposite. Since $E°$ is positive, $\Delta G°$ is negative. (This is the case for a spontaneous reaction.)

20.8

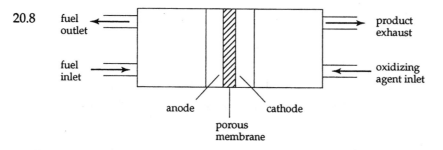

The main difference between a fuel cell and a battery is that a fuel cell is not self-contained. That is, there is a continuous supply of fuel (reductant) and oxidant to the cell, and continuous exhaust of products. The fuel cell produces electrical current as long as reactants are supplied. It never goes "dead."

20.9   Zinc, $E°_{red} = -0.763\,V$, is more easily oxidized than iron, $E°_{red} = -0.440\,V$. If conditions are favorable for oxidation, zinc will be preferentially oxidized, preventing iron from corroding. The protection lasts until all the Zn coating has reacted.

Oxidation - Reduction Reactions

20.12   (a)   *Reduction* is the gain of electrons.

         (b)   The electrons appear on the reactants side (left side) of a reduction half-reaction.

         (c)   The *reductant* is the reactant that is oxidized; it provides the electrons that are gained by the substance being reduced.

         (d)   A *reducing agent* is the substance that promotes reduction. It donates the electrons gained by the substance that is reduced. It is the same as the reductant.

20.14   (a)   False. If something is reduced, it gains electrons.

         (b)   True.

         (c)   True. Oxidation can be thought of as a gain of oxygen atoms. Looking forward, this view will be useful for organic reactions, Chapter 25.

20.16   (a)   No oxidation-reduction

         (b)   I is oxidized from –1 to +5; Cl is reduced from +1 to –1.

         (c)   S is oxidized from +4 to +6; N is reduced from +5 to +2.

         (d)   S is reduced from +6 to +4; Br is oxidized from –1 to 0.

Balancing Oxidation-Reduction Reactions

20.18 (a) $2N_2H_4(g) + N_2O_4(g) \rightarrow 3N_2(g) + 4H_2O(g)$

(b) $N_2H_4(g)$ is oxidized; $N_2O_4(g)$ is reduced.

(c) $N_2O_4(g)$ serves as the oxidizing agent; it is itself reduced. $N_2H_4(g)$ serves as the reducing agent; it is itself oxidized.

20.20 (a) $Mo^{3+}(aq) + 3e^- \rightarrow Mo(s)$, reduction

(b) $H_2SO_3(aq) + H_2O(l) \rightarrow SO_4^{2-}(aq) + 4H^+(aq) + 2e^-$, oxidation

(c) $NO_3^-(aq) + 4H^+(aq) + 3e^- \rightarrow NO(g) + 2H_2O(l)$, reduction

(d) $O_2(g) + 4H^+(aq) + 4e^- \rightarrow 2H_2O(l)$, reduction

(e) $Mn^{2+}(aq) + 4OH^-(aq) \rightarrow MnO_2(s) + 2H_2O(l) + 2e^-$, oxidation

(f) $Cr(OH)_3(s) + 5OH^-(aq) \rightarrow CrO_4^{2-}(aq) + 4H_2O(l) + 3e^-$, oxidation

(g) $O_2(g) + 2H_2O(l) + 4e^- \rightarrow 4OH^-(aq)$, reduction
(O_2 is reduced to OH^-, not H_2O, in basic solution)

20.22 (a)

$$3[NO_2^-(aq) + H_2O(l) \rightarrow NO_3^-(aq) + 2H^+(aq) + 2e^-]$$

$$\underline{Cr_2O_7^{2-}(aq) + 14H^+(aq) + 6e^- \rightarrow 2Cr^{3+}(aq) + 7H_2O(l)}$$

Net: $3NO_2^-(aq) + Cr_2O_7^{2-}(aq) + 8H^+(aq) \rightarrow 3NO_3^-(aq) + 2Cr^{3+}(aq) + 4H_2O(l)$

oxidizing agent, $Cr_2O_7^{2-}$; reducing agent, NO_2^-

(b) The oxidation half-reaction involves S, and is listed in Appendix E. The reduction half-reaction involves N, and must be written and balanced, according to the procedure in Section 20.2.

$$HNO_3(aq) \rightarrow N_2O(g)$$
$$2HNO_3(aq) \rightarrow N_2O(g)$$
$$2HNO_3(aq) \rightarrow N_2O(g) + 5H_2O(l)$$
$$2HNO_3(aq) + 8H^+(aq) \rightarrow N_2O(g) + 5H_2O(l)$$
$$2HNO_3(aq) + 8H^+ + 8e^- \rightarrow N_2O(g) + 5H_2O(l)$$

$$2[S(s) + 3H_2O(l) \rightarrow H_2SO_3(aq) + 4H^+ + 4e^-]$$

$$\underline{2HNO_3(aq) + 8H^+ + 8e^- \rightarrow N_2O(g) + 5H_2O(l)}$$

$$2HNO_3(aq) + 2S(s) + H_2O(l) \rightarrow 2H_2SO_3(aq) + N_2O(g)$$

oxidizing agent, HNO_3; reducing agent, S.

(c)

$$2[Cr_2O_7^{2-}(aq) + 14H^+(aq) + 6e^- \rightarrow 2Cr^{3+}(aq) + 7H_2O(l)]$$

$$\underline{3[CH_3OH(aq) + H_2O(l) \rightarrow HCO_2H(aq) + 4H^+(aq) + 4e^-]}$$

Net: $2Cr_2O_7^{2-}(aq) + 3CH_3OH(aq) + 16H^+(aq) \rightarrow 4Cr^{3+}(aq) + 3HCO_2H(aq) + 11H_2O(l)$

oxidizing agent, $Cr_2O_7^{2-}$; reducing agent, CH_3OH

(d)

$$2[MnO_4^-(aq) + 8H^+(aq) + 5e^- \rightarrow Mn^{2+}(aq) + 4H_2O(l)]$$

$$5[2Cl^-(aq) \rightarrow Cl_2(aq) + 2e^-]$$

Net: $\overline{2MnO_4^-(aq) + 10Cl^-(aq) + 16H^+(aq) \rightarrow 2Mn^{2+}(aq) + 5Cl_2(g) + 8H_2O(l)}$

oxidizing agent, MnO_4^-; reducing agent, Cl^-

(e) Write and balance each half-reaction, and then sum to get the overall reaction. Follow the procedure in Sample Exercise 20.3 for reactions in basic solution.

oxidation: $Al(s) \rightarrow AlO_2^-(aq)$

$$Al(s) + 2H_2O(l) \rightarrow AlO_2^-(aq) + 4H^+(aq) + 3e^-$$

$$Al(s) + 4OH^-(aq) \rightarrow AlO_2^-(aq) + 2H_2O(l) + 3e^-$$

reduction: $NO_2^-(aq) \rightarrow NH_4^+(aq)$

$$NO_2^-(aq) + 8H^+(aq) + 6e^- \rightarrow NH_4^+(aq)) + 2H_2O(l)$$

$$NO_2^-(aq) + 6H_2O(l) + 6e^- \rightarrow NH_4^+(aq) + 8OH^-(aq)$$

$$NO_2^-(aq) + 6H_2O(l) + 6e^- \rightarrow NH_4^+(aq) + 8OH^-(aq)$$

$$2[Al(s) + 4OH^-(aq) \rightarrow AlO_2^-(aq) + 2H_2O(l) + 3e^-]$$

$$NO_2^-(aq) + 6H_2O(l) + 6e^- \rightarrow NH_4^+(aq) + 8OH^-(aq)$$

$$\overline{NO_2^-(aq) + 2Al(s) + 2H_2O(l) \rightarrow NH_4^+(aq) + 2AlO_2^-(aq)}$$

(f) $H_2O_2(aq) + 2e^- \rightarrow O_2(g) + 2H^+(aq)$

Since the reaction is in base, the H^+ can be "neutralized" by adding $2OH^-$ to each side of the equation to give $H_2O_2(aq) + 2OH^-(aq) + 2e^- \rightarrow O_2(g) + 2H_2O(l)$. The other half reaction is $2[ClO_2(aq) + e^- \rightarrow ClO_2^-(aq)]$.

Net: $H_2O_2(aq) + 2ClO_2(aq) + 2OH^-(aq) \rightarrow O_2(g) + 2ClO_2^-(aq) + 2H_2O(l)$

oxidizing agent, ClO_2; reducing agent, H_2O_2

Voltaic Cells

20.24 (a) The porous glass dish in Figure 20.4 provides a mechanism by which ions not directly involved in the redox reaction can migrate into the anode and cathode compartments to maintain charge neutrality of the solutions. Ionic conduction within the cell, through the glass disk, completes the cell circuit.

(b) In the anode compartment of Figure 20.5, Zn atoms are oxidized to Zn^{2+} cations, increasing the number of positively charged particles in the compartment. NO_3^- anions migrate into the compartment to maintain charge balance as Zn^{2+} ions are produced.

20.26 (a) $Al(s)$ is oxidized, $Ni^{2+}(aq)$ is reduced.

(b) $Al(s) \rightarrow Al^{3+}(aq) + 3e^-$; $Ni^{2+}(aq) + 2e^- \rightarrow Ni(s)$

(c) $Al(s)$ is the anode; $Ni(s)$ is the cathode.

(d) $Al(s)$ is negative $(-)$; $Ni(s)$ is positive $(+)$.

(e)   Electrons flow from the Al(–) electrode toward the Ni(+) electrode.

(f)   Cations migrate toward the Ni(s) cathode; anions migrate toward the Al(s) anode.

Cell EMF under Standard Conditions

20.28   (a)   In a voltaic cell, the anode has the higher potential energy for electrons. To achieve a lower potential energy, electrons flow from the anode to the cathode.

(b)   The units of electrical potential are volts. A potential of one volt imparts one joule of energy to one coulomb of charge.

(c)   A *standard* cell potential describes the potential of an electrochemical cell where all components are present at standard conditions: elements in their standard states, gases at 1 atm pressure and 1 M aqueous solutions.

20.30   (a) $H_2(g) \rightarrow 2H^+(aq) + 2e^-$

(b)   The platinum electrode serves as a reaction surface; the greater the surface area, the more H_2 or H^+ that can be adsorbed onto the surface to facilitate the flow of electrons.

(c)

20.32   (a)   It is not possible to measure the standard reduction potential of a single half-reaction because each voltaic cell consists of two half-reactions and only the potential of a complete cell can be measured.

(b)   The standard reduction potential of a half-reaction is determined by combining it with a reference half-reaction of known potential and measuring the cell potential. Assuming the half-reaction of interest is the reduction half-reaction:

$E^{\circ}_{cell} = E^{\circ}_{red}(cathode) - E^{\circ}_{red}(anode) = E^{\circ}_{red}(unknown) - E^{\circ}_{red}(reference);$

$E^{\circ}_{red}(unknown) = E^{\circ}_{cell} + E^{\circ}_{red}(reference).$

20.34   (a) $PdCl_4^{2-}(aq) + 2e^- \rightarrow Pd(s) + 4Cl^-$ cathode $E^{\circ}_{red} = ?$

$Cd(s) \rightarrow Cd^{2+}(aq) + 2e^-$ anode $E^{\circ}_{red} = -0.403\,V$

(b) $E^{\circ}_{cell} = E^{\circ}_{red}(cathode) - E^{\circ}_{red}(anode); 1.03\,V = E^{\circ}_{red} - (-0.403\,V);$

$E^{\circ}_{red} = 1.03\,V - 0.403 = 0.63\,V$

(c)

20.36 (a) $F_2(g) + 2e^- \rightarrow 2F^-(aq)$ $E^\circ_{red} = 2.87$ V

$H_2(g) \rightarrow 2H^+(aq) + 2e^-$ $E^\circ_{red} = 0.00$ V

$E^\circ = 2.87$ V $- 0.00$ V $= 2.87$ V

(b) $Cu^{2+}(aq) + 2e^- \rightarrow Cu(s)$ $E^\circ_{red} = 0.337$ V

$Ca(s) \rightarrow Ca^{2+}(aq) + 2e^-$ $E^\circ_{red} = -2.87$ V

$E^\circ = 0.337$ V $- (-2.87$ V$) = 3.21$ V

(c) $Fe^{2+}(aq) + 2e^- \rightarrow Fe(s)$ $E^\circ_{red} = -0.440$ V

$2[Fe^{2+}(aq) \rightarrow Fe^{3+}(aq) + 1e^-]$ $E^\circ_{red} = 0.771$ V

$E^\circ = -0.440$ V $- 0.771$ V $= -1.211$ V

(d) $Hg_2^{2+}(aq) + 2e^- \rightarrow 2Hg(l)$ $E^\circ_{red} = 0.789$ V

$2[Cu^+(aq) \rightarrow Cu^{2+}(aq) + 1e^-]$ $E^\circ_{red} = 0.153$ V

$E^\circ = 0.789$ V $- 0.153$ V $= 0.636$ V

20.38 (a)

$2[Au(s) + 4Br^-(aq) \rightarrow AuBr_4^-(aq) + 3e^-]$ $E^\circ_{red} = -0.858$ V

$3[2e^- + IO^-(aq) + H_2O(l) \rightarrow I^-(aq) + 2OH^-(aq)]$ $E^\circ_{red} = 0.49$ V

$2Au(s) + 8Br^-(aq) + 3IO^-(aq) + 3H_2O(l) \rightarrow 2AuBr_4^-(aq) + 3I^-(aq) + 6OH^-(aq)$

$E^\circ = 0.49 - (-0.858) = 1.35$ V

(b)

$2[Eu^{2+}(aq) \rightarrow Eu^{3+}(aq) + 1e^-]$ $E^\circ_{red} = -0.43$ V

$Sn^{2+}(aq) + 2e^- \rightarrow Sn(s)$ $E^\circ_{red} = -0.14$ V

$2Eu^{2+}(aq) + Sn^{2+}(aq) \rightarrow 2Eu^{3+}(aq) + Sn(s)$ $E^\circ = -0.14 - (-0.43) = 0.29$ V

20.40 (a) The two half-reactions are:

$Cd^{2+}(aq) + 2e^- \rightarrow Cd(s)$ $E^\circ = -0.403$ V

$Cl_2(g) + 2e^- \rightarrow 2Cl^-(aq)$ $E^\circ = 1.359$ V

Because E° for the reduction of Cl_2 is greater, Cl_2 is reduced at the cathode, the Pt electrode. $Cd(s)$ is oxidized at the anode, the Cd electrode.

(b) The Cd anode loses mass as $Cd^{2+}(aq)$ is produced.

(c) $Cl_2(g) + Cd(s) \rightarrow Cd^{2+}(aq) + 2Cl^-(aq)$

(d) $E° = 1.359 \text{ V} - (-0.403 \text{ V}) = 1.762 \text{ V}$

Strengths of Oxidizing and Reducing Agents

20.42   Follow the logic in Sample Exercise 20.8. In each case, choose the half-reaction with the more positive reduction potential and with the given substance on the left.

(a) $Cl_2(g)$ (1.359 V vs. 1.065 V)

(b) $Cd^{2+}(aq)$ (–0.403 V vs. –0.763 V)

(c) $BrO_3^-(aq)$ (1.52 V vs. 1.195 V)

(d) $O_3(g)$ (2.07 V vs. 1.776 V)

20.44   If the substance is on the left of a reduction half-reaction, it will be an oxidant; if it is on the right, it will be a reductant. The sign and magnitude of the $E_{red}^°$ determine whether it is strong or weak.

(a) $Ce^{3+}(aq)$: very weak reductant (on the right, $E_{red}^° = 1.61 \text{ V}$)

(b) $Ca(s)$: strong reductant (on the right, $E_{red}^° = -2.87 \text{ V}$)

(c) $ClO_3^-(aq)$: strong oxidant (on the left, $E_{red}^° = 1.47 \text{ V}$)

(d) $N_2O_5(g)$: oxidant (N has maximum oxidation number, +5; can only be reduced and act as oxidant.)

20.46   (a)   The strongest oxidizing agent is the species most readily reduced, as evidenced by a large, positive reduction potential. That species is H_2O_2. The weakest oxidizing agent is the species that least readily accepts an electron. We expect that it will be very difficult to reduce Zn(s); indeed, Zn(s) acts as a comparatively strong **reducing** agent. No potential is listed for reduction of Zn(s), but we can safely assume that it is less readily reduced than any of the other species present.

(b)   The strongest reducing agent is the species most easily oxidized (the largest negative reduction potential). Zn, $E_{red}^° = -0.76 \text{ V}$, is the strongest reducing agent and F^-, $E_{red}^° = 2.87 \text{ V}$, is the weakest.

20.48   Any oxidized species from Appendix E with a reduction potential greater than 0.59 V will oxidize RuO_4^{2-} to RuO_4^-. From the list of possible oxidants in the exercise, $Br_2(l)$ and $BrO_3^-(aq)$ will definitely oxidize RuO_4^{2-} to RuO_4^-. $Sn^{2+}(aq)$ will not, and $O_2(g)$ depends on conditions. In base, it will not, but in strongly acidic solution, it will.

Free Energy and Redox Reactions

20.50   (a)

$$2I^-(aq) \rightarrow I_2(s) + 2e^- \qquad E_{red}^° = 0.536 \text{ V}$$
$$\underline{Hg_2^{2+}(aq) + 2e^- \rightarrow 2Hg(l) \qquad E_{red}^° = 0.789 \text{ V}}$$
$$2I^-(aq) + Hg_2^{2+}(aq) \rightarrow I_2(s) + 2Hg(l) \qquad E° = 0.789 - 0.536 = 0.253 \text{ V}$$

$$\Delta G^\circ = -nFE^\circ = -2 \text{ mol e}^- \times \frac{96.5 \text{ kJ}}{\text{V - mol e}^-} \times 0.253 \text{ V} = -48.829 = -48.8 \text{ kJ}$$

$$\ln K = \frac{-(-4.8829 \times 10^4 \text{ J})}{(8.314 \text{ J/mol - K})(298 \text{ K})} = 19.708 = 19.7; \quad K = e^{19.7} = 3.61 \times 10^8 = 3.6 \times 10^8$$

(b)

$$3[Cu^+(aq) \rightarrow Cu^{2+}(aq) + 1e^-] \qquad E^\circ_{red} = 0.153 \text{ V}$$

$$\frac{NO_3^-(aq) + 4H^+(aq) + 3e^- \rightarrow NO(g) + H_2O(l) \qquad E^\circ_{red} = 0.96 \text{ V}}{3Cu^+(aq) + NO_3^-(aq) + 4H^+(aq) \rightarrow 3Cu^{2+}(aq) + NO(g) + 2H_2O(l)}$$

$$E^\circ = 0.96 - 0.153 = 0.81 \text{ V}; \Delta G^\circ = -3(96.5)(0.81) = -2.345 \times 10^2 \text{ kJ} = -2.3 \times 10^5 \text{ J}$$

$$\ln K = \frac{-(-2.345 \times 10^5 \text{ J})}{(8.314 \text{ J/mol - K})(298 \text{ K})} = 94.65 = 95; \quad K = e^{95} = 1.3 \times 10^{41} = 10^{41}$$

(c)

$$2[Cr(OH)_3(s) + 5OH^-(aq) \rightarrow CrO_4^{2-}(aq) + 4H_2O(l) + 3e^-] \quad E^\circ_{red} = -0.13 \text{ V}$$

$$\frac{3[ClO^-(aq) + H_2O(l) + 2e^- \rightarrow Cl^-(aq) + 2OH^-(aq)] \qquad E^\circ_{red} = 0.89 \text{ V}}{2Cr(OH)_3(s) + 3ClO^-(aq) + 4OH^-(aq) \rightarrow 2CrO_4^{2-}(aq) + 3Cl^-(aq) + 5H_2O(l)}$$

$$E^\circ = 0.89 - (-0.13) = 1.02 \text{ V}; \Delta G^\circ = -6(96.5)(1.02) = -590.58 \text{ kJ} = -5.91 \times 10^5 \text{ J}$$

$$\ln K = \frac{-(-5.9058 \times 10^5 \text{ J})}{(8.314 \text{ J/mol - K})(298 \text{ K})} = 238.37 = 238; \quad K = 3.3 \times 10^{103} = 10^{103}$$

This is an unimaginably large number.

20.52 $K = 8.7 \times 10^4$; $\Delta G^\circ = -RT \ln K$; $E^\circ = -\Delta G^\circ/nF$; $n = 1$; $T = 298 \text{ K}$

$$\Delta G^\circ = -8.314 \text{ J/mol-K} \times 298 \text{ K} \times \ln(8.7 \times 10^4) = -2.818 \times 10^4 \text{ J} = -28.2 \text{ kJ}$$

$$E^\circ = -\Delta G^\circ/nF = \frac{-(-28.18 \text{ kJ})}{1e^- \times 96.5 \text{ kJ/V - mol e}^-} = 0.292 \text{ V}$$

20.54 $E^\circ = \dfrac{0.0592 \text{ V}}{n} \log K$; $\log K = \dfrac{nE^\circ}{0.0592 \text{ V}}$. See Solution 20.53 for a more complete explanation.

(a) $E^\circ = 0.799 \text{ V} - 0.337 \text{ V} = 0.462 \text{ V}$; $n = 2$ $(2Ag^+ + 2e^- \rightarrow 2Ag)$

$$\log K = \frac{2(0.462 \text{ V})}{0.0592 \text{ V}} = 15.6081 = 15.6; \quad K = 4.056 \times 10^{15} = 4 \times 10^{15}$$

(b) $E^\circ = 1.61 \text{ V} - 0.32 \text{ V} = 1.29 \text{ V}$; $n = 3$ $(3Ce^{4+} + 3e^- \rightarrow 3Ce^{3+})$

$$\log K = \frac{3(1.29)}{0.0592} = 65.372 = 65.4; K = 2.35 \times 10^{65} = 2 \times 10^{65}$$

(c) $E^\circ = 0.36 \text{ V} - (-0.23 \text{ V}) = 0.59 \text{ V}$; $n = 4$ $(4Fe(CN)_6^{3-} + 4e^- \rightarrow 4Fe(CN)_6^{4-})$

$$\log K = \frac{4(0.59)}{0.0592} = 39.865 = 40; \quad K = 7.3 \times 10^{39} = 10^{40}$$

20.56 $E^\circ = \dfrac{0.0592 \text{ V}}{n} \log K$; $n = \dfrac{0.0592 \text{ V}}{E^\circ} \log K$. See Solution 20.53 for a more complete development.

$$n = \frac{0.0592 \text{ V}}{0.17 \text{ V}} \log (5.5 \times 10^5); n = 2$$

Cell EMF under Nonstandard Conditions

20.58 **(a)** No. As the spontaneous chemical reaction of the voltaic cell proceeds, the concentrations of products increase and the concentrations of reactants decrease, so standard conditions are not maintained.

(b) Yes. The Nernst equation is applicable to cell EMF at nonstandard conditions, so it must be applicable at temperatures other than 298 K. There are two terms in the Nernst Equation. First, values of $E°$ at temperatures other than 298 K are required. Then, in the form of Equation [20.14], there is a variable for T in the second term. In the short-hand form of Equation [20.16], the value 0.0592 assumes 298 K. A different coefficient would apply to cells at temperatures other than 298 K.

(c) If concentration of products increases, Q increases, and E decreases.

20.60 $Al(s) + 3Ag^+(aq) \rightarrow Al^{3+}(aq) + 3Ag(s); \quad E = E° - \dfrac{0.0592}{n} \log Q; \quad Q = \dfrac{[Al^{3+}]}{[Ag^+]^3}$

Any change that causes the reaction to be less spontaneous (that causes Q to increase and ultimately shifts the equilibrium to the left) will result in a less positive value for E.

(a) Increases E by decreasing $[Al^{3+}]$ on the right side of the equation, which decreases Q.

(b) No effect; the "concentrations" of pure solids and liquids do not influence the value of K for a heterogeneous equilibrium.

(c) No effect; the concentration of Ag^+ and the value of Q are unchanged.

(d) Decreases E; forming AgCl(s) decreases the concentration of Ag^+, which increases Q.

20.62 **(a)**

$$3[Ce^{4+}(aq) + 1e^- \rightarrow Ce^{3+}(aq)] \qquad E°_{red} = 1.61 \text{ V}$$
$$\underline{Cr(s) \rightarrow Cr^{3+}(aq) + 3e^- \qquad\qquad E°_{red} = -0.74 \text{ V}}$$
$$3Ce^{4+}(aq) + Cr(s) \rightarrow 3Ce^{3+}(aq) + Cr^{3+}(aq) \qquad E° = 1.61 - (-0.74) = 2.35 \text{ V}$$

(b)

$$E = E° - \frac{0.0592}{n} \log \frac{[Ce^{3+}]^3 [Cr^{3+}]}{[Ce^{4+}]^3}; \quad n = 3$$

$$E = 2.35 - \frac{0.0592}{3} \log \frac{(0.10)^3 (0.010)}{(3.0)^3} = 2.35 - \frac{0.0592}{3} \log (3.704 \times 10^{-7})$$

$$E = 2.35 - \frac{0.0592 (-6.431)}{3} = 2.35 + 0.127 = 2.48 \text{ V}$$

(c)

$$E = 2.35 - \frac{0.0592}{3} \log \frac{(1.75)^3 (2.5)}{(0.10)^3} = 2.35 - 0.0814 = 2.27 \text{ V}$$

20.64   (a) $2[Fe^{3+}(aq) + 1e^- \rightarrow Fe^{2+}(aq)]$ $E^\circ_{red} = 0.771$ V

$$\frac{H_2(g) \rightarrow 2H^+(aq) + 2e^- \qquad\qquad E^\circ_{red} = 0.000 \text{ V}}{2Fe^{3+}(aq) + H_2(g) \rightarrow 2Fe^{2+}(aq) + 2H^+(aq) \quad E^\circ = 0.771 - 0.000 = 0.771 \text{ V}}$$

     (b)

$$E = E^\circ - \frac{0.0592}{n} \log \frac{[Fe^{2+}]^2[H^+]^2}{[Fe^{3+}]^2 P_{H_2}}; \quad [H^+] = 10^{-pH} = 1.0 \times 10^{-5}, \quad n = 2$$

$$E = 0.771 - \frac{0.0592}{2} \log \frac{(0.0010)^2 (1.0 \times 10^{-5})^2}{(2.50)^2 (0.85)} = 0.771 - \frac{0.0592}{2} \log(1.9 \times 10^{-17})$$

$$E = 0.771 - \frac{0.0592(-16.73)}{2} = 0.771 + 0.4951 = 1.266 \text{ V}$$

20.66   (a)   The compartment with 0.0150 M Cl$^-$ (aq) is the cathode.

     (b) $E^\circ = 0$ V

     (c) $E = E^\circ - \dfrac{0.0592}{n} \log Q; \quad Q = [Cl^-, \text{dilute}] / [Cl^-, \text{conc.}]$

$$E = 0 - \frac{0.0592}{1} \log \frac{(0.0150)}{(2.55)} = -0.13204 = -0.1320 \text{ V}$$

     (d)   In the anode compartment, [Cl$^-$] will decrease from 2.55 M. In the cathode, [Cl$^-$] will increase from 0.0150 M.

20.68   (a) $E^\circ = -0.136$ V $- (-0.126$ V$) = -0.010$ V; $n = 2$

$$0.22 = -0.010 - \frac{0.0592}{2} \log \frac{[Pb^{2+}]}{[Sn^{2+}]} = -0.010 - \frac{0.0592}{2} \log \frac{[Pb^{2+}]}{1.00}$$

$$\log[Pb^{2+}] = \frac{-0.23(2)}{0.0592} = -7.770 = -7.8; \quad [Pb^{2+}] = 1.7 \times 10^{-8} = 2 \times 10^{-8} \text{ } M$$

     (b)   For PbSO$_4$(s), K$_{sp}$ = [Pb^{2+}] [SO$_4{}^{2-}$] = $(1.0)(1.7 \times 10^{-8}) = 1.7 \times 10^{-8}$

Batteries and Fuel Cells

20.70   (a)   First, H$_2$O is a reactant in the cathodic half-reaction, so it must be present in some form. Additionally, liquid water enhances mobility of the hydroxide ion in the alkaline battery. OH$^-$ is produced in the cathode compartment and consumed in the anode compartment. It must be available at all points where Zn(s) is being oxidized. If the Zn(s) near the separator is mostly reacted, OH$^-$ must diffuse through the gel until it reaches fresh Zn(s). A small amount of H$_2$O(l) mobilizes OH$^-$ so that redox can continue until reactants throughout the battery are depleted.

     (b)   Highly concentrated or solid reactants offer a large amount of reactant in a small sample volume. The more available reactant, the longer the cell produces a voltage. A voltaic cell with solid or highly concentrated reactants has the advantages of small size and long operational lifetime.

20.72   The overall cell reaction is:

$$2MnO_2(s) + Zn(s) + 2H_2O(l) \rightarrow 2MnO(OH)(s) + Zn(OH)_2(s)$$

$$4.50 \text{ g Zn} \times \frac{1 \text{ mol Zn}}{65.39 \text{ g Zn}} \times \frac{2 \text{ mol } MnO_2}{1 \text{ mol Zn}} \times \frac{86.94 \text{ g } MnO_2}{1 \text{ mol } MnO_2} = 12.0 \text{ g } MnO_2 \text{ reduced}$$

20.74   (a) $HgO(s) + Zn(s) \rightarrow Hg(l) + ZnO(s)$

        (b) $E°_{cell} = E°_{red} \text{ (cathode)} - E°_{red} \text{ (anode)}$

              $E°_{red} \text{ (anode)} = E°_{red} - E°_{cell} = 0.098 - 1.35 = -1.25 \text{ V}$

        (c) $E°_{red}$ is different from $Zn^{2+}(aq) + 2e^- \rightarrow Zn(s)$ (–0.76 V) because in the battery the process happens in the presence of base and Zn^{2+} is stabilized as $ZnO(s)$. Stabilization of a reactant in a half-reaction decreases the driving force, so $E°_{red}$ is more negative.

20.76   (a)   The alkali metal Li has much greater metallic character than Zn, Cd, Pb or Ni. The reduction potential for Li is thus more negative, leading to greater overall cell emf for the battery. Also, Li is less dense than the other metals, so greater total energy for a battery can be achieved for a given total mass of material. One disadvantage is that Li is very reactive and the cell reactions are difficult to control.

        (b)   Li has a much smaller molar mass (6.94 g/mol) than Ni (58.69 g/mol). A Li-ion battery can have many more charge-carrying particles than a Ni-based battery with the same mass. That is, Li-ion batteries have a greater *energy density* than Ni-based batteries.

20.78   (a)   Both batteries and fuel cells are electrochemical power sources. Both take advantage of spontaneous oxidation-reduction reactions to produce a certain voltage. The difference is that batteries are self-contained (all reactants and products are present inside the battery casing) while fuel cells require continuous supply of reactants and exhaust of products.

        (b)   No. The fuel in a fuel cell must be fluid, either gas or liquid. Because fuel must be continuously supplied to the fuel cell, it must be capable of flow; the fuel cannot be solid.

Corrosion

20.80   (a)   Calculate $E°_{cell}$ for the given reactants at standard conditions.

$O_2(g) + 4H^+(aq) + 4e^- \rightarrow 2H_2O(l)$	$E°_{red} = 1.23 \text{ V}$
$\underline{2[Cu(s) \rightarrow Cu^{2+}(aq) + 2e^-]}$	$\underline{E°_{red} = 0.337 \text{ V}}$
$2Cu(s) + O_2(g) + 4H^+(aq) \rightarrow 2Cu^{2+}(aq) + 2H_2O(l)$	$E° = 1.23 - 0.337 = 0.89 \text{ V}$

            At standard conditions with $O_2(g)$ and $H^+(aq)$ present, the oxidation of Cu(s) has a positive E° value and is spontaneous. Cu(s) will oxidize (corrode) in air in the presence of acid.

(b) Fe^{2+} has a more negative reduction potential (–0.440 V) than Cu^{2+} (+0.337 V), so Fe(s) is more readily oxidized than Cu(s). If the two metals are in contact, Fe(s) would act as a sacrificial anode and oxidize (corrode) in preference to Cu(s); this would weaken the iron support skeleton of the statue. The teflon spacers prevent contact between the two metals and insure that the iron skeleton doesn't corrode when the Cu(s) skin comes in contact with atmospheric O_2(g) and H^+(aq).

20.82   No. To afford cathodic protection, a metal must be more difficult to reduce (have a more negative reduction potential) than Fe^{2+}. E°_{red} Co^{2+} = –0.28 V, E°_{red} Fe^{2+} = –0.44 V.

20.84   The principal metallic component of steel is Fe. E°_{red} for Fe, –0.763 V, is more negative than that of Cu, 0.337 V. When the two are in contact, Fe acts as the sacrificial anode and corrodes (oxidizes) preferentially in the presence of O_2(g).

$2Fe(s) + O_2(g) + 4H^+(aq) \rightarrow 2Fe^{2+}(aq) + 2H_2O(l)$

$E^\circ = 1.23\ V - (-0.440\ V) = 1.67\ V$

$2Cu(s) + O_2(g) + 4H^+(aq) \rightarrow 2Cu^{2+}(aq) + 2H_2O(l)$

$E^\circ = 1.23\ V - (0.337\ V) = 0.893\ V$

Both reactions are spontaneous, but the corrosion of Fe has the larger E° value and happens preferentially.

Electrolysis; Electrical Work

20.86   (a)   An *electrolytic cell* is the vessel in which electrolysis occurs. It consists of a power source and two electrodes in a molten salt or aqueous solution.

(b)   It is the cathode. In an electrolysis cell, as in a voltaic cell, electrons are consumed (via reduction) at the cathode. Electrons flow from the negative terminal of the voltage source and then to the cathode.

(c)   A small amount of H_2SO_4(aq) present during the electrolysis of water acts as a change carrier, or supporting electrolyte. This facilitates transfer of electrons through the solution and at the electrodes, speeding up the reaction. (Considering H^+(aq) as the substance reduced at the cathode changes the details of the half-reactions, but not the overall E° for the electrolysis. SO_4^{2-}(aq) cannot be oxidized.)

20.88   Coulombs = **amps-s**; since this is a 2e⁻ reduction, each mole of Mg(s) requires 2 Faradays.

(a) $4.55\ A \times 3.50\ d \times \dfrac{24\ h}{1\ d} \times \dfrac{60\ min}{1\ h} \times \dfrac{60\ s}{1\ min} \times \dfrac{1\ C}{1\ amp\text{-}s} \times \dfrac{1\ F}{96,500\ C}$

$\times \dfrac{1\ mol\ Mg}{2\ F} \times \dfrac{24.31\ g\ Mg}{1\ mol\ Mg} = 173\ g\ Mg$

(b) $10.00\ g\ Mg \times \dfrac{1\ mol\ Mg}{24.31\ g\ Mg} \times \dfrac{2\ F}{1\ mol\ Mg} \times \dfrac{96,500\ C}{F} \times \dfrac{1\ amp\text{-}s}{C} \times \dfrac{1\ min}{60\ s} \times \dfrac{1}{3.50\ A}$

$= 378\ min$

20.90   For this cell at standard conditions, $E° = 1.10$ V.

$$w_{max} = \Delta G° = -nFE° = -2(96.5)(1.10) = -212.3 = -212 \text{ kJ/mol Cu}$$

$$50.0 \text{ g Cu} \times \frac{1 \text{ mol Cu}}{63.55 \text{ g Cu}} \times \frac{-212.3 \text{ kJ}}{\text{mol Cu}} = -167 \text{ kJ} = -1.67 \times 10^5 \text{ J}$$

20.92   (a) $7.5 \times 10^3 \text{ A} \times 48 \text{ h} \times \dfrac{3600 \text{ s}}{1 \text{ h}} \times \dfrac{1 \text{ C}}{1 \text{ amp-s}} \times \dfrac{1 \text{ F}}{96,500 \text{ C}} \times \dfrac{1 \text{ mol Ca}}{2 \text{ F}}$

$$\times \ 0.68 = 4566 = 4.6 \times 10^3 \text{ mol Ca}$$

$$4566 \text{ mol Ca} \times \frac{40.0 \text{ g Ca}}{1 \text{ mol Ca}} = 1.830 \times 10^5 = 1.8 \times 10^5 \text{ g Ca}$$

(b)   If the cell is 68% efficient, $\dfrac{96,500 \text{ C}}{\text{F}} \times \dfrac{2 \text{ F}}{0.68 \text{ mol Ca}} = 2.838 \times 10^5$

$$= 2.8 \times 10^5 \text{ C/mol Ca required}$$

$$5.00 \text{ V} \times 4566 \text{ mol Ca} \times \frac{2.838 \times 10^5 \text{ C}}{\text{mol Ca}} \times \frac{1 \text{ J}}{\text{C-V}} \times \frac{1 \text{ kWh}}{3.6 \times 10^6 \text{ J}} = 1799.85 = 1.8 \times 10^3 \text{ kWh}$$

Additional Exercises

20.94   (a)

(c) $MnO_4^-(aq) + 8H^+(aq) + 5e^- \rightarrow Mn^{2+}(aq) + 4H_2O(l)$ $E_{red}^° = 1.51$ V
$5[Fe^{2+}(aq) \rightarrow Fe^{3+}(aq) + 1e^-]$ $E_{red}^° = 0.771$ V

$E° = 1.51 \text{ V} - 0.771 \text{ V} = 0.74 \text{ V}$

(d) $E = E° - \dfrac{0.0592}{5} \log \dfrac{[Fe^{3+}]^5[Mn^{2+}]}{[Fe^{2+}]^5[MnO_4^-][H^+]^8}$; $pH = 0.0, [H^+] = 1.0$

$E = 0.74\ V - \dfrac{0.0592}{5} \log \dfrac{(2.5 \times 10^{-4})^5 (0.010)}{(0.10)^5 (1.50)(1.0)^8}$; $Q = 6.510 \times 10^{-16} = 6.5 \times 10^{-16}$

$E = 0.74\ V - \dfrac{0.0592(-15.1864)}{5} = 0.74\ V + 0.18\ V = 0.92\ V$

20.95   (a)

$Fe(s) \rightarrow Fe^{2+}(aq) + 2e^-$

$\dfrac{2Ag^+(aq) + 2e^- \rightarrow 2Ag(s)}{Fe(s) + 2Ag^+(aq) \rightarrow Fe^{2+}(aq) + 2Ag(s)}$

(b)

$Zn(s) \rightarrow Zn^{2+}(s) + 2e^-$

$\dfrac{2H^+(aq) + 2e^- \rightarrow H_2(g)}{Zn(s) + 2H^+(aq) \rightarrow Zn^{2+}(aq) + H_2(g)}$

(c) $Cu \,|\, Cu^{2+} \,\|\, ClO_3^-, Cl^- \,|\, Pt$ Here, both the oxidized and reduced forms of the cathode solution are in the same phase, so we separate them by a comma, and then indicate an inert electrode.

20.97   (a)   The reduction potential for $O_2(g)$ in the presence of acid is 1.23 V. $O_2(g)$ cannot oxidize $Au(s)$ to $Au^+(aq)$ or $Au^{3+}(aq)$, even in the presence of acid.

(b)   The possible oxidizing agents need a reduction potential greater than 1.50 V. These include $Co^{3+}(aq)$, $F_2(g)$, $H_2O_2(aq)$, and $O_3(g)$. Marginal oxidizing agents (those with reduction potential near 1.50 V) from Appendix E are $BrO_3^-(aq)$, $Ce^{4+}(aq)$, $HClO(aq)$, $MnO_4^-(aq)$, and $PbO_2(s)$.

(c) $4Au(s) + 8NaCN(aq) + 2H_2O(l) + O_2(g) \rightarrow 4Na[Au(CN)_2](aq) + 4NaOH(aq)$

$Au(s) + 2CN^-(aq) \rightarrow [Au(CN)_2]^- + 1e^-$

$$O_2(g) + 2H_2O(l) + 4e^- \rightarrow 4OH^-(aq)$$

Au(s) is being oxidized and $O_2(g)$ is being reduced.

(d) $2[Na[Au(CN)_2](aq) + 1e^- \rightarrow Au(s) + 2CN^-(aq) + Na^+(aq)]$

$$\underline{\quad\quad Zn(s) \rightarrow Zn^{2+}(aq) + 2e^- \quad\quad}$$

$$2Na[Au(CN)_2](aq) + Zn(s) \rightarrow 2Au(s) + Zn^{2+}(aq) + 2Na^+(aq) + 4CN^-(aq)$$

Zn(s) is being oxidized and $[Au(CN)_2]^-(aq)$ is being reduced. While $OH^-(aq)$ is not included in the redox reaction above, its presence in the reaction mixture probably causes $Zn(OH)_2(s)$ to form as the product. This increases the driving force (and E°) for the overall reaction.

20.98   The total charge, nF, depends on moles of electrons and thus the amount of reactants. According to the Nernst equation, cell potential depends on concentrations of reactants (and products).

20.100 (a) $I_2(s) + 2e^- \rightarrow 2I^-(aq)$ $E_{red}^{\circ} = 0.536$ V

$$\underline{\quad 2[Cu(s) \rightarrow Cu^+(aq) + 1e^-] \quad\quad E_{red}^{\circ} = 0.521 \text{ V}}$$

$$I_2(s) + 2Cu(s) \rightarrow 2Cu^+(aq) + 2I^-(aq) \quad E^{\circ} = 0.536 - 0.521 = 0.015 \text{ V}$$

$$E = E^{\circ} - \frac{0.0592}{n}\log Q = 0.015 - \frac{0.0592}{2}\log [Cu^+]^2 [I^-]^2$$

$$E = +0.015 - \frac{0.0592}{2}\log (0.25)^2(3.5)^2 = +0.015 + 0.0034 = 0.018 \text{ V}$$

(b)   Since the cell potential is positive at these concentration conditions, the reaction as written in part (a) is spontaneous in the forward direction. Cu is oxidized and Cu(s) is the anode.

(c)   Yes. E° is positive, so Cu is oxidized and Cu(s) is the anode at standard conditions.

(d) $E = 0; \quad +0.015 = \dfrac{0.0592}{2}\log (0.15)^2 [I^-]^2; \quad \dfrac{2(0.015)}{0.0592} = \log (0.15)^2 + 2\log [I^-];$

$\log[I^-] = 1.0773 = 1.08; \quad [I^-] = 10^{1.08} = 11.95 = 12 M\ I^-$

20.101 Both E° and K are related to ΔG°. (See Solution 20.53.)

ΔG° = –nFE°; ΔG° = –RT lnK

$-nFE^{\circ} = -RT \ln K, E^{\circ} = \dfrac{RT}{nF}\ln K$

In terms of base 10 logs, lnK = 2.303 log K.

$$E^{\circ} = \frac{2.303\, RT}{nF}\log K$$

From the development of the Nernst equation,

$$\frac{2.303\, RT}{F} = 0.0592, E^{\circ} = \frac{0.0592}{n}\log K$$

20.103   (a) In discharge: $Cd(s) + 2NiO(OH)(s) + 2H_2O(l) \rightarrow Cd(OH)_2(s) + 2Ni(OH)_2(s)$
In charging, the reverse reaction occurs.

        (b) $E° = 0.49\ V - (-0.76\ V) = 1.25\ V$

        (c) The 1.25 V calculated in part (b) is the standard cell potential, $E°$. The concentrations of reactants and products inside the battery are adjusted so that the cell output is greater than $E°$. Note that most of the reactants and products are pure solids or liquids, which do not appear in the Q expression. It must be $[OH^-]$ that is other than 1.0 M, producing an emf of 1.30 rather than 1.25.

        (d) $E° = \dfrac{0.0592}{n} \log K; \log k = \dfrac{nE°}{0.0592}$

$$\log K = \frac{2 \times 1.25}{0.0592} = 42.23 = 42.2; K = 1.7 \times 10^{42} = 2 \times 10^{42}$$

20.104   (a) The battery capacity expressed in units of mAh indicates the total amount of electrical charge that can be delivered by the battery.

        (b) Quantity of electrical charge is measured in coulombs, C. $C = A \cdot s$

$$2850\ m\ Ah \times \frac{1\ A}{1000\ m\ A} \times \frac{3600\ s}{h} \times \frac{1\ C}{1\ A \cdot s} = 10,260\ C$$

The battery can deliver 10,260 C. Work, electrical or otherwise, is measured in J. $J = V \times C$. If the battery voltage decreases linearly from 1.55 V to 0.80 V, assume an average voltage of 1.175 = 1.2 V.

$10,260\ C \times 1.175\ V = 12,055.5 = 12 \times 10^3\ J = 12\ kJ$

The total maximum electrical work of the battery is 12 kJ.

(This is 3.3×10^{-3} kWh or 3.3 Wh.)

20.106   It is well established that corrosion occurs most readily when the metal surface is in contact with water. Thus, moisture is a requirement for corrosion. Corrosion also occurs more readily in acid solution, because O_2 has a more positive reduction potential in the presence of $H^+(aq)$. SO_2 and its oxidation products dissolve in water to produce acidic solutions, which encourage corrosion. The anodic and cathodic reactions for the corrosion of Ni are:

$$Ni(s) \rightarrow Ni^{2+}(aq) + 2e^- \qquad E°_{red} = -0.28\ V$$
$$O_2(g) + 4H^+(aq) + 4e^- \rightarrow 2H_2O(l) \qquad E°_{red} = 1.23\ V$$

Nickel(ll) oxide, $NiO(s)$, can form by the dry air oxidation of Ni. This NiO coating serves to protect against further corrosion. However, NiO dissolves in acidic solutions such as those produced by SO_2 or SO_3, according to the reaction:

$NiO(s) + 2H^+(aq) \rightarrow Ni^{2+}(aq) + H_2O(l)$. This exposes Ni(s) to further wet corrosion.

20.108   (a) The work obtainable is given by the product of the voltage, which has units of J/C, times the number of Coulombs of electricity produced:

$$w_{max} = 300\ amp \cdot h \times \frac{3600\ s}{1\ hr} \times \frac{1\ C}{1\ amp \cdot s} \times \frac{6\ J}{1\ C} \times \frac{1kWh}{3.6 \times 10^6\ J} = 1.8\ kWh \approx 2\ kWh$$

(b)   This maximum amount of work is never realized because some of the electrical energy is dissipated in overcoming the internal resistance of the battery, because the cell voltage does not remain constant as the reaction proceeds, and because the systems to which the electrical energy is delivered are not capable of completely converting electrical energy into work.

20.109   (a) $7 \times 10^8 \text{ mol } H_2 \times \dfrac{2 F}{1 \text{ mol } H_2} \times \dfrac{96,500 \text{ C}}{1 F} = 1.35 \times 10^{14} = 1 \times 10^{14} \text{ C}$

(b)

$$2H_2O(l) \rightarrow O_2(g) + 4H^+(aq) + 4e^- \qquad E^\circ_{red} = 1.23 \text{ V}$$

$$2[2H^+(aq) + 2e^- \rightarrow H_2(g)] \qquad\qquad E^\circ_{red} = 0 \text{ V}$$

$$\overline{2H_2O(l) \rightarrow O_2(g) + 2H_2(g) \qquad\qquad E^\circ = 0.00 - 1.23 = -1.23 \text{ V}}$$

$P_t = 300 \text{ atm} = P_{O_2} + P_{H_2}$. Since $H_2(g)$ and $O_2(g)$ are generated in a 2:1 mole ratio, $P_{H_2} = 200 \text{ atm}$ and $P_{O_2} = 100 \text{ atm}$.

$$E = E^\circ - \dfrac{0.0592}{4} \log (P_{O_2} \times P_{H_2}^2) = -1.23 \text{ V} - \dfrac{0.0592}{4} \log [100 \times (200)^2]$$

$E = -1.23 \text{ V} - 0.100 \text{ V} = -1.33 \text{ V}; E_{min} = 1.33 \text{ V}$

(c) $\text{Energy} = nFE = 2(7 \times 10^8 \text{ mol})(1.33 \text{ V})\dfrac{96,500 \text{ J}}{\text{V-mol}} = 1.80 \times 10^{14} = 2 \times 10^{14} \text{ J}$

(d) $1.80 \times 10^{14} \text{ J} \times \dfrac{1 \text{ kWh}}{3.6 \times 10^6 \text{ J}} \times \dfrac{\$0.85}{\text{kWh}} = \$4.24 \times 10^7 = \$4 \times 10^7$

It would cost more than \$40 million for the electricity alone.

Integrative Exercises

20.111 $N_2(g) + 3H_2(g) \rightarrow 2NH_3(g)$

(a)   The oxidation number of $H_2(g)$ and $N_2(g)$ is 0. The oxidation number of N in NH_3 is –3, H in NH_3 is +1. H_2 is being oxidized and N_2 is being reduced.

(b)   Calculate ΔG° from ΔG°_f values in Appendix C. Use $\Delta G^\circ = -RT \ln K$ to calculate K.

$\Delta G^\circ = 2\Delta G^\circ_f NH_3(g) - \Delta G^\circ_f N_2(g) - 3\Delta G^\circ_f H_2(g)$

$\Delta G^\circ = 2(-16.66 \text{ kJ}) - 0 - 3(0) = -33.32 \text{ kJ}$

$\Delta G^\circ = -RT\ln K, \ln K = \dfrac{-\Delta G^\circ}{RT} = \dfrac{-(-33.32 \times 10^3 \text{ J})}{(8.314 \text{ J/mol-K})(298 \text{ K})} = 13.4487 = 13.45$

$K = e^{13.4487} = 6.9 \times 10^5$

(c) $\Delta G^\circ = -nfE^\circ; \quad E^\circ = \dfrac{-\Delta G^\circ}{nF}; \quad n = ?$

2 N atoms change from 0 to –3, or 6 H atoms change from 0 to +1. Either way, n = 6.

$E^\circ = \dfrac{-(-33.32 \text{ kJ})}{6 \times 96.5 \text{ kJ/V}} = 0.05755 \text{ V}$

20.112   The redox reaction is: $2Ag^+(aq) + H_2(g) \rightarrow 2Ag(s) + 2H^+(aq)$.   n = 2 for this reaction.

$$E^{\circ}_{cell} = E^{\circ}_{red} \text{ cathode} - E^{\circ}_{red} \text{ anode} = 0.799 \text{ V} - 0 \text{ V} = 0.799$$

$$E = E^{\circ} - \frac{0.0592}{n} \log \frac{[H^+]^2}{[Ag^+]^2 P_{H_2}}$$

$[H^+]$ in the cell is held essentially constant by the benzoate buffer.

$$C_6H_5COOH(aq) \rightleftharpoons H^+(aq) + C_6H_5COO^-(aq) \quad K_a = ?$$

$$K_a = \frac{[H^+][C_6H_5COO^-]}{[C_6H_5COOH]}; [H^+] = \frac{K_a[C_6H_5COOH]}{[C_6H_5COO^-]} = \frac{0.10 \text{ M}}{0.050 \text{ M}} \times K_a = 2K_a$$

Solve the Nernst expression for $[H^+]$ and calculate K_a and pK_a as shown above.

$$1.030 \text{ V} = 0.799 \text{ V} - \frac{0.0592}{n} \log \frac{[H^+]^2}{(1.00)^2(1.00)}$$

$$0.231 \times \frac{2}{0.0592} = -\log [H^+]^2 = -2 \log[H^+]$$

$$\frac{0.231}{0.0592} = -\log [H^+] = pH; pH = 3.9020 = 3.90; [H^+] = 10^{-3.902} = 1.253 \times 10^{-4} = 1.3 \times 10^{-4};$$

$$[H^+] = 2K_a, K_a = [H^+]/2 = 6.265 \times 10^{-5} = 6.3 \times 10^{-5}; pK_a = 4.20$$

Check. According to Appendix D, K_a for benzoic acid is 6.3×10^{-5}.

20.114   (a) $NO_3^-(aq) + 4H^+(aq) + 3e^- \rightarrow NO(g) + 2H_2O(l)$ $E^{\circ}_{red} = 0.96$ V

            $Au(s) \rightarrow Au^{3+}(aq) + 3e^-$ $E^{\circ}_{red} = 1.498$ V

     $Au(s) + NO_3^-(aq) + 4H^+(aq) \rightarrow Au^{3+}(aq) + NO(g) + 2H_2O(l)$

$E^{\circ} = 0.96 - 1.498 = -0.54$ V; E° is negative, the reaction is not spontaneous.

(b) $3[2H^+(aq) + 2e^- \rightarrow H_2(g)]$ $E^{\circ}_{red} = 0.000$ V

        $2[Au(s) + 4Cl^-(aq) \rightarrow AuCl_4^-(aq) + 3e^-]$ $E^{\circ}_{red} = 1.002$ V

     $2Au(s) + 6H^+(aq) + 8Cl^-(aq) \rightarrow 2AuCl_4^-(aq) + 3H_2(g)$

$E^{\circ} = 0.000 - 1.002 = -1.002$ V; E° is negative, the reaction is not spontaneous.

(c) $NO_3^-(aq) + 4H^+(aq) + 3e^- \rightarrow NO(g) + 2H_2O(l)$ $E^{\circ}_{red} = 0.96$ V

            $Au(s) + 4Cl^-(aq) \rightarrow AuCl_4^-(aq) + 3e^-$ $E^{\circ}_{red} = 1.002$ V

     $Au(s) + NO_3^-(aq) + 4Cl^-(aq) + 4H^+(aq) \rightarrow AuCl_4^-(aq) + NO(g) + 2H_2O(l)$

$E^{\circ} = 0.96 - 1.002 = -0.04$; E° is small but negative, the process is not spontaneous.

(d) $E = E^{\circ} - \frac{0.0592}{3} \log \frac{[AuCl_4^-] P_{NO}}{[NO_3^-][Cl^-]^4 [H^+]^4}$

If $[H^+]$, $[Cl^-]$ and $[NO_3^-]$ are much greater than 1.0 *M*, the log term is negative and the correction to E° is positive. If the correction term is greater than 0.042 V, the value of E is positive and the reaction at nonstandard conditions is spontaneous.

20.116   (a) $\Delta H° = 2\Delta H° \, H_2O(l) - 2\Delta H° \, H_2(g) - \Delta H° \, O_2(g) = 2(-285.83) - 2(0) - 0 = -571.66$ kJ

               $\Delta S° = 2S° \, H_2O(l) - 2S° \, H_2(g) - \Delta S° \, O_2(g)$

                    $= 2(69.91) - 2(130.58) - (205.0) = -326.34$ J

   (b)   Since $\Delta S°$ is negative, $-T\Delta S$ is positive and the value of ΔG will become more positive as T increases. The reaction will become nonspontaneous at a fairly low temperature, because the magnitude of $\Delta S°$ is large.

   (c) $\Delta G = w_{max}$. The larger the negative value of ΔG, the more work the system is capable of doing on the surroundings. As the magnitude of ΔG decreases with increasing temperature, the usefulness of H_2 as a fuel decreases.

   (d)   The combustion method increases the temperature of the system, which quickly decreases the magnitude of the work that can be done by the system. Even if the effect of temperature on this reaction could be controlled, only about 40% of the energy from any combustion can be converted to electrical energy, so combustion is intrinsically less efficient than direct production of electrical energy via a fuel cell.

20.117   First balance the equation:

       $4CyFe^{2+}(aq) + O_2(g) + 4H^+(aq) \rightarrow 4CyFe^{3+}(aq) + 2H_2O(l);$ E = +0.60 V; n = 4

   (a)   From Equation [20.11] we can calculate ΔG for the process under the conditions specified for the measured potential E:

$$\Delta G = -nFE = -(4 \text{ mol } e^-) \times \frac{96.5 \text{ kJ}}{1 \text{ V - mol } e^-}(0.60 \text{ V}) = -231.6 = -232 \text{ kJ}$$

   (b)   The moles of ATP synthesized per mole of O_2 is given by:

$$\frac{231.6 \text{ kJ}}{O_2 \text{ molecule}} \times \frac{1 \text{ mol ATP formed}}{37.7 \text{ kJ}} = \text{approximately 6 mol ATP/mol } O_2$$

20.119   The reaction can be written as a sum of the steps:

$$Pb^{2+}(aq) + 2e^- \rightarrow Pb(s) \qquad\qquad E°_{red} = -0.126 \text{ V}$$
$$\underline{PbS(s) \rightarrow Pb^{2+}(aq) + S^{2-}(aq) \qquad \text{"}E°\text{"} = ?}$$
$$PbS(s) + 2e^- \rightarrow Pb(s) + S^{2-}(aq) \qquad E°_{red} = ?$$

"E°" for the second step can be calculated from K_{sp}.

$$E° = \frac{0.0592}{n}\log K_{sp} = \frac{0.0592}{2}\log(8.0 \times 10^{-28}) = \frac{0.0592}{2}(-27.10) = -0.802 \text{ V}$$

E° for the half-reaction = -0.126 V + (-0.802 V) = -0.928 V

Calculating an imaginary $E°$ for a nonredox process like step 2 may be a disturbing idea. Alternatively, one could calculate K for step 1 (5.4×10^{-5}), K for the reaction in question (K = $K_1 \times K_{sp} = 4.4 \times 10^{-32}$) and then $E°$ for the half-reaction. The result is the same.

20.120   (a)

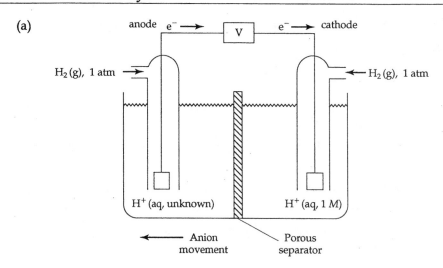

(b)
$$2H^+(aq, 1\,M) + 2e^- \rightarrow H_2(g) \qquad\qquad E^{\circ}_{red} = 0$$

$$\underline{H_2(g) \rightarrow 2H^+(aq, 1\,M) + 2e^- \qquad E^{\circ}_{red} = 0}$$

$$2H^+(aq, 1\,M) + H_2(g) \rightarrow 2H^+(aq, 1\,M) + H_2(g) \quad E^{\circ} = 0$$

(c)   At standard conditions, $[H^+] = 1\,M$, pH = 0

(d)
$$E = E^{\circ} - \frac{0.0592}{2} \log \frac{[H^+ \text{(unknown)}]^2\, P_{H_2}}{[H^+ (1\,M)]^2\, P_{H_2}}$$

$$E = 0 - \frac{0.0592}{2} \log [H^+]^2 = \frac{0.0592}{2} \times 2(-\log [H^+]) = 0.0592(pH) = 0.0592(5.0) = 0.30\ V$$

(e) E_{cell} changes $0.0592(0.01) = 0.000592 = 0.0006$ V for each 0.01 pH unit. The voltmeter would have to be precise to at least 0.001 V to detect a change of 0.01 pH units.

20.121   The two half-reactions in the electrolysis of $H_2O(l)$ are:

$$2[2H_2O(l) + 2e^- \rightarrow H_2(g) + 2OH^-]$$

$$\underline{2H_2O(l) \rightarrow O_2(g) + 4H^+ + 4e^-}$$

$$2H_2O(l) \rightarrow 2H_2(g) + O_2(g)$$

4 mol $e^-/2$ mol $H_2(g)$ or 2 mol $e^-/$ mol $H_2(g)$

Using partial pressures and the ideal-gas law, calculate the mol $H_2(g)$ produced, and the current required to do so.

$P_t = P_{H_2} + P_{H_2O}$. From Appendix B, P_{H_2O} at $25.5\,^{\circ}C$ is approximately 24.5 torr.

$P_{H_2} = 768$ torr $- 24.5$ torr $= 743.5 = 744$ torr

$$n = PV/RT = \frac{(743.5/760)\ atm\ \times\ 0.0123\ L}{298.5\ K\ \times\ 0.08206\ L\text{-}atm/mol\text{-}K} = 4.912 \times 10^{-4} = 4.91 \times 10^{-4}\ mol\ H_2$$

$$4.912 \times 10^{-4}\ mol\ H_2 \times \frac{2\ mol\ e^-}{mol\ H_2} \times \frac{96{,}500\ C}{1\ mol\ e^-} \times \frac{1\ amp\text{-}s}{1\ C} \times \frac{1\ min}{60\ s} \times \frac{1}{2.00\ min} = 0.790\ amp$$

21 Nuclear Chemistry

Visualizing Concepts

21.2   *Analyze/Plan.* From the diagram, determine the atomic number (number of protons) and mass number (number of protons plus neutrons) of the two nuclides involved. Based on the relationship between the two nuclides, decide whether the reaction is α or β-decay, positron emission or electron capture. Complete the nuclear reaction, balancing atomic numbers and mass numbers.

Solve. The two nuclides in the diagram are $^{109}_{46}$Pd and $^{109}_{47}$Ag so the second product is a β-particle. The balanced reaction is:

$$^{109}_{46}\text{Pd} \rightarrow {}^{109}_{47}\text{Ag} + {}^{0}_{-1}\text{e}$$

Check. Atomic number and mass number balance.

21.3   *Analyze/Plan.* Determine the number of protons and neutrons present in the two heavy nuclides in the reaction. Draw a graph with appropriate limits and plot the two points. Draw an arrow from reactant to product.   *Solve.*

Bi: 83 p, 128 n; Tl: 81 p, 126 n.

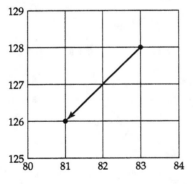

Check. An α-particle has 2 p and 2 n. The diagram shows a decrease in 2 p and 2 n for the reaction.

21.5   (a)   The difference in mass between a nuclide and its component nucleons is the mass defect. It corresponds to the energy required to separate the nuclide into individual nucleons, according to the relationship $E = \Delta mc^2$.

(b)   On Figure 21.13, ^{56}Fe has the largest binding energy per nucleon. ^{100}Ru is near this maximum and has more total nucleons than ^{56}Fe, so we expect the mass defect for ^{100}Ru to be significant.

(c)   *Plan.* Calculate the total mass of the separate nucleons. Subtract the mass of the nuclide to get Δm, the mass defect. Convert Δm to binding energy, divide by 100 to get the binding energy per nucleon.

Δm = mass of individual protons and neutrons – mass of nuclide

$\Delta m = 44(1.0072765 \text{ amu}) + 56(1.0086649 \text{ amu}) - 99.90422 \text{ amu}$

$\Delta m = 0.9011804 = 0.90118 \text{ amu}$

$$\Delta E = (2.9979246 \times 10^8 \text{ m/s})^2 \times 0.9011804 \text{ amu} \times \frac{1 \text{ g}}{6.022 \times 10^{23} \text{ amu}} \times \frac{1 \text{ kg}}{1000 \text{ g}}$$

$$= 1.34497 \times 10^{-10} \text{ J}$$

$$1.34497 \times 10^{-10} \frac{\text{J}}{\text{nuclide}} \times \frac{1 \text{ nuclide}}{100 \text{ nucleons}} = 1.34497 \times 10^{-12} \text{ J/nucleon}$$

21.6   *Analyze/Plan.* Express the particles in the diagram as a nuclear reaction. Determine the mass number and atomic number of the unknown particle by balancing these quantities in the nuclear reaction.   *Solve.*

(a) $^{239}_{94}\text{Pu} + ^{1}_{0}\text{n} \rightarrow ^{95}_{40}\text{Zr} + ? + 2\,^{1}_{0}\text{n}$

The unknown particle has an atomic number of $(94-40) = 54$; it is Xe. The mass number of the nuclide is $[(239 + 1) - (95 + 2)] = 143$. The unknown particle is $^{143}_{54}\text{Xe}$.

(b) ^{95}Zr: 40 p, 55 n is stable. ^{143}Xe: 54 p, 89 n is above the belt of stability and is not stable; it will probably undergo β-decay.

Radioactivity

21.8   p = protons, n = neutrons, e = electrons; number of protons = atomic number; number of neutrons = mass number – atomic number

(a) $^{126}_{55}\text{Cs}: 55\text{p}, 71\text{n}$ (b) $^{119}\text{Sn}: 50\text{p}, 69\text{n}$ (c) $^{141}\text{Ba}: 56\text{p}, 85\text{n}$

21.10   (a) $^{1}_{1}\text{p}$ or $^{1}_{1}\text{H}$ (b) $^{0}_{-1}\text{e}$ or $^{0}_{-1}\beta$ (c) $^{0}_{1}\text{e}$

21.12   (a) $^{191}_{79}\text{Au} + ^{0}_{-1}\text{e} \text{ (orbital electron)} \rightarrow ^{191}_{78}\text{Pt}$ (b) $^{201}_{79}\text{Au} \rightarrow ^{201}_{80}\text{Hg} + ^{0}_{-1}\beta$

         (c) $^{198}_{79}\text{Au} \rightarrow ^{198}_{80}\text{Hg} + ^{0}_{-1}\text{e}$ (d) $^{188}_{79}\text{Au} \rightarrow ^{188}_{78}\text{Pt} + ^{0}_{1}\text{e}$

21.14   (a) $^{24}_{11}\text{Na} \rightarrow ^{24}_{12}\text{Mg} + ^{0}_{-1}\text{e}$; a β particle is produced

         (b) $^{188}_{80}\text{Hg} \rightarrow ^{188}_{79}\text{Au} + ^{0}_{1}\text{e}$; a positron is produced

         (c) $^{122}_{53}\text{I} \rightarrow ^{122}_{54}\text{Xe} + ^{0}_{-1}\text{e}$; a β particle is produced

         (d) $^{242}_{94}\text{Pu} \rightarrow ^{238}_{92}\text{U} + ^{4}_{2}\text{He}$; an α particle is produced

21.16   This decay series represents a change of (232–208 =) 24 mass units. Since only alpha emissions change the nuclear mass, and each changes the mass by four, there must be a total of 6 α emissions. Each alpha emission causes a decrease of two in atomic number.

Therefore, the 6 alpha emissions, by themselves, would cause a decrease in atomic number of 12. The series as a whole involves a decrease of 8 in atomic number. Thus, there must be a total of 4 β emissions, each of which increases atomic number by one. Overall, there are 6 α emissions and 4 β emissions.

Nuclear Stability

21.18   (a) $^{66}_{32}$Ge - low neutron/proton ratio, positron emission

(b) $^{105}_{45}$Rh - high neutron/proton ratio, beta emission

(c) $^{137}_{53}$I - high neutron/proton ratio, beta emission

(d) $^{133}_{58}$Ce - low neutron/proton ratio, positron emission

21.20   Use criteria listed in Table 21.3.

(a) $^{112}_{48}$Cd   even, even more abundant

(b) $^{27}_{13}$Al   odd proton, even neutron more abundant

(c) $^{106}_{46}$Pd   even, even more abundant

(d) $^{128}_{54}$Xe   even proton, even neutron much more abundant than odd proton; odd neutron

21.22   The atomic number of tin, Sn, is 50, while the atomic number of antimony, Sb, is 51. According to Table 21.3, nuclei with even numbers of protons and/or even numbers of neutrons are more stable than nuclei with odd numbers of protons and/or neutrons. Isotopes of Sn always have an even number of protons and are more likely to be stable than isotopes of Sb, with an odd number of protons. Also, 50 is a magic number; isotopes with a magic number of protons and/or neutrons have special stability, so this makes isotopes of Sn even more likely to be stable.

21.24   The criterion employed in judging whether the nucleus is likely to be radioactive is the position of the nucleus on the plot shown in Figure 21.3. If the neutron/proton ratio is too high or low, or if the atomic number exceeds 83, the nucleus will be radioactive.

Radioactive: $^{58}_{29}$Cu —odd proton, odd neutron, low neutron/proton ratio

^{206}Po —high atomic number

Stable: $^{62}_{28}$Ni —even proton, even neutron, stable neutron/proton ratio

$^{108}_{47}$Ag —stable neutron/proton ratio, (one of 5 stable odd proton/odd neutron nuclides)

^{184}W —even proton, even neutron, stable neutron/proton ratio

Nuclear Transmutations

21.26 A major difference is that the charge on the nitrogen nucleus, +7, is much smaller than on the gold nucleus, +79. Thus, the alpha particle could more easily penetrate the coulomb barrier (that is, the repulsive energy barrier due to like charges) to make contact with the nitrogen nucleus than the gold nucleus. Rutherford used alpha particles that were being emitted from some radioactive source. He did not have access to machines that can accelerate particles to very high energy. It would be necessary to do just that to observe reaction of an alpha particle with a gold nucleus.

21.28 (a) $^{32}_{16}S + ^{1}_{0}n \rightarrow ^{1}_{1}p + ^{32}_{15}P$ (b) $^{7}_{4}Be + ^{0}_{-1}e$ (orbital electron) $\rightarrow ^{7}_{3}Li$

        (c) $^{187}_{75}Re \rightarrow ^{187}_{76}Os + ^{0}_{-1}e$ (d) $^{98}_{42}Mo + ^{2}_{1}H \rightarrow ^{1}_{0}n + ^{99}_{43}Tc$

        (e) $^{235}_{92}U + ^{1}_{0}n \rightarrow ^{135}_{54}Xe + ^{99}_{38}Sr + 2\,^{1}_{0}n$

21.30 (a) $^{238}_{92}U + ^{1}_{0}n \rightarrow ^{239}_{92}U + ^{0}_{0}\gamma$ (b) $^{14}_{7}N + ^{1}_{1}H \rightarrow ^{11}_{6}C + ^{4}_{2}He$

        (c) $^{18}_{8}O + ^{1}_{0}n \rightarrow ^{19}_{9}F + ^{0}_{-1}e$

Rates of Radioactive Decay

21.32 (a) The suggestion is not reasonable. The energies of nuclear states are very large relative to ordinary temperatures. Thus, merely changing the temperature by less than 100 K would not be expected to significantly affect the behavior of nuclei with regard to nuclear decay rates.

      (b) No. Radioactive decay has no activation energy like a chemical reaction. Activation energy is the minimum amount of energy required to initiate a chemical reaction. Radioactive decay is a spontaneous nuclear transformation from a less stable to a more stable nuclear configuration. Radioisotopes are by definition in a "transition state," prone to nuclear change or decay. Changes in external conditions such as temperature, pressure or chemical state provide insufficient energy to either excite or relax an unstable nucleus.

21.34 Calculate the decay constant, k, and then $t_{1/2}$.

$$k = \frac{-1}{t} \ln \frac{N_t}{N_o} = \frac{-1}{5.2 \text{ min}} \ln \frac{0.250 \text{ g}}{1.000 \text{ g}} = 0.2666 = 0.27 \text{ min}^{-1}$$

Using Equation [21.20], $t_{1/2} = 0.693/k = 0.693/0.02666 \text{ min}^{-1} = 2.599 = 2.6 \text{ min}$

21.36 Follow the logic in Sample Exercise 21.7. In this case, we are given initial sample mass as well as mass at time t, so we can proceed directly to calculate k (Equation [21.20], and then t (Equation [21.19]). *Solve.*

$$k = 0.693 / t_{1/2} = 0.693/27.8 \text{ d} = 0.02493 = 0.0249 \text{ d}^{-1}$$

$$t = \frac{-1}{k} \ln \frac{N_t}{N_o} = \frac{-1}{0.02493 \text{ d}^{-1}} \ln \frac{0.75}{6.25} = 85.06 = 85 \text{ d}$$

21.38   (a)   Proceeding as in Solution 21.37, calculate k in s^{-1}.

$$5.26 \text{ yr} \times \frac{365 \text{ d}}{1 \text{ yr}} \times \frac{24 \text{ hr}}{1 \text{ d}} \times \frac{3600 \text{ sec}}{1 \text{ hr}} = 1.659 \times 10^8 = 1.66 \times 10^8 \text{ s}$$

$$k = \frac{0.693}{t_{1/2}} = \frac{0.693}{1.659 \times 10^8} = 4.178 \times 10^{-9} = 4.18 \times 10^{-9} \text{ s}^{-1}$$

$$\ln \frac{N_t}{N_o} = -kt = -(4.178 \times 10^{-9} \text{ s}^{-1})(180 \text{ s}) = -7.520 \times 10^{-7} = -7.52 \times 10^{-7}$$

$$\frac{N_t}{N_o} = e^{-7.52 \times 10^{-7}} = (1.000 - 7.52 \times 10^{-7}); N_t = 3.75 \times 10^{-3} \text{ g} (1.000 - 7.52 \times 10^{-7})$$

[$e^{-7.52 \times 10^{-7}}$ is a number very close to 1. In this calculation, it is conveneint to express the number as $(1 - 7.52 \times 10^{-7})$].

The amount that decays is $N_o - N_t$:

$$3.75 \times 10^{-3} \text{ g} - [3.75 \times 10^{-3} \text{ g} (1.000 - 7.52 \times 10^{-7})] = 3.75 \times 10^{-3} \text{ g} (7.52 \times 10^{-7})$$

$$= 2.820 \times 10^{-9} = {\sim}3 \times 10^{-9} \text{ g Co}$$

(In terms of sig figs, $[N_o - N_t]$ is a very small number, found by subtracting two numbers known only to three sig figs and two decimal place. At best, we can express the result as an order of magnitude.)

$$N_o - N_t = 2.820 \times 10^{-9} \text{ g Co} \times \frac{1 \text{ mol Co}}{60 \text{ g Co}} \times \frac{6.022 \times 10^{23} \text{ Co atoms}}{1 \text{ mol Co}} \times \frac{1\beta}{1 \text{ Co atom}}$$

$$= 2.830 \times 10^{13} = {\sim}3 \times 10^{13}$$

Between 10^{13} and 10^{14} β particles are emitted in 180 seconds (3 minutes).

(b) $$\frac{2.830 \times 10^{13} \text{ dis}}{180 \text{ s}} \times \frac{1 \text{ Bq}}{1 \text{ dis/s}} = 1.57 \times 10^{11} = {\sim}10^{11} \text{ Bq}$$

The activity of the sample is approximately 10^{11} Bq.

21.40   Follow the logic in Sample Exercise 21.7.

$$t = \frac{-1}{k} \ln \frac{N_t}{N_o}; k = 0.693/5715 \text{ yr} = 1.213 \times 10^{-4} \text{ yr}^{-1}$$

$$t = \frac{-1}{1.213 \times 10^{-4} \text{ yr}^{-1}} \ln \frac{38.0}{58.2} = 3.52 \times 10^3 \text{ yr}$$

21.42   Follow the procedure outlined in Sample Exercise 21.7. The original quantity of ^{238}U is 75.0 mg plus the amount that gave rise to 18.0 mg of ^{206}Pb. This amount is 18.0(238/206) = 20.8 mg.

$$k = 0.693/4.5 \times 10^9 \text{ yr} = 1.54 \times 10^{-10} = 1.5 \times 10^{-10} \text{ yr}^{-1}$$

$$t = \frac{-1}{k} \ln \frac{N_t}{N_o} = \frac{-1}{1.54 \times 10^{-10} \text{ yr}^{-1}} \ln \frac{75.0}{95.8} = 1.59 \times 10^9 \text{ yr}$$

Energy Changes

21.44 $\Delta E = c^2 \Delta m = (3.0 \times 10^8 \text{ m/s})^2 \times 0.1 \text{ mg} \times \dfrac{1 \text{ g}}{1000 \text{ mg}} \times \dfrac{1 \text{ kg}}{1000 \text{ g}} \times \dfrac{1 \text{ kJ}}{1000 \text{ J}} = 9 \times 10^6 \text{ kJ}$

21.46 Δm = mass of individual protons and neutrons – mass of nucleus

$\Delta m = 10(1.0072765 \text{ amu}) + 11(1.0086649 \text{ amu}) - 20.98846 \text{ amu} = 0.1796189 = 0.17962 \text{ amu}$

$\Delta E = (2.9979246 \times 10^8 \text{ m/s})^2 \times 0.1796189 \text{ amu} \times \dfrac{1 \text{ g}}{6.0221421 \times 10^{23} \text{ amu}} \times \dfrac{1 \text{ kg}}{1000 \text{ g}}$

$= 2.680664 \times 10^{-11} = 2.6807 \times 10^{-11} \text{ J/}^{21}\text{Ne nucleus required}$

$2.680664 \times 10^{-11} \dfrac{\text{J}}{\text{nucleus}} \times \dfrac{6.0221421 \times 10^{23} \text{ nuclei}}{\text{mol}}$

$= 1.6143 \times 10^{13} \text{ J/mol } ^{21}\text{Ne binding energy}$

21.48   In each case, calculate the mass defect, total nuclear binding energy and then binding energy per nucleon.

(a) $\Delta m = 7(1.0072765) + 7(1.0086649) - 13.999234 = 0.1123558 = 0.112356 \text{ amu}$

$\Delta E = 0.1123558 \text{ amu} \times \dfrac{1 \text{ g}}{6.0221421 \times 10^{23} \text{ amu}} \times \dfrac{1 \text{ kg}}{1000 \text{ g}} \times \dfrac{8.987551 \times 10^{16} \text{ m}^2}{\text{s}^2}$

$= 1.676817 \times 10^{-11} = 1.67682 \times 10^{-11} \text{ J}$

binding energy/nucleon = 1.676817×10^{-11} J / 14 = 1.19773×10^{-12} J/nucleon

(b) $\Delta m = 22(1.0072765) + 26(1.0086649) - 47.935878 = 0.4494924 = 0.449492 \text{ amu}$

$\Delta E = 0.4494924 \text{ amu} \times \dfrac{1 \text{ g}}{6.0221421 \times 10^{23} \text{ amu}} \times \dfrac{1 \text{ kg}}{1000 \text{ g}} \times \dfrac{8.987551 \times 10^{16} \text{ m}^2}{\text{s}^2}$

$= 6.708304 \times 10^{-11} = 6.70830 \times 10^{-11} \text{ J}$

binding energy/nucleon = 6.708304×10^{-11} J / 48 = 1.39756×10^{-12} J/nucleon

(c)   Calculate the nuclear mass by subtracting the electron mass from the atomic mass. $128.904779 - 54(5.485799 \times 10^{-4} \text{ amu}) = 128.875156 \text{ amu}$

$\Delta m = 54(1.0072765) + 75(1.0086649) - 128.875156 = 1.1676428 = 1.167643 \text{ amu}$

$\Delta E = 1.1676428 \text{ amu} \times \dfrac{1 \text{ g}}{6.0221421 \times 10^{23} \text{ amu}} \times \dfrac{1 \text{ kg}}{1000 \text{ g}} \times \dfrac{8.987551 \times 10^{16} \text{ m}^2}{\text{s}^2}$

$= 1.742610 \times 10^{-10} \text{ J}$

binding energy/nucleon 1.742610×10^{-10} J/129 = 1.350860×10^{-12} J/nucleon

21.50   The calculated Δm is for one group of single nuclides involved in a reaction, labeled Δm/'atomic reaction'. Multiplying by Avogadro's number changes the quantity to 'mol of reaction'. Since energy is released, the sign of ΔE is negative.

(a) $\Delta m = 4.00260 + 1.0086649 - 3.01605 - 2.01410 = -0.0188851 = -0.01889$ amu

$$\Delta E = \frac{-0.0188851\,\text{amu}}{\text{'atomic reaction'}} \times \frac{1\,\text{g}}{6.022 \times 10^{23}\,\text{amu}} \times \frac{6.022 \times 10^{23}\,\text{'atomic reaction'}}{\text{mol of reaction}}$$

$$\times \frac{1\,\text{kg}}{10^3\,\text{g}} \times (2.99792458 \times 10^8\,\text{m/sec})^2 = -1.697 \times 10^{12}\,\text{J/mol}$$

(b) $\Delta m = 3.01605 + 1.0086649 - 2(2.01410) = -3.4851 \times 10^{-3} = -3.49 \times 10^{-3}$ amu

$\Delta E = -3.13 \times 10^{11}$ J/mol

(c) $\Delta m = 4.00260 + 1.00782 - 3.01605 - 2.01410 = -1.973 \times 10^{-2}$ amu

$\Delta E = -1.773 \times 10^{12}$ J/mol

21.52   According to Figure 21.13, the most stable nucleus on the periodic chart, the one with the greatest binding energy per nucleon, is iron-56.

Effects and Uses of Radioisotopes

21.54   (a)   Add ^{36}Cl to water as a chloride salt. Then dissolve ordinary CCl_3COOH. After a time, distill the volatile materials away from the salt; CCl_3COOH is volatile, and will distill with water. Count radioactivity in the volatile material. If chlorine exchange has occurred, there will be radioactivity.

       (b)   Prepare a saturated solution of $BaCl_2$ containing a small amount of solid $BaCl_2$. Add to this solution solid $BaCl_2$ containing ^{36}Cl. If the solid-solution equilibrium is dynamic, some of the ^{36}Cl in the solid will find itself in solution as chloride ion. After allowing some time for equilibrium to become established, filter the solution, measure radioactivity in the solution that is separated from the solid. If there were no dynamic equilibrium, the ^{36}Cl$^-$ would remain in the added solid, since the solution is already saturated before the addition of more solid.

       (c)   Utilize ^{36}Cl in soils of various pH values; grow plants for a given period of time. Remove plants, and directly measure radioactivity in samples from stems, leaves, and so forth, or reduce the volume of plant sample by some form of digestion and evaporation of solution to give a dry residue that can be counted.

21.56   (a)   Control rods control neutron flux so that there are enough neutrons to sustain the chain reaction but not so many that the core overheats.

       (b)   A moderator slows neutrons so that they are more easily captured by fissioning nuclei.

21.58   *Analyze/Plan.* Use conservation of A and Z to complete the equations, keeping in mind the symbols and definitions of various decay products.   *Solve.*

       (a) $^{235}_{92}U + ^{1}_{0}n \rightarrow ^{160}_{62}Sm + ^{72}_{30}Zn + 4\,^{1}_{0}n$

       (b) $^{239}_{94}Pu + ^{1}_{0}n \rightarrow ^{144}_{58}Ce + ^{94}_{36}Kr + 2\,^{1}_{0}n$

21.60   (a)   If the spent fuel rods are more radioactive than the original rods, the products of fission must lie outside the belt of stability and be radioactive themselves.

(b)   The heavy ($Z > 83$) nucleus has a high neutron/proton ratio. The lighter radioactive fission products, (e.g., barium-142 and krypton-91) also have high neutron/proton ratios, since only 2 or 3 free neutrons are produced during fission. The preferred decay mode to reduce the neutron/proton ratio is β decay, which has the effect of converting a neutron into a proton. Both barium-142 (86 n, 56 p) and krypton-91 (55 n, 36 p) undergo β decay.

21.62

H_2O^+ is a free radical because it contains seven valence electrons. Around the central O atom there are two bonding and one nonbonding electron pairs and a single unpaired electron ($3(2) + 1 = 7$ valence electrons).

21.64   (a) $1 \text{ Ci} = 3.7 \times 10^{10}$ dis/s; $1 \text{ Bq} = 1$ dis/s

$$15 \text{ mCi} \times \frac{1 \text{ Ci}}{1000 \text{ mCi}} \times 3.7 \times 10^{10} \text{ dis/s} = 5.55 \times 10^8 = 5.6 \times 10^8 \text{ dis/s} = 5.6 \times 10^8 \text{ Bq}$$

(b) $1 \text{ Gy} = 1 \text{ J/kg}$; $1 \text{ Gy} = 100$ rad

$$5.55 \times 10^8 \text{ dis/s} \times 240 \text{ s} \times 0.075 \times \frac{8.75 \times 10^{-14} \text{ J}}{\text{dis}} \times \frac{1}{65 \text{ kg}} = 1.345 \times 10^{-5} = 1.3 \times 10^{-5} \text{ J/kg}$$

$$1.3 \times 10^{-5} \text{ J/kg} \times \frac{1 \text{ Gy}}{1 \text{ J/kg}} = 1.3 \times 10^{-5} \text{ Gy}; \quad 1.3 \times 10^{-5} \text{ Gy} \times \frac{100 \text{ rad}}{1 \text{ Gy}} = 1.3 \times 10^{-3} \text{ rad}$$

(c)   rem = rad (RBE); Sv = Gy (RBE)

$$1.3 \times 10^{-3} \text{ rad} \,(1.0) = 1.3 \times 10^{-3} \text{ rem} \times \frac{1000 \text{ mrem}}{1 \text{ rem}} = 1.3 \text{ mrem}$$

$$1.3 \times 10^{-5} \text{ Gy} \,(1.0) = 1.3 \times 10^{-5} \text{ Sv}$$

(d)   The mammogram dose of 300 mrem is ~230 times as much radiation as the dose absorbed by the 65 kg person described above.

Additional Exercises

21.66   (a) $_0^1 n \rightarrow {}_1^1 p + {}_{-1}^0 e$ (or $_{-1}^0 \beta$)

The other product of neutron decay is a β particle (with the mass and charge of an electron).

(b)   Neutrons in atomic nuclei do not decay at this rate because they are stabilized by the strong forces among subatomic particles in a nucleus. Evidence for strong forces in the nucleus includes nuclear binding energies and the coexistence of like-charged protons in the very small volume of the nucleus.

21.67   (a) $_{95}^{241} \text{Am} \rightarrow {}_{93}^{237} \text{Np} + {}_2^4 \text{He}$

(b)   Alpha particles are much less penetrating than other types of nuclear emissions. They are easily contained inside the smoke detector so that users (humans) are not exposed to damaging radiation.

(c)   Each ^{241}Am emits one alpha particle. Find the number of atoms in 0.2 μg ^{241}Am, the mass of this number of alpha particles, and then the energy equivalent of this mass using $E = mc^2$. atomic mass in amu = molar mass in g

$$0.2 \, \mu g \; ^{241}Am \times \frac{1 \times 10^{-6} \, g}{\mu g} \times \frac{1 \, mol \; ^{241}Am}{241.056829 \, g} \times \frac{6.022 \times 10^{23} \, atoms}{mol}$$

$$= 4.996 \times 10^{14} = 5 \times 10^{14} \; ^{241}Am \; atoms = 5 \times 10^{14} \; \alpha \; particles$$

$$4.996 \times 10^{14} \; \alpha \; particles \times \frac{6.64 \times 10^{-24} \, g}{\alpha \; particle} = 3.3176 \times 10^{-9} = 3 \times 10^{-9} \; g \; \alpha \; particles$$

$$E = mc^2 = 3.3176 \times 10^{-9} \, g \times \frac{1 \, kg}{1000 \, g} \times \frac{(2.99792458 \times 10^8)^2 \, m^2}{s^2}$$

$$= 2.9817 \times 10^5 \, J = 3 \times 10^5 \, J = 300 \, kJ$$

(d) ^{240}Am, with a half-life of 2.12 days, would decay quickly and the smoke detector would have a very short shelf and operating life. ^{241}Am, with the much longer half-life of 432 years, will last longer than the detector. It does require attention at disposal time.

21.68   (a) $^{36}_{17}Cl \rightarrow ^{36}_{18}Ar + ^{0}_{-1}e$

         (b)   According to Table 21.3, nuclei with even numbers of both protons and neutrons, or an even number of one kind of nucleon, are more stable. ^{35}Cl and ^{37}Cl both have an odd number of protons **but** an even number of neutrons. ^{36}Cl has an odd number of protons and neutrons (17 p, 19 n), so it is less stable than the other two isotopes. Also, ^{37}Cl has 20 neutrons, a nuclear closed shell.

21.70

Time (hr)	N_t (dis/min)	$\ln N_t$
0	180	5.193
2.5	130	4.868
5.0	104	4.644
7.5	77	4.34
10.0	59	4.08
12.5	46	3.83
17.5	24	3.18

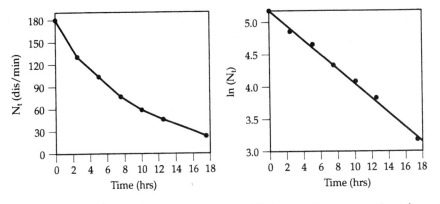

The plot on the left is a graph of activity (disintegrations per minute) vs. time. Choose $t_{1/2}$ at the time where $N_t = 1/2\, N_o = 90$ dis/min. $t_{1/2} \approx 6.0$ hr.

Rearrange Equation [21.19] to obtain the linear relationship shown on the right.

$\ln(N_t / N_o) = -kt$; $\ln N_t - \ln N_o = -kt$; $\ln N_t = -kt + \ln N_o$

The slope of this line $= -k = -0.11$; $t_{1/2} = 0.693/0.11 = 6.3$ hr.

21.71 1×10^{-6} curie $\times \dfrac{3.7 \times 10^{10}\text{ dis/s}}{\text{curie}} = 3.7 \times 10^4$ dis/s

rate $= 3.7 \times 10^4$ nuclei/s $= kN$

$k = \dfrac{0.693}{t_{1/2}} = \dfrac{0.693}{28.8\text{ yr}} \times \dfrac{1\text{ yr}}{365 \times 24 \times 3600\text{ sec}} = 7.630 \times 10^{-10} = 7.63 \times 10^{-10}\text{ s}^{-1}$

3.7×10^4 nuclei/s $= (7.63 \times 10^{-10}/\text{s})$ N; N $= 4.849 \times 10^{13} = 4.8 \times 10^{13}$ nuclei

mass ^{90}Sr $= 4.849 \times 10^{13}$ nuclei $\times \dfrac{89.907738\text{ g Sr}}{6.022 \times 10^{23}\text{ nuclei}} = 7.2 \times 10^{-9}$ g Sr

(mass of ^{90}Sr from webelements.com)

21.72   First calculate k in s^{-1}

$k = \dfrac{0.693}{2.4 \times 10^4\text{ yr}} \times \dfrac{1\text{ yr}}{365 \times 24 \times 3600\text{ s}} = 9.16 \times 10^{-13} = 9.2 \times 10^{-13}\text{ s}^{-1}$

Now calculate N:

$N = 0.385$ g Pu $\times \dfrac{1\text{ mol Pu}}{239.052\text{ g Pu}} \times \dfrac{6.022 \times 10^{23}\text{ Pu atoms}}{1\text{ mol Pu}} = 9.699 \times 10^{20}$

$= 9.7 \times 10^{20}$ Pu atoms

rate $= (9.16 \times 10^{-13}\text{ s}^{-1})(9.699 \times 10^{20}\text{ Pu atoms}) = 8.880 \times 10^8 = 8.9 \times 10^8$ dis/s

(mass of ^{239}Pu from webelements.com)

21.74   Assume that no depletion of iodide from the water due to plant uptake has occurred. Then the activity after 30 days would be:

$k = 0.693/t_{1/2} = 0.693/8.02$ d $= 0.0864$ d^{-1}

$$\ln\frac{N_t}{N_o} = -(0.0864\ d^{-1})\ (30\ d) = -2.586 = -2.6; \frac{N_t}{N_o} = 0.07485 = 0.07$$

We thus expect $N_t = 0.07485(214) = 16.0 = 2 \times 10^1$ counts/min. The measured value of 15.7 counts/min is within experimental uncertainty of the expected value, 2×10^1 (16.0) counts/minute, for no iodine uptake by the plant. We conclude that no iodine is absorbed by the plant.

21.76   First, calculate k in s^{-1}

$$k = \frac{0.693}{12.3\ yr} \times \frac{1\ yr}{365\ d} \times \frac{1\ d}{24\ hr} \times \frac{1\ hr}{3600\ sec} = 1.7866 \times 10^{-9} = 1.79 \times 10^{-9}\ s^{-1}$$

From Equation [21.18], $1.50 \times 10^3\ s^{-1} = (1.7866 \times 10^{-9}\ s^{-1})(N)$;

$N = 8.396 \times 10^{11} = 8.40 \times 10^{11}$. In 26.00 g of water, there are

$$26.00\ g\ H_2O \times \frac{1\ mol\ H_2O}{18.02\ g\ H_2O} \times \frac{6.022 \times 10^{23}\ H_2O}{1\ mol\ H_2O} \times \frac{2\ H}{1\ H_2O} = 1.738 \times 10^{24}\ H\ atoms$$

The mole fraction of ^3_1H atoms in the sample is thus

$8.396 \times 10^{11}/1.738 \times 10^{24} = 4.831 \times 10^{-13} = 4.83 \times 10^{-13}$

21.77   (a) $\Delta m = \Delta E/c^2; \Delta m = \dfrac{3.9 \times 10^{26}\ J/s}{(3.00 \times 10^8\ m/s)^2} \times \dfrac{1\ kg\text{-}m^2/s^2}{1\ J} = 4.3 \times 10^9\ kg/s$

The rate of mass loss is 4.3×10^9 kg/s.

(b)   The mass loss arises from fusion reactions that produce more stable nuclei from less stable ones, e.g., Equations [21.26-21.29].

(c)   Express the mass lost by the sun in terms of protons per second consumed in fusion reactions like Equations [21.26], [21.27] and [21.29].

$$\frac{4.3 \times 10^9\ kg}{s} \times \frac{1\ proton}{1.673 \times 10^{-24}\ g} \times \frac{1000\ g}{1\ kg} = 2.594 \times 10^{36} = 3 \times 10^{36}\ protons/s$$

21.78 $1000\ Mwatts \times \dfrac{1 \times 10^6\ watts}{1\ Mwatt} \times \dfrac{1\ J}{1\ watt\text{-}s} \times \dfrac{1\ ^{235}U\ atom}{3 \times 10^{-11}\ J} \times \dfrac{1\ mol\ U}{6.02 \times 10^{23}\ atoms}$

$\times \dfrac{235\ g\ U}{1\ mol} \times \dfrac{3600\ s}{1\ hr} \times \dfrac{24\ hr}{1\ d} \times \dfrac{365\ d}{1\ yr} \times \dfrac{100}{40}(\text{efficiency}) = 1.03 \times 10^6 = 1 \times 10^6\ g\ U/yr$

21.79 $2 \times 10^{-12}\ curies \times \dfrac{3.7 \times 10^{10}\ dis/s}{1\ curie} = 7.4 \times 10^{-2} = 7 \times 10^{-2}\ dis/s$

$$\frac{7.4 \times 10^{-2}\ dis/s}{75\ kg} \times \frac{8 \times 10^{-13}\ J}{dis} \times \frac{1\ rad}{1 \times 10^{-2}\ J/g} \times \frac{3600\ s}{hr} \times \frac{24\ hr}{1\ d}$$

$$\times \frac{365\ d}{1\ yr} = 2.49 \times 10^{-6} = 2 \times 10^{-6}\ rad/yr$$

Recall that there are 10 rem/rad for alpha particles.

$$\frac{2.49 \times 10^{-6} \text{ rad}}{1 \text{ yr}} \times \frac{10 \text{ rem}}{1 \text{ rad}} = 2.49 \times 10^{-5} = 2 \times 10^{-5} \text{ rem/yr}$$

Integrative Exercises

21.80   Calculate the molar mass of $NaClO_4$ that contains 29.6% ^{36}Cl. Atomic mass of the enhanced Cl is $0.296(36.0) + 0.704(35.453) = 35.615 = 35.6$. The molar mass of $NaClO_4$ is then $(22.99 + 35.615 + 64.00) = 122.605 = 122.6$. Calculate N, the number of ^{36}Cl nuclei, the value of k in s^{-1}, and the activity in dis/s.

$$53.8 \text{ mg } NaClO_4 \times \frac{1 \text{ g}}{1000 \text{ mg}} \times \frac{1 \text{ mol } NaClO_4}{122.605 \text{ g } NaClO_4} \times \frac{1 \text{ mol } Cl}{1 \text{ mol } NaClO_4} \times \frac{6.022 \times 10^{23} \text{ Cl atoms}}{\text{mol Cl}}$$

$$\times \frac{29.6 \ ^{36}Cl \text{ atoms}}{100 \text{ Cl atoms}} = 7.822 \times 10^{19} = 7.82 \times 10^{19} \ ^{36}Cl \text{ atoms}$$

$$k = 0.693/t_{1/2} = \frac{0.693}{3.0 \times 10^5 \text{ yr}} \times \frac{1 \text{ yr}}{365 \times 24 \times 3600 \text{ s}} = 7.32 \times 10^{-14} = 7.3 \times 10^{-14} \text{ s}^{-1}$$

rate = $kN = (7.32 \times 10^{-14} \text{ s}^{-1})(7.822 \times 10^{19} \text{ nuclei}) = 5.729 \times 10^6 = 5.7 \times 10^6 \text{ dis/s}$

21.82   (a) $0.18 \text{ Ci} \times \dfrac{3.7 \times 10^{10} \text{ dis/s}}{\text{Ci}} \times \dfrac{3600 \text{ s}}{\text{hr}} \times \dfrac{24 \text{ hr}}{\text{d}} \times 245 \text{ d} = 1.41 \times 10^{17} = 1.4 \times 10^{17} \text{ } \alpha \text{ particles}$

         (b) $P = nRT/V = 1.41 \times 10^{17} \text{ He atoms} \times \dfrac{1 \text{ mol He}}{6.022 \times 10^{23} \text{ atoms}} \times \dfrac{295 \text{ K}}{0.0250 \text{ L}} \times \dfrac{0.08206 \text{ L-atm}}{\text{K-mol}}$

$$= 2.27 \times 10^{-4} = 2.3 \times 10^{-4} \text{ atm} = 0.17 \text{ torr}$$

21.83   Calculate N_t in dis/min/g C from 1.5×10^{-2} dis/0.788 g $CaCO_3$. $N_o = 15.3$ dis/min/g C. Calculate k from $t_{1/2}$, calculate t from $\ln(N_t / N_o) = -kt$.

$$C(s) + O_2(g) \rightarrow CO_2(g) + Ca(OH_2)(aq) \rightarrow CaCO_3(s) + H_2O(l)$$

1 C atom → 1 $CaCO_3$ molecule

$$\frac{1.5 \times 10^{-2} \text{ Bq}}{0.788 \text{ g } CaCO_3} \times \frac{1 \text{ dis/s}}{1 \text{ Bq}} \times \frac{60 \text{ s}}{1 \text{ min}} \times \frac{100.1 \text{ g } CaCO_3}{12.01 \text{ g C}} = 9.52 = 9.5 \text{ dis/min/g C}$$

$$k = 0.693/t_{1/2} = 0.693/5.715 \times 10^3 \text{ yr} = 1.213 \times 10^{-4} = 1.21 \times 10^{-4} \text{ yr}^{-1}$$

$$t = -\frac{1}{k} \ln \frac{N_t}{N_o} = \frac{-1}{1.213 \times 10^{-4} \text{ yr}^{-1}} \ln \frac{9.52 \text{ dis/min/g C}}{15.3 \text{ dis/min/g C}} = 3.91 \times 10^3 \text{ yr}$$

21.84   Determine the wavelengths of the photons by first calculating the energy equivalent of the mass of an electron or positron. (Since **two** photons are formed by annihilation of **two** particles of equal mass, we need to calculate the energy equivalent of just one particle.) The mass of an electron is 9.109×10^{-31} kg.

$$\Delta E = (9.109 \times 10^{-31} \text{ kg}) \times (2.998 \times 10^8 \text{ m/s})^2 = 8.187 \times 10^{-14} \text{ J}$$

Also, $\Delta E = h\nu; \ \Delta E = hc/\lambda; \ \lambda = hc/\Delta E$

$$\lambda = \frac{(6.626 \times 10^{-34} \text{ J-s})(2.998 \times 10^8 \text{ m/s})}{8.187 \times 10^{-14} \text{ J}} = 2.426 \times 10^{-12} \text{ m} = 2.426 \times 10^{-3} \text{ nm}$$

This is a very short wavelength indeed; it lies at the short wavelength end of the range of observed gamma ray wavelengths (see Figure 6.4).

21.85   (a) $Ba(NO_3)_2(aq) + Na_2SO_4(aq) \rightarrow BaSO_4(s) + 2NaNO_3(aq)$

        (b) $1.25 \text{ mmol } Ba^{2+} + 1.25 \text{ mmol } SO_4^{2-} \rightarrow 1.25 \text{ mmol } BaSO_4$

Neither reactant is in excess, so the activity of the filtrate is due entirely to $[SO_4^{2-}]$ from dissociation of $BaSO_4(s)$. Calculate $[SO_4^{2-}]$ in the filtrate by comparing the activity of the filtrate to the activity of the reactant.

$$\frac{0.050 \ M \ SO_4^{2-}}{1.22 \times 10^6 \text{ Bq/mL}} = \frac{x \ M \text{ filtrate}}{250 \text{ Bq/mL}}$$

$[SO_4^{2-}]$ in the filtrate = $1.0246 \times 10^{-5} = 1.0 \times 10^{-5} \ M$

$K_{sp} = [Ba^{2+}][SO_4^{2-}]; [SO_4^{2-}] = [Ba^{2+}]$

$K_{sp} = (1.0246 \times 10^{-5})^2 = 1.0498 \times 10^{-10} = 1.0 \times 10^{-10}$

22 Chemistry of the Nonmetals

Visualizing Concepts

22.1 C_2H_4, the structure on the left, is the stable compound. Carbon, with a relatively small covalent radius owing to its location in the second row of the periodic chart, is able to closely approach other atoms. This close approach enables significant π overlap, so carbon can form strong multiple bonds to satisfy the octet rule. Silicon, in the third row of the periodic table, has a covalent radius too large for significant π overlap. Si does not form stable multiple bonds and Si_2H_4 is unstable.

22.3 *Analyze.* The structure is a trigonal bipyramid where one of the five positions about the central atom is occupied by a lone pair, often called a see-saw.

Plan A: Count the valence electrons in each molecule, draw a correct Lewis structure, and count the electron domains about the central atom.

Plan B: Molecules (a)–(d) each contain four F atoms bound to a central atom through a single bond (F is unlikely to form multiple bonds because of its high electronegativity). This represents 16 electron pairs; the fifth position is occupied by a lone pair, for a total of 17 e^- pairs. A valence e^- count for (a)–(d) will tell us which molecules are likely to have the designated structure. Molecule (e), $HClO_4$, is not exactly of the type AX_4, so a Lewis structure will be required.   *Solve.*

(a) XeF_4 36 e^-, 16 e^- pairs. Plan B predicts that this molecule **will not** adopt the see-saw structure.

6 e^- domains about the Xe
octahedral domain geometry
square planar structure

(b) BrF_4^- 34 e^-, 17 e^- pairs; structure **will** be see-saw.

5 e^- domains about Br trigonal
bipyramidal domain geometry
see-saw structure

(c) SiF_4 32 e^-, 16 e^- pairs; structure **will not** be see-saw

4 e^- domains about Si
tetrahedral domain geometry and structure

(d) $TeCl_4$ 34 e^-, 17 e^- pairs; structure **will be** see-saw

5 e^- domains about Te
trigonal bipyramidal domain geometry
see-saw structure

(e) $HClO_4$ 32 e^-, 16 e^- pairs; no prediction

($HClO_4$ is an oxyacid, so H is bound to O, not Cl.
Other Lewis structures that optimize formal charges
are possible; structure predictions are the same.)

22.4   *Analyze.* Equation [22.24] is: $O_3 \rightarrow O_2 + O$ $\Delta H = 105$ kJ

Plan. Use $\Delta H°$ and the dissociation energy barrier to draw the profile.

Solve. $\Delta H°$ tells us that the products are 105 kJ higher in energy than the reactants. The dissociation energy barrier is the total energy difference between reactants and the activated complex (the top of the peak).

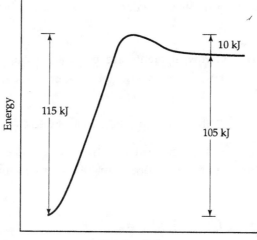

Reaction Pathway

22.6   The graph is applicable only to (c) density. Density depends on both atomic mass and volume (radius). Both increase going down a family, but atomic mass increases to a greater extent. Density, the ratio of mass to volume, increases going down the family; this trend is consistent with the data in the figure.

According to periodic trends, (a) electronegativity and (b) first ionization energy both decrease rather than increase going down the family. According to Table 22.5 both (d) X—X single bond enthalpy and (e) electron affinity are somewhat erratic, with the trends decreasing from S to Po, and anomalous values for the properties of O, probably owing to its small covalent radius.

22.7   (a)   Atomic radius increases moving downward in a group because the principal quantum number (n) of the valence electrons increases. As n increases, the average distance of an electron from the nucleus increases and so does atomic radius.

(b)   Anionic radii are greater than atomic radii because of increased electrostatic repulsions among electrons. Additional electrons in the same principle quantum level lead to additional electrostatic repulsion. This increases the energy of the electrons, and their average distance from the nucleus; the anionic radii are thus greater than the atomic radii.

(c)   The anion that is the strongest base in water is the conjugate base of the weakest conjugate acid. The conjugate acids are OH^-, SH^-, and SeH^-. According to trends in binary hydrides, the acid with the longest X—H bond will be the most readily ionized and the strongest acid. SeH^- is thus the strongest acid and OH^- the weakest. Therefore O^{2-} is the strongest base in water.

22.9   White phosphorus consists of tetrahedral P_4 molecules. This molecular geometry requires al P—P—P bond angles to be much smaller than the idealized 109°. This angular strain is relieved when P_4 reacts with another substance, which it does readily. The chains in red phosphorus require much less extreme bond angles, and there is less need to relieve steric repulsion by chemical reaction.

22.10   *Analyze/Plan.* The structure shown is a diatomic molecule or ion, depending on the value of n. Each species has 10 valence electrons and 5 electron pairs.

Solve.

(a)   Only second row elements are possible, because of the small covalent radius required for multiple bonding. Likely candidates are CO, N_2, NO^+, CN^-, and C_2^{2-}.

:C≡O: :N≡N: $[:N≡O:]^+$ $[:C≡N:]^-$ $[:C≡C:]^{2-}$

(b)   Since C_2^{2-} has the highest negative charge, it is likely to be the strongest H^+ acceptor and strongest Brønsted base. This is confirmed in Section 22.9 under "Carbides."

Periodic Trends and Chemical Reactions

22.12   Metals: (a) Ga, (b) Mo, (f) Cd   nonmetals: (e) Xe   metalloid: (c) Te, (d) As

22.14   (a)   Cl   (b)   K

(c)   K in the gas phase (lowest ionization energy), Li in aqueous solution (most positive E° value)

(d)   Ne; Ne and Ar are difficult to compare to the other elements because they do not form compounds and their radii are not measured in the same way as other elements. However, Ne is several rows to the right of C and surely has a smaller atomic radius. The next smallest is C.

(e)   C

22.16   (a)   Nitrogen is a highly electronegative element. In HNO_3 it is in its highest oxidation state, +5, and thus is more readily reduced than phosphorus, which forms stable P—O bonds.

(b)   The difference between the third row element and the second lies in the smaller size of C as compared with Si, and the fact that Si has 3d orbitals available to form an sp^3d^2 hybrid set that can accommodate more than an octet of electrons.

(c)   Two of the carbon compounds, C_2H_4 and C_2H_2, contain C—C π bonds. Si does not readily form π bonds (to itself or other atoms), so Si_2H_4 and Si_2H_2 are not known as stable compounds.

22.18   (a) $NaOCH_3(s) + H_2O(l) \rightarrow NaOH(aq) + CH_3OH(aq)$

(b) $CuO(s) + 2HNO_3(aq) \rightarrow Cu(NO_3)_2(aq) + H_2O(l)$

(c) $WO_3(s) + 3H_2(g) \rightarrow W(s) + 3H_2O(g)$

(d) $4NH_2OH(l) + O_2(g) \rightarrow 6H_2O(l) + 2N_2(g)$

(e) $Al_4C_3(s) + 12H_2O(l) \rightarrow 4Al(OH)_3(s) + 3CH_4(g)$

Hydrogen, the Noble Gases, and the Halogens

22.20   Tritium is radioactive. $^{3}_{1}H \rightarrow ^{3}_{2}He + ^{0}_{-1}e$

22.22   Hydrogen is unlike the group 1A elements because they are all solid metals and it is a gaseous nonmetal. It is more electronegative than the group 1A metals and can gain an electron to form an anion. Chemically, hydrogen can combine with goup 1A metals to form MH ionic compounds, where H^- is the anion.

Hydrogen is unlike the group 7A nonmetals because it is less electronegative than they are and can lose an electron to form a cation. Chemically, hydrogen can combine with the group 7A nonmetals to form polar covalent HX molecules, where H is the positive end of the dipole.

22.24   (a)   Electrolysis of brine; reaction of carbon with steam; reaction of methane with steam; by-product in petroleum refining

(b)   Synthesis of ammonia; synthesis of methanol; reducing agent; hydrogenation of unsaturated vegetable oils

22.26   (a) $2Al(s) + 6H^+(aq) \rightarrow 2Al^{3+}(aq) + 3H_2(g)$

(b) $Mg(s) + H_2O(g) \rightarrow MgO(s) + H_2(g)$

(c) $MnO_2(s) + H_2(g) \rightarrow MnO(s) + H_2O(g)$

(d) $CaH_2(s) + 2H_2O(l) \rightarrow Ca(OH)_2(aq) + 2H_2(g)$

22.28   (a)   molecular   (b)   ionic   (c)   metallic

22.30   Cracking is a process where higher molecular weight hydrocarbons are broken down into smaller molecules (Section 25.3). One model reaction is

$$CH_4(g) \xrightarrow{\text{catalyst}} C(s) + 2H_2(g) \quad \Delta H \sim 75 \text{ kJ}.$$

An analogous reaction can be written for "cracking" ethanol, where a third product is $O_2(g)$. The overall process is endothermic and requires energy. The question remains, what energy source drives the cracking? If energy to drive the cracking can be provided by some nonpetroleum fuel (such as grain-based ethanol) or power source (such as wind, geothermal, etc.), this could provide a satisfactory basis for a hydrogen economy.

Additionally, cracking would produce C(s) or soot and other C-based pollutants, which would have to be treated. This is probably easier at the power-plant level than the individual car level, but not insignificant.

22.32   Noble gas elements have very stable electron configurations with complete s and p subshellls. They have very large positive ionization energies; they do not lose electrons easily. They have positive electron affinities; they do not attract electrons to themselves. They do not easily gain, lose or share electrons, so they do not readily form the chemical bonds required to create compounds.

22.34   (a) $Ca(OBr)_2$, +1   (b) $HBrO_3$, +5   (c) XeO_3; Xe, +6

         (d) ClO_4^-, +7   (e) HIO_2, +3   (f) IF_5; I, +5; F, –1

22.36   (a)   potassium chlorate   (b)   calcium iodate

         (c)   aluminum chloride   (d)   bromic acid

         (e)   paraperiodic acid   (f)   xenon tetrafluoride

22.38   (a)   The more electronegative the central atom, the greater the extent to which it withdraws charge from oxygen, in turn making the O—H bond more polar, and enhancing ionization of H^+.

         (b)   HF reacts with the silica which is a major component of glass:

                $6HF(aq) + SiO_2(s) \rightarrow SiF_6^{2-}(aq) + 2H_2O(l) + 2H^+(aq)$

         (c)   Iodide is oxidized by sulfuric acid, as shown in Figure 22.12.

         (d)   The major factor is size; there is not room about Br for the three chlorides plus the two unshared electron pairs that would occupy the bromine valence shell orbitals.

22.40   Perchlorate salts are highly soluble, so the reaction is probably not a precipitation reaction. Perchloric acid, $HClO_4$, is a strong acid and ClO_4^- is a negligible base in water so the decomposition is probably not an acid-base reaction. Because oxyanions of the halogens are known to participate in oxidation-reduction reactions, the microorganisms probably destroy ClO_4^- via a redox reaction or series of reactions. While several lower oxidation states are accessible to Cl in ClO_4^-, intermediate species ClO_3^-, ClO_2^-, ClO^-, and Cl_2 are quite reactive, so the logical fate of the perchlorate anion is $Cl^-(aq)$ and $O_2(g)$ or $H_2O(l)$ with some oxidized organic compound.

Oxygen and the Group 6A Elements

22.42

Ozone has two resonance forms (Section 8.6); the molecular structure is bent, with an O—O—O bond angle of approximately 120°. The π bond in ozone is delocalized over the entire molecule; neither individual O—O bond is a full double bond, so the observed O—O distance of 1.28 Å is greater than the 1.21 Å distance in O_2, which has a full O—O double bond.

22.44　(a)　$CaO(s) + H_2O(l) \rightarrow Ca^{2+}(aq) + 2OH^-(aq)$

　　　　(b)　$Al_2O_3(s) + 6H^+(aq) \rightarrow 2Al^{3+}(aq) + 3H_2O(l)$

　　　　(c)　$Na_2O_2(s) + 2H_2O(l) \rightarrow 2Na^+(aq) + 2OH^-(aq) + H_2O_2(aq)$

　　　　(d)　$N_2O_3(g) + H_2O(l) \rightarrow 2HNO_2(aq)$

　　　　(e)　$2KO_2(s) + 2H_2O(l) \rightarrow 2K^+(aq) + 2OH^-(aq) + O_2(g) + H_2O_2(aq)$

　　　　(f)　$NO(g) + O_3(g) \rightarrow NO_2(g) + O_2(g)$

22.46　(a)　Mn_2O_7 (higher oxidation state of Mn)

　　　　(b)　SnO_2 (higher oxidation state of Sn)

　　　　(c)　SO_3 (higher oxidation state of S)

　　　　(d)　SO_2 (more nonmetallic character of S)

　　　　(e)　Ga_2O_3 (more nonmetallic character of Ga)

　　　　(f)　SO_2 (more nonmetallic character of S)

22.48　(a)　SCl_4, +4　　　　b)　SeO_3, +6　　　　(c)　$Na_2S_2O_3$, +2

　　　　(d)　H_2S, −2　　　　(e)　H_2SO_4, +6

　　　　Oxygen (a group 6A element) is in the −2 oxidation state in compounds (b), (c), and (e).

22.50　An aqueous solution of SO_2 contains H_2SO_3 and is acidic. Use H_2SO_3 as the reducing agent and balance assuming acid conditions.

　　　　(a)

$$2[MnO_4^-(aq) + 8H^+(aq) + 5e^- \rightarrow Mn^{2+}(aq) + 4H_2O(l)]$$
$$\underline{5[H_2SO_3(aq) + H_2O(l) \rightarrow SO_4^{2-}(aq) + 4H^+(aq) + 2e^-]}$$
$$2MnO_4^-(aq) + 5H_2SO_3(aq) \rightarrow 2MnSO_4(aq) + 3SO_4^{2-}(aq) + 3H_2O(l) + 4H^+(aq)$$

　　　　(b)

$$Cr_2O_7^{2-}(aq) + 14H^+(aq) + 6e^- \rightarrow 2Cr^{3+}(aq) + 7H_2O(l)$$
$$\underline{3[H_2SO_3(aq) + H_2O(l) \rightarrow SO_4^{2-}(aq) + 4H^+(aq) + 2e^-]}$$
$$Cr_2O_7^{2-}(aq) + 3H_2SO_3(aq) + 2H^+(aq) \rightarrow 2Cr^{3+}(aq) + 3SO_4^{2-}(aq) + 4H_2O(l)$$

　　　　(c)

$$Hg_2^{2+}(aq) + 2e^- \rightarrow 2Hg(l)$$
$$\underline{H_2SO_3(aq) + H_2O(l) \rightarrow SO_4^{2-}(aq) + 4H^+(aq) + 2e^-}$$
$$Hg_2^{2+}(aq) + H_2SO_3(aq) + H_2O(l) \rightarrow 2Hg(l) + SO_4^{2-}(aq) + 4H^+(aq)$$

22.52　　SF_4, 34 e⁻　　　　　　　　　SF_5^-, 42 e⁻

　　　trigonal bipyramidal　　　　　　octahedral
　　electron pair geometry　　　　electron pair geometry

　　　　　see-saw　　　　　　　　square pyramidal
　　molecular geometry　　　　　molecular geometry

22.54   (a) $Al_2Se_3(s) + 6H^+(aq) \rightarrow 2Al^{3+}(aq) + 3H_2Se(g)$

   (b) $Cl_2(aq) + S_2O_3^{2-}(aq) + H_2O(l) \rightarrow 2Cl^-(aq) + S(s) + SO_4^{2-}(aq) + 2H^+(aq)$

Nitrogen and the Group 5A Elements

22.56   (a)   NO, +2   (b) N_2H_4, –2   (c)   KCN, –3

   (d) $NaNO_2$, +3   (e) NH_4Cl, –3   (f) Li_3N, –3

22.58   (a)

$$\left[\begin{array}{c} H \\ | \\ H-N-H \\ | \\ H \end{array} \right]^+$$

tetrahedral

   (b) $\left[\ddot{O}=N-\ddot{O}\colon \right]^- \longleftrightarrow \left[\colon\ddot{O}-N=\ddot{O} \right]^-$

The ion is bent with a 120° O–N–O angle.

   (c) $\colon\ddot{N}=N=\ddot{O}\colon \longleftrightarrow \colon N\equiv N-\ddot{O}\colon \longleftrightarrow \colon\ddot{N}-N\equiv O\colon$

The molecule is linear. Again, the third resonance form makes less contribution to the structure because of the high formal charges involved.

   (d) $\ddot{O}=\dot{N}-\ddot{O}\colon \longleftrightarrow \colon\ddot{O}-\dot{N}=\ddot{O}\colon$

The molecule is bent (nonlinear). The odd electron resides on N because it is less electronegative than O.

22.60   (a) $4Zn(s) + 2NO_3^-(aq) + 10H^+(aq) \rightarrow 4Zn^{2+}(aq) + N_2O(g) + 5H_2O(l)$

   (b) $4NO_3^-(aq) + S(s) + 4H^+(aq) \rightarrow 4NO_2(g) + SO_2(g) + 2H_2O(l)$
         (or $6NO_3^-(aq) + S(s) + 4H^+(aq) \rightarrow 6NO_2(g) + SO_4^{2-}(aq) + 2H_2O(l)$

   (c) $2NO_3^-(aq) + 3SO_2(g) + 2H_2O(l) \rightarrow 2NO(g) + 3SO_4^{2-}(aq) + 4H^+(aq)$

   (d) $N_2H_4(g) + 5F_2(g) \rightarrow 2NF_3(g) + 4HF(g)$

   (e) $4CrO_4^{2-}(aq) + 3N_2H_4(aq) + 4H_2O(l) \rightarrow 4Cr(OH)_4^-(aq) + 4OH^-(aq) + 3N_2(g)$

22.62   (a) $NO_3^-(aq) + 4H^+(aq) + 3e^- \rightarrow NO(g) + 2H_2O(l)$

   (b) $HNO_2(aq) \rightarrow NO_2(g) + H^+(aq) + 1e^-$

22.64   (a) PO_4^{3-}, +5   (b) H_3AsO_3, +3   (c) Sb_2S_3, +3
   (d) $Ca(H_2PO_4)_2$, +5   (e) K_3P, –3

22.66   (a)   Only two of the hydrogens in H_3PO_3 are bound to oxygen. The third is attached directly to phosphorus, and not readily ionized, because the H—P bond is not very polar.

(b)   The smaller, more electronegative nitrogen withdraws more electron density from the O—H bond, making it more polar and more likely to ionize.

(c)   Phosphate rock consists of $Ca_3(PO_4)_2$, which is only slightly soluble in water. The phosphorus is unavailable for plant use.

(d) N_2 can form stable π bonds to complete the octet of both N atoms. Because phosphorus atoms are larger than nitrogen atoms, they do not form stable π bonds with themselves and must form σ bonds with several other phosphorus atoms (producing P_4 tetrahedral or sheet structures) to complete their octets.

(e)   In solution Na_3PO_4 is completely dissociated into Na^+ and PO_4^{3-}. PO_4^{3-}, the conjugate base of the very weak acid HPO_4^{2-}, has a K_b of 2.4×10^{-2} and produces a considerable amount of OH^- by hydrolysis of H_2O.

22.68   (a) $PCl_5(l) + 4H_2O(l) \rightarrow H_3PO_4(aq) + 5HCl(aq)$

(b) $2H_3PO_4(aq) \overset{\Delta}{\rightarrow} H_4P_2O_7(aq) + H_2O(l)$

(c) $P_4O_{10}(s) + 6H_2O(l) \rightarrow 4H_3PO_4(aq)$

Carbon, the Other Group 4A Elements, and Boron

22.70   (a) H_2CO_3 (b)   NaCN   (c) $KHCO_3$ (d) C_2H_2

22.72   (a) $CO_2(g) + OH^-(aq) \rightarrow HCO_3^-(aq)$

(b) $NaHCO_3(s) + H^+(aq) \rightarrow Na^+(aq) + H_2O(l) + CO_2(g)$

(c) $2CaO(s) + 5C(s) \overset{\Delta}{\rightarrow} 2CaC_2(s) + CO_2(g)$

(d) $C(s) + H_2O(g) \overset{\Delta}{\rightarrow} H_2(g) + CO(g)$

(e) $CuO(s) + CO(g) \rightarrow Cu(s) + CO_2(g)$

22.74   (a) $2Mg(s) + CO_2(g) \rightarrow 2MgO(s) + C(s)$

(b) $6CO_2(g) + 6H_2O(l) \overset{h\nu}{\rightarrow} C_6H_{12}O_6(aq) + 6O_2(g)$

(c) $CO_3^{2-}(aq) + H_2O(l) \rightarrow HCO_3^-(aq) + OH^-(aq)$

22.76   (a) SiO_2, +4   (b) $GeCl_4$, +4   (c) $NaBH_4$, +3

(d) $SnCl_2$, +2   (e) B_2H_6, +3

22.78   (a)   carbon   (b)   lead   (c)   germanium

22.80   (a)

(b) $Si_3O_9^{6-}$ $Si_6O_{18}^{12-}$

22.82   (a) $B_2H_6(g) + 6H_2O(l) \rightarrow 2H_3BO_3(aq) + 6H_2(g)$

       (b) $4H_3BO_3(s) \overset{\Delta}{\rightarrow} H_2B_4O_7(s) + 5H_2O(g)$

       (c) $B_2O_3(s) + 3H_2O(l) \rightarrow 2H_3BO_3(aq)$

Additional Exercises

22.83   (a)   *Allotropes* are different structural forms of the same element. They are composed of atoms of a single element bound into different structures. For example, graphite, diamond, and buckey balls are all allotropes of carbon.

       (b)   *Disproportionation* is an oxidation-reduction process where the same element is both oxidized and reduced.

       (c)   *Interhalogen* is a compound formed from atoms of two or more halogens.

       (d)   An *acidic anhydride* is a neutral molecule that is the oxide of a nonmetal. It reacts with water to produce an acid.

       (e)   A *condensation reaction* is the combination of two molecules to form a large molecule and a small one such as H_2O or HCl.

22.84   (a)   React an ionic nitride with D_2O, e.g.,

           $Mg_3N_2(s) + 6D_2O(l) \rightarrow 2ND_3(aq) + 3Mg(OD)_2(s)$

       (b)   React SO_3 with D_2O: $SO_3(g) + D_2O(l) \rightleftharpoons D_2SO_4(aq)$

       (c)   React Na_2O with D_2O: $Na_2O(s) + D_2O(l) \rightarrow 2NaOD(aq)$

       (d)   Dissolve $N_2O_5(g)$ in D_2O: $N_2O_5(g) + D_2O(l) \rightarrow 2DNO_3(aq)$

       (e)   React CaC_2 with D_2O: $CaC_2(s) + 2D_2O(l) \rightarrow Ca^{2+}(aq) + 2OD^-(aq) + C_2D_2(g)$

       (f)   Add NaCN to the D_2SO_4 solution prepared in (b):

           $NaCN(s) + D^+(aq) \overset{\Delta}{\rightarrow} DCN(aq) + Na^+(aq)$

           The DCN can be removed as gas from the reaction.

22.86   (a) $SO_2(g) + H_2O(l) \rightleftharpoons H_2SO_3(aq)$

       (b) $Cl_2O_7(g) + H_2O(l) \rightleftharpoons 2HClO_4(aq)$

       (c) $Na_2O_2(s) + 2H_2O \rightarrow H_2O_2(aq) + 2NaOH(aq)$

       (d) $BaC_2(s) + 2H_2O(l) \rightarrow Ba^{2+}(aq) + 2OH^-(aq) + C_2H_2(g)$

       (e) $2RbO_2(s) + 2H_2O(l) \rightarrow 2Rb^+(aq) + 2OH^-(aq) + O_2(g) + H_2O_2(aq)$

       (f) $Mg_3N_2(s) + 6H_2O(l) \rightarrow 3Mg(OH)_2(s) + 2NH_3(g)$

       (g) $NaH(s) + H_2O \rightarrow NaOH(aq) + H_2(g)$

22.88   Sulfur has a total of six valence electrons; its possible oxidation states range from -2 to $+6$. SO_3 has sulfur in its maximum $+6$ oxidation state; it cannot lose electrons and serve as a reducing agent. SO_2 has sulfur in the $+4$ oxidation state. It can lose electrons, be oxidized and serve as a reducing agent.

22.89

$$S(g) + O_2(g) \rightarrow SO_2(g) \qquad \Delta H = -296.9 \text{ kJ} \qquad (1)$$
$$SO_2(g) + 1/2\,O_2(g) \rightarrow SO_3(g) \qquad \Delta H = -98.3 \text{ kJ} \qquad (2)$$
$$\underline{SO_3(g) + H_2O(l) \rightarrow H_2SO_4(aq) \qquad \Delta H = -130 \text{ kJ} \qquad (3)}$$
$$S(g) + 3/2\,O_2(g) + H_2O(l) \rightarrow H_2SO_4(aq) \qquad \Delta H = -525 \text{ kJ}$$

$$1 \text{ ton } H_2SO_4 \times \frac{2000 \text{ lb}}{\text{ton}} \times \frac{453.6 \text{ g}}{1 \text{ lb}} \times \frac{1 \text{ mol } H_2SO_4}{98.09 \text{ g}} \times \frac{-525 \text{ kJ}}{\text{mol } H_2SO_4}$$

$$= -4.86 \times 10^6 \text{ kJ of heat/ton } H_2SO_4$$

22.91 (a) Although P_4, P_4O_6 and P_4O_{10} all have four P atoms in a tetrahedral arrangement, the bonding **between** P atoms and **by** P atoms is not the same in the three molecules. In P_4, the 4 P atoms are bound only to each other by P—P single bonds and strained bond angles of approximately 60°. In the two oxides, the 4 P atoms are directly bound to oxygen atoms, not to each other. Bonding by P atoms in P_4O_6 and P_4O_{10} is very similar. Each contains the P_4O_6 cage, formed by four P_3O_3 rings which share a P—O—P edge. Phosphorus bonding to oxygen maintains the overall P_4 tetrahedron but allows the P atoms to move away from each other so that the angle strain is relieved relative to molecular P_4. The P—O—P and O—P—O angles in both oxides are near the ideal 109°. In P_4O_6, each P is bound to 3 O atoms and has a lone pair completing its octet. In P_4O_{10}, the lone pair is replaced by a terminal O atom and each P is bound to 3 bridging and 1 terminal O atom.

(b)

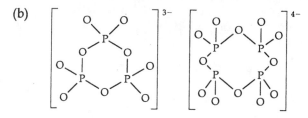

In both structures there are unshared pairs on all oxygens to give octets and the geometry around each P is approximately tetrahedral.

22.92 (a)

Si Si
 \ /
 Si

To complete their octets, the two terminal Si atoms each require three H atoms and the central Si requires two, for a total of 8 H atoms. The molecular formula is Si_3H_8.

(b) $Si_3H_8 + 5O_2 \rightarrow 3SiO_2 + 4H_2O$

22.94 (a)

$$2[5e^- + MnO_4^-(aq) + 8H^+(aq) \rightarrow Mn^{2+}(aq) + 4H_2O(l)]$$
$$\underline{5[H_2O_2(aq) \rightarrow O_2(g) + 2H^+(aq) + 2e^-]}$$
$$2MnO_4^-(aq) + 5H_2O_2(aq) + 6H^+(aq) \rightarrow 2Mn^{2+}(aq) + 5O_2(g) + 8H_2O(l)$$

(b)

$$2[Fe^{2+}(aq) \rightarrow Fe^{3+}(aq) + e^-]$$

$$\frac{H_2O_2(aq) + 2H^+(aq) + 2e^- \rightarrow 2H_2O(l)}{2Fe^{2+}(aq) + H_2O_2(aq) + 2H^+(aq) \rightarrow 2Fe^{3+}(aq) + 2H_2O(l)}$$

(c)

$$2I^-(aq) \rightarrow I_2(s) + 2e^-$$

$$\frac{H_2O_2(aq) + 2H^+(aq) + 2e^- \rightarrow 2H_2O(l)}{2\,I^-(aq) + H_2O_2(aq) + 2H^+(aq) \rightarrow I_2(s) + 2H_2O(l)}$$

(d)

$$Cu(s) \rightarrow Cu^{2+}(aq) + 2e^-$$

$$\frac{H_2O_2(aq) + 2H^+(aq) + 2e^- \rightarrow 2H_2O(l)}{Cu(s) + H_2O_2(aq) + 2H^+(aq) \rightarrow Cu^{2+}(aq) + 2H_2O(l)}$$

(e)

$$2I^-(aq) \rightarrow I_2(s) + 2e^-$$

$$\frac{O_3(g) + H_2O(l) + 2e^- \rightarrow O_2(g) + 2OH^-(aq)}{2I^-(aq) + O_3(g) + H_2O(l) \rightarrow O_2(g) + I_2(s) + 2OH^-(aq)}$$

22.96   (a) $Li_3N(s) + 3H_2O(l) \rightarrow 3Li^+(aq) + 3OH^-(aq) + NH_3(aq)$

      (b) $NH_3(aq) + H_2O(l) \rightleftharpoons NH_4^+(aq) + OH^-(aq)$

      (c) $3NO_2(g) + H_2O(l) \rightarrow NO(g) + 2H^+(aq) + 2NO_3^-(aq)$

      (d) $4NH_3(g) + 5O_2(g) \xrightarrow{\text{catalyst}} 4NO(g) + 6H_2O(g)$

      (e) $H_2CO_3(aq) \xrightarrow{\Delta} H_2O(g) + CO_2(g)$

      (f) $Ni(s) + CO(g) \rightarrow NiO(s) + C(s)$

      (h) $CS_2(g) + O_2(g) \rightarrow CO_2(g) + S_2(g)$

      (i) $CaO(s) + SO_2(g) \rightarrow CaSO_3(s)$

      (j) $CH_4(g) + H_2O(g) \xrightarrow{\Delta} CO(g) + 3H_2(g)$

      (k) $LiH(s) + H_2O(l) \rightarrow LiOH(aq) + H_2(g)$

      (l) $Fe_2O_3(s) + 3H_2(g) \rightarrow 2Fe(s) + 3H_2O(g)$

Integrative Exercises

22.98   From Appendix C, we need only ΔH_f° for F(g), so that we can estimate ΔH for the process:

$$F_2(g) \rightarrow F(g) + F(g); \qquad \Delta H^\circ = 160 \text{ kJ}$$

$$\frac{XeF_2(g) \rightarrow Xe(g) + F_2(g) \qquad -\Delta H_f^\circ = 109 \text{ kJ}}{XeF_2(g) \rightarrow Xe(g) + 2F(g) \qquad \Delta H^\circ = 269 \text{ kJ}}$$

The average Xe—F bond enthalpy is thus $269/2 = 134$ kJ. Similarly,

$$XeF_4(g) \rightarrow Xe(g) + 2F_2(g) \qquad -\Delta H_f^\circ = 218 \text{ kJ}$$

$$\underline{2F_2(g) \rightarrow 4F(g) \qquad\qquad \Delta H^\circ = 320 \text{ kJ}}$$

$$XeF_4(g) \rightarrow Xe(g) + 4F(g) \qquad \Delta H^\circ = 538 \text{ kJ}$$

Average Xe—F bond energy = 538/4 = 134 kJ

$$XeF_6(g) \rightarrow Xe(g) + 3F_2(g) \qquad -\Delta H_f^\circ = 298 \text{ kJ}$$

$$\underline{3F_2(g) \rightarrow 6F(g) \qquad\qquad \Delta H^\circ = 480 \text{ kJ}}$$

$$XeF_6(g) \rightarrow Xe(g) + 6F(g) \qquad \Delta H^\circ = 778 \text{ kJ}$$

Average Xe—F bond energy = 778/6 = 130 kJ

The average bond enthalpies are: XeF_2, 134 kJ; XeF_4, 134 kJ; XeF_6, 130 kJ. They are remarkably constant in the series.

22.99 (a) $H_2(g) + 1/2 \, O_2(g) \rightarrow H_2O(l); \Delta H = -285.83 \text{ kJ}$

$CH_4(g) + 2O_2(g) \rightarrow CO_2(g) + 2H_2O(l)$

$\Delta H = 2(-285.83) - 393.5 - (-74.8) = -890.4 \text{ kJ}$

(b) for H_2: $\dfrac{-285.83 \text{ kJ}}{1 \text{ mol } H_2} \times \dfrac{1 \text{ mol } H_2}{2.0159 \text{ g } H_2} = -141.79 \text{ kJ/g } H_2$

for CH_4: $\dfrac{-890.4 \text{ kJ}}{1 \text{ mol } CH_4} \times \dfrac{1 \text{ mol } CH_4}{16.043 \text{ g } CH_4} = -55.50 \text{ kJ/g } CH_4$

(c) Find the number of moles of gas that occupy 1 m^3 at STP:

$$n = \frac{1 \text{ atm} \times 1 \text{ m}^3}{273 \text{ K}} \times \frac{1 \text{ K-mol}}{0.08206 \text{ L-atm}} \times \left[\frac{100 \text{ cm}}{1 \text{ m}}\right]^3 \times \frac{1 \text{ L}}{10^3 \text{ cm}^3} = 44.64 \text{ mol}$$

for H_2: $\dfrac{-285.83 \text{ kJ}}{1 \text{ mol } H_2} \times \dfrac{44.64 \text{ mol } H_2}{1 \text{ m}^3 \, H_2} = 1.276 \times 10^4 \text{ kJ/m}^3 \, H_2$

for CH_4: $\dfrac{-890.4 \text{ kJ}}{1 \text{ mol } CH_4} \times \dfrac{44.64 \text{ mol } CH_4}{1 \text{ m}^3 \, CH_4} = 3.975 \times 10^4 \text{ kJ/m}^3 \, CH_4$

22.101 (a) $2NH_4ClO_4(s) \xrightarrow{\Delta} N_2(g) + 2HCl(g) + 3H_2O(g) + 5/2 \, O_2(g)$

$NH_4ClO_4(s) \xrightarrow{\Delta} 1/2 \, N_2(g) + HCl(g) + 3/2 \, H_2O(g) + 5/4 \, O_2(g)$

(b) $\Delta H^\circ = \Sigma \, \Delta H_f^\circ \text{ prod} - \Sigma \, \Delta H \text{ react}$

$\Delta H^\circ = \Delta H_f^\circ \, HCl(g) + 3/2 \, \Delta H_f^\circ \, H_2O(g) + 1/2 \, \Delta H_f^\circ \, N_2(g) + 5/4 \, \Delta H_f^\circ \, O_2(g) - \Delta H_f^\circ \, NH_4ClO_4 \Delta H^\circ$

$= -92.30 \text{ kJ} + 3/2(-241.82 \text{ kJ}) + 1/2 \, (0 \text{ kJ}) + 5/4 \, (0 \text{ kJ}) - (-295.8 \text{ kJ})$

$= -159.2 \text{ kJ/mol } NH_4ClO_4$

(c) The aluminum reacts exothermically with $O_2(g)$ and $HCl(g)$ produced in the decomposition, providing additional heat and thrust.

22.102 (a) $N_2H_4(g) + O_2(g) \rightarrow N_2(g) + 2H_2O(l)$

(b) $\Delta H° = \Delta H_f° \, N_2(g) + 2\Delta H_f° \, H_2O(l) - \Delta H_f° \, N_2H_4(aq) - \Delta H_f° \, O_2(g)$

$= 0 + 2(-285.83) - 95.40 - 0 = -667.06 \text{ kJ}$

(c) $\dfrac{9.1 \text{ g } O_2}{1 \times 10^6 \text{ g } H_2O} \times \dfrac{1.0 \text{ g } H_2O}{1 \text{ mL } H_2O} \times \dfrac{1000 \text{ mL}}{1 L} \times 3.0 \times 10^4 \text{ L} = 273 = 2.7 \times 10^2 \text{ g } O_2$

$2.73 \times 10^2 \text{ g } O_2 \times \dfrac{1 \text{ mol } O_2}{32.00 \text{ g } O_2} \times \dfrac{1 \text{ mol } N_2H_4}{1 \text{ mol } O_2} \times \dfrac{32.05 \text{ g } N_2H_4}{1 \text{ mol } N_2H_4} = 2.7 \times 10^2 \text{ g } N_2H_4$

22.104 *Plan.* vol air → kg air → g H$_2$S → g FeS. Use the ideal-gas equation to change volume of air to mass of air, (assuming 1.00 atm, 298 K and an average molar mass (MM) for air of 29.0 g/mol. Use (20 mg H$_2$S/kg) air to find the mass of H$_2$S in the given mass of air. *Solve.*

$V_{air} = 2.7 \text{ m} \times 4.3 \text{ m} \times 4.3 \text{ m} \times \dfrac{(100)^3 \text{ cm}^3}{1 \text{ m}^3} \times \dfrac{1 L}{1000 \text{ cm}^3} = 4.9923 \times 10^4 = 5.0 \times 10^4 \text{ L}$

$g_{air} = \dfrac{PV \, MM}{RT}$; assume P = 1.00 atm, T = 298 K, MM$_{air}$ = 29.0 g/mol

$g_{air} = \dfrac{1.00 \text{ atm} \times 4.9923 \times 10^4 \text{ L} \times 29.0 \text{ g/mol}}{298 \text{ K}} \times \dfrac{\text{K-mol}}{0.08206 \text{ L-atm}} = 59{,}204 = 5.9 \times 10^4 \text{ g air}$

$5.9203 \times 10^4 \text{ g air} \times \dfrac{1 \text{ kg}}{1000 \text{ g}} \times \dfrac{20 \text{ mg } H_2S}{1 \text{ kg air}} \times \dfrac{1 g}{1000 \text{ mg}} = 1.184 = 1.2 \text{ g } H_2S$

$FeS(s) + 2HCl(aq) \rightarrow FeCl_2(aq) + H_2S(g)$

$1.184 \text{ g } H_2 \times \dfrac{1 \text{ mol } H_2}{34.08 \text{ g } H_2S} \times \dfrac{1 \text{ mol FeS}}{1 \text{ mol } H_2S} \times \dfrac{87.91 \text{ g FeS}}{1 \text{ mol FeS}} = 3.054 = 3.1 \text{ g FeS}$

22.105 The reactions can be written as follows:

$H_2(g) + X(\text{std state}) \rightarrow H_2X(g) \qquad \Delta H_f°$

$2H(g) \rightarrow H_2(g) \qquad \Delta H_f°(H-H)$

$\underline{X(g) \rightarrow X(\text{std state}) \qquad \Delta H_3}$

$\text{Add}: 2H(g) + X(g) \rightarrow H_2X(g) \qquad \Delta H = \Delta H_f° + \Delta H_f° (H-H) + \Delta H_3$

These are all the necessary ΔH values. Thus,

Compound	ΔH	D H−X
H$_2$O	ΔH = –242 kJ – 436 kJ – 248 kJ = –926 kJ	463 kJ
H$_2$S	ΔH = –20 kJ – 436 kJ – 277 kJ = –733 kJ	367 kJ
H$_2$Se	ΔH = +30 kJ – 436 kJ – 227 kJ = –633 kJ	316 kJ
H$_2$Te	ΔH = +100 kJ – 436 kJ – 197 kJ = –533 kJ	266 kJ

The average H−X bond energy in each case is just half of ΔH. The H−X bond energy decreases steadily in the series. The origin of this effect is probably the increasing size of the orbital from X with which the hydrogen 1s orbital must overlap.

22.107 $N_2H_5^+(aq) \rightarrow N_2(g) + 5H^+(aq) + 4e^- \qquad E_{red}^{\circ} = -0.23 \text{ V}$

Reduction of the metal should occur when E_{red}° of the metal ion is more positive than about -0.15 V. This is the case for (b) Sn^{2+} (marginal), (c) Cu^{2+} and (d) Ag^+.

22.108   First write the balanced equation to give the number of moles of gaseous products per mole of hydrazine.

(A) $(CH_3)_2NNH_2 + 2N_2O_4 \rightarrow 3N_2(g) + 4H_2O(g) + 2CO_2(g)$

(B) $(CH_3)HNNH_2 + 5/4\,N_2O_4 \rightarrow 9/4\,N_2(g) + 3H_2O(g) + CO_2(g)$

In case (A) there are nine moles gas per one mole $(CH_3)_2NNH_2$ plus two moles N_2O_4. The total mass of reactants is $60 + 2(92) = 244$ g. Thus, there are

$$\frac{9 \text{ mol gas}}{244 \text{ g reactants}} = \frac{0.0369 \text{ mol gas}}{1 \text{ g reactants}}$$

In case (B) there are 6.25 moles of gaseous product per one mole $(CH_3)HNNH_2$ plus 1.25 moles N_2O_4. The total mass of this amount of reactants is $46.0 + 1.25(92.0) = 161$ g.

$$\frac{6.25 \text{ mol gas}}{161 \text{ g reactants}} = \frac{0.0388 \text{ mol gas}}{1 \text{ g reactants}}$$

Thus the methylhydrazine (B) has marginally greater thrust.

22.110   BN has the same number of valence electrons per formula unit as carbon. (Three from B, five from N, for an average of four per atom.) To the extent that we can neglect the difference in nuclear charges between B and N, we can think of BN as carbon-like. Indeed, BN takes on the same structural forms as carbon. However, because the B—N bonds are somewhat polar, BN is in fact even harder than diamond.

23 Metals and Metallurgy

Visualizing Concepts

23.2 (a) $MgCl_2(l)$

(b) The carbon electrode is the anode. Oxidation occurs at the anode, so formation of $Cl_2(g)$ from $2Cl^-(l)$ is the anode process.

The steel electrode is the cathode. Reduction occurs at the cathode, so formation of $Mg(l)$ from $Mg^{2+}(l)$ is the cathode process.

Check. Note that this electrolysis cell does not comply with the convention of drawing anode on the left and cathode on the right.

(c)
$$MgCl_2(l) \rightarrow Mg(s) + Cl_2(g) \qquad \text{overall}$$
$$2Cl^- \rightarrow Cl_2(g) + 2e^- \qquad \text{anode (oxidation)}$$
$$Mg^{2+} + 2e^- \rightarrow Mg(s) \qquad \text{cathode (reduction)}$$

(d) Magnesium is an active metal. It must be separated from the $Cl_2(g)$ that is also formed by electrolysis (see screen in diagram) or $MgCl_2$ will spontaneously reform. The $Mg(s)$ also should not come in contact with air (O_2) or moisture (H_2O).

23.4 Count fractions of each atom, depending on atomic position in unit cell. Corner = 1/8, face = 1/2, body center = 1. Tally red and gray separately for each compound. Use the empirical formula to identify the two compounds.

A. gray: 8 corners × 1/8 = 1

red: 2 each on top and bottom faces, one each on four side faces, one body center; $(8 \times 1/2) + 1 = 5$

empirical formula: GR_5

Compound A is $SmCo_5$ (Sm = gray, Co = red)

B: gray: 8 corners + 1 body center = $(8 \times 1/8) + 1 = 2$

red: 2 on each of 6 faces = $12 \times 1/2 = 6$

empirical formula = GR_3

Compound B is Nb_3Sn (Sn = gray, Nb = red)

23.6 (a) Sc: $[Ar]4s^23d^1$; Sc^{3+}: $[Ar]$

Scandium has a total of 3 valence electrons; loss of all 3 electrons leads to the stable noble gas configuration of Ar.

376

(b) Sr^{2+} has the noble gas electron configuration of Kr and is diamagnetic. The oxidation state of Cr in $CrO_4{}^{2-}$ is +6. If we think of the bonding between Cr and oxygen as totally ionic, Cr(VI) has lost all its valence electrons and has the electron configuration of Ar. There are no unpaired electrons associated with either Sr or Cr, and the ions are diamagnetic.

Assuming perfectly covalent bonding between Cr and oxygen, we can draw the Lewis structure below for $CrO_4{}^{2-}$.

$CrO_4{}^{2-}$, valence electrons: 6 from Cr + 24 from O + 2 from charge = 32 e^-, 16 e^- pairs

In the Lewis structure of $CrO_4{}^{2-}$, all valence electrons are paired; Cr is in the diamagnetic state. Note that both extreme bonding models predict a diamagnetic state for $CrO_4{}^{2-}$. $SrCrO_4$ is diamagnetic.

(c)   "Inorganic" ions are those not based on carbon. $ClO_4{}^-$ has the same formula type and charge as and $MnO_4{}^-$. Cl has 7 valence electrons like Mn. $ClO_4{}^-$ is tetrahedral, so we predict that $MnO_4{}^-$ is also tetrahedral.

Metallurgy

23.8   (a)   +4

(b) $MnO_2(s) + 4H^+(aq) + 2e^- \rightarrow Mn^{2+}(aq) + 2H_2O(l)$ $E^\circ_{red} = 1.23$ V

$Mn^{2+}(aq) + 2e^- \rightarrow Mn(s)$ $E^\circ_{red} = -1.18$ V

Standard reduction potentials indicate that a very strong reducing agent is required to reduce the ore to Mn(s), at least if $Mn^{2+}(aq)$ is an intermediate product. According to Appendix E, only Group I and Group II metals (Li, Na, Mg, etc.) are strong enough to reduce Mn^{2+} to Mn. In practice, $MnO_2(s)$ is reduced by coke in blast furnaces, into which MnO_2 is added to incorporate Mn into steel.

23.10   (a)   *Calcination* is heating an ore to decompose the mineral of interest into a simple solid and volatile compound. Calcination usually produces a metal oxide and a gas that is a nonmetal oxide.

(b)   *Leaching* is dissolving the mineral of interest to remove it from an ore. The solvent is usually water or an aqueous solution of acid, base or salt.

(c)   *Smelting* is heating an ore, often in a reducing atmosphere, to a very high temperature so that two immiscible liquid layers form. The layers are usually the molten metal or metals of interest and slag.

(d)   *Slag* is the unwanted layer of the smelting process. It contains molten silicate, aluminate, phosphate or fluoride compounds.

23.12   (a) $2PbS(s) + 3O_2(s) \overset{\Delta}{\rightarrow} 2PbO(s) + 2SO_2(g)$

         (b) $CoCO_3(s) \overset{\Delta}{\rightarrow} CoO(s) + CO_2(g)$

         (c) $WO_3(s) + 3H_2(g) \overset{\Delta}{\rightarrow} W(s) + 3H_2O(g)$

         (d) $VCl_3(g) + 3K(l) \rightarrow V(s) + 3KCl(s)$

         (e) $3BaO(s) + P_2O_5(l) \rightarrow Ba_3(PO_4)_2(l)$

23.14   (a)   Spontaneous reactions have negative values for ΔG. $\Delta G = \Delta H - T\Delta S$. The effect of temperature change on ΔG depends on the sign of ΔS. For reactions where ΔS is positive, ΔG becomes more negative as temperature increases. The generic calcination reaction has all solids in the reactants, but one mole of gas among the products, so ΔS is definitely positive. Calcination becomes more spontaneous as temperature increases.

         (b) $\Delta G° = \Delta G°\ PbO(s) + \Delta G°\ CO_2(g) - \Delta G°\ PbCO_3(s)$

                $\Delta G° = -187.9\ kJ - 394.4\ kJ - (-625.5\ kJ) = 43.2\ kJ$

                The reaction is not spontaneous at 25°C and standard conditions. Calculate $\Delta H°$ and $\Delta S°$ and use $\Delta G° = \Delta H° - T\Delta S°$ to find the temperature where the reaction becomes spontaneous.

                $\Delta H° = \Delta H°\ PbO(s) + \Delta H°\ CO_2(g) - \Delta H°\ PbCO_3(s)$

                $= -217.3\ kJ + (-393.5\ kJ) - (-699.1\ kJ) = 88.3\ kJ$

                $\Delta S° = S°\ PbO(s) + S°\ CO_2(g) - S°\ PbCO_3(s)$

                $= 68.7\ J/K + 213.6\ J/K - 131.0\ J/K = 151.3\ J/K$

                The reaction becomes spontaneous when $\Delta G° = 0$ and $\Delta H° = T\Delta S°$.

                $88.3\ kJ = T\ (0.1513\ kJ/K); \quad T = \dfrac{88.3\ kJ}{0.1513\ J/K} = 583.61 = 584\ K\ (311°C)$

23.16   (a)   Use Equation 23.5 as a model.

                $TiO_2(s) + 2CO(g) \rightarrow Ti(l) + 2CO_2(g)$

         (b)   Thermodynamic data for Ti(l) is not available in Appendix C, so use data for Ti(s). This is reasonable, because the melting point of Ti is 1941 K (1668°C).

                $\Delta G° = \Delta G°\ Ti(s) + 2\Delta G°\ CO_2(g) - \Delta G°\ TiO_2(s) - 2\Delta G°\ CO$

                $= 0 + 2(-394.4\ kJ) - (-889.4\ kJ) - 2(-137.2\ kJ) = 375.0\ kJ$

                $\Delta H° = 0 + 2(-393.5\ kJ) - (-944.7\ kJ) - 2(-110.5\ kJ) = 378.7\ kJ$

                $\Delta S° = 30.7\ J/K + 2(213.6\ J/K) - 50.29\ J/K - 2(197.9\ J/K) = 11.81\ J/K$

                $\Delta G°$ is large and positive, so the reaction is not spontaneous at standard conditions and 25°C.

(c)   Reactions with $+\Delta H°$ and $+\Delta S°$, like this one, can become spontaneous at elevated temperatures. This reaction would become spontaneous at

$$T = \frac{\Delta H°}{\Delta S°} = \frac{378.7 \text{ kJ}}{0.01181 \text{ kJ/K}} = 32,066 = 32,070 \text{ K}$$

From the NIST Chemistry WebBook (webbook.nist.gov /chemistry), thermodynamic data for Ti(l) are $\Delta H° = 13.65$ kJ, $S° = 39.18$ J/mol-K.

Using these values in calculations similar to those above: $\Delta H° = 392.35$ kJ; $\Delta S° = 20.29$ J/K. The reaction that produces Ti(l) will become spontaneous at 19,340 K, still an unimaginably high temperature. Whether the product is Ti(s) or Ti(l), this process is not practical.

23.18 $FeO(s) + H_2(g) \rightarrow Fe(s) + H_2O(g)$

$FeO(s) + CO(g) \rightarrow Fe(s) + CO_2(g)$

$Fe_2O_3(s) + 3H_2(g) \rightarrow 2Fe(s) + 3H_2O(g)$

$Fe_2O_3(s) + 3CO(g) \rightarrow 2Fe(s) + 3CO_2(g)$

23.20   (a)   In the *converter*, oxidation of C, Si and metals by O_2 are exothermic reactions that raise the temperature.

         (b) $2C(s) + O_2(g) \rightarrow 2CO(g)$; $S(s) + O_2(g) \rightarrow SO_2(g)$; $Si(s) + O_2(g) \rightarrow SiO_2(l)$

23.22 $SnO_2(s) + C(s) \xrightarrow{\Delta} Sn(l) + CO_2(g)$

$Sn(s) \rightarrow Sn^{2+}(aq) + 2e^-$ (anode)

$Sn^{2+}(aq) + 2e^- \rightarrow Sn(s)$ (cathode)

Metals and Alloys

23.24   Since silicon has the same crystal structure as diamond, it is a covalent-network solid with its bonding electrons localized between Si atoms. Since there is no significant delocalization in Si, it is not likely to have the metallic properties of malleability, ductility and high electrical and thermal conductivity. It is likely to be hard and high-melting like other covalent-network solids.

23.26   (a)   Cr: $[Ar]4s^13d^5$, $Z = 24$; Se: $[Ar]4s^23d^{10}4p^4$, $Z = 34$

         Both elements have the [Ar] core configuration and both have six valence electrons. The orbital locations of the six valence electrons are different in the two elements, because Se has more total electrons.

         (b)   Different Z and Z_{eff} for the two elements, and the different orbital locations of the valence electrons, are the main factors that lead to the differences in properties. In Cr, the 4s and 3d electrons are the valence electrons. Its Z and Z_{eff} are smaller than those of Se and it is not likely to gain enough electrons to achieve a noble-gas configuration. Thus, Cr loses electrons when it forms ions, acting like a metal. Se is in the same row of the periodic table as Cr, but its 3d subshell is filled, so its valence electrons are in 4s and 4p. Because Se has a larger Z and Z_{eff}, it is more likely to hold its own valence electrons and gain other electrons when it forms ions. That Se needs only two additional electrons to achieve the noble-gas configuration of Kr is also a driving force for it gaining electrons when it forms ions, acting like a nonmetal.

23.28   Moving across the fifth period from Y to Mo, the melting points of the metals increase. The number of valence electrons also increases, from 3 for Y to 6 for Mo. More valence electrons (up to 6) mean increased occupancy of the bonding molecular orbital band, and increased strength of metallic bonding. Melting requires that atoms are moving relative to each other. Stronger metallic bonding requires more energy to break bonds and mobilize atoms, resulting in higher melting points from Y to Mo.

23.30   Metals, with few valence electrons, cannot satisfy their "bonding needs" by localized covalent bonds. Metals form structures with several nearest neighbor atoms. (Na and K have body-centered cubic lattices where each atom has eight nearest neighbors.) Bonding electrons are delocalized throughout the lattice. Although bonding electrons are associated with individual atoms, their locations are not restricted. Changes in atomic position due to deformation like cutting (or pounding or shaping) are at least partly accommodated by shifting of the delocalized bonding electrons. In solids with rigid lattices formed by ionic or network covalent bonding, electrons are locked into localized bonds. A small "crack" (shift of atom positions) breaks bonds and leads to total destruction of the lattice and catastrophic failure of the material. Metals, on the other hand, are able to "adjust" to atom dislocation (cuts, cracks, dents, etc.) because of their unique delocalized bonding characteristics.

23.32   Electrical conductivity is related to the extent of valence electron delocalization in the material.

                In the hexagonal close-packed structure of titanium, each Ti atom has twelve nearest neighbors. The four valence electrons of a Ti atom are delocalized over bonding interactions with twelve neighbors. In the diamond structure of silicon, each Si atom has four nearest neighbors and four valence electrons. These four valence electrons are essentially localized in four covalent (sigma) bonds to the four nearest neighbors. The much more extensive electron delocalization in Ti leads to its significantly greater electrical conductivity.

23.34   Substitutional and interstitial alloys are both solution alloys. In a *substitutional* alloy, the atoms of the "solute" take positions normally occupied by the "solvent." Substitutional alloys tend to form when solute and solvent atoms are of comparable size and have similar bonding characteristics. In an *interstitial* alloy, the atoms of the "solute" occupy the holes or interstitial positions between "solvent" atoms. Solute atoms are necessarily much smaller than solvent atoms.

23.36   (a) $Cu_{0.66}Zn_{0.34}$, substitutional alloy; similar atomic radii, substantial amounts of both components

         (b) Ag_3Sn, intermetallic compound; set stoichiometric ratio of components

         (c) $Ti_{0.99}O_{0.01}$, interstitial alloy; very different atomic radii, tiny amount of smaller component

Transition Metals

23.38 (b) NiCo alloy and (c) W will have metallic properties. The lattices of these substances are composed of neutral metal atoms. Delocalization of valence electrons via the MO model of metallic bonding produces metallic properties.

(d) Ge is a metalloid, not a metal. (a) $TiCl_4$ is an ionic compound and (e) Hg_2^{2+} is a metal ion. In ions and ionic compounds, electrons are localized on the individual ions, precluding metallic properties.

23.40 *Analyze/Plan.* Define lanthanide contraction (Section 23.7). Based on the definition, list properties related to atomic radius. *Solve.*

The *lanthanide contraction* is the name given to the decrease in atomic size due to the build-up in effective nuclear charge as we move through the lanthanides (elements 58–71) and beyond them. This effect offsets the expected increase in atomic size going from the second to the third transition series. The lanthanide contraction affects size-related properties such as ionization energy, electron affinity, and density.

23.42 (a) CdO (b) TiO (c) Nb_2O_5 (d) NiO_2

23.44 V: $[Ar]4s^2 3d^3$, (V^{2+}: $[Ar]3d^3$); Sc: $[Ar]4s^2 3d^1$; (Sc^{2+}: $[Ar]3d^1$)

V has a slightly larger Z (23) and Z_{eff} than Sc (Z > 21), so the 3d electrons in V are more tightly held than those in Sc. Also, losing the lone 3d electron in Sc^{2+} leads to the stable noble-gas configuration of $[Ar]$ for Sc^{3+}.

23.46 (a) Ti^{2+}: $[Ar]3d^2$ (b) Co^{3+}: $[Ar]3d^6$ (c) Pd^{2+}: $[Kr]4d^8$

(d) Mo^{3+}: $[Kr]4d^3$ (e) Ru^{3+}: $[Kr]4d^5$ (f) Ni^{4+}: $[Ar]3d^6$

23.48 The stronger reducing agent is more easily oxidized; Cr^{2+} is more easily oxidized (see Solution 23.47), so it is the stronger reducing agent.

23.50 (a) Chromate ion, CrO_4^{2-}, is bright yellow. Dichromate, $Cr_2O_7^{2-}$, is orange.

(b) $Cr_2O_7^{2-}$ is more stable in acid solution than CrO_4^{2-}.

(c) Their interconversion in solution involves the acid-base equilibrium

$$2CrO_4^{2-}(aq) + 2H^+(aq) \rightleftharpoons Cr_2O_7^{2-}(aq) + H_2O(l).$$

23.52 (a) $MnO_2(s) + 4HCl(aq) \rightarrow MnCl_2(aq) + Cl_2(g) + 2H_2O(l)$

(b) Yes. MnO_2 is the oxidizing agent; HCl is the reducing agent.

25.54 On the atomic level, each atom in an antiferromagnetic material has an unpaired spin, but spins on coupled atoms are paired. The material is diamagnetic below its Neel temperature and paramagnetic above it.

In a diamagnetic material, each atom has only paired spins, and the material is diamagnetic at all temperatures.

23.56 Fe_2O_3 has all Fe atoms in the +3 oxidation state, while Fe_3O_4 contains Fe atoms in both the +2 and +3 states. Each Fe_3O_4 formula unit has one Fe(II) and two Fe(III).

     (a)   In an antiferromagnetic material, spins on coupled atoms are oppositely aligned, producing a net spin of zero. This is only possible for Fe_2O_3, where all Fe atoms have the same oxidation state, d-electron configuration and number of unpaired electrons. In Fe_3O_4, Fe(II) and Fe(III) atoms have different d-electron configurations and different numbers of unpaired electrons. Assuming an Fe(II) is coupled to an Fe(III), even if spins on coupled centers are oppositely aligned, their spins do not fully cancel and the material is ferrimagnetic.

     (b)   In principle, it should be possible to use magnetic fields to separate the two materials. The ferrimagnetic Fe_3O_4 has a permanent magnetic moment and will be somewhat attracted to a magnet. At temperatures below its Neel temperature, the antiferromagnetic FeO will not.

Additional Exercises

23.58 Al^{3+}, Mg^{2+}, and Na^+ all have large negative reduction potentials (Al, Mg, and Na are very active metals). A substance with a more negative reduction potential would have to be used to chemically reduce them. All such substances are more expensive and difficult to obtain than Al, Mg, and Na. Electrolysis is thus the most cost-efficient way to reduce Al^{3+}, Mg^{2+}, and Na^+ to their metallic states.

23.60   (a) $2VCl_3(s) + O_2(g) \rightarrow 2VOCl_3(s)$

     (b) $Nb_2O_5(s) + 5H_2(g) \rightarrow 2Nb(s) + 5H_2O(l)$

     (c) $2Fe^{3+}(aq) + Zn(s) \rightarrow 2Fe^{2+}(aq) + Zn^{2+}(aq)$

     (d) $NbCl_5(s) + 3H_2O(l) \rightarrow HNbO_3(s) + 5HCl(aq)$

23.61   (a) $NiO(s) + 2H^+(aq) \rightarrow Ni^{2+}(aq) + H_2O(l)$

     (b)   The simple answer is that the solid is subjected to acid hydrolysis:

        $CuCo_2S_4(s) + 8H^+(aq) \rightarrow Cu^{2+}(aq) + 2Co^{3+}(aq) + 4H_2S(g)$

        However, in the absence of a strong complexing ligand, Co^{3+} is not stable in water. It oxidizes water according to the following reaction:

        $4Co^{3+}(aq) + 2H_2O(l) \rightarrow 4Co^{2+}(aq) + O_2(g) + 4H^+(aq)$

     (c) $TiO_2(s) + C(s) + 2Cl_2(g) \rightarrow TiCl_4(g) + CO_2(g)$

     (d)   In this reaction O_2 is reduced and sulfide is oxidized. Writing the sulfur product as S_8, the balanced equation is:

        $8ZnS(s) + 4O_2(g) + 16H^+(aq) \rightarrow 8Zn^{2+}(aq) + S_8(s) + 8H_2O(l)$

23.63   All transition metals have the generic electron configuration $ns^2(n-1)d^x$. Regardless of the number of d electrons, each transition metal has 2 ns valence electrons that are the first electrons lost when metal ions are formed. Thus, almost every transition metal has a stable +2 oxidation state.

After the 2 ns electrons are lost, a varying number of $(n–1)d$ electrons can be lost, depending on the identity of the transition metal. The availability of different numbers of d electrons leads to a wide variety of accessible oxidation states for the transition metals.

23.64   Assuming that SO_2 and N_2 are the nonmetallic products, two half-reactions can be written:

$$5[MoS_2(s)+7H_2O(l) \rightarrow MoO_3(s)+2SO_2(g)+14H^+(aq)+14e^-]$$

$$\underline{7[12H^+(aq)+2NO_3^-(aq)+10e^- \rightarrow N_2(g)+6H_2O(l)]}$$

$$5MoS_2(s)+14H^+(aq)+14NO_3^-(aq) \rightarrow 5MoO_3(s)+10SO_2(g)+7N_2(g)+7H_2O(l)$$

$$MoO_3(s) + 2NH_3(aq) + H_2O(l) \rightarrow (NH_4)_2MoO_4(s)$$

$$(NH_4)_2MoO_4(s) \xrightarrow{\Delta} 2NH_3(g)+H_2O(g)+MoO_3(s)$$

$$MoO_3(s)+3H_2(g) \xrightarrow{\Delta} Mo(s)+3H_2O(g)$$

23.65   Niobium is a transition metal with 5 valence electrons ($5s^24d^3$), while antimony is a metalloid ($5s^25p^3$). Based strictly on their positions in the periodic chart, we expect Nb to be a better conductor of electricity.

In terms of strength of metallic bonding, Nb has a lower Fermi level and stronger metallic bonding than Sb. However, trends in conductivity (conductivity = 1/resistivity) as shown on <u>webelements.com</u> indicate that conductivity does not vary in the same pattern as metallic bond strength.

Rather, conductivity (and resistivity) are more strongly correlated to structure. According to Figure 7.7, Nb and Sb have nearly equal atomic radii. Nb adopts a body-centered cubic lattice (like Na and K) and each Nb atom has 8 nearest neighbors. Sb exists in a trigonal lattice and each Sb atom has 6 nearest neighbors. With equal atomic radii and 25% more close contacts (pathways for electron mobility), Nb has the greater conductivity.

23.66   The metallic properties of malleability, ductility, and high electrical and thermal conductivity are results of the delocalization of valence electrons throughout the lattice. Delocalization occurs because metal atom valence orbitals of nearest-neighbor atoms interact to produce nearly continuous molecular orbital energy bands. When C atoms are introduced into the metal lattice, their valence orbitals do not have the same energies as metal orbitals, and their interaction is different. This causes a discontinuity in the band structure and limits delocalization of electrons. The properties of the carbon-infused metal begin to resemble those of a covalent-network lattice with localized electrons (Solution 23.24). The substance is harder and less conductive than the pure metal.

23.68   (a) Nb^{5+}: [Ar]; diamagnetic, no unpaired electrons

(b) Cr^{2+}: $[Ar]3d^4$ paramagnetic, unpaired electrons

(c) Cu^+: $[Ar]3d^{10}$; diamagnetic, no unpaired electrons

(d) Ru^{8+}: [Ar]; diamagnetic, no unpaired electrons

(e) Ni^{2+}: $[Ar]3d^8$; paramagnetic, unpaired electrons

23.69 In a ferromagnetic solid, the magnetic centers are coupled such that the spins of all unpaired electrons are parallel. As the temperature of the solid increases, the average kinetic energy of the atoms increases until the energy of motion overcomes the force aligning the electron spins. The substance becomes paramagnetic; it still has unpaired electrons, but their spins are no longer aligned.

23.70 Antiferromagnetic materials have unpaired electrons (spin) on individual metal centers. Below the Neél temperature, spins on coupled (usually adjacent, sometimes coupled by bridging atoms) metals centers are oppositely aligned, producing zero net spin. Above the Neél temperature, atoms in the lattice have sufficient energy to undo the coupling interaction, spins on individual atoms are no longer aligned, and the substance is paramagnetic.

Diamagnetic materials have all electrons on individual atoms paired. There is no spin (or magnetic moment) on individual atoms. Thermal energy is unsufficient to cause electron transitions leading to unpaired electrons The material remains diamagnetic upon heating.

23.72 (a) $2NiS(s) + 3O_2(g) \rightarrow 2NiO(s) + 2SO_2(g)$

(b) $2C(s) + O_2(g) \rightarrow 2CO(g); C(s) + H_2O(g) \rightarrow CO(g) + H_2(g)$

$NiO(s) + CO(g) \rightarrow Ni(s) + CO_2(g); NiO(s) + H_2(g) \rightarrow Ni(s) + H_2O(g)$

(c) $Ni(s) + 2HCl(aq) \rightarrow NiCl_2(aq) + H_2(g)$

(d) $NiCl_2(aq) + 2NaOH(aq) \rightarrow Ni(OH)_2(s) + 2NaCl(aq)$

(e) $Ni(OH)_2(s) \overset{\Delta}{\rightarrow} NiO(s) + H_2O(g)$

Integrative Exercises

23.75 Recall from the discussion in Chapter 13 that like substances tend to be soluble in one another, whereas unlike substances do not. Molten metal consists of atoms that continue to be bound to one another by metallic bonding, even though the substance is liquid. In a slag, on the other hand, the attractive forces are those between ions. The slag phase is a highly polar, ionic medium, whereas the metallic phase is nonpolar, and the attractive interactions are due to metallic bond formation. There is little driving force for materials with such different characteristics to dissolve in one another.

23.77 (a) $\Delta G° = \Delta H° - T\Delta S°$ (assume $\Delta H°$ and $S°$ are constant with changes in temperature)

$Si(s) + 2MnO(s) \rightarrow SiO_2(s) + 2Mn(s)$

$\Delta H° = \Delta H_f° \ SiO_2(s) + 2\Delta H_f° \ Mn(s) - 2\Delta H_f° \ MnO(s) - \Delta H_f° \ Si(s)$

$\Delta H° = -910.9 + 2(0) - 2(-385.2) + 0 = -140.5$ kJ

$\Delta S° = S° \ SiO_2(s) + 2S° \ Mn(s) - 2S° \ MnO(s) - S° \ Si(s)$

$= 41.84 + 2(32.0) - 2(59.7) - 18.7 = -32.26 = -32.3$ J/K

$\Delta G° = -140.5$ kJ $- 1473$ K$(-0.03226$ kJ/K$) = -93.0$ kJ

    (b)   At 1473 K, the reactants and products are all solids, so they are in their standard states. Since $\Delta G°$ is negative at this temperature, the reaction should be spontaneous and thus feasible.

23.78   (a)   According to Section 20.8, the reduction of O_2 during oxidation of Fe(s) to Fe_2O_3 requires H^+. Above pH 9, iron does not corrode. At the high temperature of the converter, it is unlikely to find H_2O or H^+ in contact with the molten Fe. Also, the basic slag (CaO(l)) that is present to remove phosphorus will keep the environment basic rather than acidic. Thus, the H^+ necessary for oxidation of Fe in air is not present in the converter.

    (b) $C + O_2(g) \rightarrow CO_2(g)$

           $S + O_2(g) \rightarrow SO_2(g)$

           $P + O_2(g) \rightarrow P_2O_5(l); P_2O_5(l) + 3CaO(l) \rightarrow Ca_3(PO_4)(l)$

           $Si + O_2(g) \rightarrow SiO_2$

           $M + O_2(g) \rightarrow M_xO_y(l); M_xO_y + SiO_2 \rightarrow$ silicates

           CO_2 and SO_2 escape as gases. P_2O_5 reacts with CaO(l) to form $Ca_3(PO_4)_2(l)$, which is removed with the basic slag layer. SiO_2 and metal oxides can combine to form other silicates; SiO_2, M_xO_y, and complex silicates are all removed with the basic slag layer.

23.80 $\Delta G° = -RT \ln K; \Delta G° = \Delta H° - T\Delta S°$

    Calculate $\Delta H°$ and $\Delta S°$ using data from Appendix C, assuming $\Delta H°$ and $\Delta S°$ remain constant with changing temperature. Then calculate $\Delta G°$ and K at the two temperatures.

    $\Delta H° = 2\Delta H_f \, CO(g) - \Delta H_f° \, C(s) - \Delta H_f° \, CO_2(g)$

    $\Delta H° = 2(-110.5) - 0 - (-393.5) = +172.5$ kJ

    $\Delta S° = 2S° \, CO(g) - S° \, C(s) - S° \, CO_2(g)$

        $= 2(197.9) - 5.69 - 213.6 = +176.5$ J/K $= 0.1765$ kJ/K

    $\Delta G_{298}° = 172.5$ kJ $- 298$ K(0.1765 kJ/K) $= +119.9$ kJ

    $\ln K = \dfrac{\Delta G°}{-RT} = \dfrac{119.9 \text{ kJ}}{-(8.314 \times 10^{-3} \text{ kJ/K})(298 \text{ K})} = -48.3942 = -48.39; K = 9.6 \times 10^{-22}$

    $\Delta G_{2000}° = 172.5$ kJ $- 2000$ K(0.1765 kJ/K) $= -180.5$ kJ

    $\ln K = \dfrac{-180.5}{-(8.314 \times 10^{-3} \text{ kJ/K})(2000 \text{ K})} = 10.8552 = 10.86; K = 5.18 \times 10^4$

    (log K has 3 decimal places, so K has 3 sig figs.)

23.81   (a) $TiO_2(s) + 2Cl_2(g) + C(s) \rightarrow TiCl_4(g) + CO_2(g)$

           $\Delta H° = \Delta H_f° \, TiCl_4(g) + \Delta H_f° \, CO_2(g) - \Delta H_f° \, TiO_2(s) - 2\Delta H_f° \, Cl_2(g) - \Delta H_f° \, C(s)$

           $\Delta H° = -763.2 - 393.5 - (-944.7) - 2(0) - 0 = -212.0$ kJ

           The reaction is exothermic, ΔH is negative.

(b) $TiO_2(s) + 2Cl_2(g) \rightarrow TiCl_4(g) + O_2(g)$

$\Delta H° = \Delta H_f° \, TiCl_4(g) + \Delta H_f° \, O_2(g) - \Delta H_f° \, TiO_2(s) - 2\Delta H_f° \, Cl_2(g)$

$\Delta H° = -763.2 - 0 - (-944.7) - 2(0) = 181.5 \text{ kJ}$

The reaction is endothermic; ΔH is positive.

23.83   (a)   The very low melting and boiling points for VF_5 indicate that it is molecular rather than ionic, and that the intermolecular forces are probably weak London-dispersion forces. In order for the molecule to experience only London-dispersion forces, it must be nonpolar covalent, which requires the symmetrical trigonal bipyramidal structure shown below. PF_5 also has this structure.

(b) $VCl_3(s) + 3HF(g) \xrightarrow{\Delta} VF_3(s) + 3HCl(g)$

(c)   V(V) has a relatively small covalent radius. F is the smallest and most electronegative halogen. The steric repulsions associated with placing five larger halogens around the small V(V) central atom would be substantial. Also, the extreme electron attracting nature of F might be required to coax V into the +5 oxidation state.

23.84   Calculate the mass of Zn(s) that will be deposited.

$$2.0 \text{ m} \times 80 \text{ m} \times \frac{(100)^2 \text{ cm}^2}{1 \text{ m}^2} \times 0.49 \text{ mm} \times \frac{1 \text{ cm}}{10 \text{ mm}} \times \frac{7.1 \text{ g}}{\text{cm}^3} \times 2 \text{ sides}$$

$$= 1.113 \times 10^6 = 1.1 \times 10^6 \text{ g Zn}$$

$$1.113 \times 10^6 \text{ g Zn} \times \frac{1 \text{ mol Zn}}{65.39 \text{ g Zn}} \times \frac{2 \text{ F}}{0.90 \text{ mol Zn}} \times \frac{96,500 \text{ C}}{\text{F}} = 3.651 \times 10^9 = 3.7 \times 10^9 \text{ C}$$

(2 F/0.90 mol Zn takes the 90% efficiency into account.)

$$3.651 \times 10^9 \text{ C} \times 3.5 \text{ V} \times \frac{1 \text{ J}}{\text{C} \cdot \text{V}} \times \frac{1 \text{ kWh}}{3.6 \times 10^6 \text{ J}} = 3,550 = 3.6 \times 10^3 \text{ kWh}$$

$$3.550 \times 10^3 \text{ kWh} \times \frac{\$0.082}{1 \text{ kWh}} = \$291.06 \rightarrow \$291$$

23.86   (a) $Ag_2S(s) \rightleftharpoons 2Ag^+(aq) + S^{2-}(aq)$ K_{sp}

$\underline{\hspace{1.5cm} 2[Ag^+(aq) + 2CN^-(aq) \rightleftharpoons Ag(CN)_2^-]   \quad K_f^2 \hspace{1cm}}$

$Ag_2S(s) + 4CN^-(aq) \rightleftharpoons 2Ag(CN)_2^-(aq) + S^{2-}(aq)$

$$K = K_{sp} \times K_f^2 = [Ag^+]^2[S^{2-}] \times \frac{[Ag(CN)_2^-]^2}{[Ag^+]^2[CN^-]^4} = (6 \times 10^{-51})(1 \times 10^{21})^2 = 6 \times 10^{-9}$$

(b) The equilibrium constant for the cyanidation of Ag_2S, 6×10^{-9}, is much less than one and favors the presence of reactants rather than products. The process is not practical.

(c)
$$AgCl(s) \rightleftharpoons Ag^+(aq) + Cl^-(aq) \qquad K_{sp}$$

$$Ag^+(aq) + 2CN^-(aq) \rightleftharpoons Ag(CN)_2^-(aq) \qquad K_f$$

$$AgCl(s) + 2CN^-(aq) \rightleftharpoons Ag(CN)_2^-(aq) + Cl^-(aq)$$

$$K = K_{sp} \times K_f = [Ag^+][Cl^-] \times \frac{[Ag(CN)_2^-]}{[Ag^+][CN^-]^2} = (1.8 \times 10^{-10})(1 \times 10^{21}) = 2 \times 10^{11}$$

Since K >> 1 for this process, it is potentially useful for recovering silver from horn silver. However, the magnitude of K says nothing about the rate of reaction. The reaction could be slow and require heat, a catalyst, or both to be practical.

23.87   (a) $M(s) \rightarrow M(g)$. The process of atomization is essentially breaking the "metallic bonds" in the solid metal and separating the particles into isolated gas-phase atoms. This requires relocalizing electrons from the solid lattice onto the individual metal atoms.

(b) ΔH_{atom} is the difference between the energy of a mole of gaseous metal atoms, isolated from one another, and a mole of the metal, with all its metal-metal bonding. The difference will be smaller if: 1) the gaseous atoms have some special stability relative to other metallic elements or 2) the metal-metal bonding in the solid is weaker.

The data indicate that Cr and Mn, in the middle of the first transition series, and Cu at the end, have smaller ΔH_{atom} than their neighbors. The electron configurations for these elements are: Cr, $[Ar]4s^13d^5$ (exception); Mn, $[Ar]4s^23d^5$; Cu, $[Ar]4s^13d^{10}$. The gaseous atoms of each of these elements have special stability due to either full or half-full subshells. Assuming relatively constant metal-metal bond strength, the special stability of the gaseous atoms reduces ΔH_{atom} for these elements, relative to their neighbors.

The lower values of ΔH_{atom} for Fe, Co, and Ni relative to the elements around V (after taking account of the variations in stability of the gaseous atoms) is likely due to decreasing metal-metal bond strength. Moving to the right across the transition series from the middle onward, effective nuclear charge increases, the radial extension of the d-orbitals decreases, and the strength of metallic bonding decreases. This is somewhat of a trend.

24 Chemistry of Coordination Compounds

Visualizing Concepts

24.2   *Analyze.* Given a ball-and-stick figure of a ligand, write the Lewis structure and answer questions about the ligand.

Plan. Assume that each atom in the Lewis structure obeys the octet rule. Complete each octet with unshared electron pairs or multiple bonds, depending on the bond angles in the ball-and-stick model. Black = C, blue = N, red = O, gray = H.

There is a second resonance structure with the double bond drawn to the second O atom.

Check. Write the molecular formula, count the valence electron pairs and see if it matches your structure. $[C_4H_9N_2O_2]^-$ (16 + 9 + 10 + 12 + 1) = 48 valence e^-, 24 e^- pair Our Lewis structure also has 24 e^- pairs.

(a)   Donor atoms have unshared electron pairs. The potential donors in this structure are the two N and two O atoms.

The ligand is tridentate. (Even though there are four possible donor atoms, the structure would be strained if all four were bound to one metal center. It is likely that only one of the two O atoms binds to the same metal as the two N atoms.)

(b)   An octahedral complex has 6 coordination sites. A single ligand has only 4 possible donors, so two ligands are needed. From a steric perspective, the likely donors would be the 2 N atoms and 1 of the carbonyl oxygen atoms. The chelate bite of a carboxyl group is relatively small and would require an O—M—O angle of less than 90°.

24.3   *Analyze.* Given a ball-and-stick structure, name the complex ion, which has a 1– charge.

Plan. Write the chemical formula of the complex ion, determine the oxidation state of the metal, and name the complex.

Solve. $[Pt(NH_3)Cl_3]^-$. Oxidation numbers: [Pt + 0 + 3(–1) = –1, Pt = +2, Pt(II)

Arrange the ligands alphabetically, followed y the metal. Since the complex is an anion, add the suffix -ate, then the oxidation state of the metal: aminotrichloroplatinate(II)

24.5   *Analyze.* Given four structures, decide which are chiral.

Plan. Chiral molecules have nonsuperimposable mirror images. Draw the mirror image of each molecule and visualize whether it can be rotated into the original molecule. If so, the complex is not chiral. If the original orientation cannot be regenerated by rotation, the complex is chiral. *Solve.*

(1)   (1) mirror

The two orientations are not superimposable and molecule (1) is chiral.

The two orientations are superimposible. Rotate the right-most structure 90° counterclockwise about the B-M-B axis to align the G's; the bidentate ligands then also overlap. Molecule (2) is not chiral.

The two orientations are not superimposable and molecule (3) is chiral.

The two orientations are not superimposable and molecule (4) is chiral.

24.7   *Analyze.* Fit the crystal field splitting diagram to the complex description in each part.

Plan. Determine the number of d-electrons in each transition metal. On the splitting diagrams match the d-orbital splitting patterns to complex geometry and electron pairing to the definition of high-spin and low-spin.

Solve. Octahedral complexes have the 3 lower, 2 higher splitting pattern, while tetrahedral complexes have the opposite 2 lower, 3 higher pattern. Low spin complexes favor electron pairing because of large d-orbital splitting. High-spin complexes have maximum occupancy because of small orbital splitting.

(a) Fe^{3+}, 5d-electrons; weak field: spins unpaired; octahedral: 3 lower, 2 higher d-splitting ∴ diagram (4)

(b) Fe^{3+}, 5d-electrons; strong field: spins paired; octahedral: 3 lower, 2 higher d-splitting ∴ diagram (1)

(c) Fe^{3+}, 5d-electrons; tetrahedral: 2 lower, 3 higher d-splitting ∴ diagram (3)

(d) Ni^{2+}, 8 d-electrons; tetrahedral: 2 lower, 3 higher d-splitting ∴ diagram (2)

Check. Diagram (2) was the remaining choice for (d) and it fits the description.

24.8   *Analyze/Plan.* Given the linear diagram and axial labels, answer the questions and predict crystal field splitting. Orbitals with lobes nearest ligand charges (or partial charges) will be highest in energy; orbitals with lobes away from charges are lowest in energy.

Solve. d_{z^2} has lobes nearest the charges. $d_{x^2-y^2}$ and d_{xy} have lobes in the xy-plane farthest from the charges. d_{xz} and d_{yz} point between the respective axes and are intermediate in energy.

Introduction to Metal Complexes

24.10   (a)   *Coordination number* is the number of atoms bound directly to the metal in a metal complex.

(b)   Ligands in a metal complex usually have a lone (unshared, nonbonding) pair of electrons. NH_3 is neutral, CN^- is negatively charged.

(c)   No, ligands with positive charges are not common. Metal atoms in a complex usually have a positive charge, so a positively charged ligand would neither be electrostatically attracted to the metal nor provide the electron density required to form a metal-ligand bond.

(d)   In $Co(NH_3)_6Cl_3$, there are two types of chemical bonds. The bonds formed between Co^{3+} and $:NH_3$ are covalent. They are the result of Lewis acid-Lewis base interactions, where $:NH_3$ is the electron pair donor (Lewis base) and Co^{3+} is the electron pair acceptor (Lewis acid). The interactions between the $[Co(NH_3)_6]^{3+}$ complex cation and the three Cl^- anions are ionic bonds.

24.12   (a)   Yes. Conductivity is directly related to the number of ions in a solution. The lower the conductivity, the more Cl^- ions that act as ligands. Conductivity measurements on a set of standard solutions with various moles of ions per mole of complex would provide a comparative method for quantitative determination of the number of free and bound Cl^- ions.

(b)   In using conductivity measurements to test which ligands are bound to the metal atom, the assumption is that the total number of ions in solution is constant. This requires that either the Cl^- in the coordination sphere are so tightly bound that exchange with ions in solution is extremely slow, or that the rates at which they leave and enter are equal.

24.14   (a)   Coordination number = 6, oxidation number = +3

(b)   4, +2   (c)   6, +4   (d)   6, +3

(e)   6, +3   (f)   5, +2

24.16   (a)   6 C (see Solution 24.15(d))

(b)   4 N

(c)   5 C, 1 Br. In CO, both C and O have an unshared electron pair. C is less electronegative and more likely to donate its unshared pair. (This is analogous to the situation in CN^-).

(d)   4 N, 2 O. en is a bidentate ligand bound through N, for a total of 4 N donors. $C_2O_4^{2-}$ is bidentate with 2 O donors.

(e)   6 N. When thiocyanate is written "NCS," it is bound through N. This makes a total of 6 N donors.

(f)   4 N, 1 I. bipy is bidentate bound through N, for a total of 4 N donors.

Polydendate Ligands; Nomenclature

24.18   (a)   2 coordination sites, 2 N donor atoms

(b)   2 coordination sites, 2 N donor atoms

(c)   2 coordination sites, 2 O donor atoms (Although there are four potential O donor atoms in $C_2O_4^{2-}$, it is geometrically impossible for more than two of these to be bound to a single metal ion.)

(d)   4 coordination sites, 4 N donor atoms

(e)   6 coordination sites, 2 N and 4 O donor atoms

24.20   (a)   6   (b)   6   (c)   6   (d)   7

24.22   (a)   Pyridine is a **mono**dentate ligand because it has one N donor atom and therefore occupies one coordination site in a metal complex.

(b)   K for this reaction will be less than one. Two free pyridine molecules are replaced by one free bipy molecule. There are more moles of particles in the reactants than products, so ΔS is predicted to be negative. Processes with a net decrease in entropy are usually nonspontaneous, have positive ΔG, and values of K less than one. This equilibrium is likely to be spontaneous in the reverse direction.

24.24   (a) $[Mn(H_2O)_4Br_2]ClO_4$ (b) $[Cd(bipy)_2]Cl_2$

(c) $K[Co(o\text{-phen})Br_4]$ (d) $Cs[Cr(NH_3)_2(CN)_4]$

(e) $[Rh(en)_3][Co(ox)_3]$

24.26   (a)   dichloroethylenediamminecadmium(II)

(b)   potassium hexacyanomanganate(II)

(c)   pentaamminecarbonatochromium(III) chloride

(d)   tetraamminediaquairidium(III) nitrate

Isomerism

24.28   (a)

coordination sphere isomerism

(b)

(c)   coordination sphere isomerism

24.30   Two geometric isomers are possible for an octahedral MA_3B_3 complex (see below). All other arrangements, including mirror images, can be rotated into these two structures. Neither isomer is optically active.

24.32

(a)

optical isomers

(b)

(c)

Color, Magnetism; Crystal-Field Theory

24.34 (a) Yes. A complex that absorbs visible light of one wavelength or color will appear as the complementary color. This complex absorbs green light and will appear red.

(b) No. A solution can appear green by transmitting or reflecting only green light (the situation stated in the exercise) **or** by absorbing red light, the complementary color of green.

(c) A visible absorption spectrum shows the amount of light absorbed at a given wavelength. It is a plot of absorbance (dependent variable, y-axis) vs. wavelength (independent variable, x-axis).

(d) $E(J/\text{photon}) = h\nu = hc/\lambda$. Change J/photon to kJ/mol.

$$E = \frac{6.626 \times 10^{-34} \, J\text{-}s}{530 \, nm} \times \frac{3.00 \times 10^8 \, m}{s} \times \frac{1 \, nm}{1 \times 10^{-9} \, m} = 3.751 \times 10^{-19} = 3.75 \times 10^{-19} \, J$$

$$3.751 \times 10^{-19} = \frac{1 \, kJ}{1000 \, J} \times \frac{6.022 \times 10^{23} \, \text{photons}}{mol} = 226 \, kJ/mol$$

24.36 Six ligands in an octahedral arrangement are oriented along the x, y, and z axes of the metal. These negatively charged ligands (or the negative end of ligand dipoles) have greater electrostatic repulsion with valence electrons in metal orbitals that also lie along these axes, the d_{z^2}, and $d_{x^2-y^2}$. The d_{xy}, d_{xz} and d_{yz} metal orbitals point between the x, y, and z axes, and electrons in these orbitals experience less repulsion with ligand electrons. Thus, in the presence of an octahedral ligand field, the d_{xy}, d_{xz} and d_{xy} metal orbitals are lower in energy than the $d_{x^2-y^2}$ and d_{z^2}.

24.38 (a) $\Delta E = hc/\lambda = \dfrac{6.626 \times 10^{-34} \, J\text{-}s \times 2.998 \times 10^8 \, m/s}{500 \times 10^{-9} \, m} = 3.973 \times 10^{-19} = 3.97 \times 10^{-19} \, J/\text{photon}$

$$\Delta = 3.973 \times 10^{-19} \, J/\text{photon} \times \frac{6.022 \times 10^{23} \, \text{photons}}{1 \, mol} \times \frac{1 \, kJ}{1000 \, J} = 239.25 = 239 \, kJ/mol$$

(b) The *spectrochemical* series is an ordering of ligands according to their ability to increase the energy gap Δ. If H_2O is replaced by NH_3 in the complex, the magnitude of Δ would increase because NH_3 is higher in the spectrochemical series and creates a stronger ligand field.

24.40 (a) $[Ni(H_2O)_6]^{2+}$ absorbs 725 nm red light and appears as the complementary color, green. $[Ni(NH_3)_6]^{2+}$ absorbs 570 nm yellow light and appears as the complementary color, violet.

(b)   The shorter the wavelength of light absorbed, the greater the value of Δ, and the stronger the ligand field. The order of increasing ligand field strength is the order of decreasing wavelength absorbed.

$H_2O < NH_3 < en < bipy$

24.42   (a) Fe^{3+}, d^5 (b) Mn^{2+}, d^5 (c) Ag^+, d^{10}

        (d) Cr^{3+}, d^3 (e) Sr^{2+}, d^0

24.44   (a)   Ru: $[Kr]5s^14d^7$ (b)   Mo: $[Kr]5s^14d^5$ (c)   Co: $[Ar]4s^23d^7$

            $Ru^{3+}: [Kr]4d^5$ $Mo^{3+}: [Kr]3d^3$ $Co^{3+}: [Ar]3d^6$

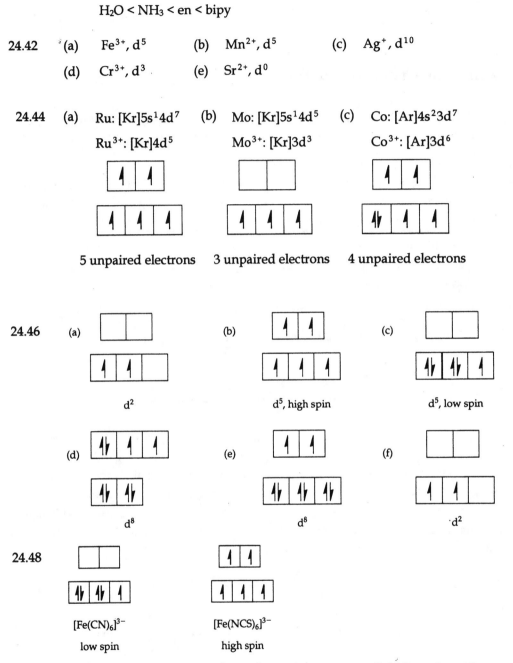

     5 unpaired electrons   3 unpaired electrons   4 unpaired electrons

24.46   (a)   (b)   (c)

               d^2 d^5, high spin d^5, low spin

   (d)   (e)   (f)

               d^8 d^8 $\cdot d^2$

24.48

   $[Fe(CN)_6]^{3-}$ $[Fe(NCS)_6]^{3-}$

    low spin   high spin

Both complexes contain Fe^{3+}, a d^5 ion. CN^-, a strong field ligand, produces such a large Δ that the splitting energy is greater than the pairing energy, and the complex is low spin. NCS^- produces a smaller Δ, so it is energetically favorable for d-electrons to be unpaired in the higher energy d-orbitals. NCS^- is a much weaker-field ligand than CN^-. It is probably weaker than NH_3 and near H_2O in the spectrochemical series.

Additional Exercises

24.50   (a)

[Ru(H₂O)₅Cl]Cl₂   ⟶   [Ru(H₂O)₆]Cl₃

24.51   (a)

octahedral   octahedral

     (c)

octahedral   octahedral

24.53   (a)   Valence electrons: $2P + 6C + 16H = 10 + 24 + 16 = 50$ e⁻, 25 e⁻ pr

Both dmpe and en are bidentate ligands. The dmpe ligand binds through P, while en binds through N. Phosphorus is less electronegative than N, so dmpe is a stronger electron pair donor and Lewis base than en. Dmpe creates a stronger ligand field and is higher on the spectrochemical series.

Structurally, P has a larger covalent radius than N, so M−P bonds are longer than M−N bonds. This is convenient because the two −CH₃ groups on each P atom in dmpe create more steric hindrance (bumping with adjacent atoms) than the H atoms on N in en.

     (b)   CO and dmpe are neutral, $2CN^- = 2^-$, $2Na^+ = 2+$. The ion charges balance, so the oxidation state of Mo is zero.

(c)   The symbol P͡P represents the bidentate dmpe ligand.

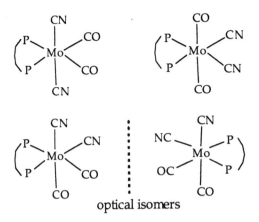

optical isomers

24.55   We will represent the end of the bidentate ligand containing the CF_3 group by a shaded oval, the other end by an open oval:

24.56   (a)   Hemoglobin is the iron-containing protein that transports O_2 in human blood.

(b)   Chlorophylls are magnesium-containing porphyrins in plants. They are the key components in the conversion of solar energy into chemical energy that can be used by living organisms.

(c)   Siderophores are iron-binding compounds or ligands produced by a microorganism. They compete on a molecular level for iron in the medium outside the organism and carry needed iron into the cells of the organism.

24.58   (a)   pentacarbonyliron(0)

(b)   Since CO is a neutral molecule, the oxidation state of iron must be zero.

(c) $[Fe(CO)_4CN]^-$ has two geometric isomers. In a trigonal bipyramid, the axial and equatorial positions are not equivalent and not superimposable. One isomer has CN in an axial position and the other has it in an equatorial position.

24.59   (a)   left shoe

(c)   wood screw

(e)   a typical golf club

24.61   (a)   Formally, the two Ru centers have different oxidation states; one is +2 and the other is +3.

   (b)

   (c)   There is extensive bonding-electron delocalization in the isolated pyrazine molecule. When pyrazine acts as a bridging ligand, its delocalized molecular orbitals provide a pathway for delocalization of the "odd" d-electron in the Creutz-Taube ion. The two metal ions appear equivalent because the odd d-electron is delocalized across the pyrazine bridge.

24.63   (a)   oxyhemoglobin   deoxyhemoglobin

   $Fe^{2+} : d^6$ $Fe^{2+} : d^6$

   low spin,   high spin,
   no unpaired electrons   4 unpaired electrons

   (b)   In deoxyhemoglobin, H_2O is bound to Fe in place of O_2.

   (c)   The two forms of hemoglobin have different colors because they absorb different wavelengths of visible light. They differ by just the H_2O or O_2 ligand, which means that the two ligands have slightly different ligand fields. Oxyhemoglobin appears red and absorbs green light, while deoxyhemoglobin appears bluish and absorbs longer wavelength yellow-green light. O_2 has a stronger ligand field than H_2O.

   (d)   According to Table 18.1, air is 20.948 mole percent O_2. This translates to 209,480 ppm O_2. This is approximately 500 times the 400 ppm concentration of CO in the experiment. If air with a CO concentration 1/500th that of O_2 converts 1/10 of the oxyhemoglobin to carboxyhemoglobin, the equilibrium constant for binding CO is much larger than that for binding O_2.

24.65   Application of pressure would result in shorter metal ionoxide distances. This would have the effect of increasing the ligand-electron repulsions, and would result in a larger splitting in the d-orbital energies. Thus, application of pressure should result in a shift in the absorption to a higher energy and shorter wavelength.

24.67

24.68   A large part of the metal-ligand interaction is electrostatic attraction between the positively charged metal and the fully or partially negatively charged ligand. For the same ligand, the greater the charge on the metal or the shorter the M−L separation, the stronger the interaction and the more stable the complex. The greater positive charge and smaller ionic radius of a metal in the 3+ oxidation state means that, for the same ligand, complexes with metals in the 3+ state are more stable than those with metals in the 2+ state. The same is true for the series of first-row transition metals in the 2+ state. Going from Mn^{2+} to Cu^{2+}, ionic radius decreases, M−L bond strength increases and complex stability increases. This is consistent with the crystal-field model that shows the energies of d-electrons on the metal to be influenced by the presence of an array of ligands, while not participating directly in metal-ligand bonding.

24.69   (a)   Only one   (b)   Two

(c)   Four; two are geometric, the other two are stereoisomers of each of these.

24.70

For a d^6 metal ion in a strong ligand field, there would be two unpaired electrons.

Integrative Exercises

24.72   (a)   Both compounds have the same general formulation, so Co is in the same (+3) oxidation state in both complexes.

(b)   Cobalt(III) complexes are generally inert; that is, they do not rapidly exchange ligands inside the coordination sphere. Therefore, the ions that form precipitates in these two cases are probably outside the coordination sphere. The dark violet compound A forms a precipitate with $BaCl_2(aq)$ but not $AgNO_3(aq)$, so it has SO_4^{2-} outside the coordination sphere and coordinated Br^-, $[Co(NH_3)_5Br]SO_4$. The red-violet compound B forms a precipitate with $AgNO_3(aq)$ but not $BaCl_2(aq)$ so it has Br^- outside the coordination sphere and coordinated SO_4^{2-}, $[Co(NH_3)_5SO_4]Br$.

Compound A, dark violet   Compound B, red-violet

(c)   Compounds A and B have the same formula but different properties (color, chemical reactivity), so they are isomers. They vary by which ion is inside the coordination sphere, so they are *coordination sphere isomers*.

(d)   Compound A is an ionic sulfate and compound B is an ionic bromide, so both are strong electrolytes. According to the solubility rules in Table 4.1, both should be water-soluble.

24.74   (a) $\Delta G° = -nFE°$. The positive E° values for both sets of complexes correspond to $-\Delta G°$ values. Negative values of $\Delta G°$ mean that both processes are spontaneous. For both o-phen and CN⁻ ligands, the Fe(II) complex is more stable than the Fe(III) complex.

(b)   That both the Fe(II) complexes are low spin means that both CN⁻ and o-phen are strong-field ligands. The negatively charged CN⁻ has a stronger electrostatic interaction with Fe^{3+} than the neutral o-phen has. This stabilizes the Fe(III) complex of CN⁻ relative to the Fe(III) complex of o-phen, which reduces the driving force for reduction of $[Fe(CN)_6]^{3-}$ relative to reduction of $[Fe(o\text{-}phen)_3]^{3-}$. The E° value and magnitude of $\Delta G°$ for the reduction of the CN⁻ complex are thus smaller than those values for the o-phen complex.

24.76   (a)   The reaction that occurs increases the conductivity of the solution by producing a greater number of charged particles, particles with higher charges, or both. It is likely that H_2O from the bulk solvent exchanges with a coordinated Br⁻ according to the reaction below. This reaction would convert the 1:1 electrolyte, $[Co(NH_3)_4Br_2]Br$, to a 1:2 electrolyte, $[Co(NH_3)_3(H_2O)Br]Br_2$.

(b) $[Co(NH_3)_4Br_2]^+(aq) + H_2O(l) \rightarrow [Co(NH_3)_4(H_2O)Br]^{2+}(aq) + Br^-(aq)$

(c)   Before the exchange reaction, there is one mole of free Br⁻ per mole of complex.
mol Br⁻ = mol Ag⁺

M = mol/L; L $AgNO_3$ = mol $AgNO_3$/M $AgNO_3$

$$\frac{3.87\,g\,complex}{0.500\,L\,soln} \times \frac{1\,mol\,complex}{366.77\,g\,complex} \times 0.02500\,L\,soln\,used =$$

$$5.276 \times 10^{-4} = 5.28 \times 10^{-4}\,mol\,complex$$

$$5.276 \times 10^{-4}\,mol\,complex \times \frac{1\,mol\,Br^-}{1\,mol\,complex} \times \frac{1\,mol\,Ag^+}{1\,mol\,Br^-} \times \frac{1\,L\,Ag^+(aq)}{0.0100\,mol\,Ag^+(aq)}$$

$$= 0.05276\,L = 52.8\,mL\,AgNO_3(aq)$$

(d)   After the exchange reaction, there are 2 mol free Br⁻ per mol of complex. Since M $AgNO_3$(aq) and volume of complex solution are the same for the second experiment, the titration after conductivity changes will require twice the volume calculated in part (c), 105.52 = 106 mL of 0.0100 M $AgNO_3$(aq).

24.78   Use Hess' law to calculate $\Delta G°$ for the desired equilibrium. Then $\Delta G° = -RTlnK$ to calculate K.

$$
\begin{array}{ll}
Hb + CO \rightarrow HbCO & \Delta G° = -80\,kJ \\
HbO_2 \rightarrow Hb + O_2 & \Delta G° = 70\,kJ \\
\hline
HbO_2 + Hb + CO \rightarrow HbCO + Hb + O_2 & \\
HbO_2 + CO \rightarrow HbCO + O_2 & \Delta G° = -10\,kJ
\end{array}
$$

$$\Delta G^\circ = -RT\ln K,\; \ln K = \frac{-\Delta G^\circ}{RT} = \frac{-(-10\text{ kJ})}{8.314\text{ J/K-mol} \times 298\text{ K}} \times \frac{1000\text{ J}}{\text{kJ}} = 4.036 = 4.04$$

$$K = e^{4.04} = 56.61 = 57$$

24.79   (a)

$[Cd(CH_3NH_2)_4]^{2+} \rightleftharpoons Cd^{2+}(aq) + 4CH_3NH_2(aq)$	$\Delta G^\circ = 37.2$ kJ
$Cd^{2+}(aq) + 2en(aq) \rightleftharpoons [Cd(en)_2]^{2+}(aq)$	$\Delta G^\circ = -60.7$ kJ
$Cd(CH_3NH_2)_4]^{2+} + 2en(aq) \rightleftharpoons [Cd(en)_2]^{2+}(aq) + 4CH_3NH_2(aq)$	$\Delta G^\circ = -23.5$ kJ

$$\Delta G^\circ = -RT\ln K;\; -23.5\text{ kJ} = -2.35 \times 10^4\text{ J}$$

$$-2.35 \times 10^4\text{ J} = \frac{-8.314\text{ J}}{\text{K-mol}} \times 298\text{ K} \times \ln K;\; \ln K = 9.485,\; K = 1.32 \times 10^4$$

(b)   The magnitude of K is large, so the reaction favors products. The bidentate chelating ligand en will spontaneously replace the monodentate ligand CH_3NH_2. This is an illustration of the chelate effect.

(c)   Using the stepwise construction from part (a),

$$\Delta H^\circ = 57.3\text{ kJ} - 56.5\text{ kJ} = 0.8\text{ kJ}$$

$$\Delta S^\circ = 67.3\text{ J/K} + 14.1\text{ J/K} = 81.4\text{ J/K}$$

$$-T\Delta S = -298\text{ K} \times 81.4\text{ J/K} = -2.43 \times 10^4\text{ J} = -24.3\text{ kJ}$$

The chelate effect is mainly the result of entropy. The reaction is spontaneous due to the increase in the number of free particles and corresponding increase in entropy going from reactants to products. The enthalpic contribution is essentially zero because the bonding interactions of the two ligands are very similar and the reaction is not "downhill" in enthalpy.

(d) ΔH° will be very small and negative. When NH_3 replaces H_2O in a complex (Closer Look Box), the tighter bonding of the NH_3 ligand causes a substantial negative ΔH° for the substitution reaction. When a bidentate amine ligand replaces a monodentate amine ligand of similar bond strength, ΔH° is very small and either positive (part (c)) or negative (Closer Look Box). In the case of NH_3 replacing CH_3NH_2, the bonding characteristics are very similar. The presence of CH_3 groups in CH_3NH_2 produces some steric hindrance in $[Cd(CH_3NH_2)_4]^{2+}$. This complex is at a slightly higher energy than $[Cd(NH_3)_4]^{2+}$, which experiences no steric hindrance, so ΔH° will have a negative sign but a very small magnitude. Relief of steric hindrance leads to a very small negative ΔH° for the substitution reaction.

24.81   The process can be written:

$$H_2(g) + 2e \rightarrow 2H^+(aq) \qquad\qquad E^{\circ}_{red} = 0.0 \text{ V}$$

$$Cu(s) \rightarrow Cu^{2+} + 2e^- \qquad\qquad E^{\circ}_{red} = 0.337 \text{ V}$$

$$\underline{Cu^{2+}(aq) + 4NH_3(aq) \rightarrow [Cu(NH_3)_4]^{2+}(aq)} \qquad "E^{\circ}_f" = ?$$

$$H_2(g) + Cu(s) + 4NH_3(aq) \rightarrow 2H^+(aq) + [Cu(NH_3)_4]^{2+}(aq) \qquad E = 0.08 \text{ V}$$

$$E = E^{\circ} - RT \ln K; \ K = \frac{[H^+]^2[Cu(NH_3)_4^{2+}]}{P_{H_2}[NH_3]^4}$$

$$P_{H_2} = 1 \text{ atm}, \ [H^+] = 1 \ M, \ [NH_3] = 1 \ M, \ [Cu(NH_3)_4]^{2+} = 1 \ M, \ Q = 1$$

$$E = E^{\circ} - RT \ln(1); \ E = E^{\circ} - RT(0); \ E = E^{\circ} = 0.08 \text{ V}$$

Since we know E° values for two steps and the overall reaction, we can calculate "E°" for the formation reaction and then K_f, using $E^{\circ} = \dfrac{0.0592}{n} \log K_f$ for the step.

$$E_{cell} = 0.08 \text{ V} = 0.0 \text{ V} - 0.337 \text{ V} + "E^{\circ}_f" \quad "E^{\circ}_f" = 0.08 \text{ V} + 0.337 \text{ V} = 0.417 \text{ V} = 0.42 \text{ V}$$

$$"E^{\circ}_f" = \frac{0.0592}{n} \log K_f; \ \log K_f = \frac{n(E^{\circ}_f)}{0.0592} = \frac{2(0.417)}{0.0592} = 14.0878 = 14$$

$$K_f = 10^{14.0878} = 1.2 \times 10^{14} = 10^{14}$$

24.82   (a)   The units of the rate constant and the rate dependence on the identity of the second ligand show that the reaction is second order. Therefore, the rate-determining step cannot be a dissociation of water, since that would be independent of the concentration and identity of the incoming ligand. The alternative mechanism, a bimolecular association of the incoming ligand with the complex, is indicated.

(b)   The relative values of rate constant are a reflection of the kinetic basicities of the three ligands: Pyridine $>$ SCN^- $>$ CH_3CN.

(c)   Ru(III) is a d^5 ion. In a low-spin d^5 octahedral complex, there is one unpaired electron.

25 The Chemistry of Life: Organic and Biological Chemistry

Visualizing Concepts

25.2 *Analyze/Plan.* Given structural formulas, decide which molecule will undergo addition. Consider which functional groups are present in the molecules, and which are most susceptible to addition. *Solve.*

Addition reactions are characteristic of alkenes. Molecule (c), an alkene, will readily undergo addition.

Molecule (a) is an aromatic hydrocarbon, which does not typically undergo addition because the delocalized electron cloud is too difficult to disrupt. Molecules (b) and (d) contain carbonyl groups (actually carboxylic acid groups) that do not typically undergo addition, except under special conditions.

25.4 *Analyze.* Given structural formulas, decide which molecules are capable of isomerism, and what type. *Plan.* Analyze each molecule for possible structural, geometric, and optical isomers/enantiomers. *Solve.*

(a) $C_5H_{11}NO_2$. Structural, geometric and optical. There are many ways to arrange the atoms in molecules with this empirical formula, so there are many structural isomers. There is one point of unsaturation in the given molecule, the C=O group; structural isomers with their point of unsaturation at a C=C group could have geometric isomers as well. The C atom to which the $-NH_3^+$ group is bound is a chiral center, so there are enantiomers. All amino acids except glycine have two possible enantiomers.

(b) $C_7H_5O_2Cl$. The most obvious isomers for this aromatic compound are ortho, meta, and para geometric isomers. Because the molecule has several points of unsaturation, the number of structural isomers is limited, but there are a few possibilities with two triple bonds. Switching the $-OH$ and $-Cl$ groups also generates a structural isomer. There are no chiral centers, so no optical isomers.

(c) C_5H_{10}. There are many structural isomers for this empirical formula. The straight-chain alkene shown also has geometric (cis-trans) isomers.

(d) C_3H_8. There are no other structural, geometric, or optical isomers for this molecule.

25.6   *Analyze/Plan.* Follow the logic in Sample Exercise 25.1 to name each compound. Decide which structures are the same compound.   *Solve.*

    (a)   2,2,4-trimethylpentane   (b)   3-ethyl-2-methylpentane

    (c)   2,3,4-trimethylpentane   (d)   2,3,4-trimethylpentane

Structures (c) and (d) are the same molecule.

Introduction to Organic Compounds; Hydrocarbons

25.8

$$N\equiv\underset{7}{C}-\underset{6}{C}-\underset{5}{C}-\underset{4}{C}=\underset{3}{C}-\underset{2}{C}-\underset{1}{C}-H$$

with H atoms: C6 (H, H), C5 (H, H), C4 (H), C3, C2 (OH), C1 (O double bond, H)

    (a)   C2, C5 and C6 have sp^3 hybridization (4 e⁻ domains around C)

    (b)   C7 has sp hybridization (2 e⁻ domains around C)

    (c)   C1, C3, and C4 have sp^2 hybridization (3 e⁻ domains around C)

25.10   From Table 8.4, the bond enthalpies in kJ/mol are: C—H, 413; C—C, 348; C—O, 358; C—Cl, 328. The bond enthalpes indicate that C—H bonds are most difficult to break, and C—Cl bonds least difficult. However, they do not explain the reactivity of C—O bonds, or stability of C—C bonds.

The reactivity of molecules containing C—O and C—Cl bonds is a result of their unequal charge distribution, which attracts reactants that are either electron deficient (electrophilic) or electron rich (nucleophilic).

25.12   All the classifications listed are hydrocarbons; they contain only the elements hydrogen and carbon.

    (a)   *Alkanes* are hydrocarbons that contain only single bonds.

    (b)   *Cycloalkanes* contain at least one ring of three or more carbon atoms joined by single bonds. Because it is a type of alkane, all bonds in a cycloalkane are single bonds.

    (c)   *Alkenes* contain at least one C=C double bond.

    (d)   *Alkynes* contain at least one C≡C triple bond.

    (e)   A *saturated hydrocarbon* contains only single bonds. Alkanes and cycloalkanes fit this definition.

    (f)   An *aromatic hydrocarbon* contains one or more planar, six-membered rings of carbon atoms with delocalized π-bonding throughout the ring.

25.14   cycloalkane, H_2C—CH_2 , C_6H_{12}, saturated

cycloalkene, HC=CH , C_6H_{10}, unsaturated

alkyne, CH_3—CH_2—C≡C—CH_2—CH_3 , C_6H_{10}, unsaturated

aromatic hydrocarbon, H—C ⬡ C—H, C_6H_6, unsaturated

25.16 C_nH_{2n-2}

25.18 CH_3—CH_2—CH_2—CH=CH_2
             pentene

CH_3—CH_2—CH=CH—CH_3
           2-pentene

CH_2=CH—CH—CH_3 with CH_3
           3-methyl-1-butene

CH_2=C—CH_2—CH_3 with CH_3
       2-methyl-1-butene

CH_3—C=CH—CH_3 with CH_3
        2-methyl-2-butene

25.20   (a)   109°   (b)   120°   (c)   180°

25.22   (a)   3,3,5-trimethylheptane

         (b)   3,4,4-trimethylheptane

         (c) $CH_3CH_2CH_2$—CH—CH—CH_2—CH_2—CH—CH_3 with CH_3, CH_3, CH_3 substituents

         (d) $CH_3CH_2CH_2CH_2CH_2$—CH—CH—CH—CH_2—CH_3 with CH_3, CH_3 substituents and $CH_2CH_2CH_2CH_2$ chain

(e)

25.24   (a)   1,4-dichlorocyclohexane

    (b)   3-chloro-1-propyne

    (c)   *trans*-2-hexene

    (d)   1-chloro-2-methyl-2-phenyl-butane or (1-chloro-2-methyl)-2-butylbenzene

    (e)   *cis*-5-chloro-1,3-pentadiene

25.26   Butene is an alkene, C_4H_8. There are two possible placements for the double bond:

$$CH_2\!=\!CHCH_2CH_3 \quad or \quad CH_3CH\!=\!CHCH_3$$
$$\text{1-butene} \qquad\qquad \text{2-butene}$$

These two compounds are *structural isomers*. For 2-butene, there are two different, noninterchangeable ways to construct the carbon skeleton (owing to the absence of free rotation around the double bond). These two compounds are *geometric isomers*.

    cis-2-butene   *trans*-2-butene

25.28

25.30   Octane number can be increased by increasing the fraction of branched-chain alkanes or
aromatics, since these have high octane numbers. This can be done by cracking. The
octane number also can be increased by adding an anti-knock agent such as tetraethyl
lead, $Pb(C_2H_5)_4$ (no longer legal); methyl t-butyl ether (MTBE); or an alcohol,
methanol, or ethanol.

Reactions of Hydrocarbons

25.32   (a)

(b)

(c)

25.34   (a)   The reaction of Br_2 with an alkene to form a colorless halogenated alkane is an
addition reaction. Aromatic hydrocarbons do not readily undergo addition
reactions, because their π-electrons are stabilized by delocalization.

(b)   *Plan.* Use a Friedel-Crafts reaction to substitute a —CH_2CH_3 onto benzene. Do
a second substitution reaction to get *para*-bromoethylbenzene.   *Solve.*

It appears that ortho, meta, and para geometric isomers of bromoethylbenzene
would be possible. However, because of electronic effects beyond the scope of
this chapter, the ethyl group favors formation of ortho and para isomers, but not
the meta. The ortho and para products must be separated by distillation or some
other technique.

25.36   The partially positive end of the hydrogen halide, $\overset{\delta^+\ \ \delta^-}{H-X}$, is attached to the π electron
cloud of the alkene cyclohexene. The electrons that formed the π bond in cyclohexene
form a sigma bond to the H atom of HX, leaving a halide ion, X^-. The intermediate is a
carbocation; one of the C atoms formerly involved in the π bond is now bound to a
second H atom. The other C atom formerly involved in the π bond carries a full positive
charge and forms only three sigma bonds, two to adjacent C atoms and one to H.

25.38

$$\begin{array}{lr} & \underline{\mathbf{\Delta H}} \\ C_{10}H_8(l) + 12O_2(g) \rightarrow 10CO_2(g) + 4H_2O(l) & -5157 \text{ kJ} \\ -[C_{10}H_{18}(l) + 29/2\,O_2(g) \rightarrow 10CO_2(g) + 9H_2O(l) & -(-6286) \text{ kJ} \\ \hline C_{10}H_8(l) + 5H_2O(l) \rightarrow C_{10}H_{18}(l) + 5/2\,O_2(g) & +1129 \text{ kJ} \\ 5/2\,O_2(g) + 5H_2(g) \rightarrow 5H_2O(l) & 5(-285.8) \text{ kJ} \\ \hline C_{10}H_8(l) + 5H_2(g) \rightarrow C_{10}H_{18}(l) & -300 \text{ kJ} \end{array}$$

Compare this with the heat of hydrogenation of ethylene:

$C_2H_4(g) + H_2(g) \rightarrow C_2H_6(g)$; $\Delta H = -84.7 - (52.3) = -137$ kJ. This value applies to just one double bond. For five double bonds, we would expect about –685 kJ. The fact that hydrogenation of napthalene yields only –300 kJ indicates that the overall energy of the napthalene molecule is lower than expected for five isolated double bonds and that there must be some special stability associated with the aromatic system in this molecule.

Functional Groups and Chirality

25.40 (a) $-\overset{\overset{\displaystyle O}{\|}}{C}-O$, ester

(b) –Cl, halocarbon; –OH, alcohol (aromatic alcohols are phenols)

(c) $-\overset{\overset{\displaystyle O}{\|}}{C}-NH-$, amide

(d) alkane

(e) $-C{=}C-$, alkene; $-\overset{\overset{\displaystyle O}{\|}}{C}H$, aldehyde

(f) $-\overset{\overset{\displaystyle O}{\|}}{C}-$, ketone

25.42 (a) C_4H_8O, $H_2C\overset{\displaystyle O}{\diagdown}CH_2$ ... H_2C-CH_2

(b) $CH_3CH_2\overset{\overset{\displaystyle O}{\|}}{C}CH_3$, $CH_3CH_2CH_2\overset{\overset{\displaystyle O}{\|}}{C}-H$

$CH_2{=}CH_2CH_2CH_2OH$, $CH_3CH{=}CHCH_2OH$, (cis and trans)
$CH_2{=}CHCH(OH)CH_3$ (enantiomers)

(Structures with the –OH group attached to an alkene carbon atom are not included. These molecules are called "vinyl alcohols" and are not the major form at equilibrium.)

25.44   (a)

$$CH_3CH_2\overset{\displaystyle O}{\overset{\|}{C}}\!-\!H$$

(b)

$$CH_3CH_2CH_2\overset{\displaystyle O}{\overset{\|}{C}}CH_3$$

(c)

$$CH_3\underset{\underset{\displaystyle CH_3}{|}}{CH}\overset{\displaystyle O}{\overset{\|}{C}}CH_3$$

(d)

$$CH_3CH_2\underset{\underset{\displaystyle CH_3}{|}}{CH}\overset{\displaystyle O}{\overset{\|}{C}}\!-\!H$$

25.46   (a)

$$CH_3CH_2CH_2\overset{\displaystyle O}{\overset{\|}{C}}\!-\!O\!-\!CH_3$$

methylbutanoate

(b)

$$\overset{\displaystyle O}{\overset{\|}{C}}\!-\!O\!-\!\underset{\underset{\displaystyle CH_3}{|}}{\overset{\overset{\displaystyle CH_3}{|}}{C}}\!-\!H$$

2-propylbenzoate

(c)

$$CH_3CH_2\overset{\displaystyle O}{\overset{\|}{C}}\!-\!\underset{\underset{\displaystyle CH}{|}}{N}\!-\!CH_3$$

N, N-dimethylpropanamide

25.48   (a)

$$CH_3CH_2CH_2CH_2OH + HO\overset{\displaystyle O}{\overset{\|}{C}}CH_2CH_3 \longrightarrow CH_3CH_2CH_2CH_2O\overset{\displaystyle O}{\overset{\|}{C}}CH_2CH_3$$

1-butanol   propionic acid   butyl proprionate
                   (propanoic acid)

(b)

$$CH_3O\overset{\displaystyle O}{\overset{\|}{C}}\!-\!\!\bigcirc\!\!-\!\! + NaOH \longrightarrow \left[\bigcirc\!\!-\!\!\overset{O}{\underset{O}{C}}\right]^{-} + Na^+ + CH_3OH$$

25.50

$$2\,CH_3COOH(l) \longrightarrow CH_3\overset{\displaystyle O}{\overset{\|}{C}}O\overset{\displaystyle O}{\overset{\|}{C}}CH_3(l) + H_2O(l)$$

$$CH_3\overset{\displaystyle O}{\overset{\|}{C}}\boxed{-OH + H}-O\!-\!\overset{\displaystyle O}{\overset{\|}{C}}CH_3 \longrightarrow \underset{\underset{\displaystyle CH_3}{}}{\overset{\overset{\displaystyle O}{\|}}{C}}O\underset{\underset{\displaystyle CH_3}{}}{\overset{\overset{\displaystyle O}{\|}}{C}} + H_2O$$

25.52 (a)

(b)

(c)

(d)

(e)

25.54

Yes, the molecule has optical isomers. The chiral carbon atom is attached to chloro, methyl, ethyl, and propyl groups. (If the root was a 5-carbon chain, the molecule would not have optical isomers because two of the groups would be ethyl groups.)

Proteins

25.56 The side chains possess three characteristics that may be of importance. They may be bulky (e.g., the phenyl group in phenylalanine) and thus impose restraints on where and how the amino acid can undergo reaction. Secondly, the side chain will be either hydrophobic, containing mostly nonpolar groups such as $(CH_3)_2 CH-$ in valine, or hydrophilic, containing a polar group such as $-OH$ in serine. The hydrophobic or hydrophilic nature of the side chain definitely influences solubility and other intermolecular interactions. Finally, the side chain may contain an acidic (e.g., the $-COOH$ group in glutamic acid) or basic (e.g., the $-NH_2$ group in lysine) functional group. These groups will be protonated or deprotonated, depending on the pH of the solution, and determine the variation of properties (including solubility) over a range of pH values. Acidic or basic side chains may also become involved in hydrogen-bonding with other amino acids.

25.58

methionine   glycine   methionylglycine

25.60   (a)   Valine, serine, glutamic acid

       (b)   Six (assuming the tripeptid contains all three amino acids):

          Gly-Ser-Glu, GSE; Gly-Glu-Ser, GES; Ser-Gly-Glu, SGE; Ser-Glu-Gly, SEG; Glu-Ser-Gly, ESG; Glu-Gly-Ser, EGS

25.62   The α-helix and β-sheet are examples of regular orientations in protein chains that are termed protein secondary structures. Both patterns are formed by hydrogen bonding. The main difference is that an α-helix is formed by hydrogen bonds between amino acids in the same chain, while a β-sheet is formed by hydrogen bonds between two chains (or a chain with a flexible loop that has bent back on itself to form hydrogen bonds).

       An α-helix is a column, with hydrogen bonds between a particular amino acid and a different amino acid several groups away. In Figure 25.26, the pitch of the helix orients hydrogen bonds between every fourth amino acid. A β-sheet has a hydrogen-bonding pattern that zippers together two uncoiled protein strands to form a "pleated" sheet. These two secondary structures are enabled by different amino acid sequences. And, many sequences form neither an α-helix nor a β-sheet.

Carbohydrates and Lipids

25.64   Glucose exists in solution as a cyclic structure in which the aldehyde function on carbon 1 reacts with the OH group of carbon 5 to form what is called a hemiacetal, Figure 25.29. Carbon atom 1 carries an OH group in the hemiacetal form; in α-glucose this OH group is on the opposite side of the ring as the CH_2OH group on carbon atom 5. In the β (beta) form the OH group on carbon 1 is on the same side of the ring as the CH_2OH group on carbon 5.

       The condensation product of two glucose units looks like this:

               α-linkage   β-linkage

25.66   (a)   In the linear form of galactose, the aldehydic carbon is C1. Carbon atoms 2, 3, 4, and 5 are chiral because they each carry four different groups. Carbon 6 is not chiral because it contains two H atoms.

(b)   The structure is best deduced by comparing
galactose with glucose, and inverting the
configurations at the appropriate carbon atoms.
Recall from Solution 25.64 that both the β-form
(shown here) and the α-form (OH on carbon 1 on
the opposite side of ring as the CH_2OH on carbon
5) are possible.

galactose

25.68   The empirical formula of glycogen is $C_6H_{10}O_5$. The six-membered ring form of glucose
is the unit that forms the basis of glycogen. The monomeric glucose units are joined by
α linkages.

25.70   Consider the fuels ethane, C_2H_6, and ethanol, C_2H_5OH, where one C–H bond has been
replaced by C–O–H, a C–O and an O–H bond. Combustion reactions for the two fuels
follow.

C_2H_6 + 7/2 O_2 → 2CO_2 + 3H_2O; C_2H_5OH + 3O_2 → 2CO_2 + 3H_2O

Both reactions have the same products, so the exothermic (negative) parts of the
reaction enthalpies are the same. The difference is in the endothermic (positive) part,
the energy required to break bonds of the reactants. According to Table 8.4, the
energies required to break bonds in the two sets of reactants follow.

C_2H_6: 6D(C–H) + D(C –C) + 7/2D(O_2) = 6(413) + 348 + 3.5(495) = 4558.5 = 4559 kJ

C_2H_5OH: 5D(C–H) + D(C–C) + D(C–O) + D(O–H) + 3D(O_2) =

5(413) + 348 + 358 + 463 + 3(495) = 4719 kJ

Since more energy is required to break bonds in the combustion of one mole of ethanol,
the reaction is less exothermic overall. Because C_2H_6 has a more exothermic
combustion reaction, we say that more energy is "stored" in C_2H_6 than in C_2H_5OH.

Nucleic Acids

25.72

25.74   In the helical structure for DNA, the strands of the polynucleotides are held together by
hydrogen-bonding interactions between particular pairs of bases. It happens that
adenine and thymine form an especially effective base pair, and that guanine and

cytosine are similarly related. Thus, each adenine has a thymine as its opposite number in the other strand, and each guanine has a cytosine as its opposite number. In the overall analysis of the double strand, total adenine must then equal total thymine, and total guanine equals total cytosine.

25.76   In terms of the "small molecule" components of DNA and RNA, there are two main differences. Both DNA and RNA contain 5-membered ring sugars. In RNA, the sugar is ribose; in DNA the substituent at C2 is –H instead of –OH, and the sugar is deoxyribose. Both include four nitrogen-containing organic bases. Three of these, adenine, guanine and cytosine, occur in both DNA and RNA. The fourth base is thymine in DNA and uracil in RNA; thymine and uracil differ by a single –CH_3 substituent.

In terms of macromolecular structure, DNA is double-stranded and exists as the famous "double helix", while RNA is single-stranded. The seemingly minor structural differences in small-molecule components results in a major difference between the macromolecular structures of DNA and RNA.

Additional Exercises

25.78   *Analyze/Plan.* We are asked the number of structural isomers for two specified carbon chain lengths and a certain number of double bonds. Structural isomers have different connectivity. Since the chain length is specified, we can ignore structural isomers created by branching. We are not asked about geometrical isomers, so we ignore those as well. The resulting question is: How many ways are there to place the specified number of double bonds along the specified C chain? *Solve.*

5 C chain with one double bond: 2 structural isomers

C=C–C–C–C   C–C=C–C–C

6 C chain with two double bonds: 6 structural isomers

C=C–C=C–C–C   C=C–C–C=C–C   C=C–C–C–C=C

C–C=C–C=C–C   C=C=C–C–C–C   C–C=C=C–C–C

25.79   Because of the strain in bond angles about the ring, cyclic alkynes with less than eight carbons are not stable. Alkyne carbon atoms preferentially have 180° bond angles; this requires a linear four-carbon group in the ring. Three additional carbons in the ring do not provide enough flexibility to make this possible without gross bond length or angle distortions. It is possible that a ring with eight or more carbons could accommodate an alkyne linkage. The (n–4) carbon atoms in the ring must provide enough flexibility for ring closure without large distortions of C–C bond lengths or angles. [You can test this with models, or by searching online data bases of known compounds.]

25.81   In alkanes, carbon forms only single, sigma bonds. Alkenes contain at least one C–C double bond, consisting of one sigma and one pi bond. Alkynes have a least one C–C triple bond, composed of one sigma and two pi bonds. In both alkenes and alkynes, C atoms are involved in pi overlap. The question is, what feature of Si prevents it from forming double or triple bonds which involve pi overlap.

According to Table 8.5, the average C—C single bond length is 1.54 Å, C=C is 1.34 Å, and C≡C is 1.20 Å. These distances show that pi overlap requires substantially closer approach of the two bonded atoms than sigma overlap alone. The bonding atomic radius of Si is 1.11 Å, while that of C is 0.77 Å (Figure 7.7). The close approach of Si atoms that is required for pi overlap is not possible because of its large bonding atomic radius. Thus, silicon analogs of alkenes and alkynes that involve multiple bonds and pi overlap are virtually unknown. Silicon analogs of alkanes with exclusively sigma overlap are known; the average Si—Si single bond length is 2.22 Å.

25.82   The suffic −ene signifies an alkene, −one a ketone. The molecule has alkene and ketone functional groups.

25.84   Two plausible decomposition reactions are:

(i) $CH_2Cl_2(l) \rightarrow C(s) + H_2(g) + Cl_2(g)$

(ii) $CH_2(NO_2)_2(l) \rightarrow N_2(g) + CO_2(g) + H_2O(g) + 1/2 O_2(g)$

Use bond dissociation energies (Table 8.4) to evaluate approximate ΔH values for each reaction.

(i)   ΔH = 2D(C—H) + 2D(C—Cl) − D(H—H) − D(Cl—Cl)

      = 2(413) + 2(328) − 436 − 242 = +804 kJ

(ii)   ΔH = 2D(C—H) + 2D(C—N) + 2D(N=O) + 2D(N—O) − D(N≡N) − 2D(C=O)

                                                           − 2D(O—H) − 1/2D(O=O)

      = 2(413) + 2(293) + 2(607) + 2(201) − 941 − 2(799) − 2(463) − 1/2(495)

ΔH = −685 kJ

Clearly, the decomposition of $CH_2(NO_2)_2$ is thermodynamically favorable, while the decomposition of CH_2Cl_2 is not. In particular, this is because of the stability of N_2 and CO_2 relative to $CH_2(NO_2)_2$. For CH_2Cl_2, no oxygen atoms are available to form stable products such as CO_2 and H_2O.

25.86   (a)

$CH_3CH_2CH_2COH$ or $(CH_3)_2CHCOH$

(b) CH_3—CH—CH_2 or CH_2—CH_2—CH_2 (with OH, OH and OH, OH groups)

25.87 Th e difference between an alcoholic hydrogen and a carboxylic acid hydrogen is two-fold. First, the electronegative carbonyl oxygen in a carboxylic acid withdraws electron density from the O—H bond, rendering the bond more polar and the H more ionizable. Second, the conjugate base of a carboxylic acid, carboxylate anion, exhibits resonance. This stabilizes the conjugate base and encourages ionization of the carboxylic acid. In an alcohol no electronegative atoms are bound to the carbon that holds the —OH group, and the H is tightly bound to the O.

25.88 In order for indole to be planar, the N atom must be sp² hybridized. The nonbonded electron pair on N is in a pure p orbital perpendicular to the plane of the molecule. The electrons that form the π bonds in the molecule are also in pure p orbitals perpendicular to the plane of the molecule. Thus, each of these p orbitals is in the correct orientation for π overlap; the delocalized π system extends over the entire molecule and includes the "nonbonded" electron pair on N. The reason that indole is such a weak base (H⁺ acceptor) is that the nonbonded electron pair is delocalized and a H⁺ ion does not feel the attraction of a full localized electron pair.

25.89 (a) None

(b) The carbon bearing the secondary —OH has four different groups attached, and is thus chiral.

(c) The carbon bearing the —NH₂ group and the carbon bearing the CH₃ group are both chiral.

25.90 In the zwitterion form of a tripeptide present in aqueous solution near pH 7, the terminal carboxyl group is deprotonated and the terminal amino group is protonated, resulting in a net zero charge. The molecule has a net charge only if a side (R) group contains a charged (protonated or deprotonated) group. The tripeptide is positively charged if a side group contains a protonated amine. According to Figure 25.23, the only amino acids with protonated amines in their side groups are arginine (Arg) and lysine (Lys). Of the tripeptides listed, only (a) Gly-Ser-Lys will have a net positive charge at pH 7.

[Note that aspartic acid (Asp) has a deprotonated carboxyl in its side group, so (c) Phe-Tyr-Asp will have a net negative charge at pH 7.]

25.92 Starch, glycogen, and cellulose are all biopolymers built by linking glucose monomers. Starch and glycogen have alpha (α) glucose linkages, where the bridging O atom is on the opposite side of the ring as the CH₂OH group. The smallest repeating unit in starch and glycogen is a single glucose unit. Starch and glycogen can have branched structures, while cellulose is always linear.

Cellulose has beta (β) glucose linkages, where the bridging O atom is on the same side of one of the rings as the CH₂OH group and on the opposite side of the CH₂OH group on the second ring. The geometry of the β linkage requires that the two linked glucose units have different orientations and that the smallest repeating unit in cellulose is two glucose units with a β linkage.

The molecular weight of a polymer is an indication of the number of monomer units present. Starch, glycogen, and cellulose all have a range of molecular weights. Glycogen has the widest range of molecular weights, 5,000–5,000,000 amu, and is potentially the largest polymer. Cellulose is intermediate in size with an average molar mass of 500,000 amu.

Starch and cellulose are produced in plants, while glycogen is produced in animals and serves as an energy storage mechanism.

25.93 Both glucose and fructose contain six C atoms, so both are hexoses. Glucose contains an aldehyde group at C1, so it is an aldohexose. Fructose has a ketone at C2, so it is a ketohexose.

25.94 No. RNA cannot form a strand complementary to DNA. DNA contains guanine, cytosine, adenine and thymine and requires these same four bases in a complementary strand. RNA contains uracil in place of thymine and cannot function as a complementary strand to DNA.

Integrative Exercises

25.95 CH_3CH_2OH $CH_3—O—CH_3$
        ethanol   dimethyl ether

Ethanol contains $—O—H$ bonds which form strong intermolecular hydrogen bonds, while dimethyl ether experiences only weak dipole-dipole and dispersion forces.

   difluoromethane   tetrafluoromethane

CH_2F_2 is a polar molecule, while CF_4 is nonpolar. CH_2F_2 experiences dipole-dipole and dispersion forces, while CF_4 experiences only dispersion forces.

In both cases, stronger intermolecular forces lead to the higher boiling point.

25.96 Determine the empirical formula of the unknown compound and its oxidation product. Use chemical properties to propose possible structures.

$$68.1\,g\,C \times \frac{1\,mol\,C}{12.01\,g\,C} = 5.6703;\ 5.6703/1.1375 = 4.98 \approx 5$$

$$13.7\,g\,H \times \frac{1\,mol\,H}{1.008\,g\,H} = 13.5913;\ 13.5913/1.1375 = 11.95 \approx 12$$

$$18.2\,g\,P \times \frac{1\,mol\,O}{16.00\,g\,O} = 1.1375;\ 1.1375/1.1375 = 1$$

The empirical formula of the unknown is $C_5H_{12}O$.

$$69.7 \, \text{g C} \times \frac{1 \, \text{mol C}}{12.01 \, \text{g C}} = 5.8035; \, 5.8035/1.1625 = 4.99 \approx 5$$

$$11.7 \, \text{g H} \times \frac{1 \, \text{mol H}}{1.008 \, \text{g H}} = 11.6071; \, 11.6071/1.1625 = 9.99 \approx 10$$

$$18.6 \, \text{g O} \times \frac{1 \, \text{mol O}}{16.00 \, \text{g O}} = 1.1625; \, 1.1625/1.1625 = 1$$

The empirical formula of the oxidation product is $C_5H_{10}O$.

The compound is clearly an alcohol. Its slight solubility in water is consistent with the properties expected of a secondary alcohol with a five-carbon chain. The fact that oxidation results in a ketone, rather than an aldehyde or a carboxylic acid, tells us that it is a secondary alcohol. Some reasonable structures for the unknown secondary alcohol are:

$$\underset{\overset{|}{\text{OH}}}{\text{CH}_3\text{CHCH}_2\text{CH}_2\text{CH}_3} \quad \underset{\overset{|}{\text{OH}}}{\text{CH}_3\text{CHCHCH}_2\text{CH}_3} \quad \underset{\overset{|}{\text{OH}}}{\text{CH}_3\text{CHCH(CH}_3)_2}$$

25.98   Determine the empirical formula, molar mass, and thus molecular formula of the compound. Confirm with physical data.

$$85.7 \, \text{g C} \times \frac{1 \, \text{mol C}}{12.01 \, \text{g C}} = 7.136 \, \text{mol C}; \, 7.136/7.136 = 1$$

$$14.3 \, \text{g H} \times \frac{1 \, \text{mol H}}{1.008 \, \text{g H}} = 14.19 \, \text{mol H}; \, 14.19/7.136 \approx 2$$

Empirical formula is CH_2. Using Equation 10.11 (MM = molar mass):

$$MM = \frac{(2.21 \, \text{g/L})(0.08206 \, \text{L-atm/mol-K})(373\text{K})}{(735/760) \, \text{atm}} = 69.9 \, \text{g/mol}$$

The molecular formula is thus C_5H_{10}. The absence of reaction with aqueous Br_2 indicates that the compound is not an alkene, so the compound is probably the cycloalkane cyclopentane. According to the *Handbook of Chemistry and Physics*, the boiling point of cyclopentane is 49°C at 760 torr. This confirms the identity of the unknown.

25.99   The reaction is: $2NH_2CH_2COOH(aq) \rightarrow NH_2CH_2CONHCH_2COOH(aq) + H_2O(l)$

$\Delta G° = (-488) + (-237.13) - 2(-369) = 12.87 = 13 \, \text{kJ}$

25.100   (a)   A = adenosine = $C_{10}H_{12}O_3N_5$

$$[A—P_3O_{10}]^{4-} + H_2O \longrightarrow [A—P_2O_6(OH)]^{2-} + HPO_4^{2-}$$

(The placement of the H^+ in these reactions is somewhat arbitrary; H^+ is attracted to the strongest base, but the equilibria are complex.)

(b)   If the hydrolysis reaction is spontaneous, the sign of ΔG must be negative.

(c)   Adenosine monophosphate (AMP) + inorganic phosphate

(The placement of the H^+ in these reactions is somewhat arbitrary; H^+ is attracted to the strongest base, but the equilibria are complex.)

25.101   (a)   At low pH, the amine and carboxyl groups are protonated.   At high pH, the amine and carboxyl groups are deprotonated.

(b)

$$CH_3C{—}OH(aq) \longrightarrow CH_3C{—}O^-(aq) + H^+(aq)$$

$K_a = 1.8 \times 10^{-5}$, $pK_a = -\log(1.8 \times 10^{-5}) = 4.74$

The conjugate acid of NH_3 is NH_4^+.

$$NH_4^+(aq) \rightleftharpoons NH_3(aq) + H^+(aq)$$

$K_a = K_w/K_b = 1.0 \times 10^{-14} / 1.8 \times 10^{-5} = 5.55 \times 10^{-10} = 5.6 \times 10^{-10}$

$pK_a = -\log(5.55 \times 10^{-10}) = 9.26$

In general, a $-COOH$ group is stronger acid than a $-NH_3^+$ group. The lower pK_a value for amino acids is for the ionization (deprotonation) of the $-COOH$ group and the higher pK_a is for the deprotonation of the $-NH_3^+$ group.

(c)

By analogy to serine, the carboxyl group near the amine will have pK_a ~2 and the amino group will have pKa ~9. By elimination, the carboxyl group in the side chain has pK_a ~4.

25.102 (a) Because the native form is most stable, it has a lower, more negative free energy than the denatured form. Another way to say this is that ΔG for the process of denaturing the protein is positive.

(b) ΔS is negative in going from the denatured form to the folded (native) form; the native protein is more ordered.

(c) The four S—S linkages are strong covalent links holding the chain in place in the folded structure. A folded structure without these links would be less stable (higher G) and have more motional freedom (more positive entropy).

(d) After reduction, the eight S—H groups will form hydrogen-bond-like interations with acceptors along the protein backbone, but these will be weaker and less specifically located than the S—S covalent bonds of the native protein. Overall, the tertiary structure of the reduced protein will be looser and less compact due to the loss of the S—S linkages.

(e) The amino acid cysteine must be present in order for −SH bonds to be found in ribonuclease A. (Methionine contains S, but no −SH functional group.)

25.103 $AMPOH^-(aq) \rightleftharpoons AMPO^{2-}(aq) + H^+(aq)$

$pK_a = 7.21; K_a = 10^{-pK_a} = 6.17 \times 10^{-8} = 6.2 \times 10^{-8}$

$K_a = \dfrac{[AMPO^{2-}][H^+]}{[AMPOH^-]} = 6.2 \times 10^{-8}$. When pH = 7.40, $[H^+] = 3.98 \times 10^{-8} = 4 \times 10^{-8}$.

Then $\dfrac{[AMPOH^-]}{[AMPO^{2-}]} = 3.98 \times 10^{-8} / 6.17 \times 10^{-8} = 0.6457 = 0.6$